# CRUSTACEAN PHYLOGENY

# CRUSTACEAN ISSUES

# 1

*General editor:*

FREDERICK R.SCHRAM

*San Diego Natural History Museum, California*

A.A.BALKEMA/ROTTERDAM/1983

# CRUSTACEAN PHYLOGENY

*Edited by*
FREDERICK R.SCHRAM
*San Diego Natural History Museum, California*

A.A.BALKEMA / ROTTERDAM / 1983

There is one radical distinction between different minds . . . that some minds are stronger and apter to mark the differences of things, others to mark their resemblances.

Francis Bacon

ISBN 90 6191 231 8

© 1983 A.A.Balkema, P.O.Box 1675, 3000 BR Rotterdam, Netherlands

Distributed in USA & Canada by: MBS, 99 Main Street, Salem, NH 03079, USA

Printed in the Netherlands

# TABLE OF CONTENTS

V

VI *Table of contents*

# PREFACE

In 1979, Dr Kristian Fauchald asked me to consider organizing a symposium on arthropod phylogeny for the Division of Invertebrate Zoology of the American Society of Zoologists. At that time, several books, both single author and edited compendia, had then recently been published which dealt with various aspects of phylogeny within arthropods; so there seemed little compelling reason to produce yet another symposium on the topic. Furthermore, in surveying the recent literature on individual arthropod groups, it appeared that almost every major arthropod taxon (insects, myriapods, trilobitomorphs, chelicerates, and pycnogonids) had its phylogeny and evolution reviewed — all that is, except the crustaceans. The last overview of phylogeny within the Crustacea had been in the 1963 proceedings of the conference organized by Drs H.B.Whittington and W.D.I.Rolfe at the Museum of Comparative Zoology, Harvard University. Obviously, a symposium on crustacean phylogeny was needed to survey the advances in the field in the last 20 years.

Originally, the invited speakers were to be urged in their symposium presentations to provide overview and not just review. To encourage 'sticking one's neck out', it was not intended to publish the symposium proceedings. However, after organization began and funding secured, Mr A.T.Balkema independently approached me with the idea of beginning a series of volumes which would survey in depth timely topics in crustacean biology. The symposium schedule was limited to one day, and thus only a limited number of speakers on a few topics could participate. The possibility of an unrestricted volume to cover the subject of crustacean evolution afforded a means by which more participants, in addition to the symposium speakers, could contribute towards providing an adequate overview. And so originated this first volume of *Crustacean Issues.*

The Harvard Symposium, *Phylogeny and Evolution of Crustacea,* is a slender tome, but its impact on the field has been enormous. Hardly a paper dealing with some aspect of crustacean evolution has been published in the last two decades that has not cited at least one of the papers in that benchmark volume. The contributors to the present volume have produced insightful, and in several instances, deliberately controversial analyses of their respective groups. If we as authors can excite the reader and stimulate research as well as the 1963 volume, we will have served our purpose well.

A few additional authorities to those included here were asked to participate, but found the press of other duties prevented them from meeting a deadline. Some taxa are not covered simply because there was no one working on their phylogeny for the present. The discussion sessions of those papers presented at the symposium in Dallas, Texas, are also included, and substantial changes added in editing the transcripts are placed in brackets.

Special acknowledgements must be made to the American Society of Zoologists and

VII

The Crustacean Society, which jointly sponsored, and the National Science Foundation, which through Grant DEB 81-07870 afforded financial support for, the international symposium, 'Phylogeny within Crustacea', held 30 December 1981 in Dallas, Texas.

Frederick R.Schram
San Diego, California

# INTRODUCTION

'How much diversity of opinion can be tolerated?' This is the underlying concern of any science, but it is a central issue of concern in the areas of systematics, phylogeny, and evolution. Phylogeny of Crustacea is no exception, especially as presented in this volume.

Crustacean studies have passed through a series of distinct periods in the course of its history. The earliest period, chiefly characterized by the taxonomic works of Latreille and H.Milne-Edwards, was a time in which the discipline was being founded and organized, basic forms were being described and the first classifications established. This initial period persisted through the 1800's until the late 19th and early 20th centuries when a synthesis of previous knowledge was achieved. This time of synthesis is best characterized by the works of W.T.Calman and C.Claus, and culminated in 1909 with Calman's volume in E.R. Lankester's *Treatise on Zoology*. That synthesis basically established the taxonomic and phylogenetic framework upon which carcinology came to be built in the next succeeding period, which extended right up to the present time.

Now a new period of synthesis is beginning, triggered in large part by the revolutionary works of S.M.Manton, on functional morphology of arthropodous locomotory mechanisms, and D.T.Anderson, on early patterns of development in arthropods. Manton and Anderson independently concluded that Arthropoda is a polyphyletic taxon, i.e. that none of the arthropodous groups (Uniramia, Crustacea, Cheliceriformes, and probably Trilobitomorpha) can be derived from immediate common ancestors, not without recourse to elaborate sequences of hypothetical forms such as to render Arthropoda as a formal taxonomic unit quite meaningless.

In addition, Manton irrevocably broke apart the union of the crustaceans and the myriapod-insect groups in a taxon Mandibulata by showing that structurally and functionally the jaws of these two groups are no more similar to each other than either is to the jaws of merostomes. It is interesting to note that while crustacean workers have generally accepted Manton's dissolution of the Mandibulata and moved onto more productive activities, myriapod and insect workers have continued to produce reams of paper attempting to refute Manton and uphold mandibulate orthodoxy.

In part, I feel the more liberal attitudes and inclinations of carcinologists as opposed to entomologists has been due to the former having had, until relatively recently, little formal adherence to the principles of phylogenetic systematics as espoused by the entomologist Willi Hennig and his disciples. The 'orthodoxy' of carcinologists has stemmed largely from Claus and Calman, who themselves were more pragmatic in their approaches to their science. Carcinology is thus not founded on a formal methodology such as cladistics, so much as on an adherence to a historic tradition of ideas bequeathed us from venerated workers of the past. Ideas, rather than method, have governed the investigations of carcinologists.

This historic tradition has produced certain grand unifying themes in crustacean phylogeny. Some of the more important of these concepts are: cephalocarids as ancestral types; the caridoid facies; the taxonomic system as essentially outlined by Calman; and the distinctness of all entomostracan types. All of these concepts have now begun to be re-examined and questioned. For some of these concepts the reconsideration is more advanced than for others, e.g. the union of some entomostracan types into a taxon Maxillopoda, while for others the controversy is just beginning, e.g. the usefulness of a taxon like 'Peracarida'.

We still do not understand how crustaceans came to be. Our understanding of earliest arthropods only seems to become more clouded with increasing knowledge, as outlined in this volume in the paper by Briggs on Cambrian forms. Manton and Anderson would have us believe in the separateness of the basic arthropodan stocks seen today. However, in the remote recesses of time we are confronted with more numerous distinctive arthropodan types. Briggs asks if we are to believe that all these arthropodous lines are separate phyla? We may never resolve this, since we can probably never have the necessarily equivalent data on the Cambrian forms to compare to the data Manton and Anderson used to evaluate the modern groups.

Attempting to elucidate phylogeny in some taxa is generating tremendous controversy. This is evidenced by the contributions herein on Maxillopoda, by Grygier, Newman and Boxshall; on higher malacostracan evolution, by Hessler, Kunze and Schram; and on Peracarida as a useful taxon, by Watling and Hessler with some additional comments by Sieg and Bousfield. Phylogenetic speculation in other groups seems to generate less controversy, such as the contributions included here by Burkenroad on dendrobranchiates, Sieg on tanaids, and to a lesser extent Bousfield on amphipods; mainly because basic data on which the phylogenies are based in these groups is just being gathered. The lack of adequate data is an especially crucial problem for a group like the ostracods in which the data base is either so staggeringly large or in other aspects almost non-existent, such that only the broadest generalities about phylogeny are offered by McKenzie, Müller and Gramm. Still other groups like the 'shrimps' discussed by Felgenhauer and Abele, or the brachyurans treated by Rice, long thought to be phylogenetic backwaters, are being subjected to a new analysis which promises to ignite controversy in taxa which we have heretofore mistakenly thought to be 'understood'. And how can we ever be prepared for a totally unexpected discovery like that of the remipede *Speleonectes lucayensis,* which may eventually completely overturn *all* previous notions about crustacean origins and early evolution.

I think it especially encouraging and refreshing to see a variety of methodologies being used by the authors in this book to elucidate phylogeny within the crustaceans. Cladistics is beginning to be employed by carcinologists, who fortunately have the advantage of having delayed in adopting cladistics long enough to be able to see its strengths and weaknesses as manifested by use in other groups. Cladistics does provide a reasonable tool, as in the case of Sieg's chapter on tanaids or Grygier's evaluation of maxillopodans, that can provide formal schemes ready to be tested by future work. General cladistic principles, if not strict methodology, however, have already brought about more reasoned and formal debate on opposite sides of important issues, such as that which involves caridoid facies and affinities within Eumalacostraca *sensu stricto.* The working scientist should never be slavishly tied to rigid methodology, as Bousfield illustrates when he creatively combines cladistic principles with numerical methods to evaluate possible amphipod relationships.

It should not be surprising that functional morphology is still in the thick of debate, since the dissent in carcinology was inspired and fostered by Manton's work in that discipline. Grygier presents a formal cladistic analysis to justify the use of Maxillopoda, yet Boxshall's counter arguments afford a basis for caution that reminds us that analysis of form has its limits without consideration of function. Felgenhauer and Abele, and Burkenroad both present prospecti for phylogeny within parts of the Decapoda that will only yield to eventual solutions when detailed analyses of functional morphology in these groups are performed.

Though developmental considerations have frequently been used to interpret phylogenetic schemes, most notably by Gurney in his classic monograph, *Larvae of Decapod Crustacea;* such has not achieved widespread use. One of the few exceptions has been Rice, who herein summarized data on larval patterns as applied to phylogeny of brachyuran decapods. Some new impetus in this regard has come from Gould's recent tome on *Ontogeny and Phylogeny,* wherein he outlines developmental phenomena applicable to understanding macroevolution in higher taxa. Thus inspired, Boxshall and especially Newman consider paedomorphosis as an agent in explaining the origin of crustacean classes.

Finally, least we overlook larger issues, study of crustacean phylogeny can afford insights applicable to evolution theory as a whole. After all, Darwin himself was first noted as a carcinologist (his monographic treatment of barnacles is still a standard working reference for cirripede workers) before he published *Origin of Species.* In this volume Marcotte poses some questions relevant to understanding the nature of adaptive radiations, and I suggest a new model for explaining the origin and evolution of major structural types.

'How much diversity of opinion can be tolerated?' I think the reply, in light of the contributions in this volume, must be, 'As much as can be gotten away with!' Procrustean adherence to orthodox traditions causes science to stagnate. This can be absolutely lethal to disciplines like phylogeny and systematics; since if theorizing is frozen, then the collection of facts becomes stereotyped. George Gaylord Simpson aptly summarized the dilemma here when he said, 'Theory without facts and facts without theory are both indefensibly worthless'. An idea unthinkable today may be in the mainstream tomorrow. If a concept is truly without foundation, it will find little application in practice. There really is something to be said for letting a thousand flowers bloom; intellectual natural selection operates soon enough.

Frederick R. Schram

DEREK E. G. BRIGGS
*Department of Geology, Goldsmiths' College, University of London, UK*

# AFFINITIES AND EARLY EVOLUTION OF THE CRUSTACEA: THE EVIDENCE OF THE CAMBRIAN FOSSILS

## ABSTRACT

The only reliable character available to identify Crustacea among the wide diversity of Cambrian arthropods is the arrangement of the cephalic appendages. These are rarely preserved, and this review is confined to a consideration of arthropods with known soft-parts, particularly those of the Middle Cambrian Burgess Shale of British Columbia and the Upper Cambrian 'Orsten' of Sweden. The majority of Cambrian arthropods do not fall into easily defined groups; only the Crustacea of the major living taxa are represented. A critical review of the fossils reveals that only two crustacean classes are present in the Cambrian, the Phyllocarida and the Ostracoda. Previous discussions of crustacean affinity considered many of the other Cambrian arthropods to belong with the trilobites in the Class Trilobitoidea, since shown to be invalid, thus obscuring their potential significance in early arthropod evolution. A consideration of all the well preserved Cambrian arthropods reveals no unequivocal evidence of crustacean affinities, but the possibility of a common origin with at least some other arthropods cannot be ruled out.

## 1 INTRODUCTION

The last decade has seen a wider acceptance of the view of Manton, that the arthropods are not monophyletic, i.e. united by descent from a single common ancestor into one large phylum. Manton (1973, 1977) recognized three arthropod phyla, Crustacea, Chelicerata and Uniramia, and left the status of the extinct Trilobita undecided. Schram (1978) retained her scheme in essence, incorporating the Pycnogonida into the Phylum Cheliceriformes, but added a fourth phylum, the Trilobitomorpha, to include the trilobites. An alternative view (Cisne 1974, Hessler & Newman 1975) while accepting an independent origin for the Uniramia, argues that the trilobites, Crustacea and Chelicerata form a single phylum.

In the course of the same decade a great deal of new information has been added to our knowledge of early arthropods from the fossil record, much of it since the publication of the papers on arthropod phylogeny referred to above. This is mainly the result of projects on soft-bodied faunas, particularly those of the Middle Cambrian Burgess Shale of British Columbia (Briggs & Whittington 1981) and the Upper Cambrian 'Orsten' of Sweden (Müller 1981). In contrast to the opinion of some more extreme adherents to the methods of phylogenetic systematics or cladistic taxonomy this information provides an important

1

base for an assessment of interrelationships between the major arthropod groups, as some of the early arthropods represent morphologies which are not represented in the living fauna. This paper is thus timely in its objective to review for neocarcinologists the fossil evidence for the affinities and early evolution of the Crustacea.

The scope of the review is deliberately restricted to the Crustacea and crustacean-like arthropods of the Cambrian. Briggs & Whittington (in preparation) will review all the arthropods of the Middle Cambrian Burgess Shale and their relationships. Schram (1982) has already considered the fossil record and evolution of the Crustacea as a whole.

## 2  WHAT CRITERIA CAN BE USED TO IDENTIFY EARLY ARTHROPODS AS CRUSTACEA?

The identification of living arthropods to phylum or superclass level is rarely a problem, as the effect of evolution through the Phanerozoic has been to eliminate all but three major morphologically distinct groups (Crustacea, Chelicerata and Uniramia). The identification of Crustacea in the fossil record is not so easy, however, due mainly to incomplete preservation — important diagnostic characters, like the appendages, commonly fail to survive. There is also ample evidence that arthropod taxa in addition to the three major living groups, and the trilobites, were present in the Palaeozoic. While the Burgess Shale may bias the record it appears that the number of such additional taxa was greatest in the Cambrian and decreased through geological time as extinct forms were replaced by diversification at lower levels in the taxonomic hierarchy. A number of metazoans have been described from the Burgess Shale which do not belong to any living phyla (Conway Morris 1979, Conway Morris & Whittington 1979, Whittington 1980).

Many features of Crustacea are shared with other arthropods. The crustacean head, however, is unique, with two preoral somites bearing antennae and three postoral bearing a mandible and two pairs of maxillae. In addition the crustacean body is usually divided into three tagmata, head, thorax and abdomen, and according to some workers (Hessler & Newman 1975, Schram 1979) the appendages are primitively polyramous and lamellate. These last two characters, however, are insufficient to define a Crustacean as they may occur in other arthropods and the segmentation of the head is the only reliable diagnostic character. It is a difficult character to use, however, in fossils. The only evidence of segmentation with much preservation potential is the arrangement of cephalic appendages, and even in the rare occurrences of soft-bodied preservation this is not easily studied. Further, as Hessler (in House 1978) has pointed out, the diagnostic segmentation may only be evident in embryology. Thus if living mystacocarids were fossilized they would have one fewer segment than other crustaceans, and in the malacostracans either some or all of the segments of the thorax, which is fused to the head, might be interpreted as cephalic. Thus the evidence for crustacean affinity, particularly in early arthropods which do not fall neatly into living taxa, may be tenuous at best. Unless the diagnostic arrangement of crustacean appendages can be demonstrated, assignment to the Crustacea must be considered equivocal — other crustacean characters may be acquired convergently.

It may be argued that adoption of the segmentation of the head as the overriding diagnostic character in fossil Crustacea is too restrictive. It is, however, included as a feature of the otherwise primitive ancestral crustacean postulated by Hessler & Newman (1975). In the absence of the other evidence available to neontologists, and in view of the presence

of a number of superficially crustacean-like arthropods in the Cambrian which are clearly
not Crustacea (e.g. *Branchiocaris,* Fig.2c, Briggs 1976) it seems appropriate to adopt such
a rigorous approach in a review of the evidence for crustacean affinities. The possibility of
interrelationships of early Crustacea *sensu stricto* (thus defined) and other crustacean-like
arthropods is considered later in the paper.

## 3 THE FOSSILS

The dominant arthropods in Cambrian fossil assemblages are the trilobites, as they have a
mineralized exoskeleton and are readily preserved; the vast majority of specimens represent
exuviae. Work on soft-bodied faunas has demonstrated that trilobites were a relatively
minor element of Cambrian communities, which were dominated by a diversity of arthropods, the majority without mineralized exoskeletons. Most of these non-trilobite arthropods
are only preserved under rare and exceptional conditions and a review of the early Crustacea
and crustacean-like arthropods may thus be confined largely to a consideration of soft-
bodied occurrences. The most important of these is undoubtedly the Middle Cambrian
Burgess Shale of British Columbia (Conway Morris & Whittington 1979, Whittington 1980).
This fauna lived in the open sea adjacent to the steep face of an algal carbonate reef about
160 m high. The assemblage appears to be exceptional only in the nature of the preservation (Conway Morris 1979, Conway Morris & Whittington 1979). There is no evidence of
a barrier to migration into the embayment in the reef front in which the locality occurs.
The shelly fossils present occur widely in 'typical' Cambrian faunas which differ from the
Burgess Shale only in lacking a soft-bodied component. The 44 genera of arthropods
account for the largest proportion of the fauna in terms of genera (37 %). Only 14 of these,
however, are the readily preserved trilobites; the majority of the other 30 which lacked
mineralized exoskeletons are therefore not normally represented in Cambrian faunas.

Almost none of the non-trilobite arthropods in the Burgess Shale falls into the three
major living taxa, Crustacea, Chelicerata and Uniramia. There are no known Cambrian
chelicerates (Briggs et al. 1979). The lobopod animal *Aysheaia pedunculata* is the only
form that shows a similarity to living uniramians, Onychophora and Tardigrada. In spite of
its striking resemblance to these two groups, *Aysheaia* lacks the ventral mouth and specialized head appendages characteristic of the Uniramia. Whittington (1978, p.195) suggests
that it 'may be regarded as the sole known representative of an early group of soft-bodied,
metamerically segmented, lobopodial animals' from which 'may have been descended both
the Tardigrada and the Onychophora'. He is nevertheless careful to emphasize (1978,
p.195) that 'it does not fit readily into any extant higher taxon'. A small number of the
Burgess Shale arthropods do, however, show affinities with the Crustacea.

3.1 The bivalved arthropod *Canadaspis perfecta* (Fig.1c,d) has been described as 'the
earliest positively identified crustacean' (Briggs 1977a). It is the second most abundant
arthropod in the Burgess Shale; over 5 000 specimens are known (Briggs 1978). The valves,
which cover the head and thorax, and all but the distal extremities of the appendages,
average about 35 mm long, but examples less than 10 mm are known. The head bears a
pair of small pedunculate eyes. Between these a pair of small appendages, interpreted as
first antennae, flank a median cephalic spine. The second antennae are larger and multisegmented, but still do not extend far beyond the carapace. The mandible (Fig.1c) consists

of a series of large spines forming both a molar and incisor process; there is no evidence of a palp. The two pairs of maxillae are similar to the appendages of the thorax, but the first pair at least bears an array of bunches of spines along the inner margin, one each on the six or seven most proximal segments. The thorax is made up of eight somites, each bearing a pair of biramous limbs. The inner ramus consists of up to 14 segments and terminates in a group of hooked claws. A single group of adaxially directed proximal spines on the basal segment of each limb presumably passed food anteriorly along a ventral food groove. The outer ramus is lamellate and includes a number of rays attached to a proximal lobe. The abdomen consists of seven somites and terminates in a telson covered in small spines. Only the last somite bears appendages – a pair of spinose projections extending ventrally beyond the telson. There is no caudal furca.

The first two pairs of biramous appendages in *Canadaspis* are interpreted as maxillae, due to their position behind the mandible, and the apparent specialization of at least the first one for feeding (Fig.1c). The thorax and abdomen then include eight and seven somites respectively, the arrangement characteristic of the Class Malacostraca. *Canadaspis* is classified in a separate Order (Canadaspidida) of the Subclass Phyllocarida (Briggs 1978).

The arthropod is primitive in two important respects. Firstly, the number of segments in the inner ramus of the biramous appendages (14) is much higher than in living crustaceans. The larger number may have been necessary to achieve an equivalent flexibility to that in Recent arthropods which have evolved specializations of the joints and podomeres. Secondly, there is no well-defined boundary between the cephalon and thorax and the maxillae are similar to the anterior trunk appendages. This lack of differentiation of the maxillae prevails only in the Cephalocarida of living crustaceans, but the evidence of embryology indicates that the second maxilla of living Crustacea is functionally part of the thorax in the earliest juvenile stage, and is a late addition to the cephalon (Anderson 1973, p.347, Manton 1977, p.249, fig.6.4d). *Canadaspis* thus appears to represent a stage in crustacean evolution prior to complete cephalization (Briggs 1979).

*Canadaspis* is retained in a slightly emended subclass Phyllocarida in spite of these primitive features. Just as shared primitive characters are not a reliable indication of affinity, their presence in only one of two groups under comparison does not preclude relationship. The lack of a rostral plate, and of appendages on the abdomen (presumably an adaptation for a benthic mode of life) are not considered sufficient justification for the erection of a

---

Figure 1. Cambrian Crustacea and crustacean-like arthropods. A, B. *Odaraia alata* Walcott, 1912. A. USNM 189232, ultraviolet illumination from top, x 1.5 (Briggs 1981, pl.7, fig.67). Dorsal view with the carapace largely removed by splitting of the slab and only evident laterally. The large eyes project anteriorly and the right series of trunk limbs is well shown; B. USNM 213812a, u.v. illumination from right, x 1.5 (Briggs 1981, pl.5, fig.42). Dorsal view of telson showing lateral 'flukes' in outline and median vertical fluke in 'section' projecting posteriorly. C, D. *Canadaspis perfecta* Walcott, 1912. C. USNM 189017, u.v. illumination from upper left, x 5 (Briggs 1978, pl.5, fig.70). Left anterior appendages seen from the right hand side (the right appendages removed); second antenna – an2, mandible – ma, first and second maxilla – mx1 and mx2; D. USNM 189018, u.v. illumination from upper left, x 2 (Briggs 1978, pl.6, fig.74). Dorsal view showing the bivalved carapace and the abdomen beyond it. The posterior spines are those of the projections borne by the pre-telson somite. E. *Perspicaris dictynna* Simonetta & Delle Cave, 1975, USNM 189245, u.v. illumination from top, x 5 (Briggs 1977, pl.68, fig.6). Oblique view showing eyes – e, bivalved carapace, and curved abdomen and telson with caudal furca.

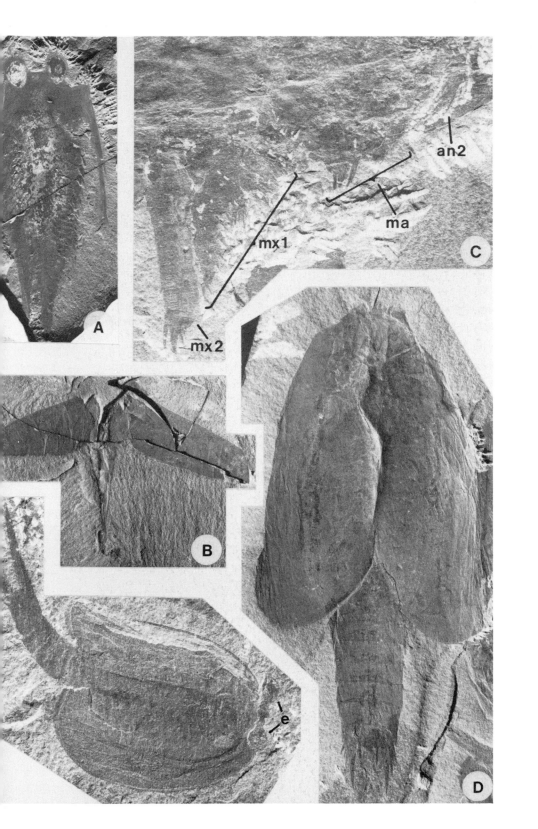

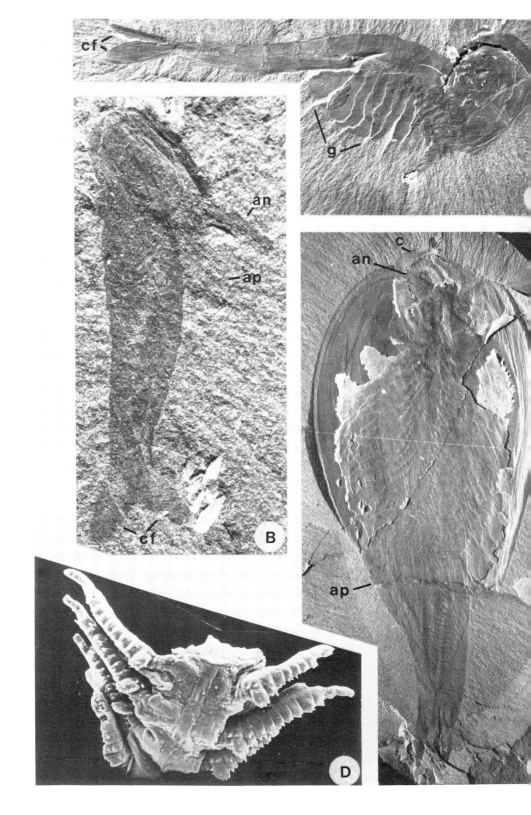

new subclass. An alternative classification, mentioned but not favoured by Briggs (1978, p.484) would elevate both the orders Leptostraca and Canadaspidida to subclass status (i.e. two subclasses, not one combining both as suggested by Schram 1982) and leave the remaining more poorly known fossil orders in the subclass Phyllocarida. I prefer, however, to include all these forms in the Phyllocarida to emphasize the similarity between the Leptostraca and the fossil forms at least until further research on the latter suggests otherwise.

3.2 A second much rarer bivalved genus, *Perspicaris* (Fig.1e) has been referred to the same order of phyllocarids as *Canadaspis* (although the evidence in this case is less satisfactory) but assigned to a new family, the Perspicarididae (Briggs 1977b, 1978). The rarity of *Perspicaris* may be due to a mainly nektonic mode of life; there is no evidence of segmented walking inner rami like those in *Canadaspis*. There are two species, the smaller *P.dictynna* and larger *P.recondita*. The eyes are pedunculate and one pair of antennae is larger than in *Canadaspis*. *P.recondita* shows some evidence of an additional smaller antenna. The mandible is unknown. Both species bear caudal furca, and seven abdominal somites which lack appendages. The trunk appendages are known only in *P.dictynna,* and they are poorly preserved. Ten pairs of similar lamellate limbs are evident, however, and the anterior two pairs are interpreted as maxillae by comparison with *Canadaspis*.

Of the Burgess Shale arthropods, only *Canadaspis* of the Canadaspidida preserves clear evidence of the diagnostic segmentation of the crustacean head. The Burgess Shale arthropods most closely related to the Canadaspidida are *Waptia* and *Plenocaris* (Briggs & Whittington 1981). An account of *Waptia* (Fig.2a) is in preparation by C.P.Hughes. Although no line along which the valves separate has been observed, there is a pronounced posterior indentation terminating at a point on the mid-line. In lateral aspect the carapace folds to reveal a slightly convex dorsal margin with a slight projection at the posterior end. There thus appears to be a functional hinge even if the flexible cuticle along the hinge line is indistinguishable as fossilized from that of the carapace as a whole. The cephalon bears a pair of long segmented antennae anteriorly and there is also evidence of a second reduced antenna. There are small pedunculate eyes. A series of short appendage-bearing somites lies beneath the carapace followed by an abdomen of five longer somites which lack limbs, and a telson bearing segmented caudal furca. The division between the posterior cephalic and thoracic appendages remains to be resolved. Beneath the carapace the appendages are

---

Figure 2. Cambrian crustacean-like arthropods. A. *Waptia fieldensis* Walcott, 1912, USNM 57682, u.v. illumination from top, x 1.5 (photograph by C.P.Hughes). Right lateral view, the carapace tilted forward; the filamentous thoracic appendages – g, the apodous abdomen and caudal furca – cf. B. *Plenocaris plena* (Walcott) 1912, USNM 57700, u.v. illumination from upper left x 10 (Whittington 1974, pl.XII, fig.1: plates X and XII are transposed in this paper). Oblique view showing the carapace, and trunk and caudal furca – cf; one antenna – an; trace of the appendages posterior to antenna – ap. C. *Branchiocaris pretiosa* (Resser) 1929, USNM 189028, u.v. illumination from right, x 1.5 (Briggs 1976, pl.3, fig.3). Dorsal view with valves partly removed to reveal multisegmented trunk and beneath it the lamellate appendages – ap. The antennae – an, and ?chelate limbs – c, of the head are evident, particularly on the left. D. *Dala peilertae* Müller 1981, x 180 (photograph provided by K.J.Müller). Ventral view of a fragment of the trunk showing endopodites of postoral appendages and two lines along the thorax possibly representing a food-groove.

all (presumably secondarily) uniramous. Four somites bearing a segmented 'walking leg' are followed by six legs with a filamentous 'gill-branch'. All ten somites may belong to the thorax, as there is evidence of up to three small appendages anterior of the 'walking legs' in addition to the antennae. Thus the very crustacean-like *Waptia* may display the diagnostic segmentation of the crustacean head. If so *Waptia* does not fall readily into any living class but shows greatest affinity with the Branchiopoda.

3.3  *Plenocaris* (Fig.2b) also has a bivalved carapace which covers the anterior part of the body. The head bears a pair of large segmented antennae, the trunk consists of 13 somites and terminates in a telson with flap-like caudal furca. Only the anterior three or four trunk somites bear limbs, which are poorly preserved, and they may constitute a thorax. *Plenocaris* is superficially phyllocarid-like and was placed tentatively in the Phyllocarida by Whittington (1974, p.20) 'because it is like some fossil forms placed in this group', although he observed that 'at the same time the trunk does not exhibit all the typical phyllocarid characteristics'. *Plenocaris* lacks the diagnostic segmentation of the malacostracan trunk and the nature of the critical cephalic appendages is not known. It is not a phyllocarid. If *Plenocaris* were a crustacean, which seems unlikely, like *Waptia* it does not fall into any living class.

3.4  Two further bivalved genera with soft parts have been described trom the Burgess Shale, *Odaraia* (Briggs 1981) and *Branchiocaris* (Briggs 1976). Of the two *Odaraia* (Fig.1a, b) shows the greater similarity to the Crustacea. The head bears a pair of very large eyes anteriorly and a mandible posteriorly. There is no evidence of antennae projecting beyond the anterior margin of the head, but two short reflective features in front of the mandible may represent limbs. The first two appendages posterior of the mandible differ from those following and may represent maxillae. Thus *Odaraia* displays some evidence of the characteristic segmentation of the crustacean head. Most of the body is enclosed in a tube-like carapace. The first four limbs of the series posterior of the possible maxillae show some specialization compared to the rest of the trunk appendages (up to 40) and may have assisted in feeding. The telson of *Odaraia* (Fig.1b) is unique and completely unlike that in any extant crustacean. It bears three large blades or 'flukes', two lateral horizontal and one median vertical, which functioned in stabilization and steering.

*Odaraia* shows greatest similarity to the Branchiopoda of the Crustacean classes. The characters it shares with the branchiopods, the bivalved carapace and large number of trunk somites, are however symplesiomorphic, i.e. shared primitive (Briggs 1981, p.578, 579). *Odaraia* lacks the caudal furca and apodous posterior trunk somites characteristic of the Branchiopoda, and this and the unique telson morphology indicate that it belongs to a new extinct class.

*Branchiocaris* (Fig.2c, Briggs 1976) also shares a bivalved carapace and large number of similar trunk somites with the branchiopods but it is not a crustacean. The head bears a pair of stout antennae anteriorly, but posteriorly there is only one pair of appendages which are large, segmented, and probably chelate and which presumably functioned in passing food directly to the mouth. The trunk consists of up to 46 somites but bears a slightly greater number of lamellate appendages. These lamellae appear to have been strengthened by a short, segmented proximal element which runs along their inner margin. The nature of these proximal structures is not known. The anterior 12 or 13 are significantly larger than the rest, which may indicate a division of the trunk into anterior and posterior tagmata. The telson bears a pair of ventral broad pointed processes, which may not have articulated proximally, and presumably functioned in stabilizing the arthropod during swimming. The similarities between *Branchiocaris* and *Odaraia* are symplesiomorphic and therefore not a reliable indication of affinity.

3.5 Collins & Rudkin (1981) have recently described a new genus and species from the Burgess Shale, *Priscansermarinus barnetti,* as a probable lepadomorph cirriped. If their identification proves correct it extends the known range of the cirripeds back by about 140 million years from the Upper Silurian (Wills 1963). The evidence for their interpretation, however, is equivocal.

The description is based on a single slab with 62 specimens probably from a level higher than that yielding most of the Burgess Shale fossils,and of a slightly inferior preservation. Specimens range from 13 to 31 mm in length. The outline of *Priscansermarinus* is strikingly reminiscent of a lepadomorph barnacle — 'a rounded, irregularly ovoid body from which protrudes a thick stalk terminating in a thickened "collar"'. The stalk was apparently flexible. The collar consits of a ; slight flaring at the distal end of the stalk which appears to terminate in a circular disc. Variations in its preserved outline can be explained by different orientations to the bedding. The stalk bears 'a rounded, irregularly ovoid body'. The margins of this 'body' are not continuous proximally with those of the stalk, but make a high angle to it. No structures are preserved within the body, but Collins & Rudkin (1981, p.1007) describe it as being a 'darker grey than the rock matrix' with 'a concentration of iron oxide at the end away from the stalk'.

*Priscansermarinus* clearly did not have a calcified shell. The presence of a thin organic 'plate-like covering on the body' is argued on the basis of a 'medium-dark-grey area covering most of the body, along with the distinct triangular shape of the (smaller) grey area' mostly covered in iron oxide, in one of the paratypes (Collins & Rudkin 1981, Text-fig.1d). They suggest (p.1001) that these grey areas may 'represent chitinous plates, similar in composition to the chitinous bivalved capitulum of *Cyprilepas*', a Silurian lepadomorph described by Wills (1963) attached to *Baltoeurypterus.* Collins & Rudkin thus consider that these plates did not reach the margins of the ovoid body, i.e. 'do not cover all of the capitulum at maturity' (p.1014), even though the bivalved capitulum of *Cyprilepas* is apparently chitinous throughout. The variability of the irregular outline of the grey area and associated iron oxide, and its apparently consistent presence in the centre of the ovoid body (Collins & Rudkin 1981, pl.1) irrespective of likely orientation provides no substantial evidence for the suggested plates. The variation in orientation to bedding indicated by the outline of specimens illustrated in Collins & Rudkin's text-figure 1C and D, and on the slab (Plate 1) suggests that in at least some cases flattening should have resulted in the folding of plates at the margin of the ovoid body. The grey area and associated iron oxide (which is unlikely to be original in the low oxygen environment postulated for the preservation of these soft-bodied organisms and therefore probably replaces pyrite) might equally be the result of precipitation in an area high in organic material, i.e. the soft tissues of the body.

Collins & Rudkin (1981) performed a series of experiments in squashing recent goose barnacles between glass plates (an invalid analogue for compression in a matrix: Briggs & Williams 1981) to show that if appendages were present they would not have been squeezed out beyond the supposed valves (an inevitable consequence of lateral sediment confining pressures anyway). If, however, valves or plates were present, they must presumably have separated as in recent barnacles to allow feeding and respiratory currents to reach the appendages. In this case it is possible that sediment would have penetrated the valves during burial as in other Burgess Shale arthropods (e.g. Briggs 1979). Thus specimens would have split between the valves or at least allowed their removal by preparation to reveal appendages beneath. Collins & Rudkin record no evidence that this is the case. They (p.1012) make the unwarranted inference that *Priscansermarinus* had two chitinous plates in the capitulum — 'knowing that if *Priscansermarinus* had had thick calcareous plates, they would probably have been preserved as iron oxide infillings, and that *Cyprilepas* had a chitinous bivalved capitulum, it is reasonable to presume that *Priscansermarinus* had two chitinous plates on the capitulum'.

The living group most similar *in outline* to *Priscansermarinus* is the lepadomorph barnacles, and like them *Priscansermarinus* appears to have been attached to the substrate by a stalk. The Burgess Shale fauna, however, includes a number of forms which cannot be assigned even to living phyla (Conway Morris 1979, Conway Morris & Whittington 1979). All, however, show at least some similarity to extant groups: *Dinomischus* Conway Morris 1977, for example, resembles an entoproct, *Opabinia* is superficially arthropod-like. Confirmation of the interpretation of *Priscansermarinus* as a barnacle therefore 'must await discovery of specimens with identifiable cirriped structures, such as cirri, within the capitulum' (Collins & Rudkin 1981, p.1011). Until then its affinities remain uncertain; it may not be an arthropod.

3.6 An important fauna of upper Cambrian arthropods from black nodular limestones in southern Sweden is presently under study by Müller (1979, 1981). In contrast to the Burgess Shale which preserves arthropods up to 300 mm long (*Tegopelte*, Simonetta & Delle Cave 1975), these upper Cambrian arthropods are microfossils, complete individuals or fragments less than 1 mm in dimension. They are collected as phosphatized residues in a plastic screen when the limestones containing them are dissolved in acetic acid. In spite of the minute size of the fossils, exquisite details of the appendages are revealed by the Scanning Electron Microscope. By far the most common arthropods in the samples are ostracods. Müller has demonstrated that the carapaces were originally phosphatic, although the appendages are secondarily so. He used this criterion to erect a new suborder, the Phosphatocopina.

The affinities and taxonomy of the Phosphatocopina have been a matter of controversy (Müller 1979, p.24). Kozur (1974), in particular, disputed Müller's (1964) contention that the carapaces are primarily phosphatic, but the validity of Müller's observation has since been demonstrated. Müller (1964), prior to his detailed study of the appendages, assigned the suborder Phosphatocopina to the Order Bradoriida Raymond 1935, retaining Raymond's name for Sylvester-Bradley's 1961 Order Archaeocopida. Sylvester-Bradley (1961) discussed the arguments for classifying the Archaeocopida (Bradoriida) as Ostracoda rather than Conchostraca (Ulrich & Bassler 1931) or Archaeostraca (Raymond 1935), based on the available evidence at that time, which was exclusively that of the carapace. He concluded (p.Q101) that these Cambrian fossils 'differ from later Ostracoda, but the differences are not profound' and therefore regarded them 'as true Ostracoda belonging to a primitive order'. Jones & McKenzie (1980) discussed the Order Bradoriida more recently. They regard the suborder Phosphatocopina as a valid taxon occupying a primitive position within the Ostracoda. Of the other suborder, the Bradoriina, they consider some families (Bradoriidae, Beyrichonidae) ancestral ostracodes, but point out that 'others (e.g. Indianidae) may be related to phyllocarid-like or branchiopod-like crustaceans'. In the absence of any evidence of soft-part morphology this is a reasonable but untestable assertion.

The appendages of three genera of phosphatocopines are now known (Müller 1979, 1981): *Falites, Vestrogothia* and *Hesslandona* (see McKenzie et al., this volume). Their discovery prompts two related questions: (1) is their assignment to the Ostracoda justified? and (2) how do the limbs differ from those of later ostracodes? Müller grouped *Falites* and *Vestrogothia* in the same family, Falitidae, but proposed a separate family Hesslandonidae for *Hesslandona* with its unique double hinge. The appendages of these two families and those of living ostracodes are compared in Table 1.

Schram (1982) comments that 'there is a serious difficulty in considering these fossils as ostracodes' pointing out that the limbs 'are actually more like maxillopodan types rather than ostracodes, especially when one considers cirripeds and naupliar copepod forms'. This observation is certainly true of the second antenna and mandible (see Sanders 1963a, figs.34, 35) and to a slightly lesser extent of the trunk limbs. The similarity, however, is simply due to the relatively primitive nature of the maxillopodan appendages (particularly the second antenna and mandible; Hessler & Newman 1975, p.441) and is thus a symplesiomorphy (shared primitive character). It is therefore of limited use in determining relationships and certainly less significant than the combination of a bivalved carapace and reduced trunk which suggests an affinity between the Phosphatocopina and other ostracodes. Schram (1982) also raises the possibility that the posterior end of *Vestrogothia spinata* 'may well have been a segmental and appendaged tagma', but there is no evidence whether or not this was the case.

In the absence of preserved appendages the true affinity of ostracode-like forms remains equivocal unless the evidence of carapace morphology is conclusive. The study of larger bivalved forms in the Burgess Shale has amply demonstrated that diverse morphologies can reside in essentially similar carapaces; there is no reason to suppose that this may not also have been the case in smaller bivalved arthropods.

Sylvester-Bradley (1961) diagnosed the soft-part morphology of the Subclass Ostracoda as 'head undifferentiated, bearing 4 pairs of cephalic appendages, 1 to 3 pairs of thoracic appendages and a pair of furcal rami, but no abdominal appendages'. The phosphatocopines are similar to ostracodes in the bivalved carapace, the enclosure of the body within it, and the small number of appendage-bearing somites (which is not simply a function of size; small Crustacea, like *Hutchinsoniella*, have many more somites). The lack of specialization of the post-antennal appendages eliminates the phosphatocopines from the ostracodes *sensu stricto* (as defined by Sylvester-Bradley, and commonly understood), as does the phosphatized carapace. Neither of these characters, however, are particularly reliable indicators of affinity. Briggs & Fortey (1982) have confirmed that the cuticle of aglaspids was also primary phosphate, but point out that this hardly implies a relationship between them and the phosphatocopines, even though phosphate is rarely used in arthropod cuticle. Conversely the phosphatic composition of the Phosphatocopina does not preclude a relationship between them and other ostracodes with calcified shells (cf. the brachiopods which include taxa with both phosphatic and calcareous shells which are nevertheless related, as convincingly argued by Rowell 1981). Some living ostracod carapaces (e.g. the Halocyprididae) are not mineralized in any way.

Similarly the lack of differentiation of the appendages, particularly in the Falitidae, is a primitive character. The concept of a progressive evolutionary specialization of arthropod appendages from the front of the head posteriorly is well established (e.g. Sanders 1963b).

Table 1. Table comparing the nature of the appendages of the two families of Upper Cambrian Phosphatocopina (Müller 1979, 1981) with those of living Ostracods (based on Howe, Kesling & Scott 1961).

|  | Falitidae | Hesslandonidae | Living Ostracods |
|---|---|---|---|
| 1st antenna | Biramous: comb-like exopod, 16 podomeres with spines; shorter endopod with large endites | Uniramous: about 9 podomeres; distal setae | Uniramous: typically 8 podomeres; may be fusion to give 5 |
| 2nd antenna | Biramous; similar to 1st antenna | Biramous: exopod of 20-23 podomeres with setae; shorter endopod with elongate endites | Biramous: variable, but usually locomotory |
| 3rd appendage | Similar to antennae, fewer exopod podomeres | Similar to 2nd antenna | Mandible: masticatory teeth on protopod; small exopod |
| 4th appendage | Similar to 3rd but smaller | Biramous: exopod with vibratory plate; endopod with projecting endites | Maxilla: variable but used in feeding; commonly endites on protopod |
| 5th appendage | Similar to 4th but smaller | Similar to 4th | First thoracic limb: may function in feeding (maxilliped) or be similar to succeeding thoracic limbs; may be dimorphic |
| 6th appendage | Poorly known | Similar to 4th | Second thoracic limb: variable, may be dimorphic |
| 7th appendage | Poorly known, possibly uniramous (further limbs may have been present) | Unknown: ?absent | Third thoracic limb: only developed in some orders, used for valve cleaning |

*Hesslandona* represents a later stage in this process than *Vestrogothia* and *Falites,* but neither approach the degree of cephalization seen in living ostracodes. Some of the Burgess Shale arthropods, such as *Canadaspis* (Briggs 1978) and *Sidneyia* (Bruton 1981) for example, indicate that trunk tagmosis may proceed more rapidly than specialization of the posterior head appendages. Thus Briggs (1978) has argued that *Canadaspis* shows the diagnostic tagmosis of the leptostracan Malacostraca even though the maxillae are still very similar to the thoracic limbs. The Phosphatocopina, similarly, have acquired the overall structure of the ostracod trunk prior to the full differentiation of the cephalic limbs. This is to imply that the nature of the bivalved carapace and segmentation of the trunk are significant at a higher taxonomic level than the detailed structure of the appendages. Müller (1979, p.26) has pointed out that the large number of podomeres in the exopods, and the presence of endites (presumably for feeding) even on the antennae are also primitive features. Although the antennules and antennae of the Falitidae have large endites used in feeding and are therefore in some sense postoral, their point of attachment in relation to the mouth is not dramatically different to that in living ostracodes where they have lost their feeding function and become preoral. This feeding function of the antennule and antennae is a further primitive feature of the phosphatocopines although the antennule of *Hesslandona* does not display it (and has presumably lost the endites secondarily). This shift of the antennae to a preoral position occurs in the embryology of living ostracodes; its limited progress in the

Cambrian forms may be partially due to the fact that 'only stages up to the preadult have been found with appendages' (Müller 1979, p.6). In conclusion, Müller's assignment of the phosphatocopines to Ostracoda appears vindicated. It may be argued that they occupy a primitive position, prior to complete cephalization, equivalent to that of *Canadaspis* within the phyllocarid Malacostraca.

There is some evidence of the soft part morphology of the other suborder of the Order Bradoriida, the Bradoriina, although the appendages are unknown. McKenzie & Jones (1979) have described a single specimen of a new bradoriid genus from the Middle Cambrian of Western Queensland which appears to show four segments. Unfortunately there is no evidence to indicate whether or not these four segments, interpreted by McKenzie & Jones as thoracic, include that bearing the maxilla or its equivalent. This appendage (the fourth) on *Hesslandona* for example (Table 1) shows greater similarity to the trunk limbs which follow it than to those appendages anterior to it. Thus the evidence provided by this unique specimen in support of the contention that 'there were originally four pairs of post-cephalic limbs' (including a second maxilla) 'and that the second maxillae have been lost' (implied but not specifically stated by McKenzie & Jones 1979, p.444) is inconclusive. The morphology revealed, however, is not inconsistent with the interpretation of this arthropod as an ostracod.

3.7 Müller's (1979, 1981) upper Cambrian fauna from southern Sweden includes other arthropods which he interprets as Crustacea, including, as might be expected in such a microscopic fauna, larval forms. These have yet to be studied and described in detail. The larval forms include metanauplii and smaller nauplii down to 0.1 mm in length excluding the appendages. One metanauplius described and figured by Müller (1981, Fig.5) has a large hypostome and a structure possibly equivalent to the lateral eyes. One pair of uniramous and two of biramous appendages are followed by a pair of partially developed limbs. Other metanauplii, of different groups, have up to five pairs of such partially developed limbs.

More interesting than these larval forms is *Dala peilertae* (ca.1.5 mm; fig.2d) which Müller interprets as a cephalocarid. In spite of its minute size the final (of two, possibly three) growth stages of this arthropod is interpreted as the adult because three pairs of reproductive 'bulbs', absent in the earlier stages, are developed at the posterior end of the trunk. Müller interprets the first two pairs of biramous appendages in *Dala* as part of a 'frontal region', although only the first is significantly differentiated from the rest. Only the proximal part of this first appendage is known. It consists of a larger coxal area than in the following appendages, and bears the basal segments of two rami which Müller assumes were antenniform and projected forwards. The basal segments bear setose projections (endites) which flank the mouth and presumably functioned in feeding. A circular disc-like structure lies between this first pair of appendages and presumably represents the labrum.

There follow eight pairs of similar, presumably postoral, biramous appendages. The first four apparently belonged to the dorsal (?cephalic) shield. Proximally these eight appendages consist of a precoxa of eight short annulae and a coxa with up to six divisions. The first segment of the coxae bears a setose pre-epipodite which Müller considers (1981, p.147) 'probably served for mastication and transport of food to the mouth' although they appear too far apart in his reconstruction (1981, fig.3) to function in either way, but particularly in mastication. Distally the coxa bears endites, one on each of the segments. The basis in all these appendages also bears two endites. The exopodite (1981, fig.4) consists of two long podomeres and terminates in a fringe of sensory setae or bristles. There are four podomeres in the endopodite which also terminates in a fringe of setae. The posterior part of the trunk comprises an abdomen of four somites and a telson with furca.

Müller concludes that *Dala* should be assigned to the Cephalocarida and comments that 'it is remarkable that the late Cambrian representative has been found in an environment of very soft sediment, similar to the modern environment . . . little change seems to have occurred in the living conditions of the group for 500 m.y.'.

The interstitial environment may have allowed a morphology such as that in *Dala* and the living Cephalocarida to survive largely unchanged. This indeed is implied by Sanders' (1963b, p.172) interpretation of the Cephalocarida as representative of 'a retention of an early stage of morphological development in the Crustacea' although Schram (1982), for example, has argued that the group is in fact specialized. There are, however, obstacles to arguing an affinity between *Dala* and the living forms. Chief among these is the structure of the tagmata in living cephalocarids which is uniform in nine species of four living genera: eight thoracic somites bearing seven or eight thoracopods, and 11 abdominal somites plus a telson. *Dala,* on the other hand, has only four thoracic somites (assuming that the anterior four appendages in the 'trunk' series belong to the cephalic shield), and four abdominal somites plus a telson. This alone is sufficient to separate it from the living Cephalocarida at class level. The similarities upon which Müller presumably based his argument for assigning *Dala* to the Cephalocarida are mainly symplesiomorphic (i.e. primitive rather than shared evolutionary novelties), and cannot be used to imply relationship. The small size, and lack of differentiation of the posterior 'cephalic' appendages from those of the thorax do not need justification as primitive features. An apodous abdomen with caudal furca also occurs, for example, in anostracan branchiopods, ostracodes, mystacocarids, and the Burgess Shale phyllocarid *Perspicaris* (Briggs 1977b, 1978). A cephalic shield is also present in the copepods and in a wide variety of Burgess Shale arthropods (Briggs & Whittington 1981). *Dala* is more primitive in some respects than the adult cephalocarid, and it is therefore not surprising that similarities to the naupliar stages of *Hutchinsoniella* (Sanders 1963b) are more striking. These include the presence of an endite in the second antenna and a mandibular palp, and the relatively larger size of the cephalic shield. These characters are likewise primitive and are not reliable indicators of relationship. The affinities of *Dala* remain uncertain although it shows a greater similarity to the Crustacea than any other major arthropod taxon.

## 4 DISCUSSION

This review of the early Crustacea and crustacean-like arthropods confirms the presence of only two of the major living taxa in the Cambrian: (1) the Class Malacostraca, represented by the bivalved phyllocarid *Canadaspis,* (2) the Class Ostracoda, represented by the Phosphatocopina (and probably other forms with calcareous as opposed to phosphatic carapaces of which the appendages are not known). A number of additional bivalved arthropods from the Burgess Shale may be Crustacea (*Odaraia, Waptia*) and show greatest similarity to the Branchiopoda of the living classes. It is doubtful that they are branchiopods, however, as many of the shared characters are symplesiomorphic.

*Dala* (Müller 1981) is not a cephalocarid and as yet there is no unequivocal fossil representative of this class. The earliest well-documented branchiopod is *Lepidocaris* (Scourfield 1926) from the Upper Devonian Rhynie Chert of Aberdeenshire. The evidence for the presence of the Cirripedia in the Cambrian is equivocal; *Priscansermarinus* (Collins & Rudkin 1981) may not even be an arthropod. The fossil record of the Class Maxillopoda dates from the Upper Silurian record of the probable lepadomorph *Cyprilepas* (Wills 1963). Of the other maxillopodan subclasses, the Branchiura and Mystacocarida have no fossil record; Cressey & Patterson (1973) have described well-preserved parasitic Copepoda from the gill chambers of Lower Cretaceous bony fish (earlier records of copepod parasitism are equivocal; Conway Morris 1981), but the earliest record of free-living forms is Miocene (Palmer 1969). Only the subclass Phyllocarida of the Malacostraca occurs in the Lower Palaeozoic.

The known Cambrian Crustacea reveal a number of features which may have wider implications for the evolution of the group (although it must be emphasized that the sample is very small!). First, the structure of the appendages is primitive. They are biramous and characterized by a large number of segments (up to 14 in the inner ramus of *Canadaspis* (Briggs 1978, fig.108) and 18 in the exopod of the phosphatocopines (Müller 1979, fig.34c)) and by several sets of medially directed proximal spines. Apart from being biramous the appendages show considerable variability. Second, the posterior cephalic appendages are little differentiated from those of the trunk. Third, the configuration of the trunk segments characteristic of both classes is well-developed (the eight-segmented thorax and seven-segmented abdomen of the malacostracan *Canadaspis;* the reduced trunk of the phosphatocopine ostracodes). This indicates that tagmosis of the trunk took place at an earlier stage in the evolution of at least these two crustacean classes than much specialization of the appendages or complete cephalization. Thus less significance should probably be attached to appendage morphology than has been to date in considerations of the affinities of the Crustacea and their classification and phylogeny at high levels in the taxonomic hierarchy. The differentiation of thorax and abdomen in *Canadaspis* may have been a response to a walking benthic mode of life as opposed to swimming. Manton (1969, p.52) suggested that the abdominal appendages 'might have disappeared had the Malacostraca remained entirely bottom living'. The specialization and reduction of the ostracode trunk was probably a response to the functional advantages of enclosing the entire body within the bivalved carapace.

Previous considerations of crustacean affinities (Manton 1977 and references therein; Cisne 1974, 1975, 1981, Hessler & Newman 1975, Schram 1978, 1982) have been largely concerned with their relation to the Chelicerata, Uniramia and the extinct trilobites, and have been mainly based, directly or indirectly, on neontological data. The vast majority of Cambrian arthropods without mineralized exoskeletons do not belong to these taxa, and as they may include morphologies intermediate between them, these additional arthropods are an obvious line of evidence for crustacean affinities and early evolution. The potential significance of these arthropods has been concealed until recently by the inclusion of many with the trilobites in the taxa Trilobitomorpha and Trilobitoidea (Størmer 1959, Hessler & Newman 1975, Schram 1978, 1982). The Class Trilobitoidea (Størmer 1959), erected to accommodate many of the non-trilobite Burgess Shale arthropods with so-called 'trilobite-like' limbs has little validity (Whittington 1979). The 'trilobitoids' include a wide variety of arthropods, most with appendages that are in no sense trilobite-like. Nevertheless, many of the recently published descriptive papers on the Burgess Shale arthropods omit any discussion of an alternative classification into higher taxa. It was considered appropriate to postpone consideration of interrelationships until the arthropod fauna was more completely known, rather than tackling the classification piecemeal and revising it as the descriptive work progressed.

Only arthropods with preserved appendages provide the critical evidence for a consideration of interrelationships and apart from Müller's (1979, 1981) upper Cambrian microfauna from Sweden, and the arthropod *Aglaspis* from the upper Cambrian of Wisconsin (Briggs, Bruton & Whittington 1979), these are confined to the Burgess Shale (Whittington 1979). The majority of the Cambrian arthropods do not fall into discrete, easily identified groups. Large morphological differences separate most of the genera, and the characters they share tend to be common to many arthropods. A 'traditional' approach to the classification of this heterogeneous group of arthropods reveals little evidence of interrelationships. This suggested a 'pattern' of evolution to Whittington (1979, fig.2) comprising a large number of independent lines, what Manton (in Manton & Anderson 1979, p.281 — following Størmer) referred to as a phylogenetic 'lawn'. The only clusters of Cambrian genera in this 'lawn' are a small group of Crustacea (Canadaspidida) and the trilobites. Otherwise individual genera are shown as parallel lines.

The completion of descriptive work on several genera between 1979 and 1981 brought the total number of well-known Cambrian arthropods with preserved appendages to over 20 and prompted Briggs & Whittington (1981) to make a preliminary attempt to identify relationships between them. While some taxa of equivalent rank to the four major ones (Trilobita, Crustacea, Chelicerata, Uniramia) may be present among these Cambrian arthropods it is clearly highly unlikely that many represent independent lines of arthropodization. The problem of identifying relationships was tackled in two main ways: (1) using numerical taxonomy (principal components analysis) to identify simple morphological similarity, and (2) by cladistic methods. The Burgess Shale arthropods are an ideal subject for a taxonomic study using either method. They can be treated as representative of a single time plane, thus avoiding any stratigraphic bias in the interpretation. They present no problem of sample selection. The genera are clearly discrete and provide far more characters for analysis than are normally available in fossil examples.

The numerical technique (Briggs & Whittington 1981) revealed a wide morphological spread with only one large cluster, around *Alalcomenaeus* (Whittington 1981). *Alalcomenaeus* occupies this central position due to its generalized morphology. A simple cephalic shield is followed by 12 uniform trunk tergites and a flattened paddle-shaped telson. There is a pair of pedunculate eyes. A single differentiated anterior appendage 'with a single long, slim distal portion' (Whittington 1981, p.353) is followed by three pairs of biramous appendages in the cephalon which are essentially similar to the 12 trunk pairs that follow. These biramous appendages comprise an outer lamellate branch fringed with filaments and an inner branch which is armored with inward facing spines. Many of the characters shared between the arthropods in the cluster around *Alalcomenaeus* are either of little significance, or common to almost all the Cambrian arthropods under consideration. The arthropods of the cluster are united, however, in the possession of a simple cephalic shield associated with a series of dorsal tergites on the trunk. The remaining genera, apart from the uniramian-like *Aysheaia* (Whittington 1978) which lacks an exoskeleton, possess a carapace that originates in the head region and extends back over the trunk, and are much more widely scattered. This carapace is bivalved in the Crustacea and crustacean-like genera which are loosely grouped; *Branchiocaris* and *Odaràia* are widely separated from the rest. The carapace of *Burgessia* (Hughes 1975) and cephalic shield of *Marrella* (Whittington

1971) are not bivalved, and these arthropods occupy different positions closer to the main cluster than to the Crustacea.

While a numerical analysis reveals basic similarities between genera it has drawbacks as a method of determining affinities. It does not differentiate between shared derived characters (evolutionary novelties) which indicate relationship, and primitive characters, which do not. Nor does it indicate the hierarchical grouping of the taxa. The formulation of a cladogram showing a nested pattern of synapomorphies achieves both of these aims, but requires an assessment of which characters are significant at which level in the taxonomic hierarchy. A character state considered derived at one level becomes primitive for those below. The high degree of convergent evolution in the arthropods, particularly in minor attributes, makes the selection and ordering of synapomorphies difficult. This difficulty is augmented by the wide morphological separation of taxa; a number of clearly derived characters occur in only one of these Cambrian genera and are therefore of no use in determining relationships. For the purposes of the study the arthropods under consideration are assumed to be monophyletic. The Uniramia are not represented in the Cambrian and the uniramian-like *Aysheaia* is omitted from the cladograms; it stands distinct from the other arthropods in lacking a jointed exoskeleton. Any grouping within the cladograms could be removed and considered to have evolved independently from a soft-bodied ancestor, but there is no objective way of demonstrating such an origin.

The cladogram presented in Figure 3 is a modification of that of Briggs & Whittington (1981, fig.2). *Emeraldella* (Bruton & Whittington, in preparation) and *Dala* (Müller 1981) have been added. The 'new genus' of Briggs & Whittington's cladogram is identified as *Sarotrocercus* Whittington,1981 and the position of *Leanchoilia* is changed on the basis of more accurate information on the cephalic appendages (Bruton & Whittington, in preparation). *Aysheaia* is omitted. This cladogram separates a group of bivalved Crustacea and crustacean-like arthropods including *Perspicaris, Canadaspis, Waptia,* the Phosphatocopina and *Odaraia* from the rest. These genera, together with *Plenocaris, Branchiocaris* and the univalved *Burgessia* are united in the possession of a carapace extending posteriorly beyond the head, rather than a cephalic shield and trunk tergites. This major dichotomy, identified by the principal components analysis (Briggs & Whittington 1981) is unsatisfactory, however, when considered from the point of view of crustacean origins and affinities, as both forms occur within the group. A bivalved carapace occurs in the Phyllocarida, Notostraca and Conchostraca, Cirripedia and Ostracoda for example, and a cephalic shield in the Cephalocarida, Copepoda, and Isopoda. Hessler & Newman (1975) were unable to agree on which condition was primitive. A bivalved carapace, however, does not occur in the Trilobita or Chelicerata and may therefore be a useful indicator of crustacean affinity in some cases (although the bivalved *Branchiocaris* is clearly not a crustacean). None of the arthropods under consideration which lack a bivalved carapace shows any obvious affinity with the Crustacea, with the exception of *Dala*. Neither *Dala* nor *Marrella* fall into the two major groups identified by this cladogram; although they possess a cephalic shield both lack trunk tergites and their relationship is therefore unresolved. Only *Dala* of the two, however, approaches the degree of cephalization diagnostic of the Crustacea with five appendages borne beneath the cephalic shield.

A second character important in classifying living Crustacea is the nature of the trunk tagmata. This, however, cannot be used as a fundamental dichotomy. Trunk tagmosis *per se* must have originated convergently in a number of groups. The arrangement of trunk tagmata is a very variable character in the living Crustacea; in the notostracan, anostracan and

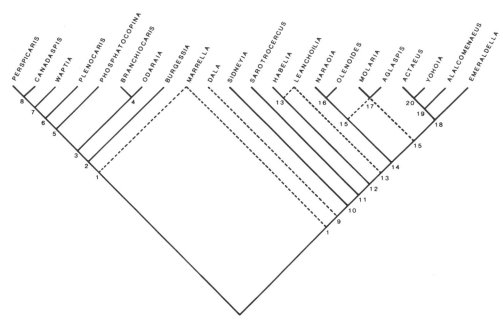

Figure 3. Cladogram illustrating one hypothesis of the interrelationships of Cambrian arthropods; the presence of either a carapace or a cephalic shield and trunk tergites is incorporated as a fundamental dichotomy. Key to synapomorphies as follows: (1) Cephalic shield (dotted lines indicate relationship is undecided), (2) Carapace (extending posteriorly beyond cephalon), tergites lacking, (3) Carapace bivalved, (4) Telson processes (in this case almost certainly convergently acquired and therefore invalid), (5) Distinction between thorax and abdomen, (6) Paddle-shaped limbs in caudal furca. (7) Incorporation of limbs (other than antennae) into head posteriorly, reduced second antenna, (8) Phyllocarid trunk segmentation, (9) Unresolved; the position of *Dala* is uncertain, (10) Tergites with pleurae following cephalic shield, (11) Incorporation of one limb (other than antennae) into head posteriorly, (12) Two posterior cephalic limbs, (13) Unresolved, (14) Three posterior cephalic limbs, (15) Unresolved, (16) Fused posterior shield or pygidium, (17) Loss of pre-telson appendages, (18) Four posterior cephalic limbs, (19) Flat paddle-like telson, (20) Loss of inner rami of trunk limbs.

diplostracan branchiopods the only evidence of tagmosis may be an apodous abdomen. The nature of the trunk tagmata may naturally be a useful character at levels higher in the cladogram, as in the presence of a malacostracan trunk in the Canadaspidida.

The only character which serves to distinguish the whole diversity of living Crustacea from the other major groups and which, unlike the nature of the mandibular mechanisms or details of embryology, can be used in these exceptionally preserved fossils, is the arrangement of the head appendages as a reflection of cephalic segmentation. The arrangement of head appendages was used at higher levels in the cladogram by Briggs & Whittington (1981) but they attempted to distinguish between anterior (supposed preoral) and posterior (postoral) cephalic limbs. This approach has the potential advantage of distinguishing between the three major groups on the basis of simple 'formulae' (simplified of necessity). Thus the Trilobita have one preoral pair of antennae and three biramous appendages in the head, i.e. 1 + 3 (Whittington 1975, 1977, Cisne 1975, 1981, Stürmer & Bergstrom 1973), the Chelicerata 1 + 5 in the cephalothorax (but see Stürmer & Bergstrom 1981) and the Crustacea 2 + 3. In the Crustacea, however, the second antenna is fundamentally postoral (Hessler &

Newman 1975) migrating forwards during ontogeny. The antennal endites are not lost until the end of the naupliar stages (e.g. see Sanders 1963a, fig.26) and this may be reflected in the adult morphology of some Cambrian forms like *Dala,* for example. In addition, the posterior cephalic limbs may be very similar to those of the trunk (in the Cephalocarida and Canadaspidida). In a consideration of crustacean affinities among the Cambrian arthropods it may therefore be more appropriate to consider the total number of cephalic appendages rather than attempting to distinguish between those which are preoral and postoral. The arrangement of cephalic appendages in the arthropods under consideration is shown in Table 2, the resulting cladogram in Figure 4. The implied homologies between cephalic appendages cannot, of course, be substantiated and clearly some limbs may not have been preserved (in *Plenocaris* Whittington 1974, for example).

The nested sets in this cladogram (Figure 4) are based on the concept of cephalization as a sequential process, commencing anteriorly and incorporating successive appendages into the head posteriorly. This approach separates the Crustacea from the other arthropods as they are the only group with five appendages in the cephalon (apart from *Emeraldella* which has six). The number of cephalic limbs, however, like the structure of the exoskeleton, has shortcomings as the main basis for the cladogram. Some taxa with a common number of cephalic appendages show no other evidence of affinity. *Branchiocaris, Sarotrocercus* and *Marrella* all have two cephalic appendages, but their morphologies are otherwise widely divergent and their relationships cannot be resolved. The Phosphatocopina share few attributes, apart from four cephalic appendages, with the arthropods with which they

Table 2. Configuration of head appendages in Cambrian arthropods used in the compilation of the cladograms (Figs.3, 4). The ratio of preoral and postoral appendages is expressed as a 'formula', the preoral number given first. Assuming that no more than two appendages will be preoral (the case in living arthropods) there are only three possible configurations for each total number of cephalic appendages. Not all the possible combinations ('formulae') actually occur. In many of the genera the arrangement is uncertain due to poor preservation.

| Appendage number | Formulae | Arthropods |
|---|---|---|
| 1 | 0 + 1 | |
| | 1 + 0 | *?Plenocaris, Sidneyia* |
| 2 | 0 + 2 | |
| | 1 + 1 | *Branchiocaris, Sarotrocercus* |
| | 2 + 0 | *Marrella* |
| 3 | 0 + 3 | |
| | 1 + 2 | *Habelia, Leanchoilia* |
| | 2 + 1 | |
| 4 | 0 + 4 | *Yohoia* |
| | 1 + 3 | *Olenoides, Naraoia, Burgessia, Molaria, ?Aglaspis, Actaeus, Alalcomenaeus* |
| | 2 + 2 | *?Phosphatocopina* |
| 5 | 0 + 5 | |
| | 1 + 4 | *Dala* |
| | 2 + 3 | *?Perspicaris, Canadaspis, ?Waptia, ?Odaraia* |
| 6 | 0 + 6 | |
| | 1 + 5 | *Emeraldella* |
| | 2 + 4 | |

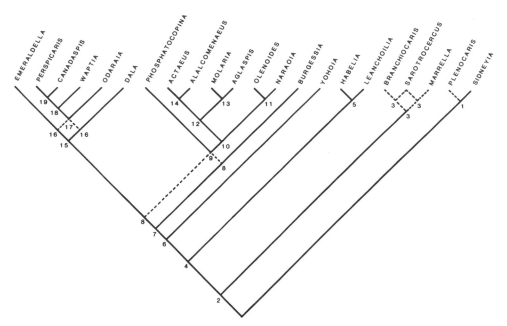

Figure 4. Cladogram illustrating an alternative hypothesis of the interrelationships of Cambrian arthropods; the main basis for nested sets is the number of cephalic limbs. Key to synapomorphies as follows: (1) One cephalic limb; the position of *Plenocaris* is uncertain — more limbs may have been present but not preserved, (2) Two cephalic limbs, (3) Unresolved, (4) Three cephalic limbs, (5) One antenna and two postoral limbs, (6) Four cephalic limbs, (7) Four cephalic limbs including one antenna, (8) Unresolved, (9) Distinction between thorax and abdomen, (10) Tergites with pleurae following cephalic shield, (11) Fused posterior shield or pygidium, (12) Telson, (13) Loss of pre-telson appendages, (14) Paddle-like telson, (15) Five cephalic limbs including one antenna, (16) Unresolved, (17) Carapace bivalved, (18) Paddle-shaped limbs in caudal furca, (19) Phyllocarid trunk segmentation.

are grouped. They differ in the possession of a bivalved carapace, two antennae, and a reduced abdomen and clearly show greater affinity with the Crustacea. A further drawback of the cephalic appendages as a basis for a cladogram is the inherent implication that the trilobites, crustaceans and chelicerates form an evolutionary progression from four through five to six cephalic appendages.

The pairs of taxa at the top of the cladograms (Figs.3, 4), such as *Olenoides* and *Naraoia* (the trilobites), *Molaria* and *Aglaspis*, and *Actaeus* and *Alalcomenaeus*, are most likely to reflect true affinity. Larger sets are progressively more likely to incorporate convergences. The diagnostic cephalic segmentation of the Crustacea clearly separates them from the other taxa. The number of head appendages, however, is not always a reliable indication of affinity among the other arthropods particularly where smaller numbers of limbs are concerned. It is not surprising that smaller numbers of cephalic limbs evolved convergently in some arthropods (as indicated by the grouping of *Branchiocaris, Sarotrocercus* and *Marrella*) as presumably a head bearing two and then three pairs of appendages must have preceded the larger numbers diagnostic of the Trilobita, Crustacea and Chelicerata. It is difficult, however, to establish the affinities of an arthropod displaying a com-

bination of primitive characters. *Marrella,* for example, has two pairs of antennae, a crustacean feature, but no further appendages in the head (i.e. posterior of the mouth) and a
large number of uniform trunk somites bearing similar biramous appendages (Whittington
1971). This lack of advanced characters makes the determination of its relationships problematic (Figs.3, 4).

Although the Crustacea form a distinct group in both cladograms (apart from the Phosphatocopina in Figure 4) their affinities with the other Cambrian arthropods are not
clearly revealed, and the interrelationships of these other arthropods remain uncertain. The
trilobites form a discrete group characterized by fused posterior somites (the pygidium)
and the arrangement of the head appendages (although the latter alone is insufficient to
to define them). The remaining arthropods do not belong to the major taxa. Apart from
the trilobites, some of those with four appendages in the head (Fig.4) could theoretically
be primitive relatives of the crustaceans or chelicerates, representing a morphology prior
to the incorporation of the diagnostic number of limbs into the head. Some of those with
three or fewer cephalic appendages could likewise be primitive relatives of the crustaceans,
chelicerates or trilobites, or some other 'new' major taxon. If the hypothesis that advanced
characters once evolved do not tend to be lost (i.e. that evolution does not usually reverse)
is accepted some of these possibilities can be tested. *Marrella* has an unusual cephalic shield
and filamentous outer rami (Whittington 1971) but in most other respects the morphology
is primitive enough to give rise to any of the three major groups. The telson morphology
of many of the other arthropods arguably precludes a relationship to the trilobites (which
are characterized by a pygidium). Although a caudal furca is common in the crustaceans
and was argued to be a primitive character by Hessler & Newman (1975) its absence does
not preclude crustacean affinity. Thus *Alalcomenaeus* (Whittington 1981) might equally
be related to either the Crustacea or chelicerates with its simple cephalic shield and uniform
trunk tergites and biramous appendages (apart from one specialized anterior pair). These
more 'generalized' Burgess Shale arthropods give an indication of possible ancestral morphologies and as such provide possible links between the major taxa. Other genera are advanced in ways which make a relationship to the major taxa difficult to demonstrate. The
cephalon of *Yohoia,* for example, bears a highly modified 'great appendage' anteriorly
(Whittington 1974) together with three other segmented limbs; *Branchiocaris* has only two
pairs of cephalic limbs, an antenna and a segmented posterior appendage which may be
chelate (Briggs 1976).

The cladograms (Figs.3, 4) highlight the problems incurred in assessing the interrelationships of these early arthropods (Briggs & Whittington 1981). The wide range of morphologies is presumably a direct result of the first major radiation of the metazoans which took
place in late PreCambrian and early Cambrian times, and is paralleled in the echinoderms,
for example (Sprinkle 1981). The agents of natural selection and mass extinction subsequently reduced the range of 'body-plans' to those which provided a basis for later radiations. The fossil record gives no indication of the origin of the Crustacea or any other
arthropod groups. Nor can it provide much evidence to test the major neontological arguments for an independent origin of the major arthropod groups — embryology, and mandibular mechanisms (see discussions in Hessler & Newman 1975, Schram 1978, 1982). The
Crustacea is the only major living taxon already established in the Cambrian. A number of
other arthropods, such as *Yohoia* and *Branchiocaris,* represent morphologies advanced to
the extent that they may represent taxa of equivalent rank to the Crustacea, and could not
have given rise to the Trilobita, Crustacea, Chelicerata or Uniramia. Others, however, such

as *Marrella* and *Alalcomenaeus,* are relatively primitive and may reflect ancestral morphologies which could have evolved into the major groups, and in that sense represent intermediates between them. It seems unlikely that many of the groups represented in the cladograms are of separate origin or have even undergone independent 'arthropodization'. The occurrence of a wide diversity of arthropods displaying similar characters to those in the major groups, but in different combinations appears to negate this possibility.

## 5 ACKNOWLEDGEMENTS

The research upon which this paper is based was funded by a Royal Society Scientific Investigations Grant, Goldsmiths' College Research Fund, and NERC Grant GR3/285 awarded to H.B.Whittington. I am grateful to C.P.Hughes, K.J.Müller and H.B.Whittington for providing information and photographs. I am indebted to the Royal Society, the Geological Survey of Canada and the Palaeontological Association for permission to reproduce figures. S.Conway Morris, H.B.Whittington and F.R.Schram kindly reviewed the manuscript.

## DISCUSSION

HESSLER: Derek, by what definition of the ostracodes could one allow that the phosphatocopines be included?
BRIGGS: Only in the sense that their derived characters are, as I understand them, the bivalved carapace, clearly reduced trunk, and the fact that they have a similar arrangement of appendages in the head and trunk. I would then argue that the differences between phosphatocopines and other ostracodes are essentially primitive, at least at class level. All this would imply phosphatocopines are ostracodes, at least it does not argue against their being ostracodes.

## REFERENCES

Anderson, D.T. 1973. *Embryology and Phylogeny in Annelids and Arthropods.* Oxford: Pergamon Press.
Briggs, D.E.G. 1976. The arthropod *Branchiocaris* n.gen., Middle Cambrian, Burgess Shale, British Columbia. *Geol. Surv. Can. Bull.* 264:1-29.
Briggs, D.E.G. 1977a. Evolutionary significance of *Canadaspis,* the earliest positively identified crustacean. *J. Paleont.* 51, suppl. to no.2:4.
Briggs, D.E.G. 1977b. Bivalved arthropods from the Cambrian Burgess Shale of British Columbia. *Palaeont.* 20:595-621.
Briggs, D.E.G. 1978. The morphology, mode of life, and affinities of *Canadaspis perfecta* (Crustacea: Phyllocarida), Middle Cambrian, Burgess Shale, British Columbia. *Phil. Trans. Roy. Soc., London* (B)281:439-487.
Briggs, D.E.G. 1979. Crustacea. *McGraw-Hill Yearbook of Science and Technology:* 144-146. New York: McGraw-Hill.
Briggs, D.E.G. 1981. The arthropod *Odaraia alata* Walcott, Middle Cambrian, Burgess Shale, British Columbia. *Phil. Trans. Roy. Soc., London* (B)291:541-584.
Briggs, D.E.G., D.L.Bruton & H.B.Whittington 1979. Appendages of the arthropod *Aglaspis spinifer,* Upper Cambrian, Wisconsin, and their significance. *Palaeont.* 22:167-180.
Briggs, D.E.G. & R.A.Fortey 1982. The cuticle of the aglaspidid arthropods, a red herring in the early history of the vertebrates. *Lethaia* 15:25-29.

Briggs, D.E.G. & H.B.Whittington 1981. Relationships of arthropods from the Burgess Shale and other Cambrian sequences. In: M.E.Taylor (ed.), *Short papers for the second international symposium on the Cambrian system.* US Dept. Interior, Geol. Surv., Open File Report 81-743:38-41.

Briggs, D.E.G. & S.H.Williams 1981. The restoration of flattened fossils. *Lethaia* 14:157-164.

Bruton, D.L. 1981. The arthropod *Sidneyia inexpectans,* Middle Cambrian Burgess Shale, British Columbia. *Phil. Trans. Roy. Soc., London* (B)295:619-653.

Cisne, J.L. 1974. Trilobites and the origin of arthropods. *Science* 186:13-18.

Cisne, J.L. 1975. Anatomy of *Triarthrus* and the relationships of the Trilobita. *Fossils and Strata* 4:45-63.

Cisne, J.L. 1981. *Triarthrus eatoni* (Trilobita): anatomy of its exoskeletal, skeletomuscular, and digestive systems. *Palaeontogr. Am.* 9(53):96-142.

Collins, D.H. & D.M.Rudkin 1981. *Priscansermarinus barnetti,* a probable lepadomorph barnacle from the Middle Cambrian Burgess Shale of British Columbia. *J.Paleont.* 55:1006-1015.

Conway Morris, S. 1977. A new entoproct-like organism from the Burgess Shale of British Columbia. *Palaeont.* 20:833-845.

Conway Morris, S. 1979. The Burgess Shale (Middle Cambrian) fauna. *Ann. Rev. Ecol. Syst.* 10:327-349.

Conway Morris, S. 1981. Parasites and the fossil record. *Parasitology* 82:489-509.

Conway Morris, S. & H.B.Whittington 1979. The animals of the Burgess Shale. *Sci. Am.* 241:122-133.

Cressey, R. & C.Patterson 1973. Fossil parasitic copepods from a Lower Cretaceous fish. *Science* 180:1283-1285.

Hessler, R.R. 1979. Relative status of arthropod groups. In: M.R.House (ed.), *The Origin of Major Invertebrate Groups:*486-487. London: Academic Press.

Hessler, R.R. & W.A.Newman 1975. A trilobitomorph origin for the Crustacea. *Fossils and Strata* 4:437-459.

Hughes, C.P. 1975. Redescription of *Burgessia bella* from the Middle Cambrian Burgess Shale, British Columbia. *Fossils and Strata* 4:415-436.

Jones, P.J. & K.G.McKenzie 1980. Queensland Middle Cambrian Bradoriida (Crustacea): new taxa, palaeobiogeography and biological affinities. *Alcheringa* 4:203-225.

Kozur, H. 1974. Die Bedeutung der Bradoriida als Vorlaufer der postkambrischen Ostracoden. *Z. geol. Wiss. Berlin* 2:823-830.

Manton, S.M. 1969. Evolution and affinities of Onychophora, Myriapoda, Hexapoda, and Crustacea. In: R.C.Moore (ed.), *Treatise on Invertebrate Paleontology, Part R, Arthropoda* 4:R15-56. Lawrence: Geol. Soc. Am. & Univ. Kansas.

Manton, S.M. 1973. Arthropod phylogeny – a modern synthesis. *J. Zool.* 171:111-130.

Manton, S.M. 1977. *The Arthropoda. Habits, Functional Morphology and Evolution.* Oxford: Clarendon Press.

Manton, S.M. & D.T.Anderson 1979. Polyphyly and the evolution of arthropods. In: M.R.House (ed.), *The Origin of Major Invertebrate Groups:*269-321. London: Academic Press.

McKenzie, K.G. & P.J.Jones 1979. Partially preserved soft anatomy of a Middle Cambrian Bradoriid (Ostracoda) from Queensland. *Search* 10:444-445.

Müller, K.J. 1964. Ostracoda (Bradorina) mit phosphatischen gehäusen and dem Oberkambrium von Schweden. *Neues Jb. Geol. Palaont. Abh.* 121:1-46.

Müller, K.J. 1979. Phosphatocopine ostracodes with preserved appendages from the Upper Cambrian of Sweden. *Lethaia* 12:1-27.

Müller, K.J. 1981. Arthropods with phosphatized soft parts from the Upper Cambrian 'Orsten' of Sweden. In: M.E.Taylor (ed.), *Short papers for the second international symposium on the Cambrian system.* US Dept. Interior, Geol. Surv., Open File Report 81-743:147-151.

Palmer, A.R. 1969. Copepoda. In: R.C.Moore (ed.), *Treatise on Invertebrate Paleontology, Part R, Arthropoda* 4:R200-203. Lawrence: Geol. Soc. Am. & Univ. Kansas.

Raymond, P.E. 1935. *Leanchoilia* and other mid-Cambrian Arthropoda. *Bull. Mus. Comp. Zool.* 76:205-230.

Rowell, A.J. 1981. The Cambrian brachiopod radiation – monophyletic or polyphyletic origins? In: M.E.Taylor (ed.), *Short papers for the second international symposium on the Cambrian System.* US Dept. Interior, Geol. Surv., Open File Report 81-743:184-187.

Sanders, H.L. 1963a. The Cephalocarida. Functional morphology, larval development, comparative external anatomy. *Mem. Conn. Acad. Arts. Sci.* 15:1-80.

Sanders, H.L. 1963b. Significance of the Cephalocarida. In: H.B.Whittington & W.D.I.Rolfe (eds), *Phylogeny and evolution of Crustacea:*163-175. Cambridge, Mass.: Mus. Comp. Zool.

Schram, F.R. 1978. Arthropods: a convergent phenomenon. *Fieldiana, Geol.* 39:61-108.

Schram, F.R. 1979. Crustacea. In: R.W.Fairbridge & D.Jablonski (eds), *The Encyclopedia of Paleontology:*238-244. Stroudsburg: Dowden, Hutchinson & Ross.

Schram, F.R. 1982. The fossil record and evolution of Crustacea. In: L.G.Abele (ed.), *The Biology of Crustacea,* Vol.1. New York: Academic Press.

Scourfield, D.J. 1926. On a new type of crustacean from the Old Red Sandstone (Rhynie Chert Bed, Aberdeenshire) – *Lepidocaris rhyniensis. Phil. Trans. Roy. Soc., London* (B)214:153-187.

Simonetta, A.M. & L.Delle Cave 1975. The Cambrian non-trilobite arthropods from the Burgess Shale of British Columbia. A study of their comparative morphology, taxonomy and evolutionary significance. *Palaeontogr. Ital.* 69 (new series 39):1-37.

Sprinkle, J. 1981. Diversity and evolutionary patterns of Cambrian echinoderms. In: M.E.Taylor (ed.), *Short Papers for the second international symposium on the Cambrian systems.* US Dept. Interior, Geol. Surv., Open File Report 81-743:219-221.

Størmer, L. 1959. Trilobitomorpha, Trilobitoidea. In: R.C.Moore (ed.), *Treatise on Invertebrate Paleontology, Part O, Arthropoda 1:*O22-37. Lawrence: Geol. Soc. Am. & Univ. Kansas.

Stürmer, W. & J.Bergström 1973. New discoveries on trilobites by X-rays. *Paläeont. Z.* 47:104-141.

Stürmer, W. & J.Bergström 1981. *Weinbergina,* a Xiphosuran arthropod from the Devonian Hunsrück Slate. *Paläont. Z.* 55:237-255.

Sylvester-Bradley, P.C. 1961. Order Archaeocopida Sylvester-Bradley, n.Order. In: R.C.Moore (ed.), *Treatise on Invertebrate Paleontology, Part Q, Arthropoda 3:*Q100-103. Lawrence: Geol. Soc. Am. & Univ. Kansas.

Ulrich, E.O. & R.S.Bassler 1931. Cambrian bivalved Crustacea of the order Conchostraca. *Proc. US Nat. Mus.* 78(4):1-130.

Whittington, H.B. 1971. Redescription of *Marrella splendens* (Trilobitoidea) from the Burgess Shale, Middle Cambrian, British Columbia. *Geol. Surv. Can. Bull.* 209:1-24.

Whittington, H.B. 1974. *Yohoia* Walcott and *Plenocaris* n.gen., arthropods from the Burgess Shale, Middle Cambrian, British Columbia. *Geol. Surv. Can. Bull.* 231:1-21 (Figs.1-6 of Plate X should be interchanged with figures 1-5 of Plate XII).

Whittington, H.B. 1975. Trilobites with appendages from the Middle Cambrian, Burgess Shale, British Columbia. *Fossils and Strata* 4:97-136.

Whittington, H.B. 1977. The Middle Cambrian trilobite *Naroia,* Burgess Shale, British Columbia. *Phil. Trans. Roy. Soc., London* (B)280:409-443.

Whittington, H.B. 1978. The lobopod animal *Aysheaia pedunculata* Walcott, Middle Cambrian, Burgess Shale, British Columbia. *Phil. Trans. Roy. Soc., London* (B)284:165-197.

Whittington, H.B. 1979. Early arthropods, their appendages and relationships. In: M.R.House (ed.), *The Origin of Major Invertebrate Groups:*253-268. London: Academic Press.

Whittington, H.B. 1980. The significance of the fauna of the Burgess Shale, Middle Cambrian. *Proc. Geol. Assoc.* 91:127-148.

Whittington, H.B. 1981. Rare arthropods from the Burgess Shale, Middle Cambrian, British Columbia. *Phil. Trans. Roy. Soc., London* (B)292:329-357.

Wills, L.J. 1963. *Cyprilepas holmi,* Wills 1962, a pedunculate cirripede from the Upper Silurian of Oesel, Esthonia. *Palaeont.* 6:161-165.

FREDERICK R. SCHRAM
*San Diego Natural History Museum, California, USA*

# REMIPEDIA AND CRUSTACEAN PHYLOGENY

ABSTRACT

A preliminary assessment is offered of the phylogenetic position of the class Remipedia in the context of conflicting theories of crustacean evolution. Some crucial information has yet to be assembled, but there would appear to be some justification for seriously reconsidering the Biramous Theory of Cannon and Manton (1927) as opposed to the Cephalocarid Theory of Sanders (1957).

## 1 INTRODUCTION

The description of *Speleonectes lucayensis* Yaeger, 1981, and its assignment to a new class of Crustacea, Remipedia, reopens the debate about the origin and early evolution of the phylum. Yaeger pointed out the various similarities *S. lucayensis* holds with other crustacean classes. It shares with the cephalocarids a cephalic shield, well-developed pleura and tergites, and caudal furca. However, *Speleonectes* has laterally directed biramous appendages, while cephalocarids have ventrally directed mixopodia. It shares with Malacostraca a biramous antennule, but lacks the tagmosis characteristic of the malacostracans. *Speleonectes* has a large number of trunk limbs like many Branchiopoda. However, it does not bear reduced antennules or maxillae, nor does it lack appendages on the posterior part of the body as generally occur in branchiopods. *Speleonectes* shares with copepods some pleural and tergal similarities, a single segment antennal exopod, and the biramous paddle appendages; but is unlike copepods in its long series of trunk limbs. Yager concluded, however, that such similarities '. . . are superficial and indicate a weak affinity, at best'.

The presence of homonomous, biramous swimming appendages on all trunk segments, a cephalic shield, and biramous antennules would seem to combine in one animal an incredible array of primitive characters. The distinctive localization of feeding activity in remipedes to the maxillae and a maxilliped, completely divorcing food gathering as a function from the trunk appendages, might at first seem to be easily interpreted as an apomorphy. However, this too might be a primitive feature in light of the issues raised by Boxshall & Newman (this volume) in regard to cephalophagy as a possibly primitive mode of feeding in crustaceans. In short, the Remipedia reopens the question of just what was the ancestral crustacean like, an issue supposedly settled by the discovery of *Hutchinsoniella macracantha* and the subsequent phyletic synthesis centered on cephalocarids as primitive crustaceans.

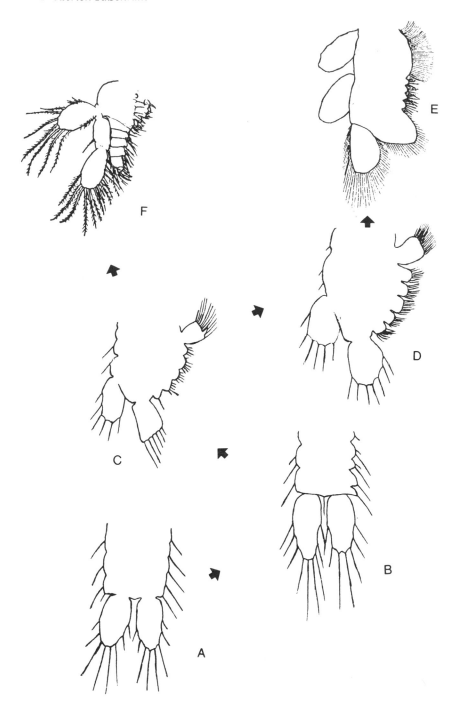

Figure 1. Biramous Theory of crustacean limb evolution. A. ancestral type with simple biramous paddle for locomotion; B. biramous appendage from the posterior trunk series of *Lepidocaris rhyniensis;* C. appendage from anterior trunk of *L.rhyniensis;* D. notostracan appendage with gnathobases on all limbs; E. anostracan polyramous foliaceous limb; F. cephalocarid polyramous mixopodium (modified from Cannon & Manton 1927).

## 2 THE BIRAMOUS THEORY

Cannon & Manton (1927) formalized the view that the ancestral crustacean limb was a simple biramous paddle which functioned primarily as a swimming organ. The conversion of such a limb to sustain a joint function of locomotion *and* feeding was facilitated by the backwash currents operable over the immediate surface of such a vibrating paddle. As the trunk limbs moved back and forth to suck in and push out water to move the animal forward, opposing backwash currents would sweep the surface. The net effect of these backwash currents was to move water towards the underside of the trunk, allowing particles and debris trapped on the appendage surface to be swept forward towards the head. Subsequent development of special setation, gnathal endites, and medially directed spines would have allowed the perfection of a filter feeding mode (Fig.1).

These ideas took great sustenance from the morphology of the peculiar Devonian fossil *Lepidocaris rhyniensis* Scourfield 1926. *Lepidocaris* has distinctive phyllopodous appendages on the anterior part of the body that were reminiscent of those seen in branchiopods (and also distantly comparable to those of cephalocarids). Posteriorly the *Lepidocaris* trunk has simple, biramous, ventrally directed paddles. Cannon & Manton felt *Lepidocaris* represented a transition phase from an ancestral, completely biramous condition to one in which polyramous foliaceous limbs were used to collect food as well as locomote.

When juveniles of *Lepidocaris* were recognized, Harding (in Scourfield 1940) pointed out that the biramous theory seemed further reinforced. Not only did *Lepidocaris* have supposedly primitive biramous limbs located on the posterior body and more highly specialized limbs anterior (this posterior-primitive pattern is found in many groups), but also, the more primitive biramous condition appeared first in the ontogeny of those anterior limbs and only became modified later in development.

## 3 THE CEPHALOCARID (OR MIXOPODIAL) THEORY

From such an apparently secure position, the Biramous Theory quickly passed out of vogue with the discovery of *Hutchinsoniella macracantha* Sanders 1955. A synthesis of a new theory of crustacean ancestry based on cephalocarids quickly evolved (Sanders 1957, 1963), and was sharply defended against even the slightest suggestion that the cephalocarids were not separate, distinct, and a representative of something close to an ancestral condition (e.g. Sanders 1961).

Sanders position (Fig.2) held that the cephalocarid mixopodium provided a central focus from which could be derived, on one hand the *Lepidocaris* anterior feeding limbs, and ultimately the phyllopium of branchiopods; and on the other the simple biramous type of *Lepidocaris* posterior limbs, and those of other forms such as copepods. In addition, a chief advantage of the Cephalocarid Theory was that it also provided an explanation for the development of leptostracan and larval eumalacostracan limbs, and thus ultimately the eumalacostracan stenopodium, by postulating an increasing specialization of the limb endopod for ambulatory activity. The Mixopodial Theory has gained wide acceptance (e.g. Hessler 1964, Hessler & Newman 1975, and Schram 1978).

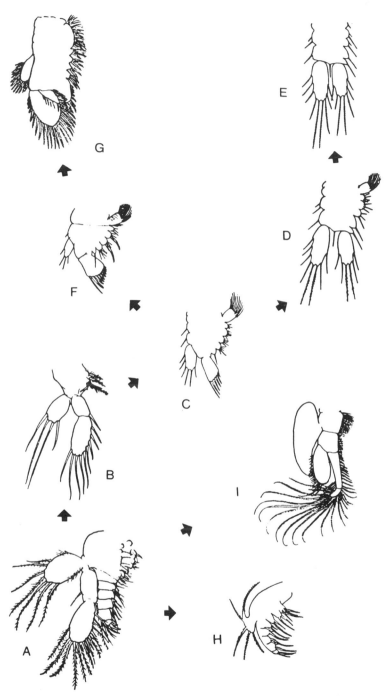

Figure 2. Mixopodial or Cephalocarid Theory of crustacean limb evolution. A. cephalocarid anterior trunk limb as an ancestral type; B. cephalocarid eighth trunk limb as an intermediate type; C. anterior trunk limb of *Lepidocaris rhyniensis;* D. intermediate trunk limb of *L.rhyniensis;* E. posterior trunk limb of *L.rhyniensis;* F. maxilla of *L.rhyniensis;* G. anostracan limb; H. maxilla of peneid larva; I. thoracic limb of leptostracan (modified from Sanders 1957).

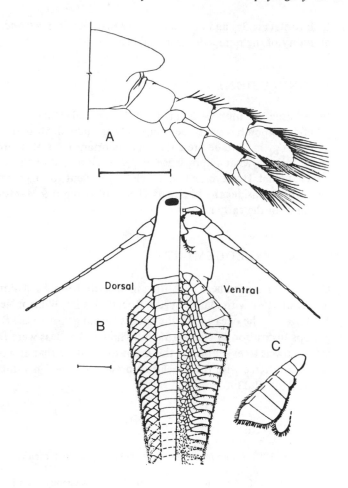

Figure 3. A. cross-section of lateral trunk of *Speleonectes lucayensis* with laterally directed biramous limb, scale 0.5 mm (from Yager 1981); B,C. Carboniferous crustacean *Tesnusocaris goldichi,* scale 10 mm (from Brooks 1955).

Dorsal   Ventral

## 4 THEORIES IN FLUX

The apparent primitive position of cephalocarids has recently been challenged. Schram (1982), inspired by Gould (1977), felt that since cephalocarids lack fully developed eyes (Burnett 1981), 'abdominal' and sometimes posterior 'thoracic' appendages, and a carapace; have a great similarity of all body appendages; and are small in size; that they could be viewed as highly specialized progenetic paedomorphs rather than as primitive animals. This would imply that the anatomy of cephalocarids is that of a derived form developed as a response to the surface, organic rich flocculent zones, interpretable as an r-selection regime, in which they live.

In the midst of this challenge, *Speleonectes lucayensis* suddenly emerges and exhibits almost the exact morphology for an ancestral condition which was outlined by Cannon & Manton 55 years ago, i.e. a form with simple biramous paddles (Fig.3a) devoted to locomotion. They postulated that food getting in such a form would have been localized in the limbs immediately behind the mouth. This same basic morphology (i.e. biramous trunk appendages with feeding apparently localized to the mouthparts) is also seen in the Carboniferous crustacean *Tesnusocaris goldichi* (Fig.3b). The fossil form exhibits variations in limb form (Fig.3c) and orientation over that seen in *Speleonectes,* but illustrates the basic antiquity of the remipede structural plan.


## 5 CONCLUSIONS

In all fairness, it must be pointed out that crucial evidence relevant to *Speleonectes'* phylogenetic position has yet to be completely assembled. More detailed studies of the external morphology have to be presented and important facts of internal anatomy are lacking, to say nothing of what a knowledge of the development of this form could contribute to the issues at hand. However, enough is already evident to suggest that future work must seriously reconsider the Biramous Theory of Cannon & Manton (1927) and the scenario it provides for the early history of crustaceans.


## 6 ACKNOWLEDGEMENTS

Concepts presented here benefitted immensely from discussions which occurred in the winter of 1981 with Mark Grygier, Robert Hessler, William Newman, George Wilson, and Jill Yager on the occasion of the latter's working visit to the San Diego Museum and Scripps Institution of Oceanography. Though the ideas were free flowing and difficult to assess now as to exact origin, this is not to construe that any or all of these people would necessarily agree with the conclusion presented here. This study was supported by NSF Grant DEB 79-03602.


## REFERENCES

Burnett, B. 1981. Compound eyes in the cephalocarid crustacean *Hutchinsoniella macracantha. J. Crust. Biol.* 1:11-15.
Cannon, H.G. & S.M.Manton 1927. On the feeding mechanism of a mysid crustacean, *Hemimysis lamornae. Trans. Roy. Soc. Edinb.* 55:219-253.
Gould, S.J. 1977. *Ontogeny and Phylogeny.* Cambridge: Belknap Press.
Hessler, R.R. 1964. The Cephalocarida, comparative skeletomusculature. *Mem. Conn. Acad. Arts Sci.* 16:1-97.
Hessler, R.R. & W.A.Newman 1975. A trilobitomorph origin for the Crustacea. *Fossils and Strata* 4:437-459.
Sanders, H.L. 1955. The Cephalocarida, a new subclass of Crustacea from Long Island Sound. *Proc. Nat. Acad. Sci.* 41:61-66.
Sanders, H.L. 1957. The Cephalocarida and crustacean phylogeny. *Syst. Zool.* 6:112-129.
Sanders, H.L. 1961. On the status of Cephalocarida. *Crustaceana* 2:1-251.
Sanders, H.L. 1963. Significance of the Cephalocarida. In: H.B.Whittington & W.D.I.Rolfe (eds), *Phylogeny and Evolution of Crustacea:* 163-175. Cambridge: Mus. Comp. Zool.
Schram, F.R. 1978. Arthropods: a convergent phenomenon. *Fieldiana: Geol.* 29:61-108.
Schram, F.R. 1982. The fossil record and evolution of Crustacea. In: L.G.Abele (ed.), *The Biology of Crustacea, Vol. I.* New York: Academic Press.
Scourfield, D.J. 1926. On a new type of crustacean from the Old Red Sandstone – *Lepidocaris rhyniensis. Phil. Trans. Roy. Soc., London* (B)214:153-187.
Scourfield, D.J. 1940. Two new and nearly complete specimens of young stages of the Devonian fossil crustacean *Lepidocaris rhyniensis. Proc. Linn. Soc. Lond.* 152:290-298.
Yager, J. 1981. Remipedia, a new class of Crustacea from a marine cave in the Bahamas. *J. Crust. Biol.* 1:328-333.

K. G. McKENZIE* , K. J. MÜLLER** & M. N. GRAMM***

   * *Riverina College, Wagga Wagga, NSW, Australia*
  ** *Institut für Paläontologie, Rhein, Freidrich-Wilhelms-Universität, Bonn, Germany*
 *** *Biological Research Institute, Vladivostock, USSR*

# PHYLOGENY OF OSTRACODA

## ABSTRACT

Ostracoda are recognized as one of the largest groups of Crustacea, with a continuous fossil record since the Early Cambrian. They are certainly the most abundant and diverse fossil crustaceans. The primitive taxa included bradoriids, phosphatocopids and leperditicopids, whose phyletic relationships with each other are unclear. Leperditicopids possibly diverged from bradoriids at the end of the Cambrian, but Phosphatocopida seem to be distinct though accepted as ostracodes. They flourished during the Middle-Late Cambrian but then died out rapidly. Leperditicopida, after initiating in the Early Ordovician, underwent a major Silurian radiation in the northern hemisphere before dying out in the Late Devonian, possibly never reaching the southern continents. Bradoriida died out in the Cambrian but before this probably gave rise to Beyrichicopida via the Beyrichonidae. Beyrichicopida were the dominant Palaeozoic ostracodes. They evolved into at least seven superfamilies — Drepanellacea, Beyrichiacea, Eurychilinacea, Nodoellinacea, Primitiopsiacea, Hollinacea and Tribolbinacea, but were extinct by the Middle Triassic. Podocopida evolved in the Ordovician, probably from a beyrichicopid hollinacean stock. The modern suborders were established in the Ordovician indicating an early radiation for the order. During the Permian-Triassic there were important radiations in the podocopid superfamilies Healdiacea and Bairdiacea. The earliest radiations on land (Permian-Triassic) involved Darwinulacea which remained the main continental group until the Middle Jurassic. During the Middle Jurassic, Cytheracea commenced the radiation which made them the dominant marine Podocopida, a status they still retain. Cytheracea replaced the Darwinulacea as the dominant land group. Cypridacea, which had evolved by the Triassic as a marine group, did not begin to dominate on land until the Purbeckian-Wealden crisis. Cenozoic podocopid faunas are dominated by Cytheracea in the sea and by Cypridacea on the continents. Myodocopida evolved in the early Palaeozoic; Cladocopida in the Silurian (Entomozoacea); and Halocyprida also in the Silurian (Entomoconchacea). Their limited fossil records makes it difficult to interpret their phylogeny in much detail.

## 1 INTRODUCTION

Ostracoda are an abundant and long ranging class in the phylum Crustacea. They are tiny animals (living species range in size from about 0.1 to over 33 mm) occupying a rich diver-

29

sity of aquatic niches — from terrestrial to oceanic — at all depths, temperatures and salinities, and encompassing a wide range of substrates.

Until recently, most biologists thought of Ostracoda as esoteric microcrustaceans characterised by extraordinarily long sperm, the longest in the animal kingdom; and most
palaeontologists recognised them as small bivalves, nearly ubiquitous in marine sediments
since the Cambrian (and in freshwater sediments since the Carboniferous) but having little
utility or intrinsic interest.

As much as anything, the revival of interest in Ostracoda stems from reawakened attention to taxonomy allied to the economic impetus given by realization of their applicability
in petroleum exploration. Major barriers to earlier progress were the existence of dual
classifications — one biological, the other palaeontological — and the depth of focus problem in ordinary microscopes which made adequate illustration frustratingly difficult.
Nevertheless, by 1961 two widely used general classifications had emerged (Orlov 1960,
Scott 1961), and some meticulous workers were obtaining good results in photography
and illustration. In 1963, the first international ostracode congress resolved to set up a com-
mittee to work towards a single general classification, though this objective remains a long
way from realization. Of equal importance has been the use, since about 1967, of the scanning electron microscope.

Table 1 provides data on the tempo of modern research. About 25 % of all known ostracodes have been described in the last 20 or so years. The results have confirmed the ecological sensitivity of Ostracoda (Oertli 1971) and have begun to underline their importance
for stratigraphy (van den Bold 1977). Meanwhile, the adoption of numerical methods has
led to publication of major environmental studies (Benson 1975a) involving the routine
handling of large data bases (McKenzie 1976) and to the provision of species lists of various
kinds, all more or less suitable for entry into such data handling systems (McKenzie, ed.
1972a, Kempf 1980). Numerical techniques have also found use in cladistic studies (Al-
Abdul-Razzaq 1973, Maddocks 1976). Nor is there any shortage of conceptual innovation
(Benson 1981). All roads now appear to lead to an improved, generally acceptable, classification and phylogeny of Ostracoda.

Table 1. New species and subspecies of Ostracoda (to the nearest higher 25) described since 1967 —
*nomina nova* not included. Data from the Zoological Record. Further records indicate the description
of over 10 000 species since 1957.

| Year | Fossil | Living | Total |
|---|---|---|---|
| 1967 | 450 | 125 | 575 |
| 1968 | 650 | 50 | 700 |
| 1969 | 450 | 100 | 550 |
| 1970 | 325 | 50 | 375 |
| 1971 | 575 | 125 | 700 |
| 1972 | 425 | 50 | 475 |
| 1973 | 525 | 25 | 550 |
| 1974 | 350 | 150 | 500 |
| 1975 | 475 | 200 | 675 |
| 1976 | 300 | 50 | 350 |
| | 4 525 | 925 | 5 450 totals |

## 2 RELATION TO OTHER CRUSTACEANS

It would be wise first to review briefly the relationships of Ostracoda to other crustaceans. Baird (1850) erected a classification of Crustacea that followed an earlier scheme of Milne Edwards and was based on the organization of the mouthparts. Baird recognised two major groups within Crustacea, which correspond to the Malacostraca and Entomostraca of modern usage. Entomostraca as a group name has fallen into disuse. It is difficult to appreciate why this should be. Probably, recent detailed research involving a greater number of taxonomic characters has led to recognition of the distinctness of numerous crustacean groups with a consequent placement of these into higher taxa. This led to introduction of the cohort by Dahl (1963) and to illustrate the point further, cohort is retained in the latest eumalacostracan phylogeny proposed by Schram (1981). If, however, we accept Manton's view (1973, 1977) supported independently by Anderson (1973) on embryological grounds, then Crustacea are to be considered a separate phylum, of equal rank to the other major arthropodan groups Uniramia (incorporating Onychophora, Myriapoda, Hexapoda), Chelicerata, and Trilobita. Elevation of Crustacea to phylum rank, incidentally, provides the desired room at the top; consequently, Entomostraca can be reintroduced at sub-phylum level since Baird's statement that '. . . Branchiopodes and Entomostraces . . . form a very natural group' remains true.

The second point raised by a review of Baird's systematic arrangement is his placement of Phyllocarida together with anostracans, notostracans and cladocerans, into the Branchiopoda. Since Calman (1909), Phyllocarida have been considered malacostracans, although several similarities to branchiopods are acknowledged in the literature. And lately they have been made to bear the burden of malacostracan ancestry (Siewing 1963), a role for which they seem poorly fitted especially because new palaeontological discoveries have pushed the origins of other malacostracans further back into the Palaeozoic.

The earliest known phyllocarid-like animal is *Canadaspis* from the celebrated mid-Cambrian Burgess Shale of Canada. *Canadaspis* is supposed to possess eight thoracic segments (Briggs 1978) with associated biramous appendages, but the illustrations suggest that it could have had ten thoracic segments with ten essentially similar biramous appendages, although the two anteriormost pairs have been interpreted as maxillae. If these ten segments were wholly thoracic, then *Canadaspis* is similar to numerous groups of entomostracous Crustacea characterised by a developmental boundary after the 11th trunk segment. Such groups include notostracans, conchostracans, and fossil as well as living anostracans (Linder 1941). Platycopid ostracodes have a posterior exoskeleton which clearly displays ten postmaxillar segments (Schulz 1976). It seems at least possible that the true maxillae of *Canadaspis* could have been either obscured in the material prepared by Briggs.

Some other characters used to support the hypothesis of phyllocaridans as ancestral to Malacostraca include the carapace fold, furca, number of abdominal segments, and so-called extension of the heart (Siewing 1963). None of these characters, however, segregate Phyllocarida from Entomostraca! The carapace has a common origin from the dorsal third cephalic segment in all carapace-bearing crustacean groups, including the entomostracan taxa. Whatever 'true' furca are [the adjective is Siewing's but the debate has been continued by Bowman (1971) and Schminke (1976)] they certainly occur in several groups of Entomostraca, e.g. Ostracoda and Conchostraca, as well as in Malacostraca. The particular significance for crustacean evolution of seven abdominal segments is not stated by Siewing, but this number of post-thoracic segments occurs in both fossil and living anostracans and there are traces of six post-thoracic exoskeletal 'segments' in saipanettid Ostracoda (Schulz 1976). In Notostraca and Conchostraca, there can be many trunk segments after the 11th – up to 16, for example, in Conchostraca (Tasch 1969). The method of heart formation is essentially the same in all arthropodans which possess one (Cannon 1924) and it is a hydrostatically maintained turgor that governs functional responses in crustaceans, including all Entomostraca, many of which are so small that they do not have (or need) a heart.

Apart from similarities in general shape, many details of the carapace are common to early phyllocaridan fossils and early entomostracans. Thus, carapace horns and median dorsal plates also occur in bradoriid and phosphatocopid ostracodes; and for the latter group, at least, an ostracode-like soft body is also known (Müller 1979a, 1982). The carapace adductor muscle scar pattern of living *Nebalia* appears to be biserial and not markedly different from an entomostracan type such as that of platycopid ostracodes.

If Phyllocarida were returned to Entomostraca, the concern to fit other malacostracans to a phyllocarid-like ancestor could be diverted profitably to searching out more direct ancestors of these groups from among the numerous available crustacean-like early Palaeozoic animals, notably those in the Burgess Shale. There should also be a correspondingly greater readiness to accept the polyphyletic origin of both Entomostraca and Malacostraca within the phylum Crustacea. Such a methodology is explicit in the latest paper on arthropodan phylogeny by Simonetta & Delle Cave (1981) although, paradoxically, these authors remain committed to the monophyletic origin of all arthropodans.

A summary taxonomy for the entomostracous crustaceans is presented below.

Phylum: Crustacea
    Subphylum: Entomostraca
        Class: Ostracoda
            Order: Myodocopida
            Order: Cladocopida
            Order: Halocyprida
            Order: Podocopida
            Order: ?Bradoriida (see Jones & McKenzie 1980)
            Order: Beyrichicopida
            Order: Leperditicopida
            Order: Phosphatocopida
        Class: Branchiopoda
            Subclass: Cladocera
            Subclass: Conchostraca
            Subclass: Cephalocarida
            Subclass: Calmanostraca
            Subclass: Sarsostraca
        Class: Cirripedia
            Order: Thoracica
            Order: Rhizocephala
            Order: Ascothoracica
            Order: Acrothoracica
        Class: Copepoda
            Order: Poecilostomatoida
            Order: Calanoida
            Order: Cyclopoida
            Order: Harpacticoida
            Order: Monstrilloida
            Order: Misophryoida
            Order: Siphonostomatoida
            Order: Mormonilloida

Class: Branchiura
Class: Phyllocarida
    Subclass: Leptostraca
    Subclass: Archaeostraca
    Subclass: Canadaspidida
    Subclass: Hymenostraca
    Subclass: Hoplostraca

## 3 NATURE OF THE FOSSIL RECORD

Ostracoda have the best fossil record of the Crustacea. Undoubtedly, this is due mainly to the readily preserved bivalved crystalline calcareous (or phosphatic) shell but another favorable factor is their generally small size. Therefore they are not merely diverse but abundant from the Cambrian onwards, with some exceptions to be discussed below.

Planktic species usually have a chitinous or weakly calcareous shell and are uncommon fossils except in restricted facies. Thus, the planktic Halocypridacea are represented in the fossil record by only two citations, both from the Cretaceous of Europe (Pokorny 1964, Kaye 1965). In the Late Devonian, however, greenish and purplish slates representing off-shore facies are widespread; and, particularly in Germany and southwestern England, entomozoid ostracodes (thought to have been planktic and ancestral to the cladocopids) are common and stratigraphically very useful fossils in them. Entomozoids occur also in the Late Devonian offshore facies of France, Spain, Poland, USSR, China and the USA (Gooday 1978). Indeed, the Late Devonian-Early Carboniferous is the time of maximum diversification among entomozoids although the group initiated much earlier (in the Silurian).

In the case of Myodocopida, which are pelagic or epibenthic ostracodes usually having calcareous shells, their poor fossil record is due to the fact that their shell calcite differs from that of other ostracodes. Sohn & Kornicker (1969) established experimentally that such shells are composed of very fine grained or amorphous calcium carbonate which crystallizes isochemically postmortem forming nodules and platelets of various kinds. After these platelets and nodules develop, the protein and chitin framework of the shell disintegrates rapidly. Thus, few myodocopid fossils are known although the modern fauna is richly diverse and widespread. Among the noteworthy exceptions are: *Philomedes homoedwardsiana* (Keij 1957) a cypridinid myodocope which is widespread in the Eocene of western Europe; a species close to the cylindroleberidid genus *Cycloleberis* from the Late Jurassic of the Upper Volga district (Dzik 1978); and *Eocypridina campbelli* a well-preserved cypridinid from the Late Devonian of Indiana. Like entomozoids, myodocopid fossils are relatively common in the Late Devonian-Carboniferous of Europe and they dominate certain assemblages. The monotypic genus *Cyprosina* is referrable to Halocyprida (Kornicker & Sohn 1976). Recently, Jones & McKenzie (1980) have suggested that the genus *Ludvigsenites* (Middle Ordovician-Silurian of north-western Canada) is also a cypridinid since it resembles the modern myodocopid *Tetragonodon* — incidentally, *Tetragonodon* although differing in shape, has a muscular scar pattern very like *Eocypridina*.

It would be wise to recognize that the ostracode fossil record is probably also biassed against very small ostracodes, such as the interstitial genera; taxa with specialized life habits, such as the parasitic forms; and those which occur in exotic habitats, such as the terrestrial, cavernicolous, arboreal, and leaf litter species. The modern representatives of these groups sometimes exhibit primitive characteristics which suggest that they have an appreciable evolutionary history although not known as fossils, e.g. Terrestricytheracea Schornikov 1969b, Bonaducecytheridae McKenzie 1977b. Nonetheless, even for such taxa, occasional fossils turn up. Thus, Colin & Danielopol (1980) argue persuasively that the Late Pliocene ostracode, *Kovalevskiella turianensis praeturianensis* was a hypogean taxon deposited by subterranean waters among a typical Parathethyan fauna. Further, interpretations for parasitic species can be based on the fossil records of their hosts (McKenzie 1973).

There remains a major bias in the ostracode fossil record, but it is one that is common to all crustaceans, indeed to all organisms with a soft anatomy: generally, the appendages are not preserved. We now realise, however, that even this bias is not invariable in the rich ostracode record and the exceptions are remarkable. Thus, Bate (1972) described *Pattersonocypris micropapillosa* from the Early Cretaceous of Brazil based on a population of 253 specimens, including 138 complete carapaces of which over 100 retained at least some appendages (and several had a complete soft anatomy). The preservation is excellent and is due to replacement of the calcite and chitin by apatite, a hydrated calcium fluorophosphate probably derived from the decaying corpse of a teleost fish on which the ostracodes were feeding. It seems, moreover, from recent work by Müller (1979b, 1982), that such occurrences are not as rare as was previously thought and that careful searches, among phosphorite rocks in particular, can be highly rewarding even though the success ratio is not as high as in the case of the Brazilian fossils.

Apart from the Cambrian phosphatocopid species discussed below, the remaining fossil record for Ostracoda is overwhelmingly confined to benthic species with crystalline calcareous shells (Langer 1973). A good idea of the gradual evolution of such forms towards the modern assemblages is obtained by reference to the recent book on stratigraphically useful British Ostracoda edited by Bate & Robinson (1978), whose treatment emphasises the importance of contemporary facies and geography in defining ostracode assemblages as well as the dominance of particular groups in the record during different geological epochs. Such considerations are often ignored by crustacean specialists interested in the evolution of other Crustacea, usually because of a lack of adequate material.

In summary, the ostracode fossil record is dominated by phosphatocopids in the Cambrian and by beyrichicopids for much of the remaining Palaeozoic, although podocopids were also important from the Ordovician, and the aberrant group Leperditicopida was widespread in the Siluro-Devonian especially of Eurasia and North America. Some myodocopids were locally common in restricted Palaeozoic facies. By the latter Palaeozoic podocopids (bairdiaceans and healdiaceans) had risen to dominance in the record. In the Mesozoic and Cenozoic only myodocopid, halocyprid, cladocopid and podocopid stocks survived — including two kirkbyocopine taxa, the genera *Manawa* and *Puncia* (Hornibrook) 1949). The relatively poor fossil record of other groups means that podocopids dominate overwhelmingly in post-Palaeozoic fossil assemblages, consequently detailed evolutionary analyses in the Late and post-Palaeozoic are virtually confined to them — for an important exception cf Kornicker & Sohn (1976).

The importance of the Cambrian phosphatocopids, as evidenced by their discussion in several other papers in this volume, demand a detailed overview of that group presented below. Not only may the phosphatocopids eventually serve to link various later ostracode groups to each other, but they are serving to elucidate ostracode relationships to other crustaceans.

# 4 PHOSPHATOCOPIDA

The occurrence of Phosphatocopida with fully preserved appendages in the Cambrian of Sweden presents an unique opportunity for observation of similarities and differences with extant ostracodes. The samples were recovered from 20 localities distributed over an area of more than 25 000 km²

The order Phosphatocopida is characterised by the development of primary phosphatic carapaces. They are smooth, black or dark amber in colour, in some cases translucent, and have a shiny, reflecting surface with a high lustre. Large carapaces measure more than 2 mm. The hinge line is straight, the hinges adont (nullidont). A distinct duplicature is developed at the ventral margin. The thin inner lamella is commonly preserved by secondary phosphatization. The body is attached to the carapace along almost its entire elngth. The isthmus is small. There are nine or less appendages depending on the growth stage. The largest specimen with preserved soft parts has a length of 1.1 mm.

Phosphatocopids are not common. Occasionally, however, they are found in large quantities and several hundred specimens may be obtained from a single kilogram sample.

The first pair of appendages, the antennules are uniramous and small or reduced. The antennae and mandibles are the most prominent appendages, with annulate exopodites and one long bristle on each annulus. Endites on protopods and endopods are well-developed. Both appendages served for swimming as well as mastication. This double function of the mandible could explain why an adductor muscle (it is regarded as a derivate of the muscles of the protopodite) or its scars cannot be observed. The postmandibular appendages are similar to each other and their endites form a filter apparatus. Maxillae and maxillulae are not specialized as additional masticatorial appendages. The upper lip (labrum) and lower lip (labium or paragnath) are well-developed and have the same general plan as in many of the non-ostracode 'orsten'-crustaceans, revealing their primordial position in crustacean evolution.

The strong development of natatory setae, particularly on the second and third appendages, indicate swimming ability. On the other hand, the carapace composed from heavy phosphatic matter and the patchy occurrence suggest that Phosphatocopida were nectobenthic and lived in large swarms on or near the bottom.

The order was recently subdivided into the suborders Hesslandonocopina Müller 1982, and Vestrogothicopina Müller 1982. Hesslandonocopina have a carapace with two valves of equal size, articulated by two adont (nullidont) hinges parallel to one another and attached to an intercalcated dorsal bar (Fig.1). The valves have a duplicature or selvage and may either close tightly or remain agape (in an undescribed genus). Dorsal bar and/or sides of valves may be adorned with conspicuous spines. The hinge structure is unique for ostracodes, and therefore some authors (Adamczak 1965, Kozur 1974) have excluded *Hesslandona* from the ostracodes, and assigned it to Cladocera (ephippia). However, based on similarities in the soft body (Fig.2), it is related to the Vestrogothicopina.

## 4.1 *Hesslandonocopina*

*Hesslandona* has a large protruding labrum which is attached to the body by softer tissue, such that it could be bent a little lengthwise. At the forehead there is a large tripartite organ on the median line; above the center its upper limit corresponds to the insertions of the first appendages. This 'frontal organ' covers about one quarter of the anterior side of the labrum. Its lengthwise subdivision into two cups with a third unpaired cup beneath them can only be observed on well-preserved specimens. The whole organ was covered by a thin membrane which on some specimens is preserved as a blown up, balloon-like, bubble, which can be explained by the development of excess pressure inside the labrum during preservation. On other specimens the membrane is wrinkled and demonstrates the primarily thin and pliable structure of this cover. The specimen illustrated in Figure 3 depicts a sheath, sucked into the labrum by low pressure. There, the chitinous bridges which separate the cups from one another become visible underneath the cover, and pits on the latter are formed above the cups. This organ is identified as the median eye because its shape and position is similar to that of the median eye in later ostracodes (for details on Recent ostracodes refer to Hartmann 1967).

The sternal plates of the posterior head segments are fused and form a prominent lower lip in the middle line of body. It extends lengthwise to the fourth-sixth appendage, depending on the growth stage and species. The mouth is in front of the labium and somewhat recessed. The center of the labium is slightly depressed into a furrow which aids transportation of food into the mouth.

As was found only recently, *Hesslandona* has a trunk region with four sternal plates, a short, limbless abdomen and short furca with rounded, annulated branches (Fig.4). However, these parts are not typically preserved. *Hesslandona* has up to nine pairs of appendages. The uniramous antennules are situated on the labrum, lateral to the dorsal rim of the eye, in distinct niches. Compared with the other appendages, they are small and short and are composed of seven segments; the last one bears four setae and is as long as all the four preceding ones together. The second pair of appendages, the antennae, are very large and biramous. The prominent protopodite is unjointed. Its laterally compressed endites bear strong spines which point to the mouth. The endopodite is comprised of two spinose elongate segments. The well-developed exopodite has 20-23 uniform annuli; each has a stiff long seta pointing posteroventrally. The terminal segment has 2-4 bristles. The third pair of limbs, the mandibles, are similar to the antennae, except that the large protopodite has developed a strong molar process. The endopodite is composed of three segments with elongated endites. The exopodite has about 25 annuli, each bearing a seta as in the second appendages. The second and third appendages obviously served for swimming as well as for mastication. Specialized maxillae and maxillulae are not developed. Appendages four to seven are all of the same general plan but decrease posteriorly in size. The exopodites lack joints, are flattened and developed as vibratory plates. The eighth and ninth pairs are small and not sufficiently preserved in the material for description.

### 4.2 *Vestrogothicopina*

Vestrogothicopina have valves of equal size with a single adont hinge line. The duplicature or selvage is similar to Hesslandonocopina. Spines may be developed in the anterior and posterior corners, on the sides of valves as well as on the ventral rim. The surface is sulcated, smooth and in the Oepikalutidae Jones & McKenzie 1980, the most highly developed family of the suborder, it may be reticulate. Of this suborder, the genera *Vestrogothia* (Fig.5) and *Falites* (Fig.6) are represented in the collections with preserved appendages. A median eye, like that in hesslandonids, is not developed.

In *Vestrogothia* the antennules are reduced. Possibly, they are developed rudimentarily in *Falites*. While all exopodites in *Vestrogothia* are composed of annuli with one seta each, in *Falites* the postmandibular ones are developed as vibratory plates similar to *Hesslandona*.

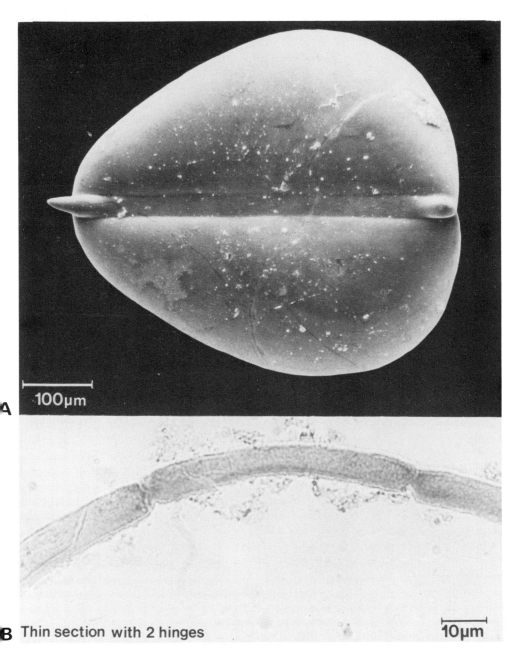

**A** 100μm

**B** Thin section with 2 hinges     10μm

Figure 1. *Hesslandona necopina* Müller, 1964. A. dorsal view with a double hinge and spined dorsal bar; B. thin section through dorsal bar and hinges.

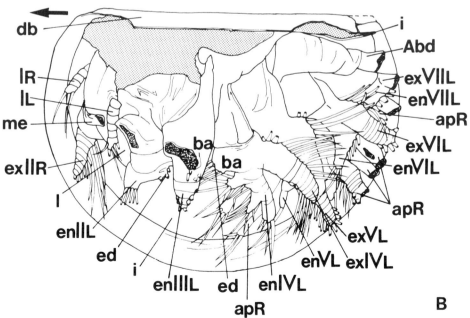

Figure 2. *Hesslandona unisulcata* Müller, 1982. A. right valve with the attached animal, showing appendages of the left side; B. explanation diagram: abd – abdomen; ba – basipodite; db – dorsal bar; i – inner lamella; l – labrum; me – median eye.

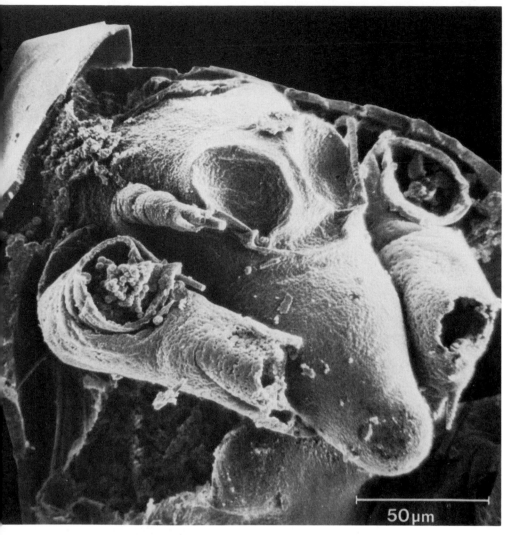

Figure 3. *Hesslandona unisulcata* Müller, 1982. Forehead with median eye, small antennules and partly preserved large antennae.

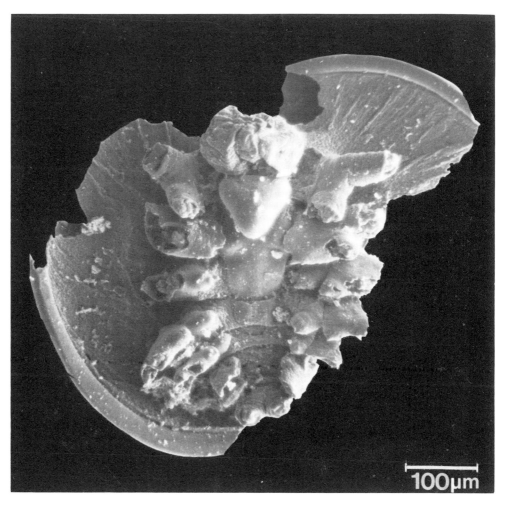

Figure 4. *Hesslandona* sp. Ventral view exposing labrum with inflated median eye, labium and trunk with sternal plates and furca.

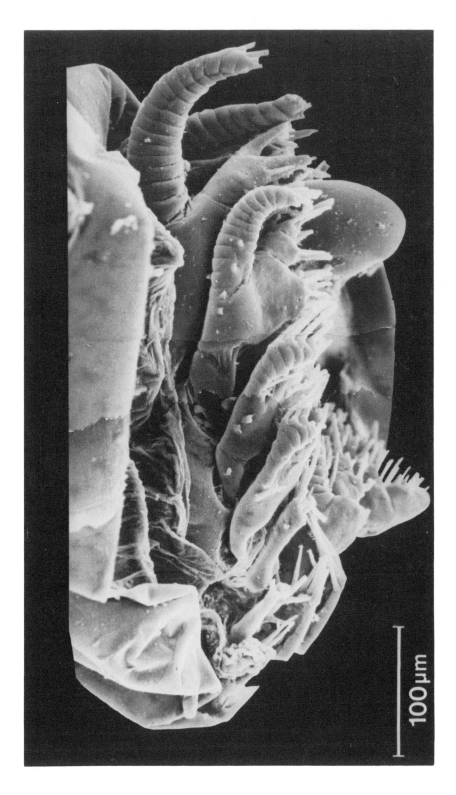

Figure 5. *Vestrogothia spinata* Müller 1964. Preadult specimen with six pairs of completely preserved appendages.

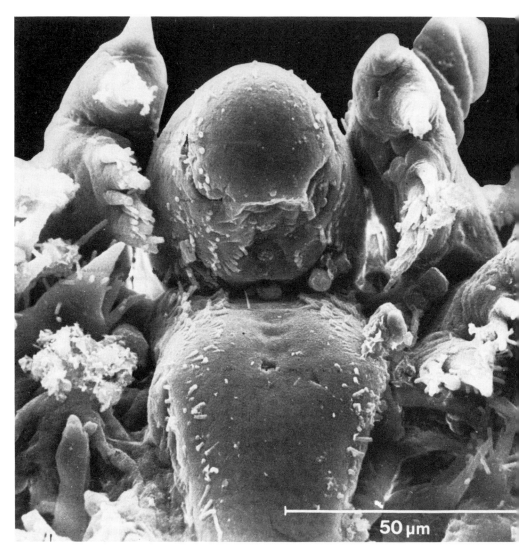

Figure 6. *Falites fala* Müller 1964. Detail with labrum, labium, mouth; and endites of antennae and mandibles pointing to mouth.

## 4.3 *Phyletic issues*

Phosphatocopida are a primordial group of Crustacea which developed a characteristic feeding apparatus with serially arranged filter appendages. Furthermore, the second and third appendages functioned as feeding organs as well as for locomotion. For that reason, the muscles of the mandibular protopodite were not available in this stage of phylogenetic development for their eventual function as valve adductor muscles, as is the case in later ostracodes. This may be also true for Beyrichicopida which also do not expose adductor scars. On the other hand, the formation of a bivalved carapace and reduction of the abdomen are already accomplished in Phosphatocopida.

As a taxonomic alternative, Phosphatocopida could be excluded from the ostracodes, and a new class could be established for them. Even if it is likely that the phosphatocopids represent a branch somewhat off the main line of development which led to the later orders of ostracodes, this taxonomic procedure would have little utility, and it would conceal the considerable general similarities between phosphatocopids and the other ostracode groups. Kozur (1974) included the genera of Vestrogothicopina into different superfamilies and families of Bradoriida and considered them ancestors for the various later ostracode orders. Although his view is not shared here, it can serve as evidence for the existence of such morphological similarity.

Landing (1980) has suggested that phosphatocopids may represent a vagrant group of crustaceans with close affinities to barnacles (class Cirripedia). However, the ontogeny of extant Cirripedia is quite different from ostracodes, in which a bivalved carapace is developed for all instars. Further evidence against a closer relationship with the 'cypris' larvae e.g. of Cirripedia has been already supplied by Jones & McKenzie (1980).

## 5 OSTRACODE EVOLUTION

The initial draft of this paper (a summary overview at that) was over 200 pages long. It was manifestly impossible to publish such a detailed discussion here, but for those readers interested in exploring the subject further, a copy of this manuscript and bibliography is available on request from the senior author. However, some pertinent facts relevant to ostracode evolution can be summarized here. A higher taxonomy of Ostracoda is presented in order to orient the reader to the diversity manifested in the class.

Phylum: Crustacea Pennant 1777 sensu Manton 1972
    Subphylum: Entomostraca O.F.Müller 1785
        Class: Ostracoda Latreille 1906
            Order: Bradoriida Raymond 1935
            Order: Phosphatocopida K.J.Müller 1964
                Suborder: Vestrogothicopina K.J.Müller 1982
                Suborder: Hesslandonocopina K.J.Müller 1982
            Order: Leperditicopida Scott 1961
                Superfamily: Leperditiacea Jones 1856
                Superfamily: Isochilinacea Swartz 1949
            Order: Beyrichicopida Pokorny 1953
                Suborder: Hollinocopina Henningsmoen 1965
                Superfamily: Eurychilinacea Ulrich & Bassler 1923
                Superfamily: Hollinacea Jaanusson 1957
                Superfamily: Nodellacea Zaspelova 1952
                Superfamily: Primitiopsacea Pokorny 1958

Suborder: Leiocopina Schallreuter 1973
  Superfamily: Aparchitacea Swarz 1945
Suborder: Binodocopina Schallreuter 1972
  Superfamily: Drepanellacea Zanina, Zaspelova & Polenova 1960
Suborder: Beyrichicopina Pokorny 1953
  Superfamily: Beyrichiacea Ulrich & Bassler 1923
Suborder: uncertain
  Superfamily: Tribolbinacea Sohn 1978
Order: Myodocopida Sars 1866
  Superfamily: Cypridinacea Baird 1850
  Superfamily: Cylindroleberidacea G.W.Müller 1906
  Superfamily: Sarsiellacea Sylvester Bradley 1961
Order: Cladocopida Sars 1866
  Superfamily: Entomozoacea Pribyl 1951
  Superfamily: Polycopacea Sars 1866
Order: Halocyprida Skogsberg 1920
  Superfamily: Entomoconchacea Sylvester Bradley 1953
  Superfamily: Thaumatocypridacea G.W.Müller 1906
  Superfamily: Halocypridacea Mertens 1958
Order: Podocopida Sars 1866
  Suborder: Metacopina Sylvester Bradley 1961
  Superfamily: Healdiacea Harlton 1933
  Superfamily: Thlipsuracea Ulrich 1894
  Superfamily: Sigilliacea Mandelstam 1960
  Superfamily: Cypridacea Baird 1845
  Superfamily: Darwinulacea Brady & Norman 1889
  Suborder: Parapodocopina Gramm 1977
  Superfamily: Jones, in preparation
  Suborder: Bairdiocopina Kozur 1972
  Superfamily: Bairdiacea Sars 1888
  Suborder: Cytherocopina Grundel 1967
  Superfamily: Cytheracea Baird 1850
  Superfamily: Terrestricytheracea Schornikov 1969
  Suborder: Paraparchitocopina Gramm 1975
  Superfamily: Paraparchitacea Gramm 1975
  Suborder: Kirkbyocopina Sohn 1961
  Superfamily: Kirkbyacea Ulrich & Bassler 1906
  Superfamily: Punciacea Hornibrook 1949
  Suborder: Platycopina Sars 1866
  Superfamily: Kloedenellacea Ulrich & Bassler 1908
  Superfamily: Cytherellacea Sars 1866

The following comments should be made on the above taxonomy. (1) Ordinal, subordinal, and superfamilial endings have been standardised but the ascriptions indicate the authorities responsible for the concepts which we have adopted in this paper. We consider Cuyancopina of Rossi de Garcia (1975) to be a junior synonym of Bairdiocopina Kozur 1972. (2) Other recent classifications of Ostracoda have been published by Kozur (1972) Hartmann & Puri (1974) and Pokorny (1978); the latter is relatively close to ours. (3) The assignment of Punciacea to Kirkbyocopina is tentative, as is the placement of Kirkbyocopina in Podocopina (Jones, personal communication). Our view of Kirkbyocopina is the concept adapted by Sohn (1961) for the superfamily Kirkbyacea, with all the qualifications included by him. The concept of Grundel (1969) is too broad. (4) The classification of Metacopina derives from McKenzie (1972b) but one of us (M.N.G.) would probably restrict Metacopina to Thlipsuracea and Healdiacea. On this matter, the views of Pokorny (1978) on metacopine groups represent an acceptable consensus (K.G.M); if preferred, it would require either a new subordinal category, or the use of Podocopina in a more restricted sense than hitherto, for Cypridacea and Darwinulacea.

Evolution in the Ostracoda is a dynamic process confined ultimately within space-time. It is now clear that global tectonic and palaeogeographic crises had profound evolutionary consequences for ostracodes matching those already well documented in numerous other groups of animals. These appear to have been the major episodes.

1. A salinity crisis during the Ediacaran (680 my BP) possibly triggered the evolution of several metazoan superphyla including an arthropodan facies. By the beginning of the Cambrian (570 my BP), Ostracoda had evolved and were represented by the Bradoriida and Phosphatocopida. The latter group was associated particularly with phosphorite facies but Bradoriida, which are thought to be polyphyletic, were less facies restricted. Phosphatocopida became extinct in the early Ordovician by which time Bradoriida were also extinct but had given rise at least to Leperditicopida and Beyrichicopida.

2. Following worldwide tectonism in the mid-Ordovician, all the modern ostracode groups became established at the ordinal and subordinal levels by the Early Silurian. Although his data are incomplete this, nevertheless, is brought out clearly in Danielopol (1980, table 1). Thus, Myodocopida were represented, possibly, by such taxa as *Ludvigsenites* (Jones & McKenzie 1980); Cladocopida by early Entomozoacea; and Halocyprida by Entomoconchacea. In parenthesis, it can be remarked that the Cambrian taxon *Isoxys* which has been assigned to several non-ostracodan groups in recent papers (Glaessner 1979, Simonetta & Delle Cave 1981) closely resembles some halocypridan ostracodes – e.g. *Conchoecia daphnoides* – although these are much smaller.

The Podocopida were represented in the Late Ordovician-Early Silurian by metacopines (Healdiacea), bairdiocopines (Bairdiacea), the paraparchitocopine Paraparchitacea, platycopines (Koedenellacea, Cytherellacea), kirkbyocopines (Kirkbyacea) and some cytherocopine Cytheracea (Bythocytheridae). The position of several other groups is uncertain or in dispute. These include Condomyridae (Schallreuter 1977) and Tricorninidae. The precise relationships of condomyrids are unclear as yet but Tricorninidae are regarded by some as early cytheraceans (Grundel 1978) related to Bythocytheridae (Grundel & Kozur 1973) and by others as having few affinities with Cytheracea (Schallreuter 1977). Finally, Cypridacea possibly evolved by this time (Schallreuter 1977). The free margin structures of these groups have been made the basis for determination of their affinities by some authors (Adamczak 1976, Gramm, in the draft version of this paper available from the authors).

Undoubtedly the most characteristic Early Palaeozoic orders, however, were the Beyrichicopida and Leperditicopida. Although a variety of hinge types has been described for beyrichicopids (Melik 1966), hingement plays little part in their classification which is based primarily on dimorphic features – cruminae, velae, lobes, sulci, loculi, histia (Henningsmoen 1965, Martinsson 1962, Jaanusson 1957, 1966). The Leperditicopida are noted for unusually large size, good preservation of numerous growth stages and highly characteristic muscle scar complexes.

3. During the rest of the Palaeozoic, worldwide tectonic crises were associated with the Middle Silurian, Middle Devonian, and Carboniferous. As a result, by the end of the Carboniferous the following groups were extinct: Beyrichiacea (Beyrichicopida); Entomoconchacea (Halocyprida); Entomozoacea (Cladocopida); some early myodocopid families (Cyprellidae, Cypridinellidae, Rhombinidae); Eridostraca and Leperditicopida. But worse was to follow.

4. The establishment of Pangaea during the Late Carboniferous-Permian was associated with profound variations in climate, sea level, and oceanic surface salinities. The net result was the Permo-Triassic faunal crisis. By the end of the Triassic the following groups were extinct: Beyrichicopida; Parapodocopina; Paraparchitocopina; Kirkbyocopina (Kirkbyacea); and Platycopina (Koedenellacea). There were brief but important radiations in the Healdiacea and Bairdiacea before the former became extinct (Jurassic) and the latter entered a prolonged evolutionary decline. There was also a radiation in the podocopid Glorianellidae (Permian-Jurassic), associated with brackish waters and transitional facies, which at least one author interprets as having important consequences for the phylogeny of Podocopida (Grundel 1978). On land, the Darwinulacea were dominant from Carboniferous-Triassic but then went into evolutionary stagnation. Intiations included the Polycopacea (Cladocopida), Thaumatocypridacea (Halocyprida) and several groups of marine Cypridacea (Podocopida).

5. Widespread marine transgressions of the Middle Jurassic and the Middle Cretaceous led to gradual establishment of the modern fauna. The former epoch followed upon the initial breakup of Gondwanaland; the latter included worldwide eustatism. The presence of Tethys as a warm sublatitudinal ocean stretching from the Americas to Australasia was a critical factor in evolution of many new groups. These included: Cylindroleberidacea (Myodocopida); Halocypridacea (Halocyprida); and, especially, most families of the modern Cytheracea (Podocopida). On land, the cytheracean limnocytherids were dominant briefly, but Cypridacea (Podocopida) soon took over this role, commencing with the worldwide *Cypridea* faunas of the Purbeckian-Wealden. During the Cenozoic, geotectonic events, associated with variations in climate and sea level continued to trigger ostracode evolution. Examples include: the development of modern deep-sea faunas following establishment of the psychrosphere; the Mediterranean salinity crisis (Messinian) leading to colonisation of the post-Messinian Mediterranean by an Atlantic fauna with relatively few Tethyan taxa remaining (mainly in the eastern relictual basins) such as *Triebelina, Cytherelloidea, Paijenborchella, Falsocythere, Tenedocythere,* and *Keijella;* the Paratethyan faunas, now relict in central Europe, the Black Sea, the Caspian, and several more easterly lake basins (McKenzie 1981); the new faunas of the East African Rift Valley lakes system dominated by *Mecnocypria* and *Allocypria* (Cypridacea) and some highly unusual cytheraceans; the Indopacific fauna which is mainly post-Neogene in origin (its evolution triggered by such events as the ultimate union of India against the Tibetan Plateau and the impingement of the Australian Block against the Indonesian Arc) although retaining numerous earlier Tethyan components from the Late Mesozoic and Paleogene. The only wholly new Cenozoic higher groups are the Punciacea (Neogene-Recent) and Terrestricytheracea (Recent). On land, the major crisis was associated with Pleistocene glaciation in the northern hemisphere which decimated local Neogene faunas and led to recolonization often via parthenogenetic populations.

The above outline history of Ostracoda establishes the fact that all main phyletic lines (orders, suborders) had developed by the Late Ordovician-Early Silurian (Parapodocopina appear to be an exception) whereas the evolution of family groups (superfamilies, families, subfamilies) was more or less continuous since the Early Palaeozoic and invariably was associated with geotectonic crises, climatic variations, and eustatism.

Such a generalization is refined by reference to particular ostracodan traits, allowing a more detailed understanding of evolution within the Class.

1. The most characteristic feature of Ostracoda is their small bivalved shell, distinguished

readily from that of conchostracans by the lack of growth lines (the Eridostraca are an exception), from the cladoceran shell by the absence of a headshield, from cirrepede 'cypris' larvae and from Ascothoracica by the occurrence of instar series (in Ostracoda). Apart from Phosphatocopida, ostracode shells are calcareous, this further distinguishing them from Conchostraca which have phosphatic shells. This calcareous bivalved shell is readily preserved and exhibits great morphologic stability (Benson 1975b); thus the classification of fossil Ostracoda has always been biased towards carapace characters of which the most important are acknowledged to be: hingement; muscle scar patterns; free margin structures; normal pore canals; features of the inner lamellae, including marginal pore canals; shape; and ornament. Conservative and homologous sculpture patterns have been identified and used to determine evolutionary trends in difficult groups (Liebau 1971).

2. There have been several finds of soft anatomy in Cambrian Phosphatocopida (Müller 1979a, 1982), Purbeckian-Wealden Cypridacea (Bate 1972), and Jurassic Cylindroleberidacea (Dzik 1978). Like the carapace features, these appear to be conservative. Uniformitarian extrapolation from living taxa to fossil ones can be made, therefore, with some confidence increasing the hope of achieving a reasonably natural classification of the Class.

3. Characteristically, Ostracoda are sensitive and reliable environmental indices. At least one author (Szczechura 1980) sees this fact, in combination with the long evolutionary history of Ostracoda, as being primarily responsible for speciation in the group. While most ostracodes are conservative in their habitat preferences — hence during episodic crises evolution is triggered in relatively 'stable' environments (McKenzie 1976) — a few groups are known to have changed their environments considerably over geological time, e.g. Cladocopida were initially planktic (Entomozoacea) but are now often interstitial. Other changes in environmental preferences or movement into new ecological niches seem to have occurred relatively recently. For example: (a) the modern psychrosphere fauna cannot predate the Eocene; and assemblages associated with phanerogams cannot predate the Palaeogene; (b) the specialised faunas of interstitial environments; high altitude lakes; caves (Danielopol 1980); terrestrial habitats; salt lakes (McKenzie 1981); and ostracode symbionts all appear to have become established within the Cenozoic, most likely since the Neogene. Such data document considerable remanent evolutionary vigour in Ostracoda. Their adaptations for such habitats include reductions (loss of eyes, limbs, segments, setae) and specializations (of shape, genitalia, setae). Of particular interest is the association of blindness with psychrospheric, cave and interstitial faunas.

4. Ostracoda are predominantly bisexual but parthenogenetic populations are known in the Podocopida (especially Cytheracea and Cypridacea). The vectors of syngamic fertilization are a diversity of sperm types ranging from simple (Cypridinacea) to highly specialised (Cypridacea), the latter including the longest known sperm (10 mm). Some continental taxa produce dessication-resistant eggs which ensure survival during periodic or prolonged aridity. Parthenogenetic populations appear to occupy the dispersal fringe of many Cypridacea and Cytheracea — for both groups parthenogenetic 'speciation' is known since the Jurassic (Purbeckian-Wealden) — including new niches such as terrestrial habitats. Darwinulacea, now represented only by parthenogenetic populations, may have used the parthenogenetic dispersal mode even earlier. Selection for parthenogenesis or syngamy is probably controlled hormonally as a response to environmental factors. Cytologically, the pathway to distinct parthenogenetic populations seems to lie via chromosomal polymorphisms which can become fixed through isolation.

5. Ostracode ontogeny (Tables 2 and 3) indicates several evolutionary trends particularly

Table 2. Ontogenetic development of Podocopida. X – structure always present in definitive form; A – anlage, at least, present in all species; a – anlage present in some species; 1 – first free living larval stage. There is no data as yet on the soft part ontogeny of Cytherellacea, Sigilliacea, Terrestricytheracea, and Punciacea. Note that the furca becomes *reduced* to its definitive form at the fourth free living larval stage'in Cytheracea and Darwinulacea. In Parvocytheridae (Cytheracea) the third thoracic limb is absent. It does not appear at any stage in the ontogeny of this family. In Xestoleberididae (Cytheracea) there are only seven free living larval stages.

| Cytheracea | Bairdiacea | Darwinulacea | Cypridacea | antennule | antenna | mandible | maxillule | PI | PII | PIII | gonads | genitalia | furca |
|---|---|---|---|---|---|---|---|---|---|---|---|---|---|
| | | | 1 | X | X | A | | | | | | | a |
| 1 | 1 | 1 | 2 | X | X | X | A | a | | | | | A |
| 2 | 2 | 2 | 3 | X | X | X | X | a | | | | | A |
| 3 | 3 | 3 | 4 | X | X | X | X | A | a | | | | A |
| 4 | 4 | 4 | 5 | X | X | X | X | X | A | a | | | X |
| 5 | 5 | 5 | 6 | X | X | X | X | X | X | A | a | | X |
| 6 | 6 | 6 | 7 | X | X | X | X | X | X | X | A | a | X |
| 7 | 7 | 7 | 8 | X | X | X | X | X | X | X | X | A | X |
| Adult 8 | 8 | 8 | 9 | X | X | X | X | X | X | X | X | X | X |

Table 3. Ontogenetic development of Cladocopida (Polycopacea); Halocyprida (Halocypridacea, Thaumatocypridacea); and Myodocopida (Cypridinacea, Cylindroleberidacea, Sarsiellacea). Note that Polycopacea lack PII, PIII. X – structure always present in definitive form; A – anlage at least present in all species; a – anlage present in some species; 1 – first free larval stage.

| Polycopacea | Thaumatocypridacea | Halocypridacea | Cypridinacea | Sarsiellacea | Cylindroleberidacea | antennule | antenna | mandible | maxillule | PI | PII | PIII | gonads | genitalia | furca |
|---|---|---|---|---|---|---|---|---|---|---|---|---|---|---|---|
| 1 | 1 | 1 | | | | X | X | X | X | a | | | | | X |
| 2 | 2 | 2 | 1 | 1 | 1 | X | X | X | X | A | a | | | | X |
| 3 | 3 | 3 | 2 | 2 | 2 | X | X | X | X | X | a | | | | X |
| 4 | 4 | 4 | 3 | 3 | 3 | X | X | X | X | X | A | a | | | X |
| 5 | 5 | 5 | 4 | 4 | 4 | X | X | X | X | X | X | A | a | | X |
| 6 | 6 | 6 | 5 | 5 | 5 | X | X | X | X | X | X | X | A | A | X |
| Adult 7 | 7 | 7 | 6 | 6 | 6 | X | X | X | X | X | X | X | X | X | X |

in such carapace characters as hingement; marginal pore canals, normal pore canals; and in such soft part features as the furca.

6. Food gathering and locomotion in Ostracoda need to be restudied since the present consensus appears to be oversimplified and ignores the physical constraints. Some nutritional modes, e.g. the various kinds of filter-feeding are not achieved with homologous structures (Table 4); thus assumptions concerning the occurrence of such modes in fossil

Table 4. Filter apparati on some filter feeding Ostracoda. Note that Schulz (1976) has illustrated prominent labral and hypostomal hair fringes, which also may operate as filters, on several cytheracean genera.

| Taxon | Oral region | Mandible endopod | Maxillule (M1) endopod | Maxilla (M2) endopod |
|---|---|---|---|---|
| Cylindroleberididae | – | – | comb | comb |
| Cytherellidae | – | comb | comb | comb ? |
| Darwinulidae | – | comb | – | – |
| Paradoxostomatidae | suction organ | – | – | – |
| *Vitjasiella* | hypostome fans | comb ? | – | – |

taxa on the basis of similar structures need to be qualified and the interpretations developed from such assumptions need to be evaluated with caution.

7. Ostracode dispersal is linked to the changes in continental and oceanic configurations discussed earlier and is a critical factor in assessing evolution within the group (McKenzie 1973b).

8. The fact that Ostracoda are oligomerized crustaceans leads to an expectation that numerous convergences occur and this is certainly the case. Very often such convergences document the attainment of a common grade of evolution by taxa which are phyletically distinct – i.e. they are examples of parallel evolution. Instances of homeomorphy need to be separated from another effect of oligomerization (and the consequent availability of relatively few characters for evolution) namely, mosaic evolution. The former (parallel evolution and convergence) tend to obscure phyletic relationships, the latter (mosaic evolution) is a characteristic phase in the evolution of numerous groups, e.g. Timiriaseviinae (Cytheracea) and *Candona* s.l. (Cypridacea).

9. With respect to other crustaceans, Ostracoda are more closely related to Entomostraca than Malacostraca as is generally acknowledged. It is suggested here on a variety of grounds that Entomostraca include Phyllocarida (Section 2 above).

10. The soft anatomies of Phosphatocopida (Müller 1979a, 1982) and Podocopida (Schulz 1976) suggest that primitive Ostracoda had at least four thoracic segments and, possibly, six abdominal segments; at least seven paired limbs and a furca; probably a nauplius eye; a relatively simple bivalved carapace; and were sexually dimorphic and small. They were benthic or epibenthic; probably included swimmers as well as crawlers; and 'filter feeders' as well as 'deposit feeders'. Such animals may well have represented an ancestral crustacean stock.

# 6 ACKNOWLEDGEMENTS

The authors wish to thank Drs F.R.Schram, for the invitation to prepare this paper, and G.Hartmann and L.S.Kornicker for reviewing the entire manuscript. It is regrettable that the tight schedule prevented more colleagues from participating, especially in regards to Meso- and Cenozoic evolution. Responsibilities for the sections are as follows: Introduction, Relationships, and Evolution (K.G.M.); Phosphatocopida (K.L.M.); and Taxonomy (K.G.M. and M.N.G.). We thank Drs J.S.Ho and P.J.Jones for advice on taxonomy.

## REFERENCES

Adamczak, F. 1965. On some Cambrian bivalved Crustacea and egg cases of the Cladocera. *Stockholm Contr. Geol.* 13:27-34.

Adamczak, F. 1976. Middle Devonian Podocopida from Poland; their morphology, systematics, and occurrence. *Senck. Lethaea* 57:265-467.

Al-Abdul-Razzaq, S. 1973. Evolution of middle Devonian species of *Euglyphella* as indicated by cladistic analysis. *Contr. Mus. Paleontol. Univ. Mich.* 24:47-65.

Anderson, D.T. 1973. *Embryology and Phylogeny in Annelids and Arthropods.* Oxford: Pergamon.

Baird, W. 1850. *Natural History of the British Entomostraca.* London: Ray Soc.

Bate, R.H. 1972. Phosphatized ostracods with appendages from the lower Cretaceous of Brazil. *Paleontol.* 15:379-393.

Bate, R.H. & E.Robinson 1978. *A Stratigraphical Index of British Ostracoda.* Liverpool: Seel House Press.

Benson, R.H. 1975a. The origin of the psychrosphere as recorded in changes of deep-sea ostracode assemblages. *Lethaia* 8:69-83.

Benson, R.H. 1975b. Morphologic stability in Ostracoda. *Bull. Am. Paleontol.* 65:13-46.

Benson, R.H. 1981. Form, function, and architecture of ostracode shells. *Ann. Rev. Earth Planetary Sci.* 9:59-80.

Bold, W.A., van den 1977. Cenozoic marine Ostracoda of the South Atlantic. In: F.M.Swain (ed.), *Stratigraphic Micropaleontology of the Atlantic Basin and Boderlands:*495-519. Amsterdam: Elsevier.

Bowman, T.E. 1971. The case of the nonubiquitous telson and the fraudulent furca. *Crustaceana* 21:165-175.

Briggs, D.E.G. 1978. The morphology, mode of life, and affinities of *Canadaspis perfecta* (Crustacea: Phyllocarida), Middle Cambrian, Burgess Shale, British Columbia. *Phil. Trans. Roy. Soc. London* (B) 281:439-487.

Calman, W.T. 1909. Crustacea. In: E.R.Lankester (ed.), *A Treatise on Zoology, Vol. 7.* London. Adam & Charles Black.

Cannon, H.G. 1924. On the development of an estherid crustacean. *Phil. Trans. Roy. Soc., London* (B) 212:395-430.

Colin, J.P. & D.L.Danielopol 1980. Sur la morphologie, la systematique, la biogéographie, et l'évolution des ostracodes Timiriaseviinae (Limnocytheridae). *Paléobiologie Continentale* 11:1-52.

Dahl. E. 1963. Main evolutionary lines among recent Crustacea. In: H.B.Whittington & W.D.I.Rolfe (eds), *Phylogeny and Evolution of Crustacea:*1-15. Cambridge: Mus. Comp. Zool.

Danielopol, D.L. 1980. An essay to assess the age of the freshwater interstitial ostracods of Europe. *Bijd. Dierk.* 50:243-291.

Dzik, J. 1978. A myodocopid ostracode with preserved appendages from the Upper Jurassic of the Volga River region. *N.Jb. Geol. Palaeontol. Monat.* 1978:393-399.

Glaessner, M.F. 1979. Lower Cambrian Crustacea and annelid worms from Kangaroo Island, South Australia. *Alcheringa* 3:21-31.

Gooday, A.J. 1978. The Devonian. In: R.H.Bate & E.Robinson (eds), *A.Stratigraphical Index of British Ostracoda:*101-120. Liverpool: Seel House Press.

Grundel, J. 1969. Neue taxonomische Einheiten der Unterklasse Ostracoda (Crustacea). *N.Jb. Geol. Palaeontol. Monat.* 1969:353-361.

Grundel, J. 1978. Die Ordnung Podocopida: Stand und Probleme der Taxonomie und Phylogenie. *Freib. Forsch.* 334:49-68.

Grundel, J. & H.Kozur 1973. Zur Phylogenie der Tricorninidae und Bythocytheridae. *Freib. Forsch.* 282:99-111.

Hartmann, G. 1967. Ostracoda 2. In: *H.G.Bronn's Klassen und Ordnungen des Tierreichs* 5(1), 2(4): 217-408. Leipzig: Acad. Verlag.

Hartmann, G. & H.S.Puri 1974. Summary of neontological and paleontological classification of Ostracoda. *Mitt. Hamb. Zool. Mus. Inst.* 70:7-73.

Henningsmoen, G. 1965. On certain features of palaeocope ostracodes. *Geol. Foren. Stockholm Forhand.* 86:329-394.

Hornibrook, N., de B. 1949. A new family of living Ostracoda with striking resemblances to some Palaeozoic Beyrichiidae. *Trans. Roy. Soc. N.Z.* 77:469-471.

Jaanusson, V. 1957. Middle Ordovician ostracodes of central Sweden. *Pub. Palaeontol. Inst., Univ. Uppsala* 17:173-442.

Jaanusson, V. 1966. Ordovician ostracodes with supravelar antra. *Pub. Palaeontol. Inst., Univ. Uppsala* 66:1-30.

Jones, P.J. & K.G.McKenzie 1980. Queensland Middle Cambrian Bradoriida (Crustacea): new taxa, paleobiogeography, and biological affinities. *Alcheringa* 4:203-225.

Kaye, P. 1965. Some new British Albian Ostracoda. *Bull. Brit. Mus. Nat. Hist. Geol.* 11:215-253.

Keij, A.J. 1957. Eocene and Oligocene Ostracoda of Belgium. *Mem. Inst. Roy. Sci. Nat. Belgique* 136: 1-210.

Kempf, E.K. 1980. *Index and Bibliography of Nonmarine Ostracoda.* Cologne.

Kornicker, L.S. & I.G.Sohn 1976. Phylogeny, ontogeny, and morphology of living and fossil Thaumato-cypridacea. *Smith. Contr. Zool.* 219:1-124.

Kozur, H. 1972. Einige Bemerkungen zur Systematik der Ostrocoden und Beschreibung neuer Platyco-pida aus der Trias Ungarns und der Slowakei. *Geol. Palaeontol. Mitt. Innsbruck* 2(10):1-27.

Kozur, H. 1974. Die Bedeutung der Bradoriida als Vorlaufer der postkambrischen Ostracoden. *Z. Geol. Wiss., Berlin* 2:823-830.

Landing, E. 1980. Late Cambrian-early Ordovician macrofaunas and phophatic microfaunas. St John Group, New Bruncwick. *J. Paleontol.* 54:752-761.

Langer, W. 1973. Zur Ultrastruktur, Mikromorphologie, und Taphonomie des Ostracod Carapax. *Palaeon-togr. Abt. A.* 144:1-54.

Liebau, A. 1971. *Homologe Sculpturmuster bei Trachyleberididen verwandten Ostrakoden.* Berlin: Diss. Techn. Univ.

Linder, F. 1941. Contributions to the morphology and taxonomy of the Branchiopoda Anostraca. *Zool. Bidrag Uppsala* 20:100-302.

Maddocks, R.F. 1976. Quest for the ancestral podocopid: numerical cladistic analysis of ostracode appendages, a preliminary report. *Abhl. Verhl. Naturwiss. Ver. Hamburg* 18/19:39-53.

Manton, S.M. 1973. Arthropod phylogeny – a modern synthesis. *J. Zool.* 171:111-130.

Manton, S.M. 1977. *The Arthropoda.* Oxford: Clarendon Press.

Martinsson, A. 1962. Ostracodes of the family Beyrichiidae from the Silurian of Gotland. *Pub. Palaeon-tol. Inst., Univ. Uppsala* 41:1-369.

McKenzie, K.G. 1972a. *Checklist of Ostracoda Recorded from the Indian Subcontinent and Ceylon (1840-1971).* Melbourne.

McKenzie, K.G. 1972b. Contribution to the ontogeny and phylogeny of Ostracoda. In: *Proceedings of the XXIIIrd International Geological Congress:*165-188. Prague.

McKenzie, K.G. 1973a. The biogeography of some Cainozoic Ostracoda. *Sp. Pap. Palaeontol.* 12:137-153.

McKenzie, K.G. 1973b. Cenozoic Ostracoda. In: A.Hallam (ed.), *Atlas of Paleobiogeography:*477-487. Amsterdam: Elsevier.

McKenzie, K.G. 1976. Sahul shelf assemblages and the evolution of post-Paleozoic Ostracoda. *Abhl. Verhl. Naturwiss. Ver. Hamburg* 18/19:215-228.

McKenzie, K.G. 1977. Bonaducecytheridae, a new family of cytheracean Ostracoda, and its phylogene-tic significance. *Proc. Biol. Soc. Wash.* 90:263-273.

McKenzie, K.G. 1981. Palaeobiogeography of some salt lake faunas. *Hydrobiol.* 82:407-418.

Melik, J.C. 1966. Hingement and contact margin structure of palaeocopid ostracodes from some middle Devonian formations of Michigan, southwestern Ontario, and western New York. *Contr. Mus. Paleontol. Univ. Mich.* 20:195-269.

Müller, K.J. 1979a. Phosphatocopine ostracodes with preserved appendages from the upper Cambrian of Sweden. *Lethaia* 12:1-27.

Müller, K.J. 1979b. Body appendages of Paleozoic ostracodes. In: *Proceedings of the VIIth International Symposium on Ostracodes:*5-8. Belgrade: Serbian Geol. Soc.

Müller, K.J. 1982. *Hesslandona unisulcata* sp.nov. with phosphatized appendages from upper Cambrian 'orsten' of Sweden. In: R.H.Bate, E.Robinson & L.M.Sheppard (eds), *Fossil and Recent Ostracoda:* Chichester.

Oertli, H.J. 1971. Paleoecologie des Ostracodes. *Bull. Centre Rech. Pau* 5:1-953.

Orlov, Yu.A. 1960. *Basic Paleontology: Trilobitomorpha, and Crustacea.* Moscow: Acad. Sci. USSR (in Russian).

Pokorny, V. 1964. *Conchoecia ?cretacea* n.sp., first fossil species of the family Halocyprididae. *Acta Univ. Carolinae Geol.* 1964:175-180.

Pokorny, V. 1978. Ostracoda. In: B.U.Haq & A.Boersma (eds), *Introduction to Micropaleontology:* 109-149. New York: Elsevier.

Rossi de Garcia, E. 1969. Cuyanocopina, un nuevo suborden de Ostracoda Paleozoico. *Actas I Cong. Argent. Paleontol. Biostrat.* 1:59-76.

Schallreuter, R.E.L. 1977. Zwei neue ordovische Podocopida und Bemerkungen zur Herkunft der Cytheracea und Cypridacea. *N.Jb. Geol. Palaeontol. Monat.* 1977:720-734.

Schminke, H.K. 1976. The ubiquitous telson and deceptive furca. *Crustaceana* 30:293-300.

Schornikov, E.I. 1969. A new family of Ostracoda from the supralittoral zone of the Kurile Islands. *Zool. Zh.* 68:494-498 (in Russian).

Schram, F.R. 1981. On the classification of Eumalacostraca. *J. Crust. Biol.* 1:1-10.

Schulz, K. 1976. *Das Chitinskelett der Podocopida und die Frage der Metamerie dieser Gruppe.* Hamburg.

Scott, H.W. 1961. Classification of Ostracoda. In: R.C.Moore (ed.), *Treatise on Invertebrate Paleontology, Part Q, Arthropoda* 3:Q74-92. Lawrence: Geol. Soc. Am. & Univ. Kansas Press.

Siewing, R. 1963. Studies in malacostracan morphology: results and problems. In: H.B.Whittington & W.D.I.Rolfe (eds.), *Phylogeny and Evolution of Crustacea:* 85-103. Cambridge: Mus. Comp. Zool.

Simonetta, A. & L.delle Cave 1981. An essay in the comparative and evolutionary morphology of Paleozoic arthropods. *Atti Convegni Lincei* 49:389-439.

Sohn, I.G. 1961. Superfamily Kirkbyacea. In: R.C.Moore (ed.), *Treatise on Invertebrate Paleontology, Part Q, Arthropoda* 3:Q163-169. Lawrence: Geol. Soc. Am. & Univ. Kansas Press.

Sohn, I.G. & L.S.Kornicker 1969. Significance of calcareous nodules in myodocopid carapaces. In: J.W. Neale (ed.), *The Taxonomy, Morphology, and Ecology of Recent Ostracoda:*99-108. Edinburgh: Oliver & Boyd.

Szczechura, J. 1980. Causes for speciation in ostracodes. *N.Jb. Geol. Palaeontol. Monat.* 1980:439-441.

Tasch, P. 1969. Branchiopoda. In: R.C.Moore (ed.), *Treatise on Invertebrate Paleontology, Part R, Arthropoda* 4, vol.1:R128-191. Lawrence: Geol. Soc. Am. & Univ. Kansas Press.

BRIAN MICHAEL MARCOTTE
*Institute of Oceanography, McGill University, Montreal, Quebec, Canada*

# THE IMPERATIVES OF COPEPOD DIVERSITY: PERCEPTION, COGNITION, COMPETITION AND PREDATION

## ABSTRACT

Perception and cognition are separate processes. They contribute fundamentally important properties to ecological interactions such as competition and predation. The interactions of competition and predation as mediated by perception and cognition may be essential to understanding the ontogeny, ecology, and phylogeny of the Copepoda.

## 1 PROLEPSIS

Astrology was not a science. With this, most 20th century scientists would tend to agree. However, by what criteria did we arrive at this demarcation of astrology from science? Astrology did have a well articulated and widely accepted central theory, i.e. '. . . that the planets exert basic forces on character, the zodiacal signs modify these and the houses "show the earthly sectors, or departments, of everyday life in which the modified forces operate" ' (Shumaker 1972). But central theory does not a science make. (One could object, however, that this theory was rashly syncretical, but so is much of what we call science, albeit 'pseudo-science', in human studies today.) Astrology did have operationally defined terms *sensu* Bridgeman (1936). Right ascendant or the position of planets above the eastern horizon and the place and time of a subject's birth could be precisely measured. It was not for want of such terms that astrology was unscientific. Furthermore, astrology made falsifiable hypotheses [Popper's (1963) often misapplied demarcation criterion] and indeed many were falsified [Pico (1572), see also relevant sections in Schumaker (1972), and Yates (1964, 1966).] Astrologers explained these failures on the inabilities of individual practitioners, insufficient or inaccurate data and on the uncertainty inherent in a very complex system — in short, on the sorts of things 20th century biologists continue to use to explain apparent failures of theory today. What then demarcates astrology from science?

Astrology was not a science. It was a craft like medicine and engineering were at the turn of the 19th century (Kuhn 1970). The tragic flaw was that astrology had no 'normal science' ruled by a constructive paradigm, the object of which was to solve puzzles (Kuhn 1962, 1970). Puzzles are normal research problems not aimed at producing major theoretical novelties (Kuhn 1962). They aim at extending the scope and precision with which a paradigm can be applied. A paradigm is, above all, concrete. It is a model, a picture and/or an analogy-drawing sequence of words in natural language (Masterman 1970). It does not have the intellectual status or force of theory. Normal science is essential to the historical reality of science (Kuhn 1962) and is a bane for purely logical interpreters (Kuhn 1970). The central theory of astrology, no matter how plausible, did not provide tools for astrologers to solve the puzzles presented by failures of their theory (Kuhn 1970).

Evolution theory is a grand, 19th century world view whether presented as natural history (e.g. Darwin 1964, Mayr 1963) or in the rigors of set theory and symbolic logic (Williams 1970, 1975). It has operationally defined terms,[1] for example, measurable 'populations' replaced idealist concepts of species. Deductions of evolution theory can be falsified, e.g. Weismann's (1889) heretability experiments. Finally, evolution theory has produced a rich and diverse normal science — from ecology to genetics. Several characteristics of evolutionary paradigms are of interest here.

1. Complex sets of interacting biological processes such as competition or predation are usually studied as discrete isolated events.[2] They are reduced to a simple set (usually two) of strongly (obviously) interacting variables — variables which are usually part of relatively simple ecosystems, e.g. intertidal Crustacea (Connell 1961). These biological processes may be further reduced by being uncoupled from interactions with their ecosystem using exclosure/inclosure cages (e.g. Paine 1974) or laboratory cultures (e.g. Gause 1934). These

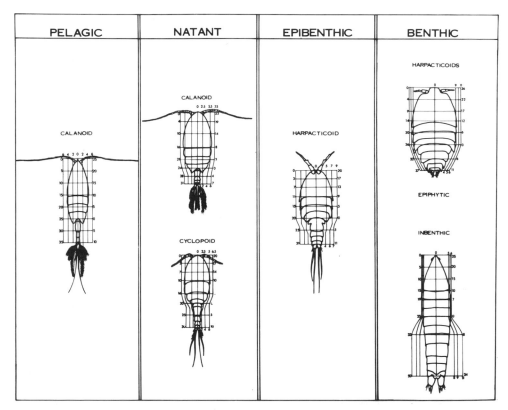

Figure 1. Representative species of three orders of free-living Copepoda. Natant are animals swimming at metres above the bottom. Epibenthic are animals living on or just above the sediment-seawater interface. Orthogonal coordinates are drawn on the pelagic calanoid. They are transformed by the morphology of the copepods to the right.

processes may also be decayed mathetically into a set of differential equations (e.g. Odum 1972, 1975). The subsequent reassembly of the ecosystem often lacks explanatory power or gives simply incorrect predictions.

2. The scientist in most evolutionary paradigms is a vestigial, Newtonian (Objective) observer over both time and space. For example, fitness is measured from his point of view, over three generations: parental, $F_1$ and $F_2$. Predator-prey interactions are defined by variables easily observed by him, e.g. gut contents, number of prey eaten/injured, and effort of the predator. These variables place the observer outside and above the interaction opining periodically on the selective value of some apparent (from the observer's point of view) adaptation: long legs, long teeth, large muscles, etc. To this observer, the decision matrix (tensor) that is the space of evolution exists absolutely without regard to organisms in it. He 'knows' the alternatives and observes the players succeeding or failing (his criteria of course).

3. Evolutionary paradigms are reactive: cause always preceeds effect. 'Preadaptation' is a locution of hindsight. To my knowledge, there is only one anticipatory paradigm in biology (see below) and no acausal, non-spatial or surreal model of evolution.

4. Evolutionary paradigms in ecology are now suffering the 'death of a thousand qualifications' (Masterman 1970). For example, population dynamics has matured from models contingent on the existence of equilibria (e.g. MacArthur 1972), to moving equilibria (Levins 1975), to sustained, bounded motion (Levins 1981). All the while, these models sacrifice precision and simplicity for generality.

An original puzzle that evolutionary paradigms were made to solve was the question of the origin and maintenance of organic diversity. 'Why are there so many kinds of animals?' (Hutchinson 1959), or its corollary, 'why aren't there more?' To this puzzle the following essay turns. The reader is cautioned not to expect a single fully formed answer here. The present essay will provide a set of spirals around the topic with the explicit intent of expressing and partially specifying holistic yet solipsistic, anticipatory yet predictive, and simple yet rich approaches to this puzzle.

## 2 INTRODUCTION

Of the free living copepods, there are about 1 500 species of calanoid, 2 000 cyclopoids and 4 500 species of harpacticoid (other groups are too few and little known to be included here). Why are there so many harpacticoids? Why aren't there more calanoids? Representatives of these three orders are pictured in Figure 1. Roughly speaking, calanoids are pelagic, cyclopoids are pelagic and near-benthic (many are parasitic), and harpacticoids are near-benthic and inbenthic. The body shapes of near-benthic (epibenthic, epiphytic, epizoic) harpacticoids vary in allometric transformations that are continuous with those of pelagic calanoids and cyclopoids. The body shapes of inbenthic harpacticoids are very different (Marcotte 1980). There are 81 genera of near-benthic harpacticoids and 222 genera of their inbenthic counterparts. Why are there so many inbenthic harpacticoids? Why is their body shape so different from that of other free living copepods? If taxonomic space is isomorphic with environment space (Hutchinson 1968), then the continuity in transformations of body form among calanoids, cyclopoids, and near-benthic harpacticoids may mean that all are experiencing similar (though continuously changing) selection pressures. Accordingly, the ecology of near-benthic harpacticoids can be used as a provisional paradigm for the ecology of the other water-dwelling copepods.

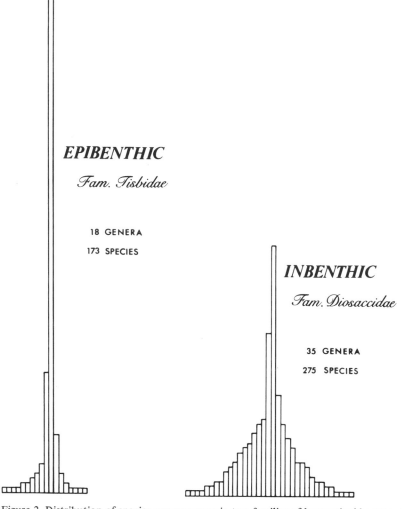

Figure 2. Distribution of species among genera in two families of harpacticoid copepod.

The evenness with which species are distributed among the genera of inbenthic and near-benthic harpacticoids differs. For example, consider the species diversity (richness and evenness) of the near-benthic harpacticoid family Tisbidae and the inbenthic family, Diosaccidae (Fig.2). There are more, morphologically defined species of diosaccid and they are more evenly distributed among their genera. Why do these differences in species diversity exist? Why aren't there more near-benthic and pelagic copepod species? To solve this puzzle, the following analyses are offered.

## 3 NEAR-BENTHIC COMPETITION FOR SPACE

The behaviour of two species of *Tisbe* was studied in the field and laboratory when they became very abundant (Marcotte, submitted 1). Adults were natant or epibenthic. Juveniles were inbenthic near the sediment-seawater interface. The swimming and feeding behaviours of the two species differed. They were identical in gross external morphology. There were substantial, species-specific differences in the morphology of their genital fields — a key to *Tisbe* taxonomy. They were sibling species.

The adults competed for space at the sediment-seawater interface. One species spent more than 94 % of its time slowly moving and feeding (on spheres of detritus) on the surface of the sediment. It was not disturbed by the approach or touch of other copepods, nematodes or mechanical devices used to simulate the vibrations created by copepods moving in water. It often allowed another copepod to walk over it without displaying a response. The second species spent fully one-third of its time swimming in the water. To feed, it would dash onto the sediment surface, grab a ball of detritus and twirl it in its mouth parts gleaning off bacteria and diatoms. It would bolt upward into the water column whenever it was approached or touched by another copepod.

These differences in behaviour became evident when copepod abundances were high. This is expected since the probability of encounter is equal to the product of copepod species' abundances. Differential sensitivity to tactile stimulation led to vertical habitat displacement for the second species, i.e. *de facto* interference competition. In ecological time, this vertical habitat separation is a form of behaviourally-defined resource allocation permitting competitors to co-exist in the same geographic location (though in different micro habitats). In evolutionary time, vertical habitat displacement might lead to (statistical) sexual isolation. The mechanism of this isolation is behavioural. The outcome need have no strong morphological consequence. Hence, sibling species can occur with 'sympatric' speciation. This example could explain how the pelagic Copepoda could have evolved from a benthic (harpacticoid-like) ancestor. It predicts that sibling species should be common among pelagic copepods. Fleminger (1975) and Fleminger & Hulsmann (1974) provide elegant demonstrations that sibling calanoids do exist.

## 4 INBENTHIC COMPETITION FOR SIZE OF FOOD-BEARING PARTICLES

For one year, the biology of meiobenthic copepods dwelling in a marsh in Nova Scotia, Canada, was studied (Marcotte, submitted 2). Two species lived in the same place at the same time and fed on bacteria and diatoms growing on spheres of organic detritus and clay floccules in the same way. These species were *Amphiascus minutus* and *Amphiascoides debilis*, both in the family Diosaccidae. Adults and juveniles of both species burrowed in the sediment. Their locomotory and feeding behaviours were similar and virtually unvarying. They seemed to be exploitatively competing for food-bearing particles, and seemed to differentiate this resource by size. 'Seemed' is a necessary caveat. Experimental verification was beyond the scope of my study, but no hypothesis other than exploitative competition accounts for my data.

The volume of the oral frustrum determined the size of the largest food-bearing particle that could be handled by these copepods. The size of this frustrum was allometrically related to the body size of the individuals. Whenever the size of the oral frustrum of the two species converged below a morphometric ratio (larger/smaller) of 1.4, the areal distributions of the two species became negatively related. When above this competitive (size) threshhold, their distributions covaried.

Basal Area of Oral Frustum

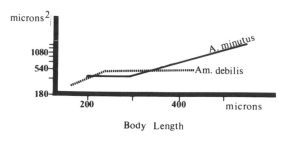

Figure 3. Schematic representation of the allometric relationship between total body length and basal area of the oral frustum of two inbenthic species of harpacticoid copepod: *Amphiascus minutus* and *Amphiascoides debilis.*

The allometric relationship between size of oral frustrum and length of body significantly differed in the two species (Fig.3). The oral frustum of *A.minutus* was constantly small over increasing body length of copepodites and enlarged rapidly with body length of adults. There were virtually no cases near the junction of the two regression lines. In *Am. debilis,* the oral frustrum enlarged rapidly with increasing body length of copepodites but was constant with body length of adults. When superimposed these two sets of data cross where *A.minutus* had the least number of cases, i.e. at its inflection point.

In the first winter when both species were present and their sizes were below the competitive threshhold, juveniles became abundant in the population of *A.minutus. Am.debilis* showed no similar demographic switch. In the following winter when *A.minutus* was without its competitor and its size decreased to the same levels as during the previous winter (when *Am.debilis* was present) no substantial demographic switch to juveniles was observed. It seemed, therefore, that during the first competitive engagement, for *A.minutus,* there was a competitive advantage to staying young, i.e. to retain a small oral frustum. To grow older and larger, would have taken the juveniles through a region of competitive overlap with *Am.debilis.* Both saltation over this region of morphometric overlap and suppression of juvenile development, and, therefore, suspension of allometric growth, seemed to be used by *A.minutus* probably both in evolutionary time (developmental saltation) and in ecological time (suppression of juvenile development) to adapt to this competition.

One can imagine from these data that had a selective advantage to staying young been sustained, heterochronic events (Gould 1977), both paedomorphosis and neoteny, could have been favourably selected had they occurred. Allometric adjustments to competition for differing sizes of food-bearing particles mediated by demographic changes could result in heterochrony. This would yield high morphological diversity, expressed as high alpha-taxonomic diversity, among these inbenthic copepods. Indeed, heterochrony has been suspected for years as a mechanism relating genera and species of inbenthic harpacticoid families (Lang 1948, Noodt 1974). The above provides an ecological explanation for such hypotheses.

These data also support the notion that a morphometric ratio of 1.4 is an important touchstone in studies of animal competition (e.g. Brown & Wilson 1956, Hutchinson 1959, Grant 1972, Fenchel 1975). Usually, however, it is recognized as the end point of morphometric divergence resulting from competition, and the phenomena is called character displacement (Brown & Wilson 1956). In the present case, it is seasonal convergence below this morphometric ratio which causes competition.[3] The result is demographic switching and/or local extinction.

The mechanism by which juvenile development was stopped in this example was un-known. This developmental pause was not equivalent to hybernation or estivation (e.g. Borutzky 1929, Donner1928).[4] No single copepodite stage seemed to numerically domi-nate the juvenile demographics.[5] This suppression of development is most like that demon-strated by Coull & Dudley (1976). In laboratory cultures they showed that naupliar deve-lopment was supressed and that the number of adults in the culture dish at one time was limited. The cue which induced this supression of juvenile development may have been chemical, i.e. pheromones excreted by copepodites or adults (Coull & Dudley 1976).[6]

## 5 PREDATION

Why don't epibenthic copepods like *Tisbe* undergo demographic adjustments to competi-tion since only their adults were observed to compete? The answer seems two pronged. First the inbenthic environment in which the juveniles of epibenthic adults live is already populated with many species of copepod that are specialized as to habitat and trophic re-sources (Marcotte, submitted 2). The juveniles probably compete with these other inbenthic copepods (including congeneric juveniles) for resources. Prolonged exposure to these com-petitors might bring about competitive exclusion.[7] Second, copepods living at and above the sediment-seawater interface are more often preyed upon than are inbenthic species (e.g. Coull 1973, Marcotte 1980). The reasons for this will be suggested below. Thus, for species with swimming, near-benthic adults, predation on this age class is more likely.[8] They are, therefore, constantly being removed, and with them any pheromones they might have produced to suppress the development of their juveniles (intraspecific competitors). Furthermore, once in the water column above the chemically stabilizing, viscous or laminar flow regime of the sediment-seawater interface, turbulence may cause pheromones to dis-perse rendering them useless. Accordingly, suppression of development may not frequently occur among these near-benthic species.

## 6 INTERACTIONS INTERACTING

The processes of ecological interactions are precisely defined. For example, competition is an interaction in which species negatively effect each others' population dynamics. Preda-tion is an interaction in which the population dynamics of the predator are positively in-fluenced and those of the prey are negatively effected. The mathematical results which accrue from these interactions interacting are well handled by one ecological paradigm, Loop Analysis (Fig.4). In this paradigm, species are viewed as existing in a network of inter-acting variables: species, functional groups, and/or abiotic factors (Levins 1970, 1973, 1975a, 1975b, 1979, Lane & Levins 1977, Lane 1981). This paradigm is rigorously holistic. To understand the mathematical outcome (persistence, fitness, etc.) of a set of interactions, one must know not only the sign of the ecological processes by which variables are directly interacting with one another, but also the sign of indirect pathways, the sign of feedback at all levels and the sign of complementary pathways.

Although the signs of the processes of ecological interactions can be used to link the variables in a loop diagram, they must not be mistaken for the sign of the mathematical outcome of these interactions interacting, which can be counter-intuitive (Levins 1975b).

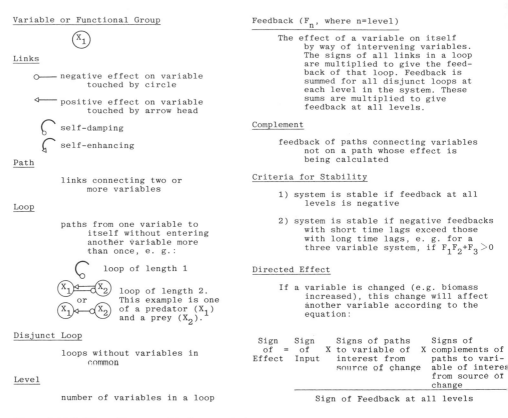

Figure 4. Definition of terms used in Loop analysis.

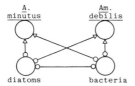

Figure 5. Example of a loop diagram in which two species (*Amphiascus minutus* and *Amphiascoides debilis* in this hypothetical example) compete through indirect, exploitative mechanisms for resources (diatoms and bacteria) which themselves compete through direct, interference mechanisms. The result is mutualism, not competition.

For example (Fig.5), two species can exploitatively compete for resources that are themselves competing by interference mechanisms. The mathematical result is mutualism not competition as the description of the ecological processes involved might have implied (Vandermeer 1980, Lane personal communication).

In short, with Loop Analysis, interactions such as competition, predation and mutualism, are not discrete processes. Rather, they are diffuse properties of a network of interacting interactions. Their mathematical meaning transcends the biology which gave rise to them. (Nature is not read in tooth and claw, but in interactions interacting.) The position of a variable in a network will determine the response of its biomass or turnover rate to a perturbation (Levins 1975b, Lane & Levins 1977). Variables which may be widely separated can have strong effects on one another. Losses of species, explosions of the population abundances of a particular species or changes in a community property (*sensu* Lane, Lauff & Levins 1975) must be taken as symptoms of network change. They may not represent the locus of a perturbation's input to the network (Lane 1981). Loss of a critical functional group in a network can lead to disintegration of the entire network [usually these critical function groups are strongly self-damping, e.g. cannibalistic copepods (Lane 1981)].

'Communities' may be networks that are not simply aggregates but substantial wholes (see footnote 9). Succession (and its ecological inverse, degradation caused by pollution) may not simply be quantitative changes involving aggregate properties. Only certain conformations of networks involving a given set of variables may be possible or stable (i.e. persistent). There may be rules relating the 'allometry' of these conformations and the number (diversity) and kinds of functional groups in the ecosystem. There may be discontinuities in these allometric transformations corresponding to what are commonly called 'communities' or 'species'. Succession (involving increasing diversity) may be constrained within a set of qualitative transformations of network structure — as was the growth (in size) of horse skulls studied by Thompson (1971). The answer to the question, 'Why aren't there more species?' may involve the nature of these transformations and the higher order principles which probably govern them.

In the explanation of copepod diversity offered above, the inbenthic and near-benthic competitors are nested variables in a complex changing ecosystem. The signs of the biological processes deduced from field data must not be taken for the mathematical outcome of these species interacting in a larger network (e.g. predation was interacting with competition). *A.minutus* and *Am.debilis* coexist across the entire boreal North Atlantic. One need not conclude therefore that there exists (and they have reached) some stable equilibrium point in their competitive interaction. Rather, their persistence together may be the result of their interaction with other variables or the signs of feedback or complementary pathways in the network.

In this qualitative model of community structure, adaptations depend for their direction and speed on selection pressures which are the product of the mathematical outcome of interactions interacting. The signal of selection pressure may be modulated by the variables directly impinging on a functional group but selection is not solely the result of these immediate biological processes.

In Loop Analysis, the whole network is important. It provides a technique to guide and test intuition and manage large data sets. It is not, however, a tool to mindlessly generate meaning from co-occurence field data. Except, perhaps, in very simple communities (few species, very strong interactions, etc.), notions about the 'localization' of biological processes, their 'control', 'regulation' or 'cause' are not so much incorrect as naive or trivial (although description of biological processes is still essential to this paradigm). 'Maladaptations', more often than not, may be a Newtonian observer's circumlocution for, 'I don't understand'.


# 7 PERCEPTION / COGNITION

We have now seen a holistic paradigm which may aid in understanding the diversity of copepods. Now we turn to a solipsistic one.

Cognition is not perception. Knowing is not by confrontation (Clarke 1971, and see epilogue below).[9] Perception connects mind to stimulus. Cognition connects mind to action. Perception is on the cutting edge of selection pressure. Cognition is on the cutting edge of adaptation. They are fundamental processes in animal biology and they are usually overlooked in evolutionary ecological paradigms.

The distinction between perception and cognition is a logical one, it is not experienced. Aristotle (*De Anima*) recognized this, asking why we do not perceive the senses when we experience external objects. The man perceiving is not aware of the senses or any mental product of them. The mental state which corresponds to an object has 'intentional existence'. The perceiver is aware only of the object. The mental state of experienced objects 'tends toward' the extramental world. Locke (*Essay*) said only primary qualities are 'objective' in this way. Others disagree saying both primary and secondary qualities are subjective. Still others say all are objective. The problems with the concept of intentionality persist and cannot be solved here.

Cognition is related to perception as syntax is to semantics (Chomsky 1957, 1965, 1967, 1968, 1978, 1980, Putnam 1975). Cognition consists of computational rules acting on symbolic representations (Fodor 1980, Pylyshyn 1980). When psychological states are being discussed, there is no need to talk about their correspondence to things in the extramental world (Niall, personal communication). Stimulus is not experience. To call psychology the science of behaviour is like calling physics the science of meter reading. Animals respond to the same stimulus in different ways and to different stimuli in the same way. Such indeterminancy reflects the representational (opaque, non-transitive) and computational (symbolic and formal) aspects of the language of the mind (Fodor 1980). The code of this language preserves only those distinctions which are relevant to execution of the computational 'program' (Niall, personal communication); it is the representation and not the stimulus that should be taken as the criterion for the act of seeing. Thus psychology can be methodologically solipsistic (Fodor 1980). Can or should ecology be solipsistic? Anticipatory?

Evolutionary paradigms use an external observer to specify a standard content of representations (see also Gombrich's 1980, study of truth in art). In building a theory of perception which is competent, it is essential to know the distinctions that mental representations make (Pylyshyn 1972). Students of human behaviour and their human subjects may understand the world in similar ways and predict behaviour toward it correctly because the observer and the subject are similar, but what of humans observing copepods, amphipods and fish interacting? These organisms perceive real things like prey, predators and competitors, but they do not perceive their attributes. Attributes are properties of representations, not external stimuli. Organisms do not adapt to an environment, they define it. Mental representations are fundamental to this act of definition. The unit of selection is not a species but rather a representation, in particular, of physical and biological causes of selection pressure. An organism cannot adapt to an unperceived cause. Without a representation of a cause, a species is condemned to compete for causeless space (e.g. predator-free or competitor-free space). Prey extinction is inevitable in regions of overlap between the perceptual field of the predator and the living (environment) space of the prey species.

Cognition may be, by its very nature, an anticipatory system (Rosen, personal communication).[10] Cognition is designed to predict — approximately — the outcome of future events (e.g. predator's bite, satiation of hunger). Present behaviour is accordingly modified by this prediction (e.g. a prey runs away, a copepod reaches out to remove a particle from the water). The discrepancy between the phenomenal world and behaviours toward it will grow with time according to the rates of change of stimuli with which cognition is not competent to deal. This is called temporal spanning (Rosen, personal communication). The result is a constant rate of extinction at least when no major new perceptual/cognitive innovations occur. Temporal spanning and its result, extinction, may help explain why so many species are now extinct.

Anticipation which causes present behaviour to change will bring an organism into a different set of extramental stimuli than previous behaviour would have. These new stimuli now act as sources of selection pressure and the recursive function is primed again. Such anticipatory characteristics make cognition 'pre-adaptive' in a hard and real sense.[11] Cognition reaches out to touch future sets of stimuli and adapt organisms to them now.[12] Let me explain these statements by expanding on the ecology of copepod diversity outlined above.

For the purposes of illustration, predators on copepods can be divided into three kinds. The first are relatively sessile predators who appear to perceive (i.e. react to) prey by contact. Examples are the cnidarian *Haliminohydra* (Swedmark 1964, Smol & Heip 1974) and the foraminiferan, *Pilulina argentea,* (Boltovskoy & Wright 1976). This first kind of predator will concern us no further.

There are two types of predators which perceive prey at a distance: tactile and sighted predators. (Chemoreception may also be used by these, and other predators, but too little is known of this for it to be included here.)

Tactile predators are sensitive to vibrations in water resulting from the movement of prey. They are also usually (presumably) chemosensitive to the presence or touch of prey. Such predators as benthic amphipods and shrimp (e.g. Bateson 1889, Bell & Coull 1978, Harrison, University of British Columbia, personal communication, personal observation) forage by scanning the sediment with their first antenna as they locomote over it. Epibenthic copepods, the usual prey of these predators, which are sensitive to vibrations in the water created by the foraging movements of these predators, bolt into the water column at the predator's approach or touch. Epibenthic copepods also sometimes 'play dead' when approached by a source of vibration in the water (Marcotte, submitted 2). They retract their head and thoracic limbs and contract their body segments along the anterior-posterior axis. The body often becomes crescent or ball shaped. This behaviour seems 'maladapted' to escaping tactile predators. The possible reasons for this behaviour are related to the activities of sighted predators.

Sighted predators are of two kinds. Some, like fish, have an eye with lens and retina. Others, like amphipods, have a compound eye composed of ocelli. Perception among these two groups of sighted predators and between tactile and sighted predators differs. Foraging strategies differ as does the resulting selection pressure on prey.

Fish are especially, visually sensitive to movement (Bateson 1889, Ingle 1968, Blaxter 1975a,b) although object orientation, forms, textures and colours may also be important (Sutherland 1968, Trevarthen 1968). We are far from knowing how or what a fish sees, let alone how it represents visual stimuli to its cognitive processes. That caveat stated, behavioural studies provide helpful hints. Anglers need to jig their bait to attract a fish's attention. Juvenile salmon, sighted predators, are attracted to prey by their movements at the sediment-seawater interface (Marcotte & Witt submitted).

Structurally, goldfish retinas have units which are sensitive to directional movements in the visual field (Jacobson & Gaze 1964, Cronley-Dillon 1964). The pattern of sensory input to the brain may also be important to a fish's perception of movement. Schwassmann & Kruger (1968) showed that some regions of the visual field of fish are expanded when projected onto the optic tectum. The non-linear projection which results is the topological field on which prey movement (speed and direction) is abstracted from the usually, kinematically active backgrounds the prey is seen against. Like a kind of (circular) log graph paper, this tectal projection may, by its very nature, differentiate particles moving at constant velocities from those accelerating, e.g. prey.

Of the copepods ingested by bottom feeding fish, epibenthic species are most numerous (e.g. Willey 1923, Marcotte 1980). These prey swim (accelerate) to escape competitors (Marcotte, submitted 2) or tactile predators. They are then visible as differential movements at the sediment-seawater interface and fish are attracted to them.[13] These prey may perceive the approach of their predator by the bow wave or other vibrations it makes in the water. 'Playing dead' causes the prey to sink into the kinematic background of its surroundings.

Many of the fish which feed on these epibenthic copepods are juveniles (e.g. Willey 1923, Karczynski, Feller & Clayton 1973, Marcotte & Witt, submitted). Juvenile fish have small retinas (Johns & Easter 1975) and lack rods (Blaxter 1975a). Their ability to see in very dim light and to discern colour contrast may be limited. They may depend to a great extent on the perception of motion to feed. Loss of kinematic contrast of the prey because of its behaviour may make the prey truly invisible. Further, Jacobson & Gaze (1964) showed that the receptive field of retinal units in fish are very large. They integrate incoming information over a large area. Loss of contrast may thus be, *a fortiori,* important to these fish.

Early quantitative studies of predator selectivity of prey emphasized the importance of prey size (Brooks & Dodson 1965, but see also Hall et al. 1976) and perhaps shape (e.g. effects of cyclomorphosis and the spines of crab zoea). Larger prey were taken preferentially (e.g. Brooks & Dodson 1965, Kerr 1974). Schoener (1979) proposed that larger prey were selected as the distance over which the predator has to pursue or transport prey increases. The usual explanation is that foraging behaviour conserves predator energy or time. Maiorana (1981) has pointed out that this behaviour may be the result of visual limitations. 'Since all objects become subjectively smaller with distance from an animal, it follows that small objects will become subjectively invisible at closer distances than larger objects' (Maiorana 1981). The size composition of the predator's diet is determined by the distance it forages and pursues prey. At short distances the predator will encounter and capture mostly small items. Such may be the case with juvenile fish which may be unable to see great distances or with adults which are feeding in dim light. At longer distances, larger items will be preferentially perceived and then taken. At intermediate distances, small, medium and large food items will be taken in equal numbers (assuming satiation does not intervene). Thus foraging strategy, a predator's gut contents and the resulting selection pressure on the prey may have less to do with energy budgets than with perception.[14]

Lack of visual contrast in aquatic environments is a result of the light scattering properties of water (Duntley 1962, Lythgoe 1966, 1968). Water is luminous. It does not provide a black background for visual perception of objects in the water.

Crustacean eyes are composed of independent ocelli, each of which is sensitive to a narrow (but overlapping) visual field. One property of compound eyes is the possibility of developing contrast. One ocellus or a small group of them may be used as a 'blank standard' against which signals from all other ocelli are compared. This 'zeroing' or 'blanking' of visual input against some arbitrary background standard may be central to the function of compound eyes and may be a critical ability for crustacean predators. Furthermore, movement of prey may appear as a pronounced flicker on the visual field of this contrast-enhancing eye. The trajectory of prey can be followed and they can be captured.

The evolutionary responses of prey to these crustacean predators will differ from their responses to predators foraging with a lens and retina eye. Contrast reducing coloration or transparent bodies may be important. Prey feeding may be limited only to periods of dim or no light so that gut contents will not add contrast and disclose their location to contrast requiring predators.

Obviously, contrast is used by all sighted predators. I argue here only that it may be more important to invertebrate predators all the time and to fish predators at various periods in their development.

In summary, the world is not seen by copepods, amphipods or fish as the observer wills. Predator perception/cognition qualifies the selection pressure it exercises on prey. Prey behaviour produced by perceptual/cognitive processes of its own alters these selection pressures and, in turn, becomes a selection pressure on the predator. Perception/cognition/behaviour of predators and prey may be anticipatory interactions (Rosen, personal communication). The predictive 'feedforward' step of cognition may cause the ecological adaptations of predator and prey to be progressively out of phase. This temporal spanning may depend on the rate of change of mental representations of external stimuli, cognitive 'programmes' and sensory organs in the interacting organisms. This temporal spanning may exist both on the evolutionary time scale of the species involved and on the ontogenetic time scale of individuals. The orderly, seasonal change in the diet of juvenile fish (i.e. increasing prey size) may be related to both changes in the perceptual abilities of the fish and to the temporal spanning of their predatory interactions with the prey.

## 8 COPEPOD FEEDING: VISCOSITY AND PERCEPTION

Do copepods forage? How do they perceive their food? Elsewhere (Marcotte, in preparation), I will review the voluminous literature concerning copepod feeding. As a result of this review and my own work (Marcotte 1975, submitted 2), it seems that the term 'filter-feeder' has been misapplied to (planktonic) copepods.

Benthic harpacticoids locomote and feed slowly. Studying their habitat preferences and feeding behaviours, four kinds of harpacticoids can be described.

1. *Ameiria* and many of its relatives live in sand with large particle diameters. Food grows in discrete clusters on these particles, i.e. food is a coarse grained resource (*sensu* Levins 1968). These animals have relatively long bodies and disproportionately long first antennae. They forage by constantly scanning the surface of the substrate with their first antennae. When a cluster of diatoms or bacteria is located, the animal moves its morphologically concentrated oral appendages over the cluster and the food is scraped from the surface. The animal then moves on.

2. Some copepods, like the laophontids, live and feed on the rectilinear edges of blades of grass (e.g. rotting *Zostera* leaves) or on cylinders of blue green algae. Their bodies are flexible with deeply indented, worm-like segmentation. The first leg is elongated into a grappling hook (Harpacticoida comes from the Greek word for grappling hook) to hold onto the linear substrate. They flex their bodies around this substrate. They feed on diatoms and bacteria living on the substrate.

3. Some copepods, like the ectinosomatids, live in sands with small particle diameters or, like *Dactylopodia* and related genera, live on the surface of algae or grass leaves. In these environments, food (diatoms, bacteria and flagellates) grows as a fine grained resource, i.e. they are spread out on the planar surfaces of the substrate, often growing in pits and crevaces, relatively homogeneously. These copepods sweep food into their oral regions using the 'elbow and forearm' of their second antennae.

4. Finally there are harpacticoids for which their world is three dimensional. Some, like some cletodids, sort food (grape-like clusters of bacteria and diatoms) from among floccules of clay. Others, like *Tisbe* and many diosaccids, glean food from the surface of spheres of detritus and clay floccules (*Tisbe* may also eat nematodes and other prey items).

There may be rigidly herbivorous or carnivorous species of harpacticoid, but I never saw them. All these copepods seem to perceive is the dimensionality of their environment. For example, if a linear substrate is not available to a 'line-dweller', they will frantically flail their appendages seeking such a substrate until they find one or die. When placed in chambers filled with only fine sand or clay floccules, or on a clean glass plate, they behave in the same manner. 'Nothing' had the same perceptual quality as 'something' too fine to be a linear substrate. The animal exhausts itself until a linear substrate is provided, whereupon the behaviour of the animal is transformed. It will grab onto the cylinder or edge, quiet its motion and eventually begin to feed.

Harpacticoids live in a bottom environment rich in suspended solids, dissolved compounds and electrically active clay crystals. Turbulence is low. Viscosity dominates. Water moves slowly in laminar flows. Chemicals excreted by microflora or the differential loss of some nutrients near colonies of microflora or chemicals left by benthic organisms as signals are probably slow to diffuse. Harpacticoids generally forage slowly and seem to use a strong chemosensory ability to locate food (Marcotte 1975, submitted 2).

Calanoids live in an apparently more turbulent environment. They feed by rapidly moving water over their first antennae and mouth parts using their second antennae (Friedman & Strickler 1975, Friedman 1980, Alcaraz et al. 1980). Because water is moved by these copepods at high speeds over short distances, a low Reynold's Number environment is generated, i.e. the water appears viscous (e.g. Zaret 1980, Alcaraz et al. 1980). Laminar flow of water over the first antennae and mouth parts occurs. An incoming particle is perceived by the copepod perhaps through chemoreception of leaked chemicals (Alcaraz et al. 1980). Because of the spatial characteristics of laminar flows, the copepod may be able to predict the position of incoming particles and grab for it using its second maxillae or other appendages. Real filtration is not only impossible at the low Reynolds Numbers (high viscosity) involved, it is not done (Alcaraz et al. 1980). Calanoids are, in the end 'raptorial' feeders, grasping at food not filtering it.

Cyclopoids are also fast moving copepods. That they are raptorial feeders has been known for a long time (Fryer 1957a,b). Their mode of perceiving food is not described.

Thus it seems that the trophic distinctions of raptorial and filter-feeding are misapplied to copepods. Harpacticoids live in a low Reynolds Number environment, calanoids and probably cyclopoids generate one. Calanoids may use their very long first antennae (as do ameirids) to forage through a dilute food environment. The length of the first antennae may also be related to its use in discriminating among vibrations of different wave lengths in the water and among directions of incoming vibrations (Strickler & Bal 1973, Friedman & Strickler 1975, Strickler 1975,b, 1977, Kerfoot et al. 1980, Friedman 1980). It seems, then, that perception of chemicals and tactile stimuli is the basis of foraging strategies among all free-living copepods. All copepods actively sense their environment. With common-held raptorial feeding mechanisms, they can accurately choose the food they wish to ingest. None are 'passive' feeders. Perception and cognition are involved (although both are too little studied for detailed description).

# 9 COPEPOD DEVELOPMENT: PERCEPTION AND ADAPTATION

Copepod development is usually anamorphic but some metamorphic events occur in parasitic groups. The first juvenile stage is the nauplius. It is usually free-living in the plankton or on the benthos. In some copepods such as the Misophrioida, the nauplius does not feed, i.e. is lecithotrophic (Gurney 1933), or the nauplius may even be passed in the egg.

The first nauplius is covered by a single body shield. The body terminates posteriorly in two, setae bearing protuberances homologous with the caudal rami (e.g. Gibbons & Ogilvie 1933). The body region carrying these protuberances is the telson. It is homologous with the pygidium of other protostome invertebrates. Body somites proliferate from the anterior margin of the telson during subsequent development.

The copepod body is embryologically composed of a cephalon or head (= acron + 4 somites), thorax (= 7 somites) and abdomen (= 4 somites + telson). The cephalon is joined by the first thoracic somite bearing a maxilliped to form a single anterior segment in the adult called the cephalothorax. Of the remaining six thoracic somites, the first and sometimes the second may later become fused to the cephalothorax. Whether or not fused, the second through the sixth thoracic somites inclusive bear locomotory legs (at least ancestrally). The seventh thoracic somite often bears a sixth pair of legs. If present these legs are reduced and resemble thoracic limb buds seen during embryological development. This last thoracic somite also bears the genital apertures of both males and females. In females, these apertures may be present as separate gonoductal and seminal receptacle pores, or as a single opening of a common atrium. In adult females and, in a very few instances, males, the last thoracic somite is fused to the first abdominal somite. Together these somites are termed the genital complex or genital segment of the adult. When fusion takes place, the genital aperture(s) remain on the portion of the genital complex corresponding to the last thoracic somite. The three remaining abdominal somites bear no appendages. The telson is the terminal segment of the adult and bears a pair of movable appendages called the caudal rami. These caudal rami are often used by free-living adults to help direct swimming or walking. The anus exists dorsally on the telson and may be covered by a variously ornamented pseudo-operculum.

To avoid confusion between the nomenclature of adult body segmentation and that of embryological somites, the adult copepod is said to be divided into a wide anterior region, the prosome, and a narrow posterior region, the urosome. The free thoracic segments of the prosome are sometimes called the metasome.

The first nauplius has a mediodorsal naupliar eye composed of three ocelli (Claus 1891, see also Esterly 1908 and references therein). These ocelli may be retained or lost in the adult. If retained, these ocelli may be untransformed, or they can be divided to form three eyes as in the calanoid *Epilabidocera* (Park 1966).

The first nauplius has three appendages: a pre-oral first antenna, a post-oral second antenna, and mandible. The first antenna is uniramous. The second antenna is biramous, and may have a large basal endite for mastication or other manipulation of food. This endite, when present, lies under a large labrum. The mandible is biramous and may have no gnatho-basis.

Five naupliar stages follow the first naupliar stage. Each of these stages can be called a nauplius or, when limb buds posterior to the mandible are externally expressed, the nau-plius can be termed a metanauplius. With the first metanaupliar molt, cephalic appendages appear as rudiments and the shield is dorsally marked off. During subsequent metanaupliar molts, the second antenna migrates to a pre-oral position with the first antenna. The man-dible is joined by a post-oral first maxilla and second maxilla.

After the six (meta) naupliar stages, there are six copepodid stages. The sixth copepodid stage is the adult. In the first copepodite, the body becomes straightened and the prosome-urosome articulation becomes clear. This articulation in the first copepodid stage is behind the third free somite, i.e. between the somites which will bear the second and third walking limbs on the adult prosome (Calman 1909). It moves back one somite with each subsequent molt. The final position of this articulation is set in the second copepodite for podoplean orders (i.e. copepods having legs on the urosome) and in the third copepodid for gymno-pleans (i.e. copepods without legs on the urosome = calanoids). During copepodid molts, all remaining limbs develop and the oral appendages including the maxilliped begin to show their adult forms. The thoracic and abdominal somites end proliferation during these molts.

Why have most free-living copepods retained this complicated, 12 stage life cycle? Not all have. The misophrioids have only one lecithotrophic nauplius which molts into the first copepodid stage (Gurney 1933). In general the sizes of all naupliar stages are very similar (Gurney 1928). If one measures growth by a ratio of the length of one stage divided by the length in the preceding stage, nauplii usually grow by a factor of 1.2 or less (Gurney 1928). Little is known of the food of these nauplii. The molt from sixth naupliar stage to first copepodid is marked by a growth of 1.4 times. Copepodid stages are generally separated by morphometric ratios greater than 1.2 especially for females, especially at early copepodid stages (Gurney 1928).

These morphometric ratios may reflect a division of trophic resources based on food (-bearing) particle size (see above). By spreading their development over a wide spectrum of body sizes, growing copepods may be able to exploit energy sources in a wide variety of particle sizes, perhaps without much inter-molt competition. By dividing this spectrum of body sizes into discrete regions occupied by individual age classes, copepods might gain the flexibility of arresting development in one of these age classes in response to competi-tors or predators which effect later age classes (see above). The fact that some free-living copepods truncate their naupliar development and/or have lecithotrophic nauplii suggests that past or present competition for food is important in understanding the biology of copepod development. Similarly, the fact that nauplii can be morphologically more diffe-rent than the adults of congeneric species which gave rise to them (Lopez 1981) suggests that important selection pressures can be acting on juveniles and not adults, or that they are qualitatively different for adults, or that adults can respond to them differently.

One much overlooked aspect of the stages in a copepod's life cycle (i.e. eggs, nauplius, copepodite, adult, hybernation, and encystment) is the ontogeny of perception and cogni-tion. Luria's (1973) introduction to neuropsychology may provide useful insights into the character of such ontogeny. Though formulated for vertebrates, his analysis may have applica-tion towards directing copepod research.

Mental processes are made possible by groups of concertedly working functional groups or areas of the nervous system and body generally. Perceptual/cognitive functions may not be localized in any one part of the network of interacting groups. For example, the perception of cylinders or rectilinear edges by 'line-dwelling' harpacticoids may not be centred or 'localized' in one seta, article, appendage or neuron. The existence of a 'line' and the apparent relaxation in movement its awareness brings, may be the product of the entire copepod's body perceiving it.

Extramental, extracorporal objects may be components in linking of one perception/cognition/behaviour complex to another. Complex sets of organism-environment feedback relationships are possible. For example, the chemical perception of food by line-dwelling harpacticoids may be cognitively linked to movements of mouth parts via the perception of linear substrates. The food may go unperceived or, if perceived, go unresponded to without a linear substrate.

During the ontogeny of a copepod, mental functions such as perception and cognition may move about the network of internal functional groups and external complements making new connections as new appendages are acquired and old connections with sense organs or neurological pathways are altered or abandoned. In the early stages of development, the central nervous system of the nauplius may have responsibility for most mental processes. As appendages and sensory organs are added, the central nervous system may divest itself of these initial responsibilities. The original perceptual/cognitive tasks may become a property of peripheral components of the network. External objects may be relied upon to initiate or direct complex actions relieving the central nervous system of memory functions. Feeding appendages from the first antenna to the maxilliped may begin to function independently of the central nervous system. New tasks might then become possible for the system. It may initiate or modulate behaviours such as migration or mating. An important task of the nervous system is the production or modulation of hormones governing molting. The rate at which the central nervous system divests itself of functions which compete with hormone production for the system's attention may pace the rate of molting. Insofar as divestment occurs through external objects such as resources or other age classes of the same species, there may be a strong environmental effect on these processes. The initiation of molting or encystment may depend to an important extent on external stimuli and how they are perceived and cognitively processed and how fast temporal spanning occurs. Feedback may exist from older age classes to earlier ones suppressing molting in the earlier age classes. Feedforward may exist from younger age classes to later ones concerving the abundance of intermediate age classes. Feedback and feedforward may need to be mediated only by perception of the abundance of the signal age class by the receptor age class. Chemical pheromones need not be produced. The extent to which memory is preserved across molts may be important in this matter, but is unstudied.

In short, naupliar and copepodid stages may be conserved both evolutionarily and ontogenetically by the rate at which mental and extramental events can be connected and effect one another. There is no reason to reject out of hand the possibility of perceptual/cognitive effects on copepod development especially when the hormonal control of molting seems so intimately related to the nervous system.


10 SUMMARY

This essay began by asking, 'Why are there so many kinds of animals?' and, 'Why aren't there more?' It applied this question to patterns of free-living copepod diversity. There are many more species of bottom-dwelling copepods than planktonic ones. There are many more species of inbenthic harpacticoids than near-benthic (epibenthic, demersal planktonic) harpacticoids. Epibenthic *Tisbe* competes for space at the sediment-seawater interface by interference (behavioural) mechanisms. It results in vertical habitat displacement of adults. Inbenthic harpacticoids seem to compete for food-bearing particles. They differentiate this resource by its size. Seasonal body (mouth) size convergence causes overlap in their use of this resource. Competition results in demographic adjustments — a selective advantage to staying young. If sustained, this selective advantage could cause heterochronic events, should they occur, to be favourably selected. This could explain repeatedly observed but unexplained instances of paedomorphosis among inbenthic species and the resulting high alpha-taxonomic diversity. The younger age classes of species with epibenthic adults (e.g. *Tisbe*) probably compete with other species in their inbenthic habitat. The epibenthic adults may be differentially preyed upon by tactile and sighted predators. The differences in predator and prey perception and cognition may have profound consequences for this interaction.

Not only energy is exchanged when passing from one trophic level to another. Trophic levels are 'doors of perception' (Huxley 1954). Competitors at the same trophic level, and predators and prey between them, are interacting perceptually and cognitively as well as economically. For example, fish may perceive and ingest second derivatives of motion not copepods or amphipods *per se.* Animals cannot adapt to an unperceived cause. An unperceived predator extinguishes its prey unless a predator-free refugium is available for it. (Such may be the case with the extinction of ammonite prey after the introduction of mosasaur predators in the Triassic.)

Cognition is distinct from perception. Cognition acts computationally on representations of external (and internal) stimuli. It may be anticipatory in function — adapting present behaviour to predicted future events. These predictions may initially be correct but their fidelity may be temporally spanned — it may decay with time. The pace of this decay may determine: ontogenetically, the pace of development, ecologically, the duration and stability of competitive or predator-prey interactions and phylogenetically, the rate of extinction.

## 11 EPILOGUE

Science proceeds methodically from empirical experience to intelligent inquiry, and from understanding to rational reflection and judgement. Intelligent understanding depends on human insight, a creative process (Lonergan 1957, 1972, 1973, Clarke 1971). Its issue is a hypothesis, concept or word. Science, 'true understanding', demands that this intelligent insight be rationally tested. Hypotheses logically precede their testing. Premature imposition of the rigorous methods appropriate to testing hypotheses on the process of making them can only stultify understanding and retard the progress of science (such as it is).

Astrology was a science in every way but in the production of normal, constructive paradigms with which to solve everyday scientific puzzles. Like a science, astrology was never really disproved. Its advocates and practitioners simply got distracted or died. Evolutionary biology is a science. It does everything right. But Evolution, like other sciences, will probably suffer the same fate as astrology. I suspect that evolution theory like Newtonian celestial physics will not be falsified (at least not convincingly so). It will simply be subsumed by some more general theory. Newtonian physics is now considered simply a special case (low speeds, short distances, etc.) of Einsteinian relativity. Darwinian evolution may become a special case (few species, strong interactions, etc.) of some as yet to be articulated theory.

A systematic re-examination of fundamental concepts in biology will presage such a revolution. It has already begun (e.g. Hutchinson 1968, Goodwin 1978, Patten et al. 1976, Webster & Goodwin 1981). Part of the purpose of the present paper was to suggest that a synthesis of biological and psychological theory may be possible and that it may precipitate or at least contribute to a revolution in biological understanding.

One must be faithful to the responsibilities engendered by the freedom of mind science provides. Now that these insights have been articulated, tests of them must begin.

## 12 FOOTNOTES

1. One must not, however, confuse primitive terms in the axioms of a theory with terms that require definition. For example, 'fitness' is a primitive term in evolution theory (Williams 1970, 1975). Time is a primitive term in physics (Cannon & Jensen 1975). Peters (1976, 1978) has shown that defining fitness leads to a tautology. It is a property of primitive terms that attempts to define them end in tautologies, an idle exercise for such terms. If primitive terms and the axioms which relate them are no longer satisfactory, they should be replaced, not infinitely regressed.

2. Perhaps this is a result of the way research is funded or scientific articles are written and published. 'Reductionism' may be less a philosophic ideal than a financial necessity. For another view, see Levins & Lewontin (1980).

3. *Am. debilis* is an arctic/boreal species. Its body size decreases in summer and increases in winter. *A. minutus* is a mediterranean/boreal species. In Nova Scotia it is probably at its cold water limit. The modal size of adult females and males remains the same all year. Body sizes of *Am. debilis* converge with those of *A. minutus* in winter.

4. Edwards (1891) showed that the marine copepod *Esola longicaudata* encysts in summer. Birge & Juday (1908) could not demonstrate that the summer encystment of *Cyclops bicuspidatus,* a freshwater copepod, was caused either by high temperatures or desiccation. Donner (1928) did show that high temperature not dryness caused encystment in a harpacticoid. Roy (1932) showed that tegumentary glands in several copepod species secrete a cyst around the copepods when desiccation occurs. Coull & Grant (1981) demonstrated that one species of marine harpacticoid encysts (as Edwards had already shown) and speculated on the existence of integumentary, cyst-producing glands. Coker (1933) demonstrated that a species of freshwater *Cyclops* arrested its development in the fourth copepodid stage in summer. Further development occurred when water temperature was lowered. This arrest in development may presage encystment since, as Coker (1933) pointed out, encystment often occurs in the fourth copepodite stage. However, his specimens never formed a cyst.

5. One result of prolongation of juvenile development may be a shift in the sex ratio of adults in favour of females. Coker (1933) showed that females only were produced from copepodites whose development was prolonged beyond 60 days. Takeda (1950) showed that prolongation of the sixth nauliar stage led to more females in the population. Sex seemed in those experiments to be determined in the sixth naupliar stage.

6. Pheromones have been invoked to explain the mating behaviour of planktonic copepods (Parker 1901, Katona 1973) and to explain the homing behaviour of intertidal harpacticoids (Bozic 1975). The existence of development-controlling pheromones may explain the development switches seen in the deep sea crustaceans Neotanaidae (Gardner 1975). Suppression of naupliar development in response to the presence of predators has also been suggested (Smoll & Heip 1974).

7. Competitive exclusion or appropriate specializations allow these juveniles to compete successfully. Lopez (1981) has shown that nauplii of *Tisbe* species are more morphologically different than the adults into which they develop. Such morphological differentiation of inbenthic juveniles is predicted by the model of copepod evolution outlined here.

8. Law (1979) has proposed a model in which predation on age classes older than some specific

age class, *i*, leads to higher rates of reproduction at age *i*. This model, though not directly applicable to copepods which molt from age class to age class and reproduce only in the last age class, does illustrate how turnover rates in age classes below that being preyed upon could be increased and could offer an explanation of *Tisbe*'s high reproductive potential.

9. Theories of knowledge are products of philosophical assumptions about the nature of the knower. It is important to understand the implications of accepting assertions about the nature of knowledge. Whole world views are logically at stake. Choices must be made, the more mindfully the better. It is so important to introduce the reader to the major issues joined in these theories that I risk appearing sophomorish by presenting the following very brief summary of these world views. I make no pretense of completeness in this endeavour, only coherence. Names of adherents are given but not references because the statements made are compiled from very many sources. The interested reader can sort these statements by examining the works of the listed adherents.

*The Milesian Greeks* (Cleve 1969) established that the critical question was how to account for the diversity of the world in the simplest possible way. Thales (624-546 BC) saw the stuff of the world as water and nothing more. The processes of rarefaction and compaction generate(d) the many from this one. Anaximander (610-546 BC) introduced 'Apeiron' – the unlimited, boundless, infinite or indefinite – as the one fundamental principle. From this one the many came by separation of opposites. Anaximines (585-528 BC) saw air more than water as the 'boundless' one. Condensation and rare-faction split the many from this one in an everlasting process of becoming.

Heraclitus of Ephesus (c.500 BC) established one of the ancient world's most important theories of elemental diversity and process. He saw the fundamental principle as the 'Logos' whose body was *Pyraeizoon* (aether), which *was* constant flux. From its pure state of fire-like change, aether dies to air which dies to water, then earth, then prester (a fiery whirlwind), then back to aether. This 'Logos' was a child playing. Bored with unity it generated a world of conflict. Then bored with conflict all was changed back to aether. There was no being, or, more properly, the categories 'being' and 'becoming' are inadequate. Both were comprised under the notion of occurrence, of temporal succession. Everything was in time. Change (becoming) is the occurrence of a succession of unlike. Permanence (being) was the occurrence of a succession of like.

Parmenides of Elea (c.510) established notions which conflicted with those of Heraclitus. For Parmenides, there was a world of seeming (change) and a world of being. The real world of being was one, devoid of any plurality. This being was undifferentiated, a one, timeless, changeless.

The Pluralists resolved the conflict between Heraclitus and Parmenides. Empedocles (c.450 BC) saw that change was a mixture or separation of fundamental, unchanging, timeless beings or elements: Fire, air, water and earth. For Anaxagoras (c.500 BC) everything consisted of an infinite number of particles or 'seeds' and in all things is a portion of everything. One object was what it was by virtue of having more seeds of that thing than any other. The mover of these seeds was the Nous (mind). The first atomist, Democratus, emerged from the Greek school of pluralists. With the atomists, we reach the first of three schools of 'materialist', philosophers who account for reality by matter and nothing else.

*Mechanistic materialism* (e.g. Democratus, Epicurus, Lucretius, Diderot, Maupertuis, Fontenelle, de Sade and with important reservations, Galileo, Descartes, Newton). Reality consists of atoms and a void. These atoms are unchangeable and eternal. Each atom is a 'one'. There is an indefinite number of these falling, bumping, and randomly associating in an unbounded, intangible, empty, infinite space. Nothing exists that is distinct from body or vacuity. Nature resolves everything into its component atoms and never reduces anything to nothing. Material objects are of two kinds: atoms and aggregates of atoms. There are no fundamental differences between living and non-living objects – both are aggregates of atoms. These atoms have only 'primary qualities': (Lucretius) size, shape, weight, (Galileo) motion, rest and position. These atoms lack 'secondary qualities': color, odor, sound, taste, heat, cold, and other objects of touch. These secondary qualities are the products of the atom's primary qualities.

*Dialectical materialism* (e.g. Marx, Engles, Lenin and with important reservations Feuerbach). Hegel taught that the material world (concrete reality) proceeded from an 'organic' totality of ideas. Feuerbach, a theologian and Hegel's student, inverted his teacher's notions and said that the material world is primary and ideas are derivative. Material being is the subject and thought the predicate. Three laws govern material reality: (1) Law of transformation of quantitative changes into qualitative leaps. A natural thing will slowly increase or decrease quantitatively (primary qualities) up to a point. Then a qualitative leap will occur through which a new species of thing will arise (e.g. quantitative lengthening of a violin string brings about qualitative changes). (2) Contradiction is the source of all change. Change is a contradiction. (3) Law of the negation of the negation. All changes negate the pre-existing being producing a new being. (These three laws of the dialectic are essential to the ecology of Levins, e.g. 1968). These laws are applicable to all nature (science), history and thought (dialectical logic).

*Naturalistic materialism* (e.g. Dewey, Santayana, Cohen, Randel, Nagel). Naturalism rejects the three laws of dialectical materialism and mechanistic materialism. Naturalists recognize a difference between living and non-living objects. They see nature giving rise to emergent qualities like mind. Naturalism denies: transcendent god, spiritual soul, and reductionisms. Naturalism rejects dualisms of (a) body and soul; (b) nature and supernaturalism; (c) man and society; (d) man and nature. Reality is a continuum which is not stable, not composed of universals. Naturalism accepts the scientific method as the only mode of knowing. A felt difficulty is located and defined. Possible solutions are suggested (ideas). Reasoning is applied to the suggestion (judgement). Further observations and experiment is performed. Thought is an instrument used by man to solve difficulties. Something is true if it helps cope with a problem.

What is the mind to a materialist? Here is a selection of their answers. Mind is a particular kind of matter (Lucretius) or a kind of internal motion of atoms (Hobbs) or a property of matter (Diderot, Lenin) (then everything has mind!) or mind is a name for a certain appearance of bodily events (modern physiology) or is a property of certain organizations of matter (Dewey and many Naturalists) or mind is a by-product or epiphenomenon of the body (Watson). In what, then, consists knowledge or awareness? Atoms can only hit each other (i.e. enter into mechanical interactions). Photons bouncing from an object with a certain wavelength strike atoms in the eye. These, or atoms they set in motion, in turn move striking atoms in the brain. This is knowledge by confrontation – atom to atom. There are no universals (e.g. 'man'). These are just names. But is a table 'aware' when I strike it? What is qualitatively different about striking atoms in a table and atoms in the brain? To solve this question (and much more) Socrates, Plato and Aristotle adopted the notion of immateriality.

*Platonism* (e.g. Plato, St Augustine, Boethius, Alan of Lille, Hugh of St Victor, Auvegne, Erigina, Duns Scotus and, with important reservations, Socrates and Descartes). Plato asked (Symposium, Alcibiades I.) 'What am I', and argued that in speaking we use words. But the user and the thing used are always different, thus the user (a person) cannot be his body (the thing used), but rather must be psyche (mind, soul). Since the user and the thing used are different and since one uses body, one cannot be either body, or body and psyche combined. Therefore, man must be psyche. For Plato, there exists a host of immaterial, really-existing, actually intelligible universals. These are pure forms in the being of which particular objects participate. These particulars are aggregates of form and matter. The really 'true' forms are in perfect act. Matter is potentially anything depending on the form with which it is associated. Because the phenomenal world is composed of forms immersed in unperfected matter, it can only be a poor copy of the 'real' world of forms. The space of this phenomenal world is pure number, a rippling pool in which perfect forms are imperfectly reflected. The 'true' forms are actually, perfectly, eternally knowable. Human knowledge consists of intellectual participation in these pure intelligibles. The senses perceive particular objects and the intellect *passively* abstracts the 'potential' residue of matter from these particulars leaving behind the immaterial form. For the Platonist, the mere presence of this immaterial substance (intellectual species) to an intellect brings about understanding, awareness in the intellect. This theory of knowing has been called knowing by duality or confrontation.

*Aristotelianism* (e.g. St Thomas Aquinas, Lonergan, Clarke and, with some reservations, Aristotle). Aristotle systematized all philosophy. He used four 'causes' to isolate particular processes and problems from the continuum of nature: formal cause, that which makes a thing to be what it is; material cause, that which is pure potential and which is actualized by form; efficient cause, that which brings form and matter together; final cause, that for which a thing is brought into being. Aristotle (*Physics*) argues that 'being' has several meanings. There are those beings which exist by themselves and are called substances, and have properties (size, color, etc.) which inhere in them. These properties are accidental and may be changed (removed, replaced, etc.) without changing the essential character of their substrate. There are two types of being: substance and accident. Accidental beings are aggregates, i.e. collections of pre-existing substances, and nothing more, e.g. a watch: a gear of a watch is the same (exhibits the same activities) before, during and after incorporation in the watch. In substantial change, new species of activity may come into or pass from being. Thus, when a man dies, a corpse comes into being. The corpse is substantially (essentially) different from the living man in that awareness and reason are lost. Essence is formed by the intersection of a form and matter, with properties (powers, activities) that are qualitatively different from form and matter alone. Thus man is a substantial whole composed of the essence 'psyche-body'. Man is not simply body as the materialists taught. Nor is man an aggregate of body and psyche (man is psyche) as Plato taught. The differences between Aristotle's notion of man as the essence psyche-body and Plato's notion that man is his psyche is subtle but very very important both for philosophy and history. The differences became pellucidly clear in late medieval philosophy with the controversies over the nature of human knowing.

For the Platonists, formal causality alone defined a man. Knowledge was passive confrontation of form with form. For the Aristotelians, man was a psyche-body. He could not, by his own nature (power) know forms as forms. He could only 'see through a glass darkly'. For the Aristotelians, man desired to know unconditionally by the dynamic of his mind's questioning, but he cannot. Man is born with a mind that is a clean slate, and experiences the phenomenal world and questions, What is it?' He has an insight and thus comes to some understanding and articulates this with a word or hypothesis. But wanting to know truly, man next asks, 'Is it what I understand it to be?' This need to test leads to a second insight. A judgement is passed. In knowing, the intellect, the object (experience), and the act of understanding are one. This is knowing by identity. Objects are only potentially intelligible. Intelligibility is in the mind of the one understanding the object. Knowledge is not passive or objective for the Aristotelians. For the Platonists, if a pupil disagrees with a teacher (who by definition possesses objective 'truth') either the pupil is ignorant and cannot learn, or is immoral and will not learn. For Aristotelians, man does not know 'The truth', he understands truly. If there is a disagreement, there is room for the parties to reason together, have new insights, arrive at a new understanding.

10. It is not necessary that cognition be more than representational and computational (i.e. distinct from perception) to make it an important element in evolutionary ecological theory. That it may also be anticipatory makes it even more important, though my adherence to this proposition is equivocal. When the issues are important and data few, opinions are legion and hotly contested. The more remote the theorist from the event, the more baroque and immovable his theory.

11. Time may not be an object of perception. Time may be embedded in the structure of cognition and projected on the phenomenal world (space may have a similar origin). The recursive function of anticipation in cognition may be the source of time (and space) and its primitive status in human understanding. If psychology is instructive, anticipation is essential for awareness of time (Dupont 1974, Hartocollis 1974, Faber 1981). The notions of cause and effect (i.e. before and after) may be categories of cognition, not perception.

12. Lamark's theory was perhaps not so much wrong as uncentred on mental processes.

13. The juvenile Coho and Chinook Salmon studied by Marcotte and Whitt (submitted) actually fed by removing copepods of the appropriate size and shape from mouthfulls of sediment taken at the sediment-seawater interface. Their repeated attack of certain areas of the bottom may have reflected the patch distribution of swimming epibenthic copepods. However, the copepods actually ingested were adults of shallow inbenthic species. The fish never ingested the inbenthic juveniles of species whose adults are epibenthic. The result of this predation could be release of inbenthic juveniles of the epibenthic adults from competition with the inbenthic adults actually eaten. The question, 'Can prey species select predators to optimize their own (groups) fitness?' is germane here but is unanswerable at this time.

14. The question, 'Why do copepods make the effort of binding their feces in a membrane?' may be solved with studies of predator perception. Such fecal pellets are large, about the size of a nauplius larva. Fish lunge at them as if they were prey (Harrison 1976). The fish takes time to fix its eyes on the pellet and to ingest it. The fish usually spits it out in disgust (literally, because of its taste) but time has been spent not persuing the prey which produced the pellet. The disgust of the fish implies that there are chemical (taste) properties to the membrane which can be recognized. Presumably, even chemosensing invertebrate predators could be fooled by the pellets. Thus fecal pellets may serve the same purpose for copepods as aluminium foil strips did for airplane pilots in World War II – producing interference on the sense organs (radar) of the predators (enemy).

## REFERENCES

Alcaraz, M., G.-A.Paffenhöfer & J.R.Strickler 1980. Catching the algae: a first account of visual observations on filter-feeding calanoids. In: W.C.Kerfoot (ed.), *Evolution and Ecology of Zooplankton Communities:* 241-248. Univ. Press of New England.

Bateson, W. 1889. Notes on the senses and habits of some Crustacea. *J. Mar. Biol. Assoc. UK* 1:211-224.

Bell, S.S. & B.C.Coull 1978. Field evidence that shrimp predation regulates meiofauna. *Oecologia* 35: 141-148.

Birge, E.A. & C.Juday 1908. A summer resting stage in the development of *Cyclops bicuspidatus* Claus. *Trans. Wisc. Acad. Sci. Arts Lett.* 16:1-9.

Blaxter, J.H.S. 1975a. The eyes of larval fish. In: M.A.Ali (ed.), *Vision in Fish: New Approaches in Research:*427-443. New York: Plenum Press.

Blaxter, J.H.S. 1975b. Fish vision and applied research. In: M.A.Ali (ed.), *Vision in Fish: New Approaches in Research:*757-773. New York: Plenum Press.

Boltovskoy, E. & R.Wright 1976. *Recent Foraminifera*. The Hague: Junk.

Borutzky, E.-W. 1929. Zur Frage über den Ruhestand der Copepoda-Harpacticoida. Dauereier bei *Canthocamptus arcticus* Lilljeborgi. *Zool. Anz.* 83:225-233.

Bozic, B. 1975. Detection actometrique d'un facteur d'interattraction chez *Tigriopus* (Crustaces, Copepodes, Harpacticoides). *Bull. Soc. zool. Fr.* 100:305-311.

Bridgeman, P.W. 1936. *The Nature of Physical Theory*. Princeton: Priceton Univ. Press.

Brooks, J.L. & S.I.Dodson 1965. Predation, body size and composition of plankton. *Science* 150:28-35.

Brown, W.L. & E.O.Wilson 1956. Character displacements. *Syst. Zool.* 5:49-64.

Cannon, W.H. & O.G.Jensen 1975. Terrestrial timekeeping and general relativity – a discovery. *Science* 188:317-328.

Chomsky, N. 1957. *Syntactic Structures*. The Hague: Mouton.

Chomsky, N. 1965. *Aspects of the Theory of Syntax*. Cambridge: M.I.T. Press.

Chomsky, N. 1967. The formal nature of language. In: E.H.Lenneberg (ed.), *Biological Foundations of Language*. New York: Wiley.

Chomsky, N. 1968. *Language and Mind*. New York: Harcourt, Brace & World.

Chomsky, N. 1978. A theory of core grammar. *Glot* 1:7-26.

Chomsky, N. 1980. Rules and representations. *Behav. Brain Sci.* 3:1-61.

Clarke. T.J. 1971. The background and implications of Duns Scotus' theory of knowing in the beatific vision. PhD Thesis, Brandeis University.

Claus, C. 1891. Das Medianauge der Crustaceen. *Arbeit. Zool. Inst. Wien* 9:225-266.

Cleve, F.M. 1969. *The Giants of Pre-sophistic Greek Philosophy. An attempt to reconstruct their thought*. The Hague: Martinus Nijhoff.

Coker, R.E. 1933. Arrêt du developpement chez les copepodes. *Bull. Biol.* 67:276-287.

Connell, J.H. 1961. The influence of interspecific competition and other factors on the distribution of the barnacle *Chthamalus stellatus*. *Ecology* 42:710-723.

Coull, B.C. 1973. Estuarine meiofauna: a review: trophic relationships and microbial interactions. In: L.H.Stevenson & R.R.Colwell (eds.), *Estuarine Microbial Ecology:*499-512. Columbia: Univ. South Carolina Press.

Coull, B.C. & B.W.Dudley 1976. Delayed naupliar development of meiobenthic copepods. *Biol. Bull.* 150:38-46.

Coull, B.C. & J.Grant 1981. Encystment discovered in a marine copepod. *Science* 212:342-344.

Cronly-Dillon, J.R. 1964. Units sensitive to direction of movement in goldfish optic tectum. *Nature* 203:214-215.

Darwin, C. 1964 reprint. *On the Origin of Species*. Cambridge: Harvard Univ. Press.

Donner, F. 1928. Die Harpaktiziden der Leipziger Umgebung und der Schneebeger Erzbergwerke. *Intern. Rev. Hydrobiol.* 20:221-353.

Duntley, S.Q. 1962. Underwater visability. In: M.N.Hill (ed.), *The Sea* 1:452-455. New York: Interscience.

Dupont, M.A. 1974. A provisional contribution to the psychoanalytical study of time. *Intern. J. Psychoanalysis* 55:483.

Easter, S.S. 1975. Retinal specializations for aquatic vision: theory and facts. In: M.A.Ali (ed.), *Vision in Fish: New Approaches in Research:*609-617. New York: Plenum Press.

Edwards, C.L. 1891. Beschreibung einiger neuen Copepoden und eines neuen copepodenahnlichen Krebes *Leuckartella paradoxa*. Berlin: Nicolaische Verlags Buchhandlung.

Esterly, C.O. 1908. The light recipient organs of the copepod *Eucalanus elongatus*. *Bull. Mus. Comp. Zool. Harvard* 53(1):1-53.

Faber, M.D. 1981. *Culture and Consciousness*. New York: The Human Sciences Press.

Fenchel, T.M. 1975. Character displacement and coexistence in mud snails (Hydrobiidae). *Oecologia* 20:19-32.

Fleminger, A. 1975. Geographical distribution and morphological divergence in American coastal-zone planktonic copepods of the genus *Labidocera*. *Estuarine Res.* 1:392-419.

Fleminger, A. & K.Hulsemann 1974. Systematics and distribution of the four sibling species comprising the genus *Pontellina* Dana (Copepoda, Calanoida). *Fish. Bull.* 74(1):63-120.

Fodor, J.A. 1980. Methodological solipsism considered as a research strategy in cognitive psychology. *Behav. Brain Sci.* 3:63-109.

Friedman, M.M. 1980. Comparative morphology and functional significance of copepod receptors and oral structures. In: W.C.Kerfoot (ed.), *Evolution and Ecology of Zooplankton Communities:*185-197. New York: Univ. Press of New England.

Fryer, G. 1957a. The feeding mechanism of some freshwater cyclopoid copepods. *Proc. Zool. Soc. Lond.* 129:1-27.

Fryer, G. 1975b. The food of some freshwater cyclopoid copepods and its ecological significance. *J. Animal Ecol.* 26:261-286.

Gardiner, L.F. 1975. The systematics, post marsupial development and ecology of the deep-sea family Neotanaidae (Crustacea: Tanaidacea). *Smith. Cont. Zool.* 170:1-265.

Gause, G.F. 1934. *The Struggle for Existence.* Baltimore: Williams & Wilkins Co.

Gibbons, S.G. & H.S.Ogilvie 1933. The development stages of *Oithona helgolandica* and *Oithona spinirostris* with a note on the occurrence of body spines in cyclopoid nauplii. *J. Mar. Biol. Assoc. UK* 23:529-550.

Gombrich, E.H. 1980. Standards of truth: the arrested image and the moving eye. *Critical Inquiry* 7:237-273.

Goodwin, B.C. 1978. A cognitive view of biological process. *J. Social Biol. Struct.* 1:117-125.

Gould, S.J. 1977. *Ontogeny and Phylogeny.* Cambridge: Harvard Univ. Press.

Grant, P.R. 1972. Convergent and divergent character displacement. *Biol. Bull. Linn. Soc.* 4:39-68.

Gurney, R. 1928. Dimorphism and rate of growth in Copopoda. *Intern. Rev. der. ges. Hydrobiol. u. Hydrographie* 21:189-207.

Gurney, R. 1933. Notes on some Copepoda from Plymouth. *J. Mar. Biol. Assoc. UK* 19:299-304.

Hartocollis, P. 1974. Origins of time. *Psychoanalytic Quart.* 43:243-261.

Hall, D.J., S.T.Threlkeld, C.W.Burns & P.H.Crowley 1976. The size-efficiency hypothesis and the size structure of zooplankton communities. *Ann. Rev. Ecol. Syst.* 7:177-208.

Harrison, B. 1976. *Feeding in the Rock Cod.* BSc Honours Thesis, Dalhousie University, Halifax, Nova Scotia, Canada.

Hutchinson, G.E. 1959. Homage to Santa Rosalia or why are there so many kinds of animals. *Am. Nat.* 93:145-159.

Hutchinson, G.E. 1968. When are species necessary? In: R.C.Lewontin (ed.), *Population Biology and Evolution:*177-186. Syracuse: Syracuse Univ. Press.

Huxley, A. 1954. *Doors of Perception.* New York: Harper & Row.

Ingle, D. 1968. Spatial dimensions of vision in fish. In: D.Ingle (ed.), *The Central Nervous System and Fish Behaviour:*51-59. Chicago: Univ. Chicago Press.

Jacobson, M. & R.M.Gaze 1964. Types of visual response from single units in the optic tectum and optic nerve of the goldfish. *Quart. J. Expt. Physiol.* 49:199-209.

Johns, P.R. & S.S.Easter 1975. Retinal growth and adult goldfish. In: M.N.Ali (ed.), *Vision in Fish: New Approaches in Research:*451-457. New York: Plenum Press.

Kaczynsky, V.W., R.J.Feller & J.Clayton 1973. Trophic analysis of juvenile Pink and Chum salmon (*Oncorhynchus gorbuscha* and *O.keta*) in Puget Sound. *J. Fish. Res. Board Canada* 30:1003-1008.

Katona, S.K. 1973. Evidence for sex pheromones in planktonic copepods. *Limnol. Oceanogr.* 18:574-583.

Kerfoot, W.C., D.L.Kellogg & J.R.Strickler 1980. Visual observations of live zooplankters: evasion, escape and chemical defenses. In: W.C.Kerfoot (ed.), *Evolution and Ecology of Zooplankton Communities:*10-27. New York: Univ. Press of New England.

Kerr, S.R. 1974. Theory of size distribution in ecological communities. *J. Fish. Res. Board Canada* 31:1859-1862.

Kuhn, T.S. 1962. The Structure of Scientific Revolutions. *Intern. Ecycl. of Unified Science* 2(2):1-210.

Kuhn, T.S. 1970. Logic of discovery of psychology of research. In: I.Lakatos & A.Musgrave (eds), *Criticism and the Growth of Knowledge:*1-23. Cambridge Univ. Press.

Lane, P.A. 1981. *Using qualitative analysis to understand perturbations to marine ecosystems in the field and laboratory. Proc. Symp.* Washington: Env. Prot. Agen.

Lane, P.A., G.H.Lauff & R.Levins 1975. The feasibility of using a holistic approach in ecosystem analysis. In: S.A.Levin (ed.), *Ecosystem Analysis and Prediction:*111-128. Philadelphia: SIAM.

Lane, P.A. & R.Levins 1977. Dynamics of aquatic systems. II. The effect of nutrient enrichment on plankton model communities. *Limnol. Oceanogr.* 22:454-471.

Lang, K. 1948. *Monographie der Harpactididen.* Koenigstein: Sven Koeltz.

Law, R. 1979. Optimal life histories under age-specific predation. *Am. Nat.* 114:399-417.

Levins, R. 1968. *Evolution in Changing Environments.* Princeton: Princeton Univ. Press.

Levins, R. 1970. Complex systems. In: C.H.Waddington (ed.), *Toward a Theoretical Biology,* 3:73-88. Chicago: Aldine Publ. Co.

Levins, R. 1973. The analysis of partially specified systems. *N. Y. Acad. Sci.* 231:123-138.

Levins, R. 1975. Evolution in communities near equilibrium. In: M.L.Cody & J.M.Diamond (eds), *Ecology and Evolution of Communities:*16-50. Cambridge: Harvard Univ. Press.

Levins, R. 1979. Coexistence in a variable environment. *Am. Nat.* 114:765-783.

Levins, R. & R.Lewontin 1980. Dialectics and reductionism in ecology. *Synthese* 43:47-78.

Lonergan, B.J.F. 1957. *Insight.* London: Darton, Longman & Todd.

Lonergan, B.J.F. 1971. *Method in Theology.* London: Darton, Longman & Todd.

Lonergan, B.J.F. 1973. *Introducting the Thought of Bernard Lonergan.* London: Darton, Longman & Todd.

Lopez, G.W. 1980. Description of the larval stages of *Tisbe cucumariae* (Copepoda: Harpacticoida) and comparative development within the genus *Tisbe. Mar. Biol.* 57:61-71.

Luria, L.A. 1973. *The Working Brain.* England: Allen Lane, The Penguin Press.

Lythgoe, J.N. 1966. Visual pigments and underwater vision. In: R.Bainbridge, G.C.Evans & O.Rackham (eds), *Light as an Ecological Factor. Br. Ecol. Soc. Symp.* 6:375-391.

Lythgoe, J.N. 1968. Visual pigments and visual range underwater. *Vision Res.* 8:997-1011.

MacArthur, R. 1972. *Geographical Ecology: Patterns in the Distribution of Species.* New York: Harper & Row.

Maiorana, V.C. 1981. Prey selection by sight: random or economical. *Am. Nat.* 118:450-451.

Marcotte, B.M. 1977. An introduction to the architecture and kinematics of harpacticoid (Copepoda) feeding: *Tisbe furcata* (Baird 1837). *Mikrofauna Meeresboden* 61:183-196.

Marcotte, B.M. 1980. The meiobenthos of fjords: a review and prospectus. In: J.Freeland, D.M.Farmer & C.D.Levings (eds), *Fjord Oceanography:*557-568. New York: Plenum Press.

Marcotte, B.M. submitted 1. Behaviourally defined econogical resources and *Tisbe.*

Marcotte, B.M. submitted 2. The ecology, functional morphology and evolution of marine meiobenthic copepods.

Marcotte, B.M. & A.Witt submitted. Juvenile Coho and Chinook Salmon eat and compete for benthic prey: a preliminary report.

Masterman, M. 1970. The nature of a paradigm. In: I.Lakatos & A.Musgrave (eds), *Criticism and the Growth of Knowledge:*59-89. Cambridge Univ. Press.

Mayr. E. 1963. *Populations, Species, and Evolution.* Cambridge: Harvard Univ. Press.

Noodt, W. 1974. Anpassung an interstitielle Bedingungen: Ein Faktor in der Evolution Höherer Taxa der Crustacea? *Fauna.-ökol. Mitt.* 4:445-452.

Odum, H.T. 1972. *An Energy Circuit Language, Its Physical Basis in a Systems Analysis and Simulation.* Vol.2. New York: Academic Press.

Odum, H.T. 1975. Combining energy laws and corollaries of the maximum power principle with visual systems mathematics. In: S.A.Levin (ed.), *Ecosystem Analysis and Prediction:*239-263. Philadelphia: SIAM.

Paine, R.T. 1974. Intertidal community structure. Experimental studies between a dominant competitor and its principle predator. *Oecologia* 15:93-120.

Park. T.S. 1966. The biology of a calanoid copepod *Epilabidocera anphitrites* McMurrich. *La Cellule* 66(2):129-256.

Parker, G.H. 1901. The reactions of copepods to various stimuli and the bearing of this on daily depth migrations. *Bull. US Fish. Comm.* 21:103-123.

Patten, B.C., R.W.Bosserman, J.T.Finn & W.G.Cale 1976. Propagation of cause in ecosystems. In: *Systems Analysis and Stimulation in Ecology* 4:457-579.

Peters, R.H. 1976. Tautology in evolution and ecology. *Am. Nat.* 110:1-12.

Peters, R.H. 1978. Predictable problems with tautology in evolution and ecology. *Am. Nat.* 112:759-786.

Pico della Mirandola, G. 1572. *Opera Omnia.* 1 Vol. Bâle.

Popper, K. 1963. *Conjectures and Refutations: The Growth of Scientific Knowledge.* London: Routledge & K.Paul.

Putnam, H. 1975. The meaning of meaning. In: K.Gundersen (ed.), *Minnesota Studies in the Philosophy of Science 7. Language, Mind and Knowledge.* Minneapolis: Univ. Minn. Press.

Pylyshyn, Z.W. 1972. Competence and psychological reality. *Am. Psychol.* 27:546-552.

Pylyshyn, Z.W. 1980. Computation and cognition: issues in the foundations of cognitive science. *Behav. Brain Sci.* 3:111-169.

Roy, J. 1932. *Copepodes et Cladocères de l'Ouest de la France.* Paris.

Schoener, T.W. 1979. Generality of the size-distance relation in models of optimal feeding. *Am. Nat.* 114:902-914.

Schwassmann, H.O. & L.Kruger 1968. Anatomy of visual centres in teleosts. In: D.Ingle (ed.), *The Central Nervous System and Fish Behaviour:*3-16. Chicago: Univ. Chicago Press.

Shumaker, W. 1972. *The Occult Sciences in the Renaissance.* Berkeley: Univ. California Press.

Smol, N. & C.Heip 1974. The culturing of some harpacticoid copepods from brackish water. *Biol. Jb. Dodonaea* 42:159-169.

Strickler, J.R. 1975a. Swimming of planktonic *Cyclops* species (Copepods, Crustacea): pattern, movements and their control. In: T.Y.-T.Wu et al. (eds), *Swimming and Flying in Nature:*599-613. New York: Plenum Press.

Strickler, J.R. 1975b. Intra- and interspecific information flow among planktonic copepods: receptors. *Int. Ver. Theor. Angew. Limnol. Verh.* 19:2951-2958.

Strickler, J.R. 1977. Observation of swimming performances of planktonic copepods. *Limnol. Oceanogr.* 22:165-170.

Strickler, J.R. & A.K.Bal 1973. Setae of the first antenna of the copepod *Cyclops scutifer* (Sars): their structure and importance. *Proc. Nat. Acad. Sci.*:2656-2659.

Sutherland, N.S. 1968. Shape discrimination in the Goldfish. In: D.Ingle (ed.), *The Central Nervous System and Fish Behaviour:*35-50. Chicago: Univ. Chicago Press.

Swedmark, B. 1964. The interstitial fauna of marine sand. *Biol. Rev.* 39:1-42.

Takeda, N. 1950. Experimental studies on the effect of external agencies on the sexuality of a marine copepod. *Physiol. Zool.* 23:288-301.

Thompson, D'A. 1971. *On Growth and Form.* Cambridge Univ. Press.

Trevarthen, C. 1968. Vision in Fish: the origins of the visual frame of action in vertebrates. In: D.Ingle (ed.), *The Central Nervous System and Fish Behaviour:*61-94. Univ. Chicago Press.

Vandermeer, J. 1980. Indirect mutualism: variations on a theme by Stephen Levine. *Am. Nat.* 116:441-448.

Webster, G. & B.Goodwin 1981. History and structure in biology. In: M.Barker, L.Birke, A.D.Muir & S.P.R.Rose (eds), *The Dialectics of Biology and Society.* London: Allison & Busby.

Weismann, A. 1889. *Essays upon Heredity and Kindred Biological Problems.* Oxford: Clarendon Press.

Willey, A. 1923. Notes on the distribution of free-living Copepoda in Canadian waters II. *Contrib. Canad. Biol.* 1:529-539.

Williams, M.B. 1970. Deducing the consequences of evolution: a mathematical model. *J. Theor. Biol.* 29:343-385.

Williams, M.B. 1975. Darwinian selection for self-limiting populations. *J. Theor. Biol.* 55:415-430.

Yates, F.A. 1964. *Giordano Bruno and the Hermetic Tradition.* Chicago: Univ. Chicago Press.

Yates, F.A. 1966. *The Art of Memory.* Chicago: Univ. Chicago Press.

Zaret, R.E. 1980. The animal and its viscous environment. In: W.C.Kerfoot (ed.), *Evolution of Zooplankton Communities:*3-9. New York: Univ. Press of New England.

MARK J. GRYGIER
*Scripps Institution of Oceanography, University of California, San Diego, USA*

# ASCOTHORACIDA AND THE UNITY OF MAXILLOPODA

## ABSTRACT

The class Maxillopoda (= Copepodoidea) originally contained the subclasses Copepoda, Mystacocarida, Branchiura, Cirripedia (including Rhizocephala), and Ascothoracida. This paper tries to answer the important, but neglected, question of whether such a grouping (now also including Ostracoda, Hansen's Y-larvae, *Basipodella harpacticola* and *Deoterthron dentatum,* and Pentastomida) is a natural one. A literature review of the class is conducted. All the previously suggested diagnostic characters, and some new ones, are analysed from a cladistic standpoint, and a new diagnosis based on the defensible synapomorphies is proposed. Archicopepoda (= Euthycarcinoidea) is excluded from the class, and doubts about the maxillopodan affinities of Bradoriida (including Phosphatocopina) are expressed. The non-parasitic ancestor of the ascothoracids could have served equally well as the progenitor of all the maxillopodan taxa (i.e. the urmaxillopodan); the modifications necessary to derive each from it are detailed. Because of the extent of convergence among the component taxa, only a partially resolved cladogram of the class is proposed. Conventional notions about the relationships among maxillopodan taxa (e.g. a Copepoda-Branchiura-Mystacocarida clade) are not supported. There is some justification for formal taxonomic recognition of the class Maxillopoda, but it does not have as solid a foundation as the other crustacean classes.

## 1 INTRODUCTION

Entomostraca, classically a major division of Crustacea including all non-malacostracan forms, is now usually divided into at least seven distinct subclasses of equal rank with Malacostraca: Cephalocarida, Branchiopoda, Cirripedia, Copepoda, Mystacocarida, Branchiura and Ostracoda. This paper critically reviews the arguments about the monophyletic nature (*sensu* Hennig 1966) of the class Copepodoidea or Maxillopoda, proposed independently by Beklemishev (1952) and Dahl (1956), respectively, for some or all of the non-cephalocarid and non-branchiopod Entomostraca. It also describes how a hypothetical ancestor of the Ascothoracida, an obscure subclass of parasitic crustaceans allied to the Cirripedia, possessed a body plan from which those of all extant maxillopodans could have been derived. The possible maxillopodan affinities of Hansen's Y-larvae, the two parasitic genera *Basipodella* and *Deoterthron,* Rhizocephala, Pentastomida, and the fossil taxa Euthycarcinoidea and Bradoriida are also discussed (see Table 1 for brief summaries of all putative maxillopodan taxa).

Table 1. List of taxa proposed for inclusion in the class Maxillopoda or Copepodoidea.

| Taxon | Proposer | Comments |
|---|---|---|
| Ascothoracida | Beklemishev (1952), Dahl (1956) | Parasites of anthozoans and echinoderms; formerly an order of Cirripedia |
| Ostracoda | Siewing (1960) | Seed shrimp; only extant forms considered herein |
| Bradoriida | Never explicitly, included with ostracods | Early Paleozoic 'ostracods'; considered herein to be of uncertain affinities; possibly not maxillopodan |
| Hansen's Y-larvae | Herein | Planktonic nauplii and 'cyprids'; adults unknown; apparently parasitic; formerly considered abberant Cirripedia or Ascothoracida |
| Branchiura | Beklemishev (1952) tentatively, Dahl (1956) | Fish lice |
| Cirripedia | Beklemishev (1952), Dahl (1956) tentatively | |
|    Thoracica | | Barnacles |
|    Acrothoracica | | Burrowing barnacles |
|    Rhizocephala | | Parasites of crustaceans; may deserve own subclass apart from Cirripedia (cf. section 6.2 herein) |
| Copepoda | Beklemishev (1952), Dahl (1956) | |
| Archicopepoda | Dahl (1956) | Non-maxillopodan fossil taxon now called Euthycarcinoidea; formerly considered primitive copepods; probably uniramian |
| Mystacocarida | Beklemishev (1952), Dahl (1956) | Members of meiofauna in sandy beaches and muddy bottoms |
| *Basipodella harpacticola* and *Deoterthron dentatum* | Bradford and Hewitt (1980) | Ectoparasites of crustaceans |
| Pentastomida | Herein | Respiratory parasites of terrestrial vertebrates; formerly a protarthropodan phylum |

## 2 HISTORICAL SURVEY

The subclass Ascothoracida played an important role in two early efforts to produce a phylogenetic history of the Entomostraca. Fowler (1890) was impressed by the similarity of ascothoracids both to cirripeds and to ostracods, and suggested that the three taxa formed a lineage independent of the copepod-phyllocarid line. Wagin (1937, 1947) compared ascothoracids to copepods as well as to cirripeds, with which they previously had been classified (Gruvel 1905). He found enough characters in common to suggest an affinity between ascothoracids and copepods, though cirripeds remained their nearest relatives.

Branchiura was long considered an order of copepods until Martin (1932). The distinctness of Mystacocarida, originally proposed as a taxon independent of Copepoda (Pennak & Zinn 1943), was also questioned (Armstrong 1949). These examples of conservative classification are important in the history of the class Maxillopoda because they have served as *a priori* guideposts for later efforts to establish a maxillopodan phylogeny (e.g. Dahl 1956, Siewing 1960).

Beklemishev (1952, not seen, but cited by Birshtein (1960) and Siewing (1960); English edition 1969) included the orders Copepoda, Mystacocarida, Cirripedia, Ascothoracida, and probably Branchiura in a new superorder, Copepodoidea. His justification was the similarity among these taxa in body segmentation and thoracopod morphology. According to Beklemishev, Cirripeds and ascothoracids exhibit the primitive state, six pairs of segmented, biramous thoracopods lacking epipods and gnathobases. The first pair has become maxillipedal in copepods and mystacocarids, and is supposedly lost in branchiurans. The sixth pair is generally reduced in copepods and is lost in mystacocarids and branchiurans. Beklemishev judged the lack of a carapace in copepods to be secondary.

Dahl (1956) independently proposed the subclass Maxillopoda to include Mystacocarida, Archicopepoda (= Euthycarcinoidea), Branchiura, Copepoda, and probably Cirripedia (including Ascothoracida). Members of each of these taxa possess six thoracomeres (number reduced in mystacocarids and branchiurans), a well-developed mandibular palp, well developed maxillules and maxillae acting as filters in filter feeders, and no gnathobases on the thoracopods. Dahl placed Mystacocarida as an early branch of the copepod line and Branchiura very close to Copepoda. The Cirripedia were included only provisionally.

Siewing (1960) recognized the Maxillopoda or Copepodoidea (he did not prefer one name over the other). He saw mystacocarids as somewhat more primitive than copepods, but closely related both by the presence of nauplius eyes and maxillipeds. Branchiurans, with their compound eyes and carapace, were considered even more primitive. Siewing derived cirripeds from an ascothoracid-like form, but did not comment explicitly on their relationship to the other maxillopodans. He suggested that ostracods had maxillopodan affinities because of their resemblance to cirriped cyprid larvae.

Ax (1960) conducted an extensive comparison of a mystacocarid with a generalized copepod, and accepted Siewing's hypothesis that Mystacocarida is a primitive offshoot of the copepod line within Maxillopoda or Copepodoidea.

Birshtein (1960), largely influenced by Wagin, included Maxillopoda in his classification of Crustacea. In addition to the characters already mentioned, he noted the reduced circulatory system in all maxillopodans. In the same volume, Novozhilov (1960) paraphrased Dahl's (1956) diagnosis of Maxillopoda.

Dahl (1963) reappraised his earlier classification of Crustacea and elevated Maxillopoda to the level of cohort. He asserted that pelagic forms are descended from benthic forms, harpacticoid copepods being secondarily benthic. He considered the lateral gut diverticula of branchiurans, cirripeds (including ascothoracids), and some ostracods; the gastric mills of some ostracods and acrothoracican cirripeds; and the sensory papilla X-organ present in a few copepods and, as frontal filaments, in *Balanus* larvae as unifying features of Maxillopoda. He tabulated data indicating that maxillopodan compound eyes may form a distinct type. Dahl was not convinced that ostracods could be included reasonably in the taxon, and suggested that the common ancestor of Copepoda and Mystacocarida had essentially maxillopodan features.

Kaestner (1967) included a taxon Maxillosa (comprising Mystacocarida, Copepoda, Branchiura, Ascothoracida, and Cirripedia) in his invertebrate treatise with no explanation or diagnosis. The English translation (Kaestner 1970) reverted to the term Maxillopoda for the same taxon.

Newman et al. (1969) revived consideration of the taxon Maxillopoda. They included a comparative table of tagmatization of the component groups and hypothesized a derivation of Cirripedia from an urascothoracid form with some copepod and branchiuran features. This work also first defined Maxillopoda in terms of a 5-6-5 body plan. Newman (1974) pointed out that all the putative maxillopodan groups (Ostracoda included) could be derived easily from an ascothoracid-like form. Hessler & Newman (1975) briefly listed the specializations of each maxillopodan group and proposed a coordinate taxon, Thoracopoda, for the rest of the Crustacea. Grygier (1980, 1981a, 1982) employed asco-thoracids to amplify and clarify contentions by Brown (1970) and Pochon-Masson (1978) that spermatological evidence supported the validity of Maxillopoda. Bradford & Hewitt (1980) explored the affinities of their enigmatic genus *Deoterthron,* as well as the closely related *Basipodella,* through a detailed tabulation of characteristics of several maxillopodan taxa. Recent Russian workers (Wagin 1976, Zevina 1981) accepted the inclusion of Ascothoracida and Cirripedia in Copepodoidea or Maxillopoda, respectively. Newman (1982) listed features characteristic of Maxillopoda, and discusses the origin of the class (this volume).

Little opposition to Beklemishev's and Dahl's formulations has arisen, even though Newman (1974, 1982, this volume, Newman et al. 1969) has been their only consistent advocate. Without mentioning the taxon by name, Lauterbach (1974) rejected Maxillopoda by classifying all carapace-bearing crustaceans into a single supposedly monophyletic taxon, Palliata, separating Ostracoda, Ascothoracida, and Cirripedia from the others (Lauterbach did not consider branchiurans to have true carapaces). Paulus (1979) stated his doubts about the monophyly of Maxillopoda, an opinion no doubt shared by many carcinologists (various personal communications), but offered no specific objections. Hessler (1982) came to an uncertain, generally pessimistic conclusion after a brief, carefully reasoned argument of both sides of the question, and recommended against formal taxonomic use of the name Maxillopoda or Copepodoidea.

The name Maxillopoda is used in this paper for the class-level taxon under consideration to conform with prevailing usage. Though the name Copepodoidea was proposed first, it was not diagnosed (according to Birshtein 1960), has an inappropriate, superfamilial suffix, and has seen much less use in the subsequent literature, even among Russians, than has Maxillopoda.

## 3 EVALUATION OF PROPOSED DIAGNOSTIC CHARACTERS OF MAXILLOPODA

For Maxillopoda to be accepted as a monophyletic taxon, it must be definable in terms of synapomorphies (shared derived characters) not found in other crustaceans except by convergence. Non-convergent features shared with other organisms are plesiomorphic (primitive) and cannot be used to define monophyletic taxa (Hennig 1966).

This section evaluates the synapomorphic status of the putative diagnostic (or at least characterstic) features that have been attributed to Maxillopoda (Table 2). Dahl (1956, 1963), Beklemishev (1969), Newman et al. (1969) and Newman (1982) have contributed most of these criteria; some have been proposed by other workers, and a few new suggestions are introduced here.

## 3.1 *Tagmosis*

The division of the body into a head with five pairs of appendages, a thorax with no more than six, and an abdomen with at most five limbless somites (including the telson) has been considered the most distinctive feature of Maxillopoda (Dahl 1956, Beklemishev 1969, Newman et al. 1969). Ascothoracids are seen as the most generalized form (Newman 1974). *Basipodella* (head appendages reduced) and copepods (first thoracomere cephalized) also adhere closely to this archetype. Cirripeds retain all the appendages, but have a reduced abdomen, as do *Deoterthron,* some Y-cyprids (but Bresciani (1965) described one with six abdominal somites), and branchiurans. Mystacocarids are interpreted as having lost the sixth pair of thoracopods, and branchiurans two pairs. Ostracods have a much reduced abdomen and thorax, except *Saipanetta,* in which the posterior end of the body is apparently divided into several limbless segments (McKenzie 1967).

It is clear that all the maxillopodan groups (except Bresciani's Y-cyprid) can be derived by reduction from an ancestor with 11 trunk segments including the telson. Is 5-6-5 the best possible representation of an urmaxillopodan body plan, and is such an arrangement apomorphic relative to other crustaceans? There is some doubt about the first question. Male ostracods generally have a pair of penes behind the thoracopods, and copepods often have paired genital appendages on the first abdominal segment (first or second urosome segment depending on the order). Ascothoracids and cirripeds have an unpaired penis, respectively on the first abdominal segment and between the last pair of cirri. In generalized ascothoracids the penis is articulated and biramous (cf. Newman 1974), which strongly suggests an origin by medial fusion of paired appendages (Newman 1982). It is likely that the penes and genital appendages of all four groups represent a seventh pair of postcephalic limbs. Thus, a 5-7-4 body plan is probably a better representation of an urmaxillopodan.

Whether the urmaxillopodan descended from a malacostracan form *sensu lato* (Gurney 1942, Siewing 1960, Newman, this volume) or separately from an urcrustacean form (e.g. Hessler & Newman 1975), its body plan (5-6-5 or 5-7-4) was modified from the ancestral state in having a smaller number of trunk segments and in lacking appendages on several posterior ones. This apomorphic state may be employed as a diagnostic feature of Maxillopoda if the variant tagmoses of the component taxa reflect secondary divergences.

## 3.2 *Cephalic sensory organs*

The structure of the nauplius eye in the five maxillopodan taxa possessing it (Ostracoda, Copepoda, Cirripedia, Branchiura and Hansen's Y-larva) was studied histologically and thoroughly reviewed by Elofsson (1966, 1971). [The reported occurrences of nauplius eyes in nauplii of the ascothoracid *Laura gerardiae* (Lacaze-Duthiers 1883) and in an ascothoracid 'cyprid' larva (Tessmann 1904) have not been confirmed, and the so-called ocelli in mystacocarids (Dahl 1952) are not eyes at all (Renaud-Mornant et al. 1977).] In each case the nauplius eye has three cups equipped with lens cells (not in cirripeds and Y-larva, in which the eyes are somewhat reduced), sensory cells, tapetal cells, and pigment. Among crustaceans, the tapetal cells and lens cells are unique to maxillopodan taxa and may be interpreted as synapomorphies. Furthermore, no maxillopodan has true frontal organs (Elofsson 1966), while other crustaceans usually do (Elofsson 1963, 1965). If this condition reflects a loss rather than a plesiomorphic lack, it may also be a synapomorphy.

The presence of the sensory organs variously called SPX-organs, sensory papilla X-organs,

Table 2. List of proposed diagnostic or characteristic features of Maxillopoda, with evaluations of their cladistic status (explanations in text).

| Feature | Proposer | Evaluation |
|---|---|---|
| *Tagmosis* (3.1) | | |
| Six thoracic segments | Beklemishev (1952), Dahl (1956) | Synapomorphic, but see below |
| Five cephalic segments, six thoracic, five abdominal including telson (5-6-5) | Newman et al. (1969) | Synapomorphic, but see below |
| Five cephalic segments, seven limb-bearing segments, four appendage-less segments including telson (5-7-4) | Herein | Synapomorphic |
| *Cephalic sensory organs* (3.2) | | |
| Nauplius eye with lens cells and tapetal cells | Elofsson (1966) | Synapomorphic |
| Frontal organs absent | Elofsson (1966) | Probably synapomorphic |
| Organs of Bellonci | Dahl (1963) | Plesiomorphic |
| Compound eye structure | Dahl (1963) | Variously autapomorphic |
| *Antennules* (3.3) | | |
| Strong development | Newman (1982) | Plesiomorphic |
| Attachment function | Herein | Probably convergent |
| *Mandible* (3.4) | | |
| Palp | Dahl (1956) | Plesiomorphic |
| Bladelike gnathobase | Herein | Synapomorphic |
| *Maxillary feeding* (3.4) | Dahl (1956) | Plesiomorphic or convergent |
| *Thoracopods* (3.5) | | |
| Lack of epipods | Beklemishev (1952) | Synapomorphic |
| Lack of gnathobases | Beklemishev (1952), Dahl (1956) | Possibly convergent |
| Paddle-like form | Herein | Possibly convergent |
| Genital appendages on seventh postcephalic segment | Newman (1982) | Synapomorphic |
| *Furca* (3.6) | Hessler (1982) | Plesiomorphic |
| *Reduced circulatory system* (3.7) | Birshtein (1960) | Possibly convergent |
| *Digestive system* (3.8) | | |
| Gastric mill | Dahl (1963) | Convergent |
| Lateral diverticula | Dahl (1963) | Synapomorphic |
| *Gonopores* (3.9) | | |
| Position anterior to that in other crustaceans | Newman (1982) | Uninterpretable cladistically |
| Male openings on seventh postcephalic segment | Herein | Synapomorphic |
| *Flagellate sperm* (3.10) | Brown (1970) | Plesiomorphic |
| *Distinctive naupliar limbs* (3.11) | Newman (1982) | Plesiomorphic |

or organs of Bellonci in cirripeds (Kauri 1966), ascothoracids (e.g. Yosii 1931), mystacocarids (Renaud-Mornant et al. 1977), ostracods (e.g. Andersson 1977) and copepods (Dahl 1963) cannot be used as a diagnostic feature of Maxillopoda because organs of Bellonci are also found in decapods and peracarids (Chaigneau 1978), and are thus plesiomorphic.

Branchiurans, some ostracods, cirriped cyprids, and Y-larva cyprids have compound eyes. Dahl (1963) and Paulus (1979) reviewed their structure and found no convincing

synapomorphic features, though the eyes in the various taxa differ from those of malacostracans and branchiopods.

### 3.3 *Antennules*

The antennules of most maxillopodans besides ostracods serve for attachment by means of
claws, an adhesive device, or a geniculate articulation. The many antennular attachment
devices probably evolved convergently, but the possibility that such a function is synapomorphic and served several times as a preadaptation for a sedentary or parasitic way of life
might also be considered. The well-developed state of most of these antennules is clearly
plesiomorphic.

### 3.4 *Mouthparts*

Possession of a mandibular palp and well-developed maxillules and maxillae by representatives of most maxillopodan taxa is certainly plesiomorphic for Crustacea as a whole, and
cannot be a diagnostic criterion of the class (cf. diagnosis by Dahl 1956). Maxillary feeding
is not found only in certain maxillopodan taxa, but in many eucarids and peracarids as
well. Its incidence is apomorphic at some level, since the ancestral crustacean is inferred to
have been a thoracic feeder (Sanders 1963a, Hessler & Newman 1975), but it is probably a
convergent condition. In referring to a non-synapomorphic character, the name Maxillopoda is thus in some sense a misnomer (Hessler 1982, Newman 1982).

   The mandibles of most non-maxillopodan crustaceans have gnathobases differentiated
into incisor and molar processes, probably the plesiomorphic state for Crustacea. The
mandibular gnathobases of non-parasitic maxillopodans are simple serrate or dentate blades.
Such a gnathobase may be a synapomorphic, diagnostic trait, the variations among parasites being secondary modifications.

### 3.5 *Thoracopods*

Dahl (1956) and Beklemishev (1969) commented on the lack of epipods and gnathobases
on the thoracopods of maxillopodan crustaceans. The limbs of groups added since the
taxon's inception (Hansen's Y-larvae, *Basipodella* and *Deoterthron,* and Ostracoda) are
also built this way. It has been suggested that the large, external, coxal setae of ascothoracids are vestigial epipods (Grygier 1981b), but this interpretation is probably unjustified.
Most other crustacean groups have members equipped with thoracic epipods, a condition
generally acknowledged as primitive (cf. Hessler & Newman 1975), so their lack in all
maxillopodan taxa may be synapomorphic and diagnostic. The paddle-like limbs of the
clearly non-maxillopodan Remipedia (Yager 1981) and Lipostraca (Scourfield 1926), which
are very similar to those of copepods, ascothoracids, and cyprid larvae, might cast doubt
on the monophyletic origin of this specific type of biramous limb in Maxillopoda.

### 3.6 *Furca*

All maxillopodans have caudal furca at some point in their lives, the form of which varies
among taxa. Furca, however, cannot be used as a diagnostic character because they are
plesiomorphic for all crustaceans. This conclusion also holds if Bowman's (1971) interpretation of furcal rami as uropods is correct.

### 3.7 *Circulatory system*

Maxillopodan crustaceans have either no special circulatory system (ascothoracids, most copepods, many ostracods, mystacocarids) or a very small heart with no more than three ostia and rudimentary arteries (calanoid copepods, myodocopid ostracods, cirripeds, branchiurans). Such circulatory systems are apomorphic compared to most other crustaceans, in which the heart is tubular with many segmentally arranged pairs of ostia and a complex arterial system (Maynard 1960). Because the positions and numbers of ostia and arteries vary among the maxillopodan taxa, heart reductions may have occurred convergently, due simply to the adequacy of diffusion and muscular agitation for transport of substances throughout a small body. The incidences of total loss are clearly convergent. Even though cephalocarids, which are the same size as many maxillopodans, have a generalized heart (Hessler 1969a), a reduced circulatory system is at best an unsatisfactory diagnostic character of Maxillopoda.

### 3.8 *Digestive system*

Dahl (1963) cited the possession of gastric mills by some ostracods and acrothoracican cirripeds as evidence for monophyly in Maxillopoda. Such a masticatory apparatus is also common throughout the Malacostraca, and its appearance here is undoubtedly convergent. The lateral gut diverticula of ascothoracids, cirripeds, branchiurans, and some ostracods are better candidates for synapomorphic status, since the gastric caeca of other crustaceans are not so arranged (Dahl 1963).

### 3.9 *Gonopore position*

Newman (1982) pointed out that the positions of the genital openings in maxillopodans are different from those in other crustaceans. The male gonopores are on the seventh postcephalic segment, or, in those taxa with reduced bodies, in an equivalent position posterior to the thoracopods. Mystacocarids, in which they are on the fourth of five limb-bearing segments, are exceptional (Dahl 1952). In malacostracans the male genital segment is the eighth postcephalic, and in branchiopods (except cladocerans) it is somewhat farther back. Only in cephalocarids, among long-bodied crustaceans, are the male openings anterior to those in most maxillopodans, being on the sixth postmaxillary segment. If the condition in mystacocarids be considered derived, male gonopores on the seventh postcephalic segment might be diagnostic for Maxillopoda.

The female genital apertures are on the fused seventh and eighth postcephalic segments in copepods, in an equivalent position relative to the thoracopods in ostracods and branchiurans, on the fourth segment in mystacocarids, and on the first in cirripeds and ascothoracids (cf. section 6.2 for disputation of position in ascothoracids). In no case do these positions correspond to those in other crustaceans, so it is not possible to state which female gonopore position is most plesiomorphic, or even whether the male and female genital segments coincided primitively. Therefore, no diagnostic position in Maxillopoda can be specified.

### 3.10 *Sperm*

Maxillopodans are unique among crustaceans in that some have flagellate sperm, specifically Ascothoracida (Wingstrand 1978, Grygier 1980, 1981a, 1982), Cirripedia (Munn & Barnes

1970, Pochon-Masson et al. 1970), Branchiura (Wingstrand 1972), and Mystacocarida (Tuzet & Fize 1958, Brown & Metz 1967). Brown (1970) and Pochon-Masson (1978) cited this structural similarity in support of Maxillopoda as a natural taxon. At a superficial level this argument is untenable, since flagellate sperm, especially those found in ascothoracids, are the most generalized type of spermatozoa (Franzén 1970, 1977) and are thus plesiomorphic. Sperm structure cannot be used in diagnoses of Maxillopoda (Wingstrand 1972, Hessler 1982), but a consideration of the variations among taxa is useful for clarifying relationships within Maxillopoda, as will be explained in section 6.2.

### 3.11 *Larval morphology*

Newman (1982) proposed that the distinctive naupliar morphology of maxillopodans serve as part of the definition of the class. This approach is reasonable, if their larvae are uniquely synapomorphic relative to a more generalized form. Out-group comparisons are not possible, at least until the soft parts of trilobite protaspis larvae became known, or the affinities of Müller's (1981) phosphatized Cambrian metanauplii are established.

In-group comparisons are of only limited utility. Sanders (1963a,b) listed numerous features distinguishing the nauplii of branchiopods from those of cephalocarids, malacostracans, and maxillopodans. Branchiopod larvae have short unsegmented antennules, an antennal protopod forming at least half the length of that limb, and uniramous mandibles. Other crustacean nauplii typically have longer, segmented antennules, a short antennal protopod, and biramous mandibles. For all three characters branchiopods seem to be apomorphic, while other crustaceans are plesiomorphic. Sander's list also includes setation differences which probably also follow this pattern.

Sanders (1963a) postulated that the nauplii of cephalocarids and mystacocarids are the most plesiomorphic among Crustacea in their retention of a primitively benthic mode of life and correlated morphological features. If this is true one cannot specify apomorphic states unique to maxillopodan nauplii. Rather, most of their common features would seem to be plesiomorphic for Crustacea. Cephalocarid nauplii are probably more plesiomorphic than those of maxillopodans and malacostracans in the larger number of antennal exopod articles and the elongate form of the antennal naupliar process, but there are no synapomorphies relative to the cephalocarid condition in maxillopodan nauplii that are not also found or accentuated in the non-feeding, gnathobase-less nauplii of malacostracans (Newman, this volume).

### 3.12 *Summary*

Table 2 lists all the proposed diagnostic criteria of Maxillopoda discussed in this section and conclusions about their value. Many can be dismissed because of their obviously plesiomorphic or convergent status, e.g. mandibular palp, furca, gastric mill, naupliar morphology. A number more may be synapomorphic, but could equally, or even more likely, be convergent; e.g. reduced circulatory system, paddle-like limbs, attachment to substrate by antennules). There remain a few characters that are not easily dismissed as synapomorphies unless one rejects the entire general history of Crustacea accepted *a priori* here. From this last set, the following diagnosis of the class Maxillopoda can be composed:

Eleven postcephalic segments (including telson), first six each with pair of biramous thoracopods (no epipods), seventh with pair of well-developed genital appendages associated with male gonopores, pos-

82 *Mark J.Grygier*

terior segments limbless. Nauplius eye with lens cells and tapetal cells; frontal organs absent. Mandibular gnathobase a simple, serrate or dentate blade. Pair of lateral midgut diverticula.

This diagnosis includes only synapomorphic features. It does not rely on ancestral (plesio-morphic) characters such as the carapace, caudal furca, compound eyes, heart, and flagellate sperm. The class Maxillopoda includes all organisms descended from those that fit this diagnosis, so it is unnecessary to qualify the diagnostic characters with such phrases as, 'no more than . . ., at least . . ., except for . . .,' etc. That no known crustacean fits this diagnosis in all particulars is an unfortunate, but not unexpected by-product of the cladistic method; it results from the diversity of modifications among the extant representatives of the class.

## 4 ORGANISMS OF DOUBTFUL MAXILLOPODAN AFFINITIES

Two groups of fossil arthropods that have been included explicitly or implicitly in the Maxillopoda, Euthycarcinoidea and Bradoriida, are actually of very uncertain affinities. They do not fit the class' diagnosis and probably have no close relationship to it.

### 4.1 *Euthycarcinoidea*

These Triassic arthropods were originally considered ancient copepods (Archicopepoda Handlirsch 1914), so Dahl (1956) included them in the Maxillopoda. It is now doubtful that they are even crustaceans, F.Schram (1971) suggested euthycarcinoids were merosto-moids. Bergström (1979) proposed a uniramian (perhaps myriopodous) affinity for them, an interpretation that's been elaborated upon (F.Schram & Rolfe, 1982). Therefore, euthycarcinoids need not be considered further.

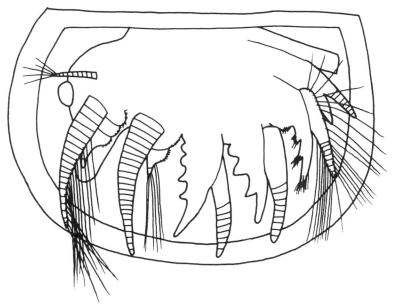

Figure 1. A reconstruction of an ostracod-like phosphatocopine bradoriid, *Hesslandona* sp., from the Upper Cambrian of Sweden (after Müller 1981).

4.2 *Bradoriida (= Archaeocopida)*

This early Paleozoic taxon (including Phosphatocopina Müller) includes a variety of bi-valved arthropods customarily treated as early ostracods (cf. Sylvester-Bradley 1961). Several recent papers have described the soft parts of phosphatocopines (bradoriids with phosphatic shells). Müller (1979, 1981) described subadult stages of three genera (Fig.1). They have up to seven pairs of nearly identical, biramous appendages (first pair uniramous in *Hesslandona*), each with a short endopod and a multiarticulate exopod bearing a seta on each article. The anterior two pairs have gnathobases in *Falites* and *Vestrogothia;* all but the first pair have them in *Hesslandona.* In none of Müller's specimens is the body visibly segmented, but Jones & McKenzie (1980) described a specimen of another genus with four distinct thoracomeres.

   The identity of bradoriids is disputed. Müller (1979) considered Phosphatocopina a dead-end ostracod line and was convinced that data on *Hesslandona* (Müller 1981) proved this point. The situation is not really so clear. Müller's original specimens were all subadult and may not yet have had a full complement of limbs. He does not say whether his recon-struction of *Hesslandona* (adapted in Figure 1), which shows the normal number of append-ages for ostracods (seven pairs), represents an adult. Aside from the antennules, the limbs of *Hesslandona* are to all appearances unlike those in adults of extant crustaceans. Landing (1980) called attention to the cirriform limbs of Müller's (1979) specimens and suggested that the organisms might have a greater affinity with cirripeds than with ostracods. Actually these limbs resemble maxillopodan naupliar limbs (cf. section 3.11) much more than they do cirri, and only substantiate Müller's inference that he was dealing with larval stages. Jones & McKenzie (1980) also regarded Phosphatocopina as an ostracodan sideline, but they distinguished the phyllocarid- or branchiopod-like bradoriids from the ostracod-like ones by the presence or absence of a posterior carapace gape. Such a gape was taken as evidence of a protruding abdomen, a character foreign to ostracods.

   The serial homogeneity of the limbs in all the phosphatocopines so far examined argues for their primitive nature (Hessler & Newman 1975), but such homogeneity is common to almost all Cambrian arthropods (cf. Bergström 1979, Briggs, this volume). The larger arthropods, for example those of the Burgess Shale, represent an incredible variety of only distantly related forms (Whittington 1980). There is no reason to expect the situation to be any different among the smaller arthropods, which are only now beginning to be studied. Jones & McKenzie (1980) may have been conservative in their estimation of widely diver-gent affinities among bradoriids; indeed, some may belong to hitherto unsuspected higher taxa. Their argument for a protruding abdomen in some bradoriids is especially interesting because it raises the possibility that some are actually representatives of an urmaxillopodan stock, members of which may have had a well-developed, probably protrudent abdomen and a bivalved carapace with a gape (cf. section 5). Only more discoveries of preserved soft parts will permit definite taxonomic assignments of these arthropods.

# 5 THE URMAXILLOPODAN AND ITS DESCENDANTS

The ability to formulate a diagnosis of Maxillopoda argues for the monophyly of the class, so the included taxa probably had a unique common ancestor. What was it like? Gurney (1942) suggested a neotenic derivation of copepods from eumalacostracan protozoea larvae.

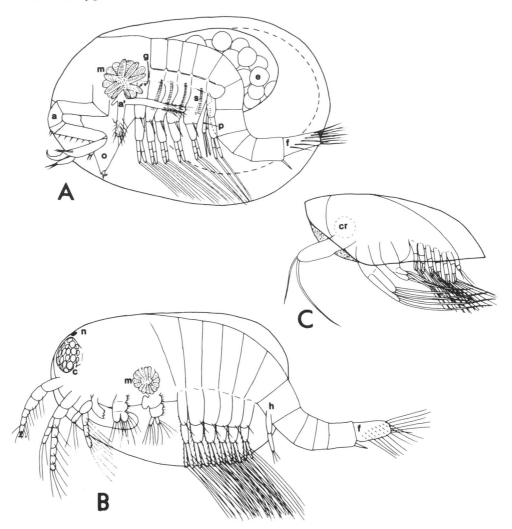

Figure 2. A. Generalized ascothoracid based on several species of *Synagoga* and incorporating features of both sexes. Only the left appendages are shown except for the right furcal ramus and medial penis. B. Hypothetical urmaxillopodan crustacean, essentially an ascothoracid without parasitic modifications. Only the left appendages are shown; they are highly stylized. It is unclear whether the carapace was univalved or bivalved. C. 'Nauplio-cyprid' larva of the rhizocephalan *Lernaeodiscus porcellanae*. Only the left appendages are shown. The drawing is traced from a photographic slide taken bythe late L. Ritchie. Abbreviations: a – antennule; a' – antenna; c – compound eye; cr – compound eye rudiment; e – eggs; f – furcal ramus; g – female gonopore; h – hemipenis (genital appendage); m – adductor muscle; n – nauplius eye; o – oral cone; s – seminal receptacles.

The body plan and distribution of appendages in copepodid larvae and protozoeas are similar. The alterations required by this hypothesis include suppression of the carapace and compound eyes, no transformation later in ontogeny of paddlelike limbs into schizopods, and the development of articulated furcal rami from the telson lobes. Gurney's proposal is quite reasonable, but, having produced a copepod, his model can go no further; other maxillopodans are clearly not derived from copepods. Too much attention has been focused on the Copepoda in discussions of maxillopodan phylogeny, reflected for example in Beklemishev's name Copepodoidea for the taxon. Dahl (1956) and Siewing (1960) compared the branchiurans and mystacocarids directly to copepods in arguing for their inclusion in Maxillopoda, and ignored the phylogenetic inferences that could be drawn from cirripeds and ascothoracids.

Newman (1974) suggested that it would prove worthwhile to look at ascothoracids for clues to the origin of the Maxillopoda, since some members of that group seem to have a very generalized body plan. This section shows in detail how ascothoracids can serve as models of the urmaxillopodan *Bauplan.* The origin of the urmaxillopodan itself is not discussed here, since Newman (this volume) examines this topic at length.

## 5.1 *Ascothoracida*

This is a small but highly diversified subclass, many of the members of which are highly modified in conjunction with their parasitism upon anthozoans and echinoderms. Wagin (1976) provided the most comprehensive review of the subclass, updating other useful summaries by Krüger (1940b) and Wagin (1947). The following description of a generalized ascothoracid (Fig.2a) is based largely on species of *Synagoga,* some of which are free-living ectoparasites; some aspects of the internal anatomy and ontogeny are based on less generalized, but more fully known ascothoracids.

The body is enclosed in a bivalved carapace, containing gut and gonad diverticula, and which is moved by a powerful postesophageal adductor muscle. The head bears a pair of large, six-segmented, raptorial antennules; and a pair of biramous, more or less plumose antennae. The labrum is produced into a conical sheath that partly surrounds the styliform mouthparts. The thorax has six segments, each with a pair of biramous, natatory appendages; each limb consists of a coxa, a basis, a triarticulate endopod (biarticulate in first and sixth pairs), and a biarticulate exopod, and bears numerous long setae. In females the coxae of the second through fifth limbs contain seminal receptacles. The five-segmented abdomen is very muscular. A biramous penis extends ventrally from its first segment; the fifth segment (telson) bears a pair of posteroventral spines and a pair of movable, blade-like furcal rami armed with natatory setae.

The internal anatomy is simple. Maxillary and possibly antennal glands are present. The gut is complete and the nervous system is highly consolidated in the thorax, without distinct ganglia. Eyes and a heart are absent. The oviduct is said to exit at the base of the first pair of thoracopods. At least some generalized species are protandric hermaphrodites (Newman 1974, Grygier 1981b).

Eggs and larvae are brooded under the carapace. There are typically two or more naupliar stages, during which posterior appendages are added and the dorsal shield becomes bivalved. These are followed by one or more bivalved larval stages with a full complement of appendages, the so-called ascothoracid larvae. Representative ontogenetic sequences are presented by Brattström (1948) and Wagin (1954).

## 5.2 *The urascothoracid*

To demonstrate how well ascothoracids serve as models for the urmaxillopodan, it is only necessary to reconstruct their non-parasitic ancestor, an organism presumably similar to

Wagin's (1976) nebulously characterized, hypothetical '*Protascothorax*', but better defined as an urascothoracid (Newman et al. 1969, Newman 1974) (Fig.2b).

The tagmosis of an urascothoracid as well as the structure of its carapace, thoracopods, and furca would have been unchanged from that seen now, but the organs altered by the adoption of parasitism would have been more generalized. For example, the antennules were smaller than in *Synagoga* and non-raptorial. The antennae were biramous and multi-articulate. The mouthparts were of the biting-chewing type rather than piercing-sucking, and were not encompassed by the labrum. The urascothoracid would have had compound and nauplius eyes, antennal and maxillary glands, a nervous system with segmental ganglia, and generalized sperm. The male genital appendages may not yet have fused to form a medial penis. There were no seminal receptacles in the thoracopods, and the position of the female gonopores cannot be guessed (cf. section 6).

A living organism organized in much this way is the so-called nauplio-cyprid larva of the rhizocephalan *Lernaeodiscus porcellanae* (Fig.2c). This abnormal larva discovered by the late L.Ritchie (in litt. 1973) resembles a typical cyprid larva posterior to the head, but it retains naupliar cephalic appendages and rudimentary compound eyes and its carapace is open along the entire ventral and anterior margins. It does not feed; yet, aside from its reduced abdomen, it shows what an urascothoracid might have been like.

The hypothetical urascothoracid designed above is almost plesiomorphic enough to serve as an urmaxillopodan. One needs only to add a heart and be less specific about the number of articles in the thoracopod rami to have an animal from which all the other maxillopodans can be derived, principally by reduction. It is uncertain whether the carapace was hinged. Newman's (this volume) urmalacostracan derivation of the maxillopodan line accepts a univalved state for the sake of simplicity. In extrapolating from extant maxillopodans, however, there is no way to choose between a bivalved or univalved carapace. The reductions and new features required to derive each maxillopodan group from an ur-maxillopodan ancestor are specified below.

## 5.3 *Ostracoda*

This discussion concerns only the extant ostracods; the problematical fossil bradoriids have been reviewed earlier. Ostracods are small, bivalved crustaceans living in nearly all aquatic environments (Fig.3a). Their typically asymmetrical valves usually have complex sculpturing, hinges, pores, and adductor muscle scars. The head is well-developed, with natatory antennules and antennae, strong mandibles, and either filtering or natatory maxillules and maxillae; some ostracods have compound eyes, and most have at least a nauplius eye. The rest of the body is reduced and unsegmented (except possibly *Saipanetta*) with at most two pairs of thoracopods (in some taxa as few as none), and in males a simple or exceedingly complex pair of penes (sometimes unpaired). The body ends with a large or small pair of unarticulated furcal rami. The nervous system is highly concentrated, and a heart is sometimes present. The body is supported by a unique endoskeletal lattice.

It would have required little, besides an extensive reduction of the thorax and abdomen, which must have taken place whatever their ancestor, to derive ostracods from the proposed urmaxillopodan. The only major new structures in the adults are the complex hinge joint and asymmetry of the valves, and the endoskeleton. The peculiar nauplius of some ostracods, which has three pairs of uniramous appendages and a bivalved carapace, is not qualitatively different from the nauplius of cirripeds like *Lepas,* whose enormous dorsal shield is strictly homologous to the bivalved shell of the cyprid larva (Newman, in preparation).

## 5.4 *Cirripedia (sensu stricto)*

Barnacles equipped with filtering cirri and shells of calcareous plates would seem at first to be far removed from the proposed urmaxillopodan. A popular scenario of the evolution of this subclass traces barnacles back to a hypothetical form resembling a permanently attached cyprid larva (Newman et al. 1969) (Fig.3b). The present analysis needs only to extend, therefore, to the cyprid larva and the most generalized lepadomorphs.

Cyprid larvae (Fig.3c) are bivalved. The valves tend to fuse ventrally, leaving anterior and posterior openings. Though the adductor muscle is anterior to the esophagus in most adult barnacles, it passes posteriorly, the plesiomorphic condition (Hessler 1964, Newman 1982), in acrothoracicans, *Ibla,* and cyprid larvae. Nauplius and compound eyes are present. The cement ducts transversing the four-segmented antennules exit on the third article (Nott & Foster 1969). Antennae and antennal glands are lacking in adults, but vestiges of the antennae may be retained for a time in the cyprid (e.g. Batham 1946). Cyprids do not feed, but adults have the usual three pairs of mouthparts. The six pairs of biramous, paddle-like thoracopods of the cyprid are modified into cirri in the adult (anterior ones often acting as maxillipeds). The abdomen has fewer than five segments in cyprids and is lost in adults; it ends with a pair of caudal appendages (furcal rami) frequently retained in the adult behind the penis, the latter lying between the last pair of cirri. The oviducts open on the first thoracomere.

An urcirriped (Newman et al. 1969) would have been very similar to the proposed ur-maxillopodan. The necessary alterations for its evolution are a reduction in the number of abdominal segments, the development of cement glands exiting through the antennules, and the loss of antennal glands. The thoracopods would have needed little or no modification to start functioning as filters.

## 5.5 *Y-larvae*

Certain unusual nauplii first described by Hansen (1899) as 'nauplius Y' have been found several times since then (summarized by Bresciani 1965, T.Schram 1970b, 1972). They have a complexly tessellated dorsal shield, a dorsocaudal organ of unknown function, and a blind gut (Elofsson 1971). Bresciani (1965) and T.Schram (1970a) described the phylogenetically interesting, apparently parasitic 'cyprid' stages of these larvae. The adults are unknown, or at least unrecognized.

Y-cyprids (Fig.3d) are minute (0.5 mm) and have compressed, univalved carapaces covering the head and thorax. Nauplius and compound eyes are present. The antennules have a hook on the second or fourth of six articles. Antennae and mouthparts are lacking, but there are spines on the large labrum. The six thoracomeres each bear a pair of biramous thoracopods constructed much like those in generalized ascothoracids and barnacle cyprid larvae. There are three abdominal segments including the telson in T.Schram's specimens, six in Bresciani's, the telson much longer than the other segments. Bresciani (1965) says the fourth and fifth segments are hard to distinguish (perhaps their boundary is illusory, and the abdomen of his Y-cyprid actually has the usual maxillopodan number of five segments). There are no abdominal appendages except for a furca. Nothing is known of the internal anatomy.

Y-cyprids can be derived easily from the proposed urmaxillopodan. They require ornamentation of the carapace and telson, development of a hook on a subterminal antennular

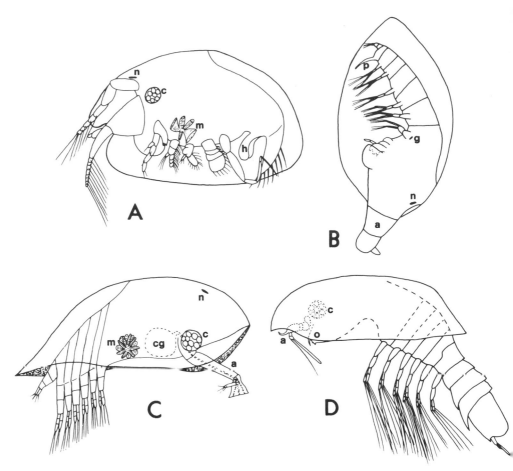

Figure 3. A. Diagrammatic representation of an ostracod combining features of several groups. Only the left appendages are shown, except for both hemipenes. B. Hypothetical ancestral barnacle (Cirripedia *s.s.*). Only the right appendages are shown except for the medial penis (after Newman et al. 1969). C. Generalized cirriped cyprid larva. Only the right appendages are shown, except for both furcal rami. D. Y-larva cyprid. Only the left appendages are shown, and the ornamentation of the carapace and abdomen is omitted. Abbreviations: cg – cement gland; otherwise cf. Figure 2.

article, and the loss of the oral and genital appendages. The unknown adult may, of course, regain these latter structures. Except for the loss of the antennae, Y-cyprids are otherwise almost unchanged.

## 5.6 *Branchiura*

The fish lice (cf. Martin 1932, Kaestner 1970) (Fig.4a) have small antennules and antennae, a piercing-sucking proboscis formed from the mandibles, and well-developed maxillules and maxillae. Many species have hooks on the antennules, a pre-oral sting, and maxillular sucking discs, but these characters are not plesiomorphic within Branchiura, being restricted to certain genera. There are nauplius and compound eyes, but no antennal glands. The flat dorsal shield, which is deeply cleft posteriorly, is usually considered a true carapace, but Lauterbach (1974) disagrees. There are four pairs of biramous natatory thoracopods, which in males may have modifications for clasping, and a bilobed unsegmented abdomen with a pair of minute furcal lobes. The newly hatched larva is similar to the adult in overall organization, but the maxillular sucking discs and posterior three pairs of appendages may not be fully developed. The gonopores of both sexes are behind the fourth pair of thoracopods.

Derivation of the branchiurans from the proposed urmaxillopodan requires the loss of two thoracomeres and their appendages, flattening and longitudinal cleavage of the carapace, loss of the antennal glands, a major reduction of the abdomen, and compression of larval development. The abdominal location of the testes and the protopodal clasping structures are unique to branchiurans.

## 5.7 *Copepoda*

Generalized copepods (Fig.4b) have a cephalosomic shield formed from the head and first thoracomere; it bears five pairs of cephalic appendages and a pair of uniramous maxillipeds. The remaining ten somites are divided into a metasome and urosome, the latter with five or six segments. The first five postcephalosomic segments bear biramous, natatory thoracopods joined across the midline to allow simultaneous beating. The sixth segment (sixth and seventh fused in females) is the genital segment, sometimes bearing reduced appendages. The telson bears a pair of uniarticulate furcal rami. Copepods lack a carapace and compound eyes, but the nauplius eye is often complexly developed. There is no antennal gland. Sperm transfer occurs by means of spermatophores, and eggs are either released directly or, more usually, carried in sacs by the female.

Copepods could have evolved from the proposed urmaxillopodan through a straightforward reduction of the body, perhaps passing through a stage like Kabata's (1979) 'archicopepod', in which there was no division of the trunk into a rigid metasome and urosome. The first thoracomere had to become incorporated into the head while its appendage became uniramous. Flexibility of the body (except at the metasome-urosome articulation) was lost, as were the antennal glands, compound eyes, and carapace. The egg sacs (necessarily plesiomorphic within the subclass?) and the medial sclerites connecting each pair of thoracopods were major new developments.

## 5.8 *Basipodella harpacticola and Deoterthron dentatum*

The unique body plan represented by these two minute (smallest crustaceans known), superficially copepod-like ectoparasites of crustaceans (harpacticoid and ostracod hosts, respectively), has only recently been recognized (Bradford & Hewitt 1980) (Fig.4c,d). The cephalic appendages include a tiny pair of grasping antennules, a piercing organ (mandibles?, labrum?), and an asymmetrically placed stylet. There are no eyes. A carapace is absent, but the cephalon is fused into a shield. The six pairs of more or less identical, biramous thoracopods are borne on free thoracomeres. The abdomen, including the telson, consists of five somites in *Basipodella* and two in *Deoterthron.* Both species have short furcal rami. and the internal anatomy of neither species is known in detail.

A derivation of these crustaceans from the proposed urmaxillopodan involves substantial changes, but could have occurred largely through reduction. The carapace and eyes were

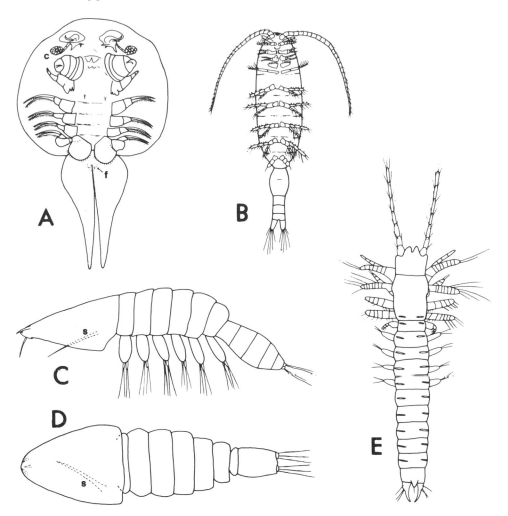

Figure 4. A. Ventral view of a generalized branchiuran, *Dolops longicauda* (after Brian 1947). B. Ventral view of a generalized, female, calanoid copepod. C. *Basipodella harpacticola*. Only left appendages are shown, except for both furcal rami (after Becker 1975). D. Dorsal view of *Deoterthron dentatum* (after Bradford & Hewitt 1980). E. Diagrammatic representation of a mystacocarid, seen in dorsal view (after Hessler 1969b). Abbreviations: c – compound eye; f – furca; s – stylet.

lost, as were the antennae and at least some mouthparts. However, the structure of the thorax, abdomen, and their appendages needed to have changed little, aside from the loss of the genital appendages, to produce a *Basipodella*-like form. Two major new features are the asymmetrically placed stylet which Bradford & Hewitt (1980) likened to a rhizocephalan kentron (cf. section 6), and the highly unusual reproductive biology of *B.harpacticola* outlined by Becker (1975). Maturation in this species involves the formation of an internal brood chamber by the expansion of the posterior end of the thorax, with subsequent loss of the thoracopods and abdomen.

### 5.9 *Mystacocarida*

These tiny interstitial dwellers of sandy beaches and muddy bottoms comprise two very similar genera, *Derocheilocaris* and *Ctenocheilocaris* (cf. Hessler & Sanders 1966, Renaud-Mornant 1976).

The body consists of 16 segments and lacks a carapace (Fig.4e). The cephalon is divided by a constriction and bears five pairs of rather generalized appendages. Next is an uncephalized thoracomere with a pair of small maxillipeds. The following four segments each bear a pair of uniarticulate vestigial thoracopods. The succeeding segments are limbless, but the last (telson) has a pair of articulated, pincer-like furcal rami. Mystacocarids are eyeless, but they do have organs of Bellonci (Renaud-Mornant et al. 1975). Most of the segments have a pair of dorsal toothed furrows. Metameric glands are scattered over the body (Pochon-Masson et al. 1975), and both antennal and maxillary glands are present (Cals et al. 1971). The nervous system is very generalized, with a pair of ganglia in each postantennal segment and the tritocerebrum not incorporated into the cephalic ganglion (Dahl 1952, Cals et al. 1971). The gonopores are on the fourth postcephalic segment (Dahl 1952, Hessler & Sanders 1966). Female mystacocarids have a highly unusual vitellarium (Dahl 1952), and males package sperm two to a spermatophore (Brown & Metz 1967).

Many alterations must be postulated to derive mystacocarids from the proposed urmaxillopodan. The cephalic appendages and nervous system, as well as the number of body segments, remained very generalized. The carapace, compound and nauplius eyes, last pair of thoracopods, genital appendages, and heart were suppressed, and the middle four pairs of thoracopods were significantly reduced. The major new features include the transverse division of the cephalon, the vitellarium, toothed furrows, and metameric glands, and the position of the gonopores.

### 5.10 *Rhizocephala and Pentastomida*

It is impractical to give detailed derivations of these highly advanced parasites directly from an urmaxillopodan form. Their precise relationship to cirripeds (Newman 1982) and branchiurans (Riley et al. 1978), respectively, will be discussed in the next section.

## 6 A CLADISTIC HYPOTHESIS

### 6.1 *Approach*

The first phylogenetic trees of Maxillopoda (Dahl 1956, Siewing 1960) seem to have been based on two considerations, overall similarity of generalized representatives of the component taxa and qualitative evaluations of their relative primitiveness. The contradictions in Dahl's and Siewing's placement of Mystacocarida and Branchiura along the lineage ancestral to Copepoda may have been due to differing judgements about whether the compound eyes and carapace of branchiurans are more or less primitive than the generalized appendages and nervous system of mystacocarids.

The fossil record does not help to confirm any hypothesis of maxillopodan phylogeny. Animals which may be ostracods (Müller 1979, 1981) and lepadomorph barnacles (Collins & Rudkin 1981) had already differentiated during the Cambrian, suggesting that the major

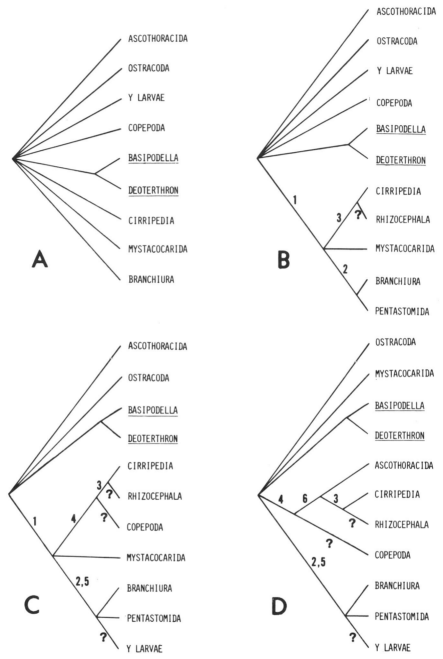

Figure 5. Cladograms of maxillopodan taxa. A. Completely unresolved cladogram (except for the *Basipodella-Deoterthron* clade) showing that all maxillopodan taxa share the synapomorphies of the class. B. Partial resolution of Figure 5a based on spermatological evidence and comparative development of kentrogonid rhizocephalans. C. Further resolution of Figures 5a and b through tentative placement of Copepoda and Hansen's Y-larvae. D. Alternative cladogram to Figures 5a and c showing the situation if most of the spermatological data are disregarded and synapomorphies are recognized between Ascothoracida and Cirripedia *s.s.*

maxillopodan radiation happened earlier than the fossil record can be traced. The other maxillopodan subclasses are not known as fossils or appear as highly specialized forms much too late to be of use in tracing phylogenies.

Two attempts have been made, involving comparative spermatology of four of the subclasses, to apply cladistic techniques to maxillopodan phylogeny (Grygier 1981a, 1982). To construct a cladogram for taxa higher than species, one must determine the most plesiomorphic state of each character as it is found within each taxon (Eldredge & Cracraft 1980). The resulting sets of character states (visualizable as 'paper animals') are then compared by conventional cladistic means, but there is a substantial uncertainty due to a lack of precision in specifying the plesiomorphic character states. This section tries to incorporate the rest of the maxillopodan taxa into a more comprehensive cladistic hypothesis.

## 6.2 *Execution*

As shown above in section 5, it is possible to derive the maxillopodan taxa from a hypothetical ancestor modelled on the Ascothoracida but incorporating plesiomorphic characters that that group no longer shows. The corresponding cladogram is a completely unresolved, multi-rayed figure (Fig.5a). The rest of this section tries to resolve some of the branch points.

Spermatological studies have provided some elegant synapomorphies. Grygier (1980, 1981a, 1982) showed that the flagellate sperm of ascothoracids closely resemble the plesiomorphic type of metazoan sperm (Franzén 1977), but differ in having a sheath of microtubules around the midpiece in some species, a more or less elongate nucleus, and more than a few simple mitochondria. The organelles are arranged sequentially: acrosome, nucleus, midpiece, flagellum. The flagellate sperm of branchiurans, cirripeds and rhizocephalans, and mystacocarids differ among themselves in the exact relationship of the basal body to the other organelles, whether or not the flagellum is free posteriorly, and in details of the mitochondria. All, however, share a trait rarely found in the animal kingdom, a basal body so far anterior that the axoneme parallels the entire length of the nucleus. Grygier (1981a, 1982) interpreted this character as a synapomorphy and, accepting *a priori* a close relationship of Ascothoracida to Cirripedia (substantiated by the diagnosibility of Maxillopoda), proposed that Ascothoracida is the sister group of the other three taxa. Unfortunately, the first cladogram he presented (Grygier 1981a) included an arbitrary branching sequence among the three taxa with advanced basal bodies. A corrected cladogram (Grygier 1982, Fig.5b, leaving out other groups) has Ascothoracida on one side of the base, and Cirripedia, Branchiura, and Mystacocarida (with a common base) on the other.

Wingstrand (1972) compared sperm of a branchiuran and a pentastomid and found them nearly identical to one another. These highly modified sperm lack an acrosome when mature, replacing it with a pseudacrosome. Other uniquely shared characters include the

---

Explanation: Question marks indicate ambiguous (Rhizocephala) or weakly justifiable placements (others). Numerals correspond to putative synapomorphies: 1 — sperm with anterior basal bodies; 2 — a suite of sperm modifications including replacement of the acrosome with a pseudacrosome and an anchored microtubule doublet in the axoneme; 3 — frontolateral horns in the nauplii and adhesive antennules in the cyprid larvae; 4 — six naupliar stages; 5 — a univalved, posteriorly indented carapace; 6 — anteroventral carapace pores and female gonopores on the first thoracomere.

dorsal and ventral ribbons, a membrane connected to one of the microtubule doublets, and the mitochondrial arrangement. Such a coincidence of abberant structures can hardly be anything but a synapomorphic condition demonstrating that Branchiura and Pentastomida are sister groups. Wingstrand (1972) and Riley et al. (1978) accepted this conclusion, and it is accepted here as well (Fig.5b). Each author described a scenario by which aquatic crustaceans could have given rise to pentastomids; the details do not bear on the present discussion.

Rhizocephalans, which as adults consist of rootlets penetrating their crustacean hosts and an external brood sac, are phylogenetically ambiguous. Because their nauplii and cyprid larvae are entirely comparable to those of barnacles, the Rhizocephala are classified as an order of Cirripedia. The classical interpretation of their origin from parasitic lepadomorphs like *Anelasma* and *Rhizolepas* (Day 1939, Newman et al. 1969) is supported by Delage's (1884) observations on host penetration in *Sacculina.* The newly settled cyprid metamorphoses into a kentrogon larva with a hypodermic device (kentron) that pierces into the host, first passing through a cyprid antennule. The peduncular rootlets in parasitic barnacles correspond broadly to the antennular region, and suggest how the rhizocephalan mode of life originated.

Ritchie & Høeg (1981) showed in *Lernaeodiscus porcellanae* that the kentron forms ventrally, in what would be the cyprid's mouth field if it were a feeding larva, and pierces the host at this position, not through an antennule. Veillet (1964) noted similar invasive behavior in other rhizocephalans. If, as seems likely, this is a more plesiomorphic mode of host penetration than in *Sacculina,* rhizocephalans may not have arisen from lepadomorph barnacles at all; they may have arisen directly from cyprid-like ancestors. Since all known barnacle cyprids are non-feeding, it is unlikely that rhizocephalans evolved progenetically from barnacles, but they may have descended from a prethoracican stock (Newman 1982). The synapomorphies of Rhizocephala and Cirripedia *s.s.* (frontolateral horns in nauplii, adhesive antennules in cyprids) make it certain that the two taxa are no farther apart than sister groups if the last hypothesis is correct (Fig.5b).

Barnacles and copepods have, plesiomorphically, six naupliar stages. Rhizocephalans have only up to four (Krüger 1940a), but since their nauplii do not feed, development may have been compressed. The mystacocarid *Derocheilocaris typicus* also has six naupliar stages, as defined by the presence of naupliar processes of the antennae (Hessler & Sanders 1966), but *D.remanei* has ten (Delamare Deboutteville 1954), so the fundamental number in mystacocarids cannot be specified. Ascothoracids have about seven naupliar stages (Wagin 1976) but it is unclear whether they are all true instars or if all occur in any one species. Other maxillopodans either have no true nauplii or their total number cannot yet be estimated. The coincidence of six naupliar stages may imply a sister group relationship between Copepoda and the cirriped clade, provided the number in Rhizocephala is reduced (Fig.5c). This is a tenuous hypothesis because it has no morphological backing; there are no structural synapomorphies unique to this clade.

Bresciani (1965) listed numerous similarities between his Y-cyprid and ascothoracids, but T.Schram (1970a), in a more detailed comparison, noted that the different antennular armament, the lack of typical ascothoracid mouthparts, and the presence of compound eyes argued against their inclusion in Ascothoracida (N.B., the last character, being plesiomorphic, cannot be used to make this distinction). Hansen (1899) and Steuer (1903) assigned the Y-nauplii to the cirriped order Apoda, but since the single apodan species, *Proteolepas bivincta,* is really a cryptoniscid isopod (Bocquet-Védrine 1972), this interpretation can be dismissed. Y-larvae might be better placed in the branchiuran clade (Newman et al. 1969). Y-cyprids share with branchiurans a posteriorly-indented, univalved carapace, but without knowledge of sperm morphology in Y-larvae, it is not possible to resolve the trichotomy between them, branchiurans, and pentastomids (Fig.5c). If a univalved carapace is plesiomorphic for Maxillopoda (Newman, this volume), the position of Y-larvae on a cladogram of the class is not resolvable.

What of *Basipodella* and *Deoterthron*? Becker (1975) considered the asymmetrically placed stylet a maxilliped, and interpreted *B. harpacticola* as a copepod with an unreduced seventh pair of trunk limbs (homologous to genital appendages in other copepods). This interpretation would allow the placement of the two genera as a sister group of Copepoda, but its corollary is that there are 12 trunk segments, too many for a copepod (or a maxillopodan). Bradford & Hewitt (1980) considered the stylet a possible homologue of the kentron in kentrogonid rhizocephalan larvae. Since its specific function is unknown, interpretations of it as anything but a mouthpart whose partner is lost are unparsimonious. In view of the lack of satisfactory synapomorphies with other maxillopodans, the position of *Basipodella* and *Deoterthron* on the cladogram of Maxillopoda must remain unresolved (Fig.5c).

Ostracods share no unambiguous, derived features with other maxillopodans, so their precise position in a cladogram can also not be resolved (Fig.5c).

The cladistic hypothesis now presented (Fig.5c) is not well resolved; some of the putative synapomorphies employed in its construction are questionable, and a cautious soul might be content with one of the earlier cladograms. Traditional conceptions of the degree of affinity of Branchiura and Mystacocarida with Copepoda and of Ascothoracida with Cirripedia have been almost totally lost. Let us examine the latter relationship in more detail.

The two characters most likely to be synapomorphies of Ascothoracida and Cirripedia are anteroventral carapace pores and the position of the female gonopores. Cirriped cyprid larvae have a pair of 'frontal glands' exiting on the anteroventral part of each carapace valve (Walley 1969). The *metacrinicola*-group of the ascothoracid genus *Synagoga* have a large pore on each valve at the anterior end of the flattened ventral side (Newman 1974). Neither sort of pore has been investigated histologically or ultrastructurally, and their functions are unknown, so their homology is not yet established. In both taxa the female gonopores are said to be on the first thoracomere, an unusually advanced position. However, the condition in ascothoracids is not firmly established. In some extremely specialized endoparasites, the eggs seem to be shed directly into the mantle cavity, not passing through an oviduct (Okada 1941). In others, organs once considered the oviducts or their orifices are now suspected to be parts of the maxillary gland (Wagin 1976) or the antennal gland (Pyefinch 1937, Grygier unpublished data). The anterior gonopore is rather firmly established in two species, *Laura gerardiae* (Lacaze-Duthiers 1883) and *Ascothorax ophioctenis* (Wagin 1947), but it is premature to claim the gonopore problem is solved in ascothoracids. It will be remembered that no plesiomorphic position was assignable to the female gonopores in maxillopodans; even if cirripeds and ascothoracids do have it in common on the first thoracomere, this might be the plesiomorphic state.

If one accepts the carapace pores and gonopore sites as synapomorphies of Ascothoracida and Cirripedia, then a different cladogram must be hypothesized (Fig.5d). All previous conclusions hold except that the spermatological data linking the cirriped, mystacocarid, and branchiuran clades as a sister group of Ascothoracida must be disregarded. The resulting cladogram is much less resolved than the previous one (Fig.5c), and therefore may be less preferable.

A variety of poorly resolved cladograms can be produced by interpreting other characters as synapomorphies: e.g. bivalved carapace in Cirripedia, Ascothoracida, and Ostracoda (assuming a univalved carapace like that of Y-cyprids is plesiomorphic); cement production in Cirripedia, Ascothoracida (Grygier, unpublished data), Ostracoda (Calman 1909), and Copepoda; spermatophores in Copepoda, Mystacocarida (Brown & Metz 1967), and Branchiura (Fryer 1958); seminal receptacles connected internally with the oviduct in Copepoda and Ostracoda (Calman 1909). Many characters are clearly convergent among the maxillopodans. Even the preferred model cladogram (Fig.5c) supposes at least four separate losses of the carapace.

Improved resolution and confidence in these cladistic hypotheses can be obtained as a result of increased knowledge of larval development, internal anatomy, and ultrastructure of some taxa, but the number of apomorphies available for analysis is unlikely to be augmented much. Since the most precise cladogram generated here (Fig.5c) is still largely unresolved, it would be unwise to assign names to each clade.

## 7 COUNTERARGUMENTS

There are a few major problems with the version of crustacean evolution proposed in this paper. Some of the challenges can be and are answered here, but some cannot be refuted with current knowledge.

The harpacticoid copepod *Limnocletodes behningi,* of the family Cletodidae, poses a challenge of the first type. Its females have what appear to be small, supernumerary limbs with setae on the posterior part of the genital segment and the succeeding segment (eighth and ninth postcephalic somites) (Fig.6a). Smirnov (1933) concluded from this unique limb configuration that the urcopepod had limbs on each segment, a condition at variance with the present hypothesis. An examination of other cletodids makes Smirnov's hypothesis seem unnecessary, for many of these species have a ring of papillae bearing apical spines on several segments, including the two in question. In a species of *Enhydrosoma* illustrated by Raibaut (1965), the ventrolateral papillae on the anterior urosome segments are larger than the others (Fig.6b). The so-called extra limbs of *L.behningi* may not be limbs at all, but enlarged, probably fused papillae homologous to those common throughout the family.

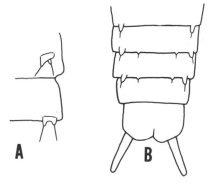

Figure 6. A possible explanation of supposed supernumerary limbs in copepods. A. Ventral view of the genital segment and the succeeding segment in a female *Limnocletodes behningi,* showing the limb-like outgrowths (after Smirnov 1933). B. Ventral view of the urosome of *Enhydrosoma caeni,* another cletodid, showing the segmental rings of papillae (including enlarged ones), some of which may have fused to give the extra pair of 'limbs' in *L.behningi* (after Raibaut 1965).

If so, the urcopepod need have had no more than the seven pairs of limbs postulated for the urmaxillopodan.

Mystacocarida is probably the weakest component of Maxillopoda as now comprised. The features that give mystacocarids a distinctly primitive facies are essentially naupliar (Delamare Deboutteville & Chappuis 1954, Hessler 1982). They may be neotonous descendants of some other type of crustacean, which did not have to be a maxillopodan because these naupliar characters are not unique to that class (cf. diagnosis section). Mystacocarids differ from other maxillopodans in the relative positions of the male gonopores and the thoracopods. In other taxa, the gonopores are posterior to the thoracopods, sometimes associated with a pair of genital appendages or a penis. Mystacocarids have no special genitalia, and the gonoduct is associated with the fourth thoracomere, the penultimate limb-bearing segment. Why have the orifices moved forward only in this group, when their position has elsewhere been conservative? Finally, mystacocarids are the most highly apomorphic taxon in Maxillopoda in terms of new structures not possessed by the hypothetical urmaxillopodan. There may be too many such structures for this hypothesis of descent to be parsimonious. An argument to remove Mystacocarida from Maxillopoda based on these considerations could plausibly be advanced.

The location of the female gonopores on the first thoracomere in cirripeds and at least some ascothoracids is disquieting because other maxillopodans have them in the same position as the male gonopores. In other crustacean classes these gonopores are static (Malacostraca, Cephalocarida) or nearly so (Branchiopoda). The instability of these orifices' positions among maxillopodans may reflect the artificiality of the group.

If any taxon accepted here as a component of Maxillopoda were shown by other criteria not to belong to it, at least some of the diagnostic characters of the class would automatically become convergent, weakening an already precarious construct. A collapse of the logical framework underlying Maxillopoda would necessitate a replacement of the present cladistic model by a phylogenetic grass (Fig. 7) composed of those clades with strong synapomorphies, such as Branchiura-Pentastomida and Cirripedia-Rhizocephala. Ascothoracida could go back to being a sister group of Cirripedia and Rhizocephala (Wagin 1947, 1976) because of the two possible synapomorphies discussed in the previous section. Hessler's (1982) complaint that the boundaries of Maxillopoda cannot be drawn with confidence would be justified, and we would be forced to return to classifying the maxillopodans in numerous independent subclasses.

## 8 CONCLUSIONS

Two nearly independent arguments have been presented for accepting Maxillopoda as a natural taxon. First, a list of diagnostic features has been drawn up. There seem to be enough defensible synapomorphies to make coincident convergence a less probable explanation than monophyly. It is not essential for all the diagnostic traits to have appeared at once. The origin of the maxillopodan clade may be defined as the point by which all had evolved. Alternatively, if a component taxon actually lacked a particular synapomorphic feature, that trait (but not the others) would cease to be diagnostic of Maxillopoda, but would remain a synapomorphy of the rest of the class, which would form the sister group of the plesiomorphic taxon. The logical derivation of this diagnosis depends on the acceptance of certain axioms about the plesiomorphic condition, within Crustacea, of the

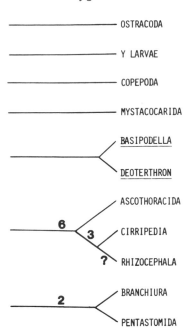

OSTRACODA

Y LARVAE

COPEPODA

MYSTACOCARIDA

BASIPODELLA

DEOTERTHRON

ASCOTHORACIDA

**6** **3** CIRRIPEDIA

**?** RHIZOCEPHALA

**2** BRANCHIURA

PENTASTOMIDA

Figure 7. A phylogenetic diagram showing the taxa that would remain bound by synapomorphies if evidence for convergence of the diagnostic characters precluded recognition of the class Maxillopoda; cf. Figure 5 for explanations of symbols.

characters examined. In this paper, the characteristics of the model urcrustacean of Hessler & Newman (1975) have been given the most credence, and Newman's paper elsewhere in this volume has supplied additional postulates about the nature of the urmaxillopodan's immediate ancestors. If a different model of the major features of crustacean evolution is favored, the maxillopodan diagnostic characters may appear plesiomorphic or convergent, instead of synapomorphic. It is left to advocates of other models to demonstrate whether or not diagnoses of Maxillopoda based on synapomorphies follow from them.

The second argument relies on the traditional practice of constructing hypothetical ancestors. This technique has fallen into disfavor, but its use here is partially justified by the existence of extant animals very similar to the required urmaxillopodan, the ascothoracids. These parasites certainly had a free-living ancestor, the characteristics of which described herein are unlikely to be controversial. The other maxillopodans can be derived from the urmaxillopodan largely through reduction.

The order of events leading to the variety of recent maxillopodan taxa is very unclear. Convergences of structure among the major taxa are numerous, and attempts to segregate rigorously the convergent from the synapomorphic characters are largely futile (F.Schram, this volume). Normally, the parsimony and utility of a cladogram depend on how well it minimizes the necessity of convergence (Eldredge & Cracraft 1980). This criterion is not applicable in the present case. Instead, the favored cladogram (Fig.5c) is the most highly resolved among those possible. This philosophy must not be taken as a preference for a false model that explains a lot over a true model that explains but little. Rather, the more comprehensive model is the one toward which future falsifying efforts should be directed.

So, is there a taxon Maxillopoda? The answer is yes, even if it only includes Ascothoracida and Cirripedia (Fig.7), which should be regarded (possibly along with Rhizocephala) as distinct subclasses within Maxillopoda (Wagin 1947, 1976, Newman 1982). The real controversy is whether Maxillopoda also includes the other taxa that have been discussed. For example, the original core of the class, a presumptive copepodan-branchiuran-mystaco-caridan clade (Dahl 1956, Siewing 1960), does not appear in any cladistic hypothesis examined in the previous section. Despite the facies resemblance among these subclasses, their phylogenetic relationships must be analysed individually. A few taxa, notably Ostra-coda and the two genera of crustacean ectoparasites, do not seem to share synapomorphies with other maxillopodans beyond those diagnostic of the class. This may imply that some taxa are being forced under the umbrella of a polyphyletic class, but it may also reflect an extensive and early radiation of maxillopodan forms from which only a few very divergent lineages survive.

Systematists who wish to recognize the class Maxillopoda and employ it formally now have grounds for doing so, although these grounds are not all the same as those invoked by the original proposers. Those who do not must falsify the favorable arguments presented here, and possible points of attack have been indicated in the previous section.

## 9 ACKNOWLEDGEMENTS

I thank Dr F.R.Schram (San Diego Natural History Museum), Dr R.R.Hessler (Scripps Institution of Oceanography), and especially Dr W.A.Newman (SIO) for inspiring me to try to resolve the maxillopodan question, for numerous discussions which helped to clarify the arguments, and for criticizing an earlier draft of the manuscript. Other carcinologists whom I thank for contributing their expertise as various aspects of the model were being formulated include Drs J.Kunze and A.Fleminger and Mr G.Wilson (all of SIO) and Dr P.Illg (University of Washington). All opinions expressed here are my own, as is any responsibility for factual errors. This is a Contribution of Scripps Institution of Oceanography, new series. Part of the work was done during the tenure of an NSF Graduate Fellowship, and partial support came from NSF Grant DEB 78-15052.

## DISCUSSION

BOXSHALL: It seems to me your spermatological data is the central thread which holds the Maxillopoda together. Yet in order to accept it you have to disregard the identical location of female gonopores in cirripedes and ascothoracids. The gonopores are a double derived character because (a) they are not on the same somite, and (b) because the female opening moves forward to the same first trunk somite. You have to reject that, a double synapomorphy, in order to accept your spermatological data.
GRYGIER: [The spermatological data give the skeleton of the cladogram, but the synapomorphies in my diagnosis are what hold the class together.] I preferred sperm because it gave me a comprehensive view of what might have gone on in the evolution of the group. However, I'm willing to entertain the other possibility. I have yet to see in the ascothoracids that I have examined more than a couple of species where I can say the gonopore probably does exit on the first thoracomere. In some genera the position where it was inferred to be is wrong. I believe that people, heretofore, have been looking at an excretory gland exit rather than an oviduct. In fact I have not been able to find an oviduct in some ascothoracids. In this respect I would agree with Schram, that when trying to assign apo-

morphic and plesiomorphic labels to characters in maxillopodans I cannot come up with a cladistic hypothesis involving a statistical analysis in terms of parsimony.

BRIGGS: I am not really clear why you are so anxious to throw the phosphatocopines out of the ostracodes. Although Jones and McKenzie demonstrated bradoriids are polyphyletic, they still prefer to maintain the phosphatocopines as ostracodes. If you are anxious to have an abdomen in your maxillopodans, then I must point out that there probably was an abdomen in phosphatocopines. Müller was just not able to demonstrate it, but there is a hole at the end of the animal. You might have had your magic four segments.

GRYGIER: If phosphatocopines are ostracodes, that would have the effect of making some of my diagnostic characters for Maxillopoda convergent by requiring the removal of the ostracods. I would then be unwilling to present any such diagnosis as I have here. I do feel I have some evidence to justify using Maxillopoda as a natural taxon as I have defined them.

ABELE: What do you mean by *statistically* falsifying a cladogram?

GRYGIER: The first thing I attempted to do was to look at all sorts of characters, not just the ones I employed here. Such an array did not fall out into patterns that would make a satisfactory cladogram. Too many of the characters would have to be convergent no matter how they were looked at. So instead, I started with a skeleton of certain groups and characters that I felt had a better chance of being synapomorphies and then added groups onto that skeleton.

WICKSTEN: I'm not at all sure why you are linking the Branchiura with the Pentastomida. They are both highly modified parasites. It seems they are more likely the result of some convergent evolution for parasitism.

GRYGIER: I'm not linking them on the basis of parasitism. I'm linking them only on the basis of their nearly identical, highly aberrant sperm.

WICKSTEN: That is the only link?

GRYGIER: That's right. Different hypotheses have been offered on how the transition occurred from one group to the other by Wingstrand, and Riley et al. I'm not convinced by either of them, but I'm not in the business here of justifying things in terms of adaptational possibilities and evolutionary stories.

[Andersson & Nilsson (1981, *Protoplasma* 107:361-374) is one of a new series of ultrastructural studies of compound and nauplius eyes in maxillopodans, which will affect the interpretation of lens cells and tapetal cells as synapomorphies. In another of these works, Hallberg (1982, *Zool. Anz.* 208:227-236) notes peculiarities common to the compound eyes of branchiurans and ostracods. I have recently seen a planktonic, ascothoracid larva with compound eyes similar to those of cyprid larvae, but with many more ommatidia (Grygier, unpublished manuscript). Boxshall, Ferrari & Tiemann (in press, *Crustaceana Suppl.*) attempt to synthesize an urcopepod by amalgamating plesiomorphic features; they do not, however, address the origin of the Copepoda. Boxshall & Lincoln (unpublished manuscript) are preparing a review of *Basipodella* and *Deoterthron* with new species and a reinterpretation of their anatomy that, if correct, makes their maxillopodan affinities a little more problematical.]

## REFERENCES

Andersson, A. 1977. The organ of Bellonci in ostracodes: An ultrastructural study of the rod-shaped, or frontal, organ. *Acta Zool. Stockh.* 58:197-204.

Armstrong, J.C. 1949. The systematic position of the crustacean genus *Derocheilocaris* and the status of the subclass Mystacocarida. *Am. Mus. Novit.* 1413:1-6.

Ax. P. 1960. *Die Entdeckung neuer Organisationstypen im Tierreich.* Wittenburg: A.Ziemsen Verlag.

Batham, E.J. 1946. *Pollicipes spinosus* Quoy and Gaimard. II. Embryonic and larval development. *Trans. Roy. Soc. N.Z.* 75:405-418.

Becker, K.H. 1975. *Basipodella harpacticola* n.gen., n.sp. (Crustacea, Copepoda). *Helgoländ. Wiss. Meeresunters.* 27:96-100.

Beklemishev, W.N. 1952. *Principles of Comparative Anatomy of Invertebrates, 2nd ed.* [in Russian]. Moscow: Nauka.

Beklemishev, W.N. 1969. *Principles of Comparative Anatomy of Invertebrates, 3rd ed., Vol.1, Promorphology.* Edinburgh: Oliver & Boyd.

Bergström, J. 1979. Morphology of fossil arthropods as a guide to phylogenetic relationships. In: A.P. Gupta (ed.), *Arthropod Phylogeny:*3-56. New York: Van Nostrand Reinhold.

Birshtein, Ya.A. 1960. Class Crustacea [in Russian]. In: Yu.A.Orlov (ed.), *Fundamentals of Paleontology Vol.8, Arthropoda – Trilobitoidea and Crustacea* (N.E.Chernysheva, ed.):201-216. Moscow: State Scientific-Technical Press.

Bocquet-Védrine, J. 1972. Supression de l'ordre des apodes (crustacés cirripèdes) et r'attachement de son unique représentant, *Proteolepas bivincta,* à la famille des Crinoniscidae (crustacé, isopode, cryptoniscien). *C.R. Acad. Sci. Paris* 275D:2145-2148.

Bowman, T.E. 1971. The case of the nonubiquitous telson and the fraudulent furca. *Crustaceana* 21: 165-175.

Bradford, J.M. & G.C.Hewitt 1980. A new maxillopodan crustacean, parasitic on a myodocopid ostracod. *Crustaceana* 38:67-72.

Brattström, H. 1948. Undersökningar över Öresund 33. On the larval development of the ascothoracid *Ulophysema öresundense* Brattström. Studies on *Ulophysema öresundense* 2. *Kungl. Fysiogr. Sällsk. Handl. N.F.* 59(5):1-70.

Bresciani, J. 1965. Nauplius 'Y' Hansen. Its distribution and relationship with a new cypris larva. *Vidensk. Medd. Dan. Naturh. Foren.* 128:245-258.

Brian, A. 1947. Los argúlidos del Museo Argentino de Ciencias Naturales (Crustacea Branchiura). *Ann. Mus. Arg. Cienc. Nat.* 47:353-370.

Brown, G.G. 1970. Some comparative aspects of selected crustacean spermatozoa and crustacean phylogeny. In: B.Baccetti (ed.), *Comparative Spermatology:*183-204. New York: Academic Press.

Brown, G.G. & C.B.Metz 1967. Ultrastructural studies on the spermatozoa of two primitive crustaceans *Hutchinsoniella macracantha* and *Derocheilocaris typicus. Z.Zellforsch.* 80:78-92.

Calman, W.T. 1909. Crustacea. In: R.Lankester (ed.), *A Treatise on Zoology, part VII, third fascicle.* London: Adam & Charles Black.

Cals, P., C.Delamare Deboutteville & J.Renaud-Mornant 1971. Caractères anatomiques des mystacocarides (Crustacea). Etude de *Derocheilocaris remanei* Delamare Deboutteville et Chappuis 1954. *C.R. Acad. Sci. Paris* 272D:3154-3157.

Chaigneau, J. 1978. L'organe de Bellonci des crustacés. Historique et état actuel des connaissances. *Arch. Zool. Exp. Gén.* 119:185-199.

Collins, D. & D.M.Rudkin 1981. *Priscansermarinus barnetti,* a probable lepadopmorph barnacle from the Middle Cambrian Burgess Shale of British Columbia. *J. Paleont.* 55:1006-1015.

Dahl. E. 1952. Reports of the Lund University Chile Expedition 1948-1949. 7. Mystacocarida. *Lunds Univ. Arsskr. N.F.* 48(6):1-40.

Dahl, E. 1956. Some crustacean relationships. In: K.G.Wingstrand (ed.), *Bertil Hanström, Zoological Papers in Honour of his 65th Birthday:*138-147. Lund: Zool. Inst.

Dahl, E. 1963. Main evolutionary lines among recent Crustacea. In: H.B.Whittington & W.D.I.Rolfe (eds), *Phylogeny and Evolution of Crustacea:*1-15. Cambridge: Mus. Comp. Zool.

Day, J.H. 1939. A new cirripede parasite – *Rhizolepas annelidicola,* nov.gen. et sp. *Proc. Linn. Soc. Lond.* 151:64-78.

Delage, Y. 1884. Évolution de la sacculine (*Sacculina carcini* Thomps.) crustacé endoparasite de l'ordre nouveau des kentrogonides. *Arch. Zool. Exp. Gén.* (2)2:417-736.

Delamare Deboutteville, C. 1954. Recherches sur les crustacés souterrains. III. Développement postembryonnaire des mystacocarides. *Arch. Zool. Exp. Gén.* 91:25-34.

Delamare Deboutteville, C. & P.A.Chappuis 1954. Recherches sur les crustacés souterrains. II. Morphologie des mystacocarides. *Arch. Zool. Exp. Gén.* 91:7-24.

Eldredge, N. & J.Cracraft 1980. *Phylogenetic Patterns and the Evolutionary Process.* New York: Columbia Univ. Press.

Elofsson, R. 1963. The nauplius eye and frontal organs in Decapoda (Crustacea). *Sarsia* 12:1-68.

Elofsson, R. 1965. The nauplius eye and frontal organs in Malacostraca (Crustacea). *Sarsia* 19:1-54.

Elofsson, R. 1966. The nauplius eye and frontal organs of the non-Malacostraca (Crustacea). *Sarsia* 25: 1-128.

Eloffson, R. 1971. Some observations on the internal morphology of Hansen's nauplius Y (Crustacea). *Sarsia* 46:23-40.

Fowler, G.H. 1890. A remarkable crustacean parasite, and its bearing on the phylogeny of the Entomostraca. *Q.J. Microsc. Sci.* 30:107-120.

Franzén, Å. 1970. Phylogenetic aspects of the morphology of spermatozoa and spermiogenesis. In: B.Baccetti (ed.), *Comparative Spermatology:* 29-46. New York: Academic Press.

Franzén, Å. 1977. Sperm structure with regard to fertilization biology and phylogenetics. *Verh. Dtsch. Zool. Ges.* 1977:123-138.

Fryer, G. 1958. Occurrence of spermatophores in the genus *Dolops* (Crustacea: Branchiura). *Nature* 181:1011-1012.

Gruvel, A. 1905. *Monographie des Cirrhipèdes où Thecostracés.* Paris, Masson et Cie.

Grygier, M.J. 1980. Comparative spermatology of Ascothoracica (Crustacea: Maxillopoda) and its phylogenetic implications. *Am. Zool.* 20(4):815.

Grygier, M.J. 1981a. Sperm of the ascothoracican parasite *Dendrogaster,* the most primitive found in Crustacea. *Int. J. Invert. Reprod.* 3:65-73.

Grygier, M.J. 1981b. *Gorgonolaureus muzikae* sp. nov. (Crustacea: Ascothoracida) parasitic on a Hawaiian gorgonian, with special reference to its protandric hermaphroditism. *J. Nat. Hist.* 15:1019-1045.

Grygier, M.J. 1982. Sperm morphology in Ascothoracida (Crustacea: Maxillopoda); confirmation of generalized nature and phylogenetic importance. *Int. J. Invert. Reprod.* 4:323-332.

Gurney, R. 1942. *Larvae of Decapod Crustacea.* London: Ray Soc.

Handlirsch, A. 1914. Ein interresante Crustaceenform aus der Trias der Vogesen. *Verh. Zool.-Botan. Ges. Wien.* 64:1-8.

Hansen, H.J. 1899. Die Cladoceren und Cirripedien der Plankton-Expedition. *Ergeb. Plankton-Exped. Humboldt-Stiftung.* 2(G.,d.):1-58.

Hennig, W. 1966. *Phylogenetic Systematics.* Urbana: Univ. Illinois Press.

Hessler, R.R. 1964. The Cephalocarida: comparative skeleto-musculature. *Mem. Conn. Acad. Arts Sci.* 16:1-97.

Hessler, R.R. 1969a. Cephalocarida. In: R.C.Moore (ed.), *Treatise on Invertebrate Paleontology, Part R, Arthropoda 4(1):* 120-128. Lawrence: Geol. Soc. Am. & Univ. Kansas Press.

Hessler, R.R. 1969b. A new species of Mystacocarida from Maine. *Vie Milieu Sér. A Biol. Mar.* 20:105-116.

Hessler, R.R. 1982. Evolution within the Crustacea. In: L.Abele (ed.), *Biology of Crustacea, Vol.1:* 149-185. New York: Academic Press.

Hessler, R.R. & W.A.Newman 1975. A trilobitomorph origin for the Crustacea. *Fossils Strata* 4:437-459.

Hessler, R.R. & H.L.Sanders 1966. *Derocheilocaris typicus* Pennak & Zinn (Mystacocarida) revisited. *Crustaceana* 11:141-155.

Jones, P.J. & K.G.McKenzie 1980. Queensland Middle Cambrian Bradoriida (Crustacea): new taxa, palaeobiogeography and biological affinities. *Alcheringa* 3:203-225.

Kabata, Z. 1979. *Parasitic Copepoda of British Fishes.* London: Ray Soc.

Kaestner, A. 1967. *Lehrbuch der Speziellen Zoologie.* Stuttgart: Gustav Fischer Verlag.

Kaestner, A. 1970. *Invertebrate Zoology.* New York: Interscience Publishers.

Kauri, T. 1966. On the sensory papilla X organ in cirriped larvae. *Crustaceana* 11:115-122.

Krüger, P. 1940a. Cirripedia. In: *Bronns Klassen und Ordnungen des Tierreichs, Vol.5, Abt.1, Buch 3, Teil 3:* 1-560.

Krüger, P. 1940b. Ascothoracida. In: *Bronns Klassen und Ordnungen des Teirreichs, Vol.5, Abt.1, Buch 3, Teil 4:* 1-46.

Lacaze-Duthiers, H., de 1883. Histoire de la *Laura gerardiae* type nouveau de crustacé parasite. *Mém. Acad. Sci. Inst. France* 42(2):1-160.

Landing, E. 1980. Late Cambrian-early Ordovician macrofauna and phosphatic microfaunas, St.John Group, New Brunswick. *J. Paleont.* 54:752-761.

Lauterbach, K.E. 1974. Über die Herkunft des Carapax der Crustaceen. *Zool. Beitr.* 20:273-327.

Martin, M.F. 1932. On the morphology and classification of *Argulus* (Crustacea). *Proc. Zool. Soc. Lond.* 1932:771-806.

Maynard, D.M. 1960. Circulation and heart function. In: T.H.Waterman (ed.), *The Physiology of Crustacea, Vol.1:*161-226. New York: Academic Press.

McKenzie, K.G. 1967. Saipanellidae: a new family of podocopid Ostracoda. *Crustaceana* 13:103-113.

Müller, K.J. 1979. Phosphatocopine ostracodes with preserved appendages from the Upper Cambrian of Sweden. *Lethaia* 46:23-40.

Müller, K.J. 1981. Arthropods with phosphatized soft parts from the Upper Cambrian 'orsten' of Sweden. In: M.E.Taylor (ed), *Short Papers for the Second International Symposium on the Cambrian System.* US Geol. Surv. Open File Rept. 81-743:147-151.

Munn, E.A. & H.Barnes 1970. The fine structure of the spermatozoa of some cirripedes. *J. Exp. Mar. Biol. Ecol.* 4:261-286.

Newman, W.A. 1974. Two new deep-sea Cirripedia (Ascothoracica and Acrothoracica) from the Atlantic. *J. Mar. Biol. Ass. UK* 54:437-456.

Newman, W.A. 1982. Cirripedia. In: L.Abele (ed.), *Biology of Crustacea, Vol.1:*197-220. New York: Academic Press.

Newman, W.A., V.A.Zullo & T.H.Withers 1969. Cirripedia. In: R.C.Moore (ed.), *Treatise on Invertebrate Paleontology, Part R, Arthropoda 4, Book 1:*206-295. Lawrence: Geol. Soc. Am. & Univ. Kansas Press.

Nott, J.A. & B.A.Foster 1969. On the structure of the antennular attachment organ of the cypris larva of *Balanus balanoides* (L.). *Phil. Trans. Roy. Soc. London* (B)256:237-280.

Novozhilov, N.I. 1960. Subclass Maxillopoda Dahl 1956 [in Russian]. In: Yu.A.Orlov (ed.), *Fundamentals of Paleontology, Vol.8, Arthropoda – Trilobitoidea and Crustacea* (N.E.Chernysheva, ed.):253. Moscow: State Scientific-Technical Press.

Okada, Y.K. 1941. Sur la construction particulière de l'organe génital femelle de *Myriocladus* et la différentiation des cellules sexuelles. *J. Fac. Sci. Imp. Univ. Tokyo Sect. IV Zool.* 5:249-263.

Paulus, H.F. 1979. Eye structure and the monophyly of Arthropoda. In; A.P.Gupta (ed.), *Arthropod Phylogeny:*299-383. New York: Van Nostrand Reinhold.

Pennak, R.W. & D.J.Zinn 1943. Mystacocarida, a new order of Crustacea from intertidal beaches in Massachusetts and Connecticut. *Smith. Misc. Coll.* 103(9):1-11.

Pochon-Masson, J. 1978. Les différentiations infrastructurales liées à la perte de la motilité chez les gamètes mâles des crustacés. *Arch. Zool. Exp. Gén.* 119:465-470.

Pochon-Masson, J., J.Bocquet-Vedrine & Y.Turquier 1970. Contribution à l'étude du spermatozoïde des crustacés cirripèdes. In: B.Baccetti (ed.), *Comparative Spermatology:*205-219. New York: Academic Press.

Pochon-Masson, J. & P.Cals 1975. Les glands metamériques: un caractère adaptif des mystacocarides (crustacés). *Cah. Biol. Mar.* 16:763-764.

Pyefinch, K.A. 1937. The anatomy of *Baccalaureus torrensis* sp.n. (Cirripedia Ascothoracica). *J.Linn. Soc. Lond. Zool.* 40:345-371.

Raibaut, A. 1965. Sur quelques Cletodidae (Copepoda, Harpacticoida) du bassin de Thau. *Crustaceana* 8:113-120.

Renaud-Morant, J. 1976. Un nouveau genre de crustacé mystacocaride de la zone néotropicale: *Ctenocheilocaris claudiae* n.g.n.sp. *C.R.Acad. Sci. Paris* 282D:863-866.

Renaud-Morant, J., J.Pochon-Masson & J.Chaigneau 1977. Mise en évidence et ultrastructure d'un organe de Bellonci chez un crustacé mystacocaride. *Ann. Sci. Nat. Zool. Biol. Anim.* 19:459-478.

Riley, J., A.A.Banajo & J.L.James 1978. The phylogenetic relationships of the Pentastomida: the case for their inclusion within Crustacea. *Int. J. Parasit.* 8:245-254.

Ritchie, L.E. & J.T.Høeg 1981. The life history of *Lernaeodiscus porcellanae* (Cirripedia: Rhizocephala) and coevolution with its porcellanid host. *J. Crust. Biol.* 1:334-347.

Sanders, H.L. 1963a. The Cephalocarida. I. Functional morphology, larval development, comparative external anatomy. *Mem. Conn. Acad. Arts Sci.* 15:1-80.

Sanders, H.L. 1963b. Significance of the Cephalocarida. In: H.B.Whittington & W.D.I.Rolfe (eds), *Phylogeny and Evolution of Crustacea:*163-175. Cambridge: Mus. Comp. Zool.

Schram, F.R. 1971. A strange arthropod from the Mazon Creek of Illinois and the trans-Permo-Triassic Merostomoidea (Trilobitoidea). *Fieldiana Geol.* 20:85-102.

Schram, F.R. & W.D.I.Rolfe 1982. New euthycarcinoid arthropods from the Upper Pennsylvanian of France and Illinois. *J. Paleontol.* 56:1434-1450.

Schram, T.A. 1970a. Marine biological investigations in the Bahamas 14. Cypris Y, a later developmental stage of nauplius Y Hansen. *Sarsia* 44:9-24.

Schram, T.A. 1970b. On the enigmatical larva nauplius Y type I Hansen. *Sarsia* 45:53-68.

Schram, T.A. 1972. Further records of nauplius Y type IV Hansen from Scandinavian waters. *Sarsia* 50:1-24.

Scourfield, D.J. 1926. On a new type of crustacean from the Old Red Sandstone (Rhynie Chert Bed, Aberdeenshire) – *Lepidocaris rhyniensis* gen. et sp.nov. *Phil. Trans. Roy. Soc. London* (B)214:153-187.

Siewing, R. 1960. Neuere Ergebnisse der Verwandschaftsforschung bei den Crustaceen. *Wiss. Zeit. Univ. Rostock Mat.-Nat. Reihe* 9:343-358.

Smirnov, S. 1933. Notiz über *Limnocletodes behningi* Borutzky. *Zool. Anz.* 102:118-129.

Steuer, A. 1903. Über eine neue Cirripedienlarve aus dem Golfe von Triest. *Arb. Zool. Inst. Univ. Wien* 15(2):1-6.

Sylvester-Bradley, P.C. 1961. Archaeocopida. In: R.C.Moore (ed.), *Treatise on Invertebrate Paleontology, Part Q, Arthropoda 3:*Q100-102. Lawrence: Geol. Soc. Am. & Univ. Kansas Press.

Tessmann, M. 1904. *Beiträge zur Entwicklungsgeschichte der Cirripedien.* Inaugural-Dissertation, Univ. of Leipzig, Germany.

Tuzet, O. & A.Fize 1958. La spermatogénèse de *Derocheilocaris remanei* Delamare et Chappuis (Crustacea Mystacocarida). *C.R. Acad. Sci. Paris* 246:3669-3671.

Veillet, A. 1964. La metamorphose des cypris femelles des rhizocéphales. *Zool. Meded. Rijksmus. Nat. Hist. Leiden* 39:573-576.

Wagin, V.L. (spelled Vagin). 1937. Die Stellung der Ascothoracida ord.nov. (Cirripedia Ascothoracica Gruvel 1905) im System der Entomostraca. *Dokl. Acad. Nauk SSSR.* 15(5):273-278.

Wagin, V.L. 1947. *Ascothorax ophioctenis* and the position of Ascothoracida Wagin in the system of the Entomostraca. *Acta Zool. Stockh.* 27:155-267.

Wagin, V.L. 1954. On the structure, larval development, and metamorphosis of Dendrogasteridae (parasitic crustaceans of the order Ascothoracida) [in Russian]. *Uchen. Zap. Leningr. Gos. Univ. Biol. Ser.* 35(172):42-89.

Wagin, V.L. 1976. *Ascothoracida* [in Russian]. Kazan: Kazan Univ. Press.

Walley, L.J. 1969. Studies on the larval structure and metamorphosis of *Balanus balanoides* (L.). *Phil. Trans. Roy. Soc. London* (B)256:237-280.

Whittington, H.B. 1980. The significance of the fauna of the Burgess Shale, Middle Cambrian, British Columbia. *Proc. Geol. Ass.* 91(3):127-148.

Wingstrand, K.G. 1972. Comparative spermatology of a pentastomid, *Raillietiella hemidactyli,* and a branchiuran crustacean, *Argulus foliaceus,* with a discussion of pentastomid relationships. *Kongel. Dan. Vidensk. Selsk. Biol. Skrift.* 19(4):1-72.

Wingstrand, K.G. 1978. Comparative spermatology of the Crustacea Entomostraca. I. Subclass Branchiopoda. *Kongel. Dan. Vidensk. Selsk. Biol. Skrift.* 22(1):1-67.

Yager, J. 1981. Remipedia, a new class of Crustacea from a marine cave in the Bahamas. *J. Crust. Biol.* 1:328-333.

Yosii, N. 1931. Notes on the organization of *Baccalaureus japonicus. Annotnes, Zool. Jap.* 13:169-187.

Zevina, G.B. 1981. *Barnacles of the Suborder Lepadomorpha (Cirripedia, Thoracica) of the World Ocean. Part I. Family Scalpellidae* [in Russian]. Leningrad: Nauka.

WILLIAM A. NEWMAN
*Scripps Institution of Oceanography, La Jolla, California, USA*

# ORIGIN OF THE MAXILLOPODA; URMALACOSTRACAN ONTOGENY AND PROGENESIS

## ABSTRACT

The Maxillopoda have been inferred to have a 5-6-5 body plan derivable from the 5-8-8 plan of the malacostracans. Since crustaceans in general have anamorphic development, it has also been inferred that the reduction was achieved by progenetic paedomorphosis; that is, precocious maturity of a larval form. The peneids and euphausiids are the only malacostracans having a relatively complete anamorphic larval sequence. It is complicated by numerous apomorphies including loss of naupliar feeding and the acquisition of the caridoid facies. Maxillopodans, however, typically have feeding nauplii, and there is no trace of the caridoid facies at any stage of development. Maxillopodans appear in the fossil record in the Cambrian, long before the first eumalacostracans, which first appear in the Devonian as caridoids. Therefore, at least a precaridoid malacostracan ancestry for the maxillopodans is indicated. Virtually all non-eucarid malacostracans, including the primitive leptostracan phyllocarids, have epimorphic or nearly epimorphic development. Therefore, the larval development of precaridoid malacostracans must be reconstructed. By removing caridoid apomorphies from the eucarid sequence, and integrating what remains with knowledge of the most anamorphic development known in crustaceans, that of the cephalocarids, a reasonable approximation of precaridoid, urmalacostracan development can be reconstructed. The urmalacostracan ontogenetic sequence suggested here consists of approximately 14 stages, beginning with a feeding nauplius and including the first adult or complete body plan. The stages are divided almost equally between naupliar and 'caripodid' sequences. No protozoeal stages appear between them because the feeding nauplii eliminate their need. The 11th or 'premancoid' caripodid of the urmalacostracan larval sequence has the essential urmaxillopodan features including a 5-6-5 body plan.

## 1 INTRODUCTION

A number of students of the crustaceans recognize relationships in various combinations between maxillopodan subclasses (Copepoda, Cirripedia, Ascothoracica, Branchiura, Mystacocarida, and Ostracoda), but there is considerable disagreement and the class has not gained general acceptance (cf. Hessler 1982). Yet the grouping has been conceptually interesting in attempts to order knowledge about these taxa, even in comparative spermatology (cf. Baccetti 1979, Grygier 1981). The Maxillopoda has been inspected cladistically and an urmaxillopodan reconstructed (Grygier, this volume). The purpose of this paper is to explore the origin of the Maxillopoda.

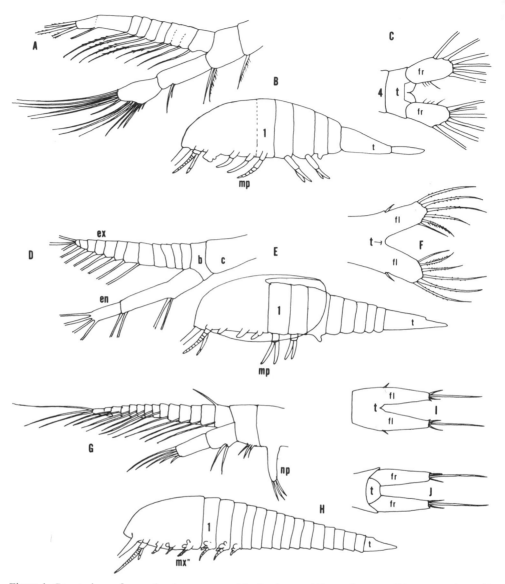

Figure 1. Comparison of second antennae, general body plans and furca of copepodids (A-C) and peneid protozoeae (D-F) (after Gurney 1942), and a cephalocarid juvenile or 'cephalopodid' (G-J) (after Sanders 1963). (A, B, D, E, & F after Gurney 1942, Figs 6, 7a, 7d, 7b and 29d, respectively; C after Hulsemann & Fleminger 1975, Fig.30; G after Sanders & Hessler 1964, Fig.3b; I and J after Sanders 1963, Figs 22 and 24 respectively; H interpreted from Sanders 1963, Fig.23).

Gurney (1942) considered that the crustaceans evolved from trilobites which had a relatively long series of similar appendage-bearing trunk segments. Hessler & Newman (1975) supported this view and proposed an urcrustacean similarly constructed, in contrast to the rather malacostracan-like urcrustacean of Siewing (1960). The anatomy of the former was deemed sufficient to account for the cephalocarid and its very generalized limbs, as well as for the existence of the relatively primitive branchiopods. It now also appears to be a satisfactory model for the Remipedia of Yager (1981). Therefore, the trilobite-cephalocarid-like urcrustacean is accepted here.

The prototype *Baupläne* can be envisaged in the course of crustacean evolution as falling into the following order: the cephalocarid-like urcrustacean (Hessler & Newman 1975), the leptostracan-like urmalacostracan (Siewing 1963), and the ascothoracican-like urmaxillopodan (Grygier, this volume). Thus we can turn our attention to the malacostracans for clues to the origin of the maxillopodans.

Likely the most provocative hypothesis for the origin of a maxillopodan group is that of Gurney (1942) concerning the copepods. He noted the marked similarity between the first copepodid larval stage and the first protozoea of the peneids in body plan and in details of their second antennae and furcal rami (Fig.1), and suggested that copepods could have evolved from malacostracans by paedomorphosis. Copepods differ from malacostracans in having a trunk of six thoracopods and an abdomen of five appendageless segments (the first may bear genital appendages and the last or telson bears furcal rami), rather than eight thorocopods and an abdomen of six appendage-bearing segments plus an appendageless one and the telson bearing furcal rami. Considering just the number of segments (including the telson), copepods are 5-6-5 while malacostracans are 5-8-8 in body plan. Following Gurney, the former is a reduction of the latter achieved through paedomorphosis.

My own interest in the Class Maxillopoda stems in good part from attempts to understand the origin of the Cirripedia (Newman et al. 1969, Newman 1982) and I concluded that the free-living ancestors of the ascothoracicans were close to the stem group of the cirripeds. The ascothoracicans, like the copepods, have a 5-6-5 plan, but in addition they have a carapace, the first thoracic segment is not fused to the head, and the first pair of thoracic limbs is similar to the following pairs. Therefore, the free-living ancestors of the ascothoracicans are potentially even closer than copepods to the stem of the maxillopodans, and a consideration of their origin becomes essentially a consideration of the origin of the Maxillopoda (Newman 1974, cf. Boxshall et al. 1983 regarding the urcopepod). If the maxillopodans were derived from malacostracans through paedomorphosis, we must take a close look at larval development. Since many maxillopodan developmental sequences include free nauplii capable of feeding, such must have existed in their ancestors.

## 2 MALACOSTRACAN LARVAL DEVELOPMENT

The only surviving phyllocarids, the nebaliaceans, have essentially direct or epimorphic development, so it is necessary to look at eumalacostracan ontogeny for clues to the anamorphy that must have underlain urmalacostracan development (Sanders 1963). The syncarids, peracarids, and pancarids are of little help, for while anaspidaceans and lophogastrids have been reported to pass through a recognizable nauplius in the egg (Siewing 1959), they lack free larval stages. Hoplocarids have free post-naupliar stages, but beyond that they are not very useful as far as early development is concerned. We are left then to consider eucarids, specifically the euphausiids and peneids, since they have a relatively complete larval sequence. However, their nauplii are non-feeding. This, in addition to their caridoid apomorphies, including the appearance of uropods before the pleopods and the pleopods before all thoracopods, leaves many gaps to be filled by inference in determining what urmalacostracan larval development may have been like. Some of the inferences required can be obtained by an out-group comparison.

The ontogeny of the sole member of the new crustacean class Remipedia (Yager 1981) is unknown. Considering the peculiarities of the adult head, including the apparently prehensile trophic apparatus, and the apparently primitive thoracoabdominal continuum of oar-feet, it does not seem likely that knowledge of its development would bear strongly on the present matter.

The Branchiopoda have anamorphic development, but their feeding nauplii are distinct from those of cephalocarids and maxillopodans, and the non-feeding nauplii of malacostracans (Sanders 1963). Furthermore, the early stages develop marked apomorphies including a ventral food-groove, reduced first antennae, and vestigial second maxillae. Therefore, other than in having feeding nauplii and anamorphic development, the branchiopods are of little use for present purposes. Left then is the Cephalocarida, and the ontogeny of two species is rather well documented (Sanders 1962, 1963, Sanders & Hessler 1964).

In cephalocarids, development is very anamorphic and the differences between *Hutchinsoniella macracantha* and *Lightiella incisa* are relatively small. *Hutchinsoniella* gains trunk segments more gradually, has a uniform distribution of pleural spines (the first eight pairs of which are replaced by pleura during ontogeny), and the mandible retains a six-segmented palp throughout all stages. Its development is therefore more anamorphic and the adult apparently more plesiomorphic than that of *Lightiella.* There are 18 stages (Table 1).

The first stage in *Hutchinsoniella* has the rudiments of two pairs of maxillae and three thoracomeres plus a telson bearing furcal lobes, in addition to the nE{appendages (two pairs of antennae plus mandibles), and it is therefore a metanauplius. Since a simple or true nauplius occurs in some branchiopods, malacostracans, and maxillopodans, the metanauplius of cephalocarids is assumed to be an apomorphy related to a benthonic existence (Sanders 1963).

By the third stage the second maxillae have come to resemble the thoracopods, and by the fifth stage the first thoracopods are well-developed and remain indistinguishable from subsequently appearing pairs. The coxal endite or naupliar process of the second antenna

Table 1. Larval development in cephalocarids, peneids, euphausiids (*Euphausia* spp.), copepods and cirripeds, and that inferred for the urmalacostracan. Cephalocarid development is divided into naupliar and juvenile ('cephalopodid') sequences, as it is in maxillopodans such as copepods. Peneid and euphausiid malacostracans have a protozoeal sequence between the naupliar and zoeal ('caripodid') sequences. The urmalacostracan is inferred to have had a developmental pattern intermediate between that of the cephalocarids and the eumalacostracans.

| | 1 | 2 | 3 | 4 | 5 | 6 | 7 | 8 | 9 | 10 | 11 | 12 | 13 | 14 | 15 | 16 | 17 | 18 |
|---|---|---|---|---|---|---|---|---|---|---|---|---|---|---|---|---|---|---|
| | | | | | nauplii | | | | | | | | | | cephalopodids | | | |
| CEPHALOCARIDA | I | II | III | IV | V | VI | VII | VIII | IX | X | XI | XII | XIII | I | II | III | IV | V |
| | | | | | | | | | | | caripodids | | | | | | | |
| URMALACOSTRACAN | I | II | III | IV | V | VI | VII | I | II | III | IV | V | VI | | | | | |
| | | | | | | | protozoeae | | | zoeae | | | | | | | | |
| PENEOIDEA | I | II | III | IV | V | VI | I | II | III | I | II | III | IV | | | | | |
| EUPHAUSIACEA | I | II | III | I | II | III | I | II | III | IV | V | VI | VII | | | | | |
| | | | | | | | copepodids | | | | | | | | | | | |
| COPEPODA | I | II | III | IV | V | VI | I | II | III | IV | V | | | | | | | |
| CIRRIPEDIA | I | II | III | IV | V | VI | cyprid I | | | | | | | | | | | |

persists into the 13th stage, marking the end of the naupliar series; feeding in the 14th stage is accomplished without it. The 14th stage ('first juvenile' of Sanders 1963, 'cephalopodid I' herein) has five thoracopods and the rudiment of the sixth. The 18th stage of podid V has seven thoracopods plus the rudiments of the eighth. This stage would be equivalent to the malacostracan manca if there were abdominal limbs, but the 12 abdominal segments lack them.

We can now attempt to develop a reasonable approximation of what urmalacostracan ontogeny may have been like by integrating cephalocarid ontogeny with that of eumalacostracans. Euphausiids and peneids are the only malacostracans having free nauplii, and, information on their structure and number, and on the number of subsequent stages, has been taken from Gurney (1942), Sanders (1963), Mauchline (1980) and Knight (1980, personal communication). Their developmental sequences are summarized comparatively in Table 1. Peneids generally have five or six nauplii and three or four zoeal stages while species of *Euphausia* have three nauplii and commonly six or seven zoeal stages (Knight, personal communication). Both have three protozoeal stages between the naupliar and zoeal sequences. Thus, the total number of stages is five or six less than in cephalocarids. The reduction has been at the expense of the naupliar series in these eumalacostracans, even if the protozoeal stages are considered equivalent to metanauplii (Gurney 1942) and included in the naupliar series.

There are at least five apomorphies by which peneids and euphausiids evidently depart from an idealized urmalacostracan ontogeny: (1) the true nauplii do not feed; (2) protozoeae feed, but lack naupliar second-antennal feeding structures; (3) the thorax posterior to the first three thoracopods is highly compressed, at least in euphausiids; (4) the uropods appear in the last protozoea, before any of the trunk limbs except the maxillipeds; and (5) the pleopods appear before the full set of thoracopods has been acquired at least in euphausiids. The first and second of these apomorphies concern feeding and herald the complete loss of the nauplius in other malacostracans. However, from the sequence, 3-6 nauplii plus 3 protozoeae, it is evident the urmalacostracan had as many as 6 or 7 naupliar stages capable of feeding.

The third difficulty, severe compression of the last five thoracic segments, is a euphausiid apomorphy. Compression is less pronounced in the relatively primitive genus, *Bentheuphausia* (Brinton, person communication); likewise, it is barely detectable in the peneid protozoea, and in part for this reason Gurney (1942) considered peneid development more primitive. The thorax in adult peneids and caridoid decapods is compressed relative to the abdomen, and this is the case in most other malacostracans including the leptostracans. Some compression is therefore likely to have existed in the adult urmalacostracan (Fig.2j).

The fourth difficulty, the very precocious appearance of the uropods in eucarids, likely did not occur in primitive eumalacostracans. As Gurney (1942) notes, the uropods in some hoplocarids appear after the last pleopods, as seems to be the case in thalassinideans (Gurney 1942), and this must have been the primitive sequence.

The fifth problem, the heterochronous appearance of pleopods before the full set of thoracopods, is certainly an apomorphy. Since it correlates, however, with the division of labor between anterior and posterior trunk limbs, and this division is basic to the malacostracans, it likely occurred in the urmalacostracan as will be enlarged upon later concerning 'mancoids'. Suffice it to say for the moment that there is ample justification for considering there was a manca in urmalacostracan ontogeny (Fig.2h).

Taking these facts and inferences into account, an integration of cephalocarid and

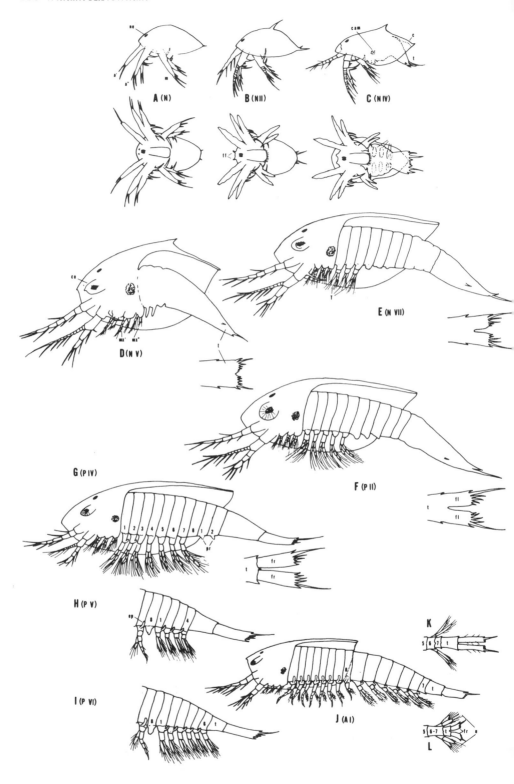

eumalacostracan ontogenies provides a reasonable basis for reconstruction of urmalacostracan development. It can be envisaged as having consisted of a naupliar series of 6 or 7 stages and a podid series of 5 or 6 stages.

In euphausiids and peneids the protozoea serves to smooth the transition from the non-feeding nauplii to the zoea on one hand, and involves the precocious aspects of the caridoid facies on the other. Since there is no trace of a protozoea in cephalocarids and there is no need to postulate one for the phyllocarids, none is indicated for the urmalacostracan (Table 1, Fig.2).

Urmalacostracan development, illustrated in Figure 2, can be summarized as follows: the first free stage illustrated (A) must have been a nauplius capable of feeding by endites of the second antennae and mandibles. The fourth stage (C) would have been a metanauplius and included a carapace, carapace adductor muscles, rudiments of the first and second maxillae and perhaps the first two pairs of thoracopods, and an undivided telson. In subsequent metanauplii (D and E) the trunk begins to segment, natatory limbs are added sequentially, and the telson begins to divide posteriorly into furcal lobes. Feeding is accomplished primarily by the two pairs of maxillae as well as by coxal endites (naupliar processes) of the second antennae (cf. D and E, Fig.1g) and of the mandibles.

Caripodids (F-I) differ from the last nauplius (E) primarily in having lost the naupliar processes of the second antennae. Primordial rudiments of the first two pairs of pleopods probably would have appeared before the last or eighth pair of thoracopods in the urmalacostracan, and certainly in the ureumalacostracan, and therefore the stage IV podid (G) would be 'premancoid'. Epipods likely also would have begun to appear, but the thoracopods themselves were probably still primarily natatory in form and function. The furcal lobes of the telson, however, had become articulated, furcal rami. In podids V and VI (H and I), pleopods appear rapidly and take over the natatory function of the thoracopods (Gurney 1942), whereby the endopods of the latter become ambulatory. Coxal epipods may have extended back onto the pleopods (Schram 1969). If the eighth pair of thoracopods remained rudimentary, mancoid stages are plesiomorphic at the urmalacostracan level rather than an apomorphy of eumalacostracans.

In the adult form (J), the eighth pair of thoracopods have appeared and the abdomen is complete. Such an urmalacostracan differs from that of Siewing (1963) in certain details including compression of the thorax relative to the abdomen, presence of epipods on the pleopods, and an appendageless rather than an appendage-bearing seventh or pretelsonic segment. The flexible carapace, provided with carapace adductor muscles associated primarily with the maxillary segment, may have been but was not necessarily bivalved; it is assumed that the bivalved carapace of conchostracan branchiopods, leptostracan malacostracans, and some maxillopodans are convergent.

---

Figure 2. Urmalacostracan larval development and the progenetic origin of the Maxillopoda. A-E, nauplii: A and B, naupliar stages I and II; C-E, metanaupliar stages IV, V and VII; F-I, caripodid stages II, IV, V and VI; G, premancoid stage or urmaxillopodan; H and I, mancoids. J, postlarval or adult stage; K, precaridoid (urmalacostracan or urphyllocarid) caudal configuration of J; and L, derived caridoid (eumalacostracan) caudal configuration, seen from above. (a' – first antenna; a" – second antenna; c – carapace; cm – carapace adductor muscle; ce – compound eye; ep – epipodite; ff – frontal filament; fl – furcal lobe; fr – furcal ramus; m – mandible; mx' – first maxilla; mx" – second maxilla; ne – nauplius eye; pr – pleopodal rudiments; r – rudiments (maxillary and thoracic); T – thoracopod; t – telson; u – uropod).

The urmalacostracan caudal configuration (K) functions primarily in balance and steering while swimming forward. The ureumalacostracan configuration (L) can also be flexed more or less rapidly under the trunk, whereby the tail fan, plesiomorphically including the last pleopods or uropods, as well as the furcal rami, effects the backward-escape reaction. This, the caridoid configuration, was apparently achieved by coalescence of the sixth and seventh abdominal segments (Manton 1928), or by arrested development following the podid VI stage (I).

## 3 MAXILLOPODAN DEVELOPMENT AND AFFINITIES

Of the maxillopodan subclasses, only the mystacocarids, copepods, cirripeds, and ascothoracicans have a relatively anamorphic sequence. Mystacocarids, as adults, are very 'juvenilized' (Hessler 1982). Their larval series begins with a metanauplius having, in *Derocheilocaris typicus,* functional first maxillae (Hessler & Sanders 1966), and there is considerable heterochrony in the development of thoracic limbs. Thus, both the developmental sequence and the end product are not particularly relevant here other than to note that the number of nauplii in *D. typicus* is six.

Copepods have a characteristic, anamorphic larval sequence of six nauplii and five podid stages preceeding the adult. The fifth and sixth naupliar stages have functional maxillae and a number of rudimentary trunk segments. The first podid stage has at least three functional thoracic limbs, and this of course is part of Gurney's (1942) comparison with the protozoea (Fig.1b, e).

In cirripeds, only the last of the six stages is a metanauplius and then the only additional limbs, the first maxillae, are rudimentary (Crisp 1962). In further contrast with copepods, there is but one podid stage, the non-feeding cyprid larva. It is achieved in a single moult; that is, the equivalent of the five copepod podids have been compressed into a single stage (Table 1).

Ascothoracican development, in which there are at least two podids (ascothoracid larvae) between the nauplii and adult (Brattström 1948), partially fills the gap between the copepod copepodid and cirriped cyprid larva. It follows that copepod development is the more plesiomorphic, and as such suggests what the urmaxillopodan larval sequence must have been like. The question is whether such an urmaxillopodan sequence is more readily derived from the cephalocarid or the malacostracan plan.

When Gurney (1942) hypothesized that copepods had evolved from malacostracans by paedomorphosis, cephalocarids were unknown and the taxonomic grouping Maxillopoda (Dahl 1956) had not been proposed. Sanders (1963), by comparison of larval appendages, concluded that similarities between copepodids and cephalocarids were greater than between copepodids and protozoeae. This does not mean that they are therefore more closely related, however, because most of the similarities are potentially plesiomorphic (cf. Fig.1). Apparent apomorphies shared by malacostracans and maxillopodans that distinguish both from cephalocarids are needed. These could include the full cephalization of the second maxillae, the structure of the second antennae, and detailed similarities of the furca. One might also include the carapace but, like the compound eyes (Burnett 1981), in my opinion it was for the most part lost in cephalocarids (Newman, in Hessler & Newman 1975).

Sanders (1963) considered the larval second antennae of cephalocarids more similar to those of copepods than to those of the protozoea, but on close inspection the similarities are primarily related to the plesiomorphies of naupliar feeding, which no longer exist in malacostracans. The second antennae of the protozoea and copepodid differ from that of the cephalocarid metanauplius in having an inferred three- rather than a two-segmented endopod and an exopod of generally eight rather than approximately 14 segments. If the cephalocarid condition proves to be the apomorphy, little more can be said; but if it is the plesiomorphic state, the differences indicate closer affinity of maxillopodans to malaco-stracans than to cephalocarids. The furcal similarity between the copepod and the proto-zoea, noted by Gurney (1942), is considered unique (Sanders 1963). It is probably an apomorphy that relates them. An unequivocal apomorphy that separates the malaco-stracans and maxillopodans from the cephalocarids is the degree of cephalization of the second maxillae. However, it does not separate them from the Branchiopoda and Remipedia, and therefore cannot be used as evidence for a close relationship between malacostracans and maxillopodans, especially since the characteristic could have quite simply been acquired convergently. Taking all of the characters at face value, the ancestral stock of the maxillo-podans appears to have been closer to that of the malacostracans than to the cephalocarids.

## 4 AN URMALACOSTRACAN ORIGIN FOR MAXILLOPODANS

As we have noted, eumalacostracan larval development contains heterochronies related to (1) the division of labor between the trunk limbs, (2) the advent of the caridoid facies, and (3) the loss of a feeding nauplius. The last problem is of no great consequence in attempt-ing to relate maxillopodan development and origins to malacostracans for we are free to assume that the ureumalacostracan had a feeding nauplius. The second is a real problem, however, unless a precaridoid ancestral stock is accepted, and this of course is the reason for having to infer what urmalacostracan ontogeny must have been like.

Inspection of the urmalacostracan larval sequence indicates that the stage IV caripodid (Table 1, Fig.2g) closely approximates the urmaxillopodan form (Grygier, this volume). It is premancoid; that is, rudiments of the pleopods appear before at least the last pair of thoracopods. As already noted, the mancoid tendency is evidently related to the division of labor between the trunk limbs which entails precocious development of the pleopods for locomotory purposes. It occurs in hoplocarid and eucarid development, and in adults of the latter, the eighth leg is frequently small and/or put to a different use relative to other pereopods. There is a manca in cumacean, tanaidacean, and isopodan peracarids, and in the pancarid *Monodella*. The adult of the pancarid *Thermosbaena* is a mancoid which, except for the presence of uropods (cf. Barker 1962), is remarkably similar to the stage IV caripodid depicted here, including having but two pairs of pleopods. Bathynellacean syn-carids are similarly and in some cases even more mancoid than *Thermosbaena* (cf. Schminke 1981). But these are all eumalacostracans; what about phyllocarids? The only surviving forms, the leptostracans, have reduced fifth and sixth pleopods and somewhat smaller last thoracic legs. Thus it appears that the mancoid tendency goes back to the roots of the malacostracans and it is therefore depicted here, in urmalacostracan development, as the stage V and VI caripodids (Fig.2h and i).

Given the premancoid stage IV caripodid, little is left to the imagination in envisioning how the maxillopodan adult body plan could have been derived through paedomorphosis. A direct cephalocarid-like ancestor has already been discounted, and an urcrustacean ances-tor like that of Hessler & Newman (1975) must be discounted for the same reason. The urcrustacean of Siewing (1960) differs little from the urmalacostracan of Siewing (1963) or that of the urmalacostracan depicted here (Fig.2j). Thus, the consensus seems to hold to an urmalacostracan that is essentially an urphyllocarid. The ascothoracicans are in many ways the most generalized of the maxillopodans, including having a carapace, a first thoracic segment free of the head and the most generalized spermatozoa known for crustaceans (Grygier 1981, 1982). They also have a strong facies similarity with phyllocarids (Norman 1888). Therefore an urmalacostracan-urphyllocarid, precaridoid origin for the maxillopodans seems the most likely.

The urmaxillopodan plan has been inferred herein to have sprung from the equivalent of the urmalacostracan podid stage IV (Fig.2g). As can be seen in podid IV, the first thoracic segment has not been incorporated into the back of the head and, while the first pair of thoracopods are so positioned and may have assisted in feeding, they are primarily natatory. There are seven pairs of thoracopods, including the first, but the seventh pair will be modified for reproductive purposes and/or lost in the maxillopodans. Thus, the urmaxillopodan plan is 5-7-4 (Grygier, this volume) since the seventh thoracic segment is homologous to the first abdominal segment. However, I have continued to use the 5-6-5 nomenclature convention for the Maxillopoda.

## 5 THE CARAPACE, THE URMALACOSTRACAN AND THE URMAXILLOPODAN

Urmalacostracan larval development outlined here (Fig.2) includes a carapace since one has been inferred to have been present in the adult (Calman 1909, Siewing 1963, Hessler & Newman 1975). However, there has been considerable disagreement in recent years over the pervasiveness of the carapace throughout the crustaceans. Lauterbach (1974) considered that the urcrustacean lacked a carapace, but treated crustaceans having one as monophyletic. Hessler (in Hessler & Newman 1975) also considered that the urcrustacean lacked a carapace but that it arose independently in the various classes. However, Hessler & Newman (1975), like Calman (1909) and Siewing (1963) before them, considered that the urmalacostracan had a carapace.

Dahl (1976) proposed that the urmalacostracan lacked a carapace. Thereby relatively primitive eumalacostracans without a carapace, such as syncarids, could be interpreted as never having had one rather than simply having lost it. Watling (1981), following Dahl's lead, suggested that the urperacarid also lacked a carapace and that the two major peracarid lines gained one independently. If one follows Watling (1981), ancestral isopods and amphipods lacked a carapace, the small carapaces and other caridoid features such as the antennal scale and thoracic exopods of tanaidaceans and spelaeogriphaceans are not vestigial structures but rather rudiments anticipating the condition in mysidaceans. Thus, the full caridoid facies of the mysidaceans and the eucarids would be convergent!

In my opinion, neither the carapace nor the caridoid facies in the strict sense (Calman 1909) were easily acquired. The virtually part-for-part correspondence between eucarids and mysidaceans cannot be put aside lightly and Watling (this volume) now recognizes this. The oldest known crustaceans, the phyllocarids (Cambrian), have a carapace, as do the earliest eucarids (Eocaridacea and Decapoda; Devonian) and peracarids (Mysidacea, Spelaeogriphacea, and Tanaidacea; Mississippian) (Schram 1981). One can distrust but not ignore the fossil evidence. Furthermore, Dahl's (1976) concerns that the phyllocarids may not have given rise to eumalacostracans have been substantially lessened by Rolfe's (1981) fossil evidence for detailed similarities between the carapaces of the certain phyllocarids and eumalacostracans, and that phyllocarids were not limited to filter feeding. Rolfe (1981) concludes that the phyllocarids were not a phyletic dead end as Dahl (1976) suggested, but readily could have given rise to the eumalacostracans. Therefore, the urmalacostracan, essentially an urphyllocarid, is envisaged as having had a carapace, one homologous with that of the maxillopodans.

# 6 PRECARIDOID ORIGIN OF MAXILLOPODA

The urmalacostracan, like that of Siewing (1963), is precaridoid *sensu* Calman (1909), not caridoid as stated by Dahl (1976). Since the ancestor of the maxillopodans is inferred to have been precaridoid, a note of explanation is required. For the urmalacostracan to become caridoid required the acquisition of two essential features: (1) conversion of the exopod of the second antenna into a scale that can be extended laterally as a diving plane, and (2) incorporation of the sixth pleopods into the tail fan as uropods at the expense of the furcal rami (Fig.2L). Both adaptations are primarily involved in the backward escape reaction, which entails flecture of the abdomen under the body.

No trace of the caridoid facies has been observed in maxillopodans. It appears that the ascothoracicans, in many ways the most generalized of the maxillopodans, use the well-developed abdomen and furca in much the same manner as precaridoids or phyllocarids. Again the inference is in agreement with the fossil record since certainly ostracods occurred in the Cambrian along with the Phyllocarida. The cirripeds apparently also appear in the Cambrian (Collins & Rudkin 1981) or Silurian (Wills 1962), millions of years before the appearance of the caridoid facies.

The maxillopodan *Bauplan* could have been acquired gradually, in a microevolutionary sense, or rapidly as a relatively macroevolutionary event (cf. Stanley 1979). Clues as to how the microevolutionary process could have progressed are found in at least two distinct lineages among the extant eumalacostracans, the bathynellacean syncarids, and the pancarids. While the trends are similar in both, all of the bathynellaceans have the first thoracic segment free of the head and its limbs not modified as maxillipeds. Bathynellaceans also have epipods on some thoracopods, a telson with furcal rami or lobes, and females that lay eggs free. While lacking these plesiomorphic characters, the pancarid has a carapace under which females brood eggs and therefore represent a very different ancestry. The similarities in reduction are convergent, but the tendency to do so in the same way must be related to the underlying fundamentals of malacostracan embryology and ontogeny.

Reduction in these malacostracans includes loss of the last one or two pairs of thoracopods and three pairs of pleopods. Except for the presence of uropods, the hallmark of eumalacostracans, both have mancoid body plans intermediate between the premancoid or podid stage IV and the mancoid or podid stage V (Fig.2g,h). Schminke (1981) considers the reduced bathynellaceans to be 'zoeal' in character and the same concept underlies Barker's (1962) analysis of the pancarid genera. In *Monodella* the telson is free of the sixth abdominal segment, but in *Thermosbaena* it is fused with the sixth abdominal segment (pleotelson) and the last two pairs of thoracopods are missing (Barker 1962). By further reduction, including the fusion of the limbless seventh and eighth thoracic segment with the first abdominal segment (actually the lack of their separation during early embryology) and loss of the two pairs of rudimentary pleopods and the uropods, but for want of furcal rami *Thermosbaena* would closely approximate the maxillopodan *Bauplan*.

While the aforegoing exercise has made it relatively easy to envisage how the urmaxillopodan could have been achieved in microevolutionary terms, the paedomorphic reduction in bathynellaceans and pancarids involves a slowing down of somatic development in connection with refugial environments specifying K-selection; that is, the forms are neotenic (Gould 1977). Both groups are apparently remnants of ancient malacostracan lineages that have survived because of adaptations to refugia. It does not seem likely that they will lead to forms that will again be important in the natural economy.

The principal maxillopodans (copepods, cirripeds, and ostracods) for the most part have r-selected life history strategies. This suggests a paedomorphic origin involving the speeding up of maturation; that is, progenesis. Evidently the maxillopodan morphology provided for diversification at lower taxonomic levels and radiation of higher taxa into niches of major importance in terms of resource allocations (space utilization, biomass accumulations, and positions in trophic systems).

The progenetic acquisition of the maxillopodan plan obviously would have required numerous small physiological, behavioral and reproductive as well as morphological adjustments in the conversion of a malacostracan larva into a functional adult. But since the 5-6-5 *Bauplan,* including the appropriate trophic and natatory appendages, can readily be inferred to have existed in the malacostracan larval sequence (Fig.2g), the adjustments could have been very rapid, to the extent that the appearance of the maxillopodan *Bauplan* could have been effectively a macroevolutionary event. Since the major maxillopodan taxa have apparently enjoyed increasing diversity since the Cambrian, it is difficult to imagine how a competitive organization could have evolved through convergence from similar stocks.

## 7 ACKNOWLEDGEMENTS

Numerous colleagues have discussed various aspects of crustacean evolution touched upon in the paper. For insights into euphausiid and peneid development I am indebted to M. Knight and E.Brinton; and into that of copepods, A.Fleminger. Thanks are due to R.Manning for references to stomatopod development, and T.Waterman for notes on the occurrence of 'mancoids' in various crustaceans and the apparent misuse of the term by previous authors in reference to nebaliaceans. Several preprints have also been useful; particularly G.Boxshall et al. (1982) on the urcopepod, and R.Hessler (1982) on crustacean evolution in general, and I thank the authors for making them available. I am especially grateful to M.Grygier, not only for the preprint of his paper on maxillopodans (this volume), but for numerous stimulating discussions that much influenced my perception of the problems of plesiomorphies and cladograms. Thanks are due to F.Schram for his editorial expertise, and also for convening this symposium, for if he had not, the desire to write this paper likely would have remained dormant. While indebted to my colleagues, responsibility for errors or other shortcomings falls on me.

## 8 ADDENDUM

Schminke's (1981) important and comprehensive study concerning the paedomorphic origin of the bathynellaceans arrived in our library a week or so before this conference was to convene. He derives the bathynellaceans from a palaeocaridoid by the same general process (paedomorphosis), as I have used to derive the Maxillopoda from a precaridoid malacostracan stock. Therefore, it is probably not surprising that the two theories run similar courses, but considering that they were headed towards quite different goals, it is remarkable that they complement each other to a significant degree. The only notable departure involves the furcal rami. Schminke (1981:584) infers that those of the bathynellaceans are homologous with the furcal lobes seen in newly hatched *Anaspides,* but he concludes that the rami became secondarily segmented and motile; that is, they are a bathynellacean apomorphy. I have inferred that the furcal rami are symplesiomorphic to the Crustacea; their retention in some primitive eumalacostracans (Eocaridacea and Bathynellacea) being an ancestral reminiscence and their loss in most eumalacostracans being a perfection of the uropodal tail fan and backward escape capabilities of the caridoid facies (cf. Hessler, this volume). While this difference in interpretation of the furcal rami does not affect consideration of the origin of the maxillopodans from a precaridoid malacostracan ancestor, it has serious implications concerning the evolution of the eumalacostracans.

## DISCUSSION

BOUSFIELD: Do I understand you correctly Bill that the ontogenetic direction of the telson is from a nearly plate-like form with a cleft to one with two separate lobes?
NEWMAN: Yes. The primitive situation is a trailing telson which develops the lobes ontogenetically, which become articulated as furcal rami. This is the situation we see in the leptostracans and bathynellaceans. Therefore, we infer the urmalacostracan had furcal rami, and that would be necessary for them to give rise to the urmaxillopodan.
BOXSHALL: Somewhere along the line the copepods must have lost compound eyes. Can you suggest why?
NEWMAN: First of all, I wonder if it is really settled that all copepods have lost their compound eyes? It is believed that the pontellids have rudiments of compound eyes [Parker, G.H. 1891. *Bull. Mus. Comp. Zool.* 21:45-141]. No, I don't know why. Copepods are very small animals; and while they are small they do have good nauplius eyes, so the explanation cannot arise from their having evolved in the deep-sea. On the other hand, ascothoracicans, which are relatively large, have lost both compound and nauplius eyes, apparently related to their deep-sea history.
WATLING: One question bothers me with respect to crustacean phylogeny. If you look at crustaceans with planktonic larvae relative to most other invertebrates with planktonic larvae, you find a lot of features that are there because of that life habit in the plankton. Why is it crustacean people are so intent on retaining those larval features in deriving whole orders of animals through paedomorphosis? Whereas, in other invertebrate groups, people are content to leave the larvae as specializations to larval existence.
NEWMAN: We are all agreed maxillopodans have a reduced body plan. The question is how does one get such a reduced plan? You can't just 'cut off their tails with a carving knife'. Progenetic paedomorphosis is one way to achieve this. Such explanations have been used in other groups. One example was the use of the tadpole larva of urochordates to explain the origin of higher chordates. While most of us do not now agree with that explanation for the evolution of chordates, I do believe paedomorphosis provides an effective explanation for maxillopodan origins.

BRIGGS: You talk about early urmaxillopodans and you made a reference to *Cyprilepas,* implying that it might indicate the origins of maxillopodans are at least earlier than the Silurian. Could you not use Muller's phosphatocopines as your urmaxillopodan?

NEWMAN: Yes, but you don't even have to go to phosphatocopines. There are supposed ostracodes known only by bivalved shells, that could be ascothoracicans. It remains to be seen just what phosphatocopines are.

HESSLER: Why didn't you use stomotopod larvae?

NEWMAN: I do in the paper. Stomatopods lack nauplii, but they have a mancoid stage and the uropods appear last in the more primitive forms.


# REFERENCES

Baccetti, B. 1979. Ultrastructure of sperm and its bearing on arthropod phylogeny. In: A.P.Gupta (ed.), *Arthropod Phylogeny:*609-644, New York: Van Nostrand Reinhold.

Boxshall, G.A., F.D.Ferrari & H.Tiemann 1983. The ancestral copepod: Towards a concensus, at the First International Conference on Copepoda. *Crustaceana:Suppl.*

Burnett, B.R. 1981. Compound eyes in the cephalocarid crustacean *Hutchinsoniella macrancantha. J. Crust. Biol.* 1:11-15.

Brattström, H. 1948. On the larval development of the ascothoracid *Ulophysema oresundense* Brattström. *Lunds Universitets Arsskrift. N.F.* Avd.2 Bd.44, Nr.5:69pp.

Calman, W.T. 1909. Crustacea: In: E.R.Lankester (ed.), *A Treatise on Zoology.* London: Adam & Charles Black.

Collins, D. & D.M.Rudkin 1981. *Priscansermarinus barnetti,* a probable lepadomorph barnacle from the Middle Cambrian Burgess Shale of British Columbia. *J.Paleont.* 55:1006-1015.

Crisp, D.J. 1962. The larval stages of *Balanus hameri* (Ascanius 1767). *Crustaceana* 4:123-130.

Dahl, E. 1956. Some crustacean relationships. In: K.G.Wingstrand (ed.), *Hanström Festschrift:*138-147. Lund: Zool. Institute.

Dahl, E. 1976. Structural plans as functional models exemplified by the Crustacea Malacostraca. *Zool. Scripta* 5:163-166.

Gould, S.J. 1977. *Ontogeny and Phylogeny.* Cambridge: Harvard University Press.

Grygier, M.J. 1981. Sperm of the ascothoracican parasite *Dendrogaster,* the most primitive found in crustaceans. *Int. J. Invert. Reprod.* 3:35-73.

Grygier, M.J. 1982. Sperm morphology in Ascotheroacida (Crustacea:Maxillopoda); confirmation of generalized nature and phylogenetic importance. *Int. J. Invert. Reprod.* 4:323-332.

Gurney, R. 1942. *Larvae of Decapod Crustacea.* London: Ray Society.

Hessler, R.H. 1982. Evolution within the Crustacea. In: L.Abele (ed.), *Biology of Crustacea, Vol.1:*149-185. New York: Academic Press.

Hessler, R.H. & W.A.Newman 1975. A trilobitomorph origin for the Crustacea. *Fossils and Strata* 4: 437-459.

Hessler, R.H. & H.L.Sanders 1966. *Derocheilocaris typicus* Pennak & Zinn (Mystacocarida) revisited. *Crustaceana* 11:141-155.

Hulsemann, K. & A.Fleminger 1975. Some aspects of copepodid development in the genus *Pontellina* Dana (Copepoda: Calanoida). *Bull. Mar. Sci.* 25:174-185.

Knight, M.D. 1980. Larval development of *Euphausia eximia* (Crustacea:Euphausiacea) with notes on its vertical distribution and morphological divergence between populations. *Fish. Bull.* 78:313-335.

Lauterbach, K.-E., von 1975. Uber die Herkunft der Malacostraca (Crustacea). *Zool. Anz. Jena* 194: 165-179.

Manton, S.M. 1928. On the embryology of a mysid crustacean, *Hemimysis lamornae. Phil. Trans. Roy. Soc. London* (B)223:363-463.

Mauchline, J. 1980. The biology of mysids and euphausids. *Adv. Mar. Biol.* 18:1-681.

Newman, W.A. 1974. Two deep-sea Cirripedia (Ascothoracica and Acrothoracica) from the Atlantic. *J. Mar. Biol. Ass. UK* 54:437-456.

Newman, W.A. 1982. Cirripedia. In: L.Abele (ed.), *Biology of Crustacea, Vol.1:*197-220. New York: Academic Press.

Newman, W.A., V.A.Zullo & T.H.Withers 1969. Cirrpedia. In: R.C.Moore (ed.), *Treatise on Invertebrate Paleontology, Part R, Arthropoda 4,* 1:R206-295. Lawrence: Geol. Soc. Amer. & Univ. Kansas Press.

Norman, A.M. 1888. Report on the occupation of the Table. *Brit. Assoc. Advan. Sci. Pret. 57th meeting* 1887:85-86.

Rolfe, W.D.I. 1981. Phyllocarida and the origin of the Malacostraca. *Geobios* 14:17-27.

Sanders, H.L. 1963a. The Cephalocarida: Functional morphology, larval development, comparative external anatomy. *Mem. Conn. Acad. Arts Sci.* 15:1-80.

Sanders, H.L. 1963b. Significance of the Cephalocarida. In: H.B.Whittington & W.D.I.Rolfe (eds.), *Phylogeny and Evolution of Crustacea:*163-175. Cambridge: Mus. Comp. Zool.

Sanders, H.L. & R.H.Hessler 1964. The larval development of *Lightiella incisa* Gooding (Cephalocarida). *Crustaceana* 7:81-97.

Schminke, H.K. 1976. The ubiquitous telson and the deceptive furca. *Crustaceana* 30:292-300.

Schminke, H.K. 1981. Adaptation of Bathynellacea (Crustacea, Syncarida) to life in the interstitial ('Zoea Theory'). *Int. Rev. ges. Hydrobiol.* 66:575-637.

Schram, F.R. 1969. Some Middle Pennsylvanian Hoplocarida (Crustacea) and their phylogenetic significance. *Fieldiana Geol.* 12:235-289.

Schram, F.R. 1981. On the classification of Eumalacostraca. *J. Crust. Biol.* 1:1-10.

Siewing, R. 1959. Syncarida. In: *H.G.Bronn, Klassen und Ordnungen des Tierreichs* 5:1-121. Leipsig: Akad. Verlag.

Siewing, R. 1960. Neuere Ergebnisse der Verwandtschaftforschung bei den Crustaceen. *Wiss. Z. Univers. Rostock, Math.-Nat. Reihe* 9:343-358.

Siewing, R. 1963. Studies in malacostracan morphology: Results and problems. In: H.B.Whittington & W.D.I.Rolfe (eds), *Phylogeny and Evolution of Crustacea:*85-103. Cambridge: Mus. Comp. Zool.

Stanley, S.M. 1979. *Macroevolution: Pattern and Process.* San Francisco: Freeman.

Watling, L. 1981. An alternative phylogeny of peracarid crustaceans. *J. Crust. Biol.* 1:201-210.

Wills, L.J. 1963. *Cyprilepas holmi* Wills 1962, a pedunculate cirripede from the Upper Silurian of Oesel, Esthonia. *Palaeont.* 6:161-165.

Yager, J. 1981. Remipedia, a new class of Crustacea from a marine cave in the Bahamas. *J. Crust. Biol.* 1:329-333.

GEOFFREY A. BOXSHALL
*Department of Zoology, British Museum (Natural History), London, UK*

# A COMPARATIVE FUNCTIONAL ANALYSIS OF THE MAJOR MAXILLOPODAN GROUPS

## ABSTRACT

The basic functional plans of the four major maxillopodan groups are compared, and the probable sequence of steps by which each evolved is analysed. It is suggested that the evolution of the copepod and cirripede/branchiuran lines was initially determined by the differentiation and specialization of the anterior part of the trunk as a locomotory center. The evolution of the mystacocarids was initially determined by the retention of both cephalic feeding and cephalic locomotory mechanisms, and evidence is presented which indicates that a specialized thoracic locomotory tagma has never been a feature of this line. Other evidence presented in support of a taxon Maxillopoda is reinterpreted. It is concluded that Maxillopoda is not a valid taxon.

## 1 INTRODUCTION

Many authors during the last century have regarded the Copepoda as a group with characters that are persistently larval. Claus (1876) first pointed out the marked parallel development of the copepodid stages of copepods and the protozoeal stages of peneid decapods. Barnard (1892) suggested that the Copepoda could be regarded as permanently larval, derived by neoteny from the developmental stages of branchiopods. Calman (1909) regarded them as a specialized group with characters that could be persistently larval rather than phylogenetically primitive. Beurlen (1930) referred to the Copepoda as representatives of the Pygocephalomorpha (?forerunners of the Decapoda, see Tiegs & Manton 1958) which have remained in the larval condition. Finally Gurney (1942) adopted and expanded the hypothesis that copepods were persistently larval and he derived them by arrested development from a larval form having the general characters of the protozoeal larva of decapods. Many subsequent authors appear to have tentatively regarded the basic copepod organization as representing a paedomorphic condition with a possible malacostracan ancestry (Tiegs & Manton 1958, Hessler & Newman 1975).

The term paedomorphosis is used here as a general term for the presence of ancestral larval characters in adult descendants. Paedomorphosis can be the result of two evolutionary processes: progenesis (precocious sexual maturation of a descendant while still in the equivalent of the larval stage of its ancestor), and neoteny (the retention of ancestral larval characters in adult descendants).

In 1956 Dahl established a new taxon, the Maxillopoda, to include the Copepoda,

121

Archicopepoda, Branchiura, Mystacocarida, and possibly the Cirripedia. The Archicopepoda have subsequently been shown to represent an independent fossil class, the Euthycarcinoidea (Gall & Grauvogel 1964) regarded by Moore (1969) as crustacean, Schram (1971) as trilobitomorphan and Schram & Rolfe (1982) as uniramian. Dahl (1963) later refined his concept of the Maxillopoda and included within it the Copepoda, Mystacocarida, Branchiura, and Cirripedia. The Ostracoda have also been considered as a possible derivative of the maxillopodan line by further reduction of the body plan (Siewing 1960, Hessler & Newman 1975). The Maxillopoda, as a taxon, has had a mixed acceptance. The two main hypotheses concerning the origins of the Copepoda, their descent by paedomorphosis from a malacostracan ancestor, and their derivation from the common ancestor of the Maxillopoda, appear to be mutually incompatible unless the urmaxillopodan was itself derived by a paedomorphic process, as indicated by Hessler & Newman (1975). Clearly any consideration of the ancestry of the Copepoda must involve comparisons with the other maxillopodan taxa.

While it is important to consider the possible processes by which evolutionary change takes place, it seems to me that the identification of paedomorphosis as a significant process in the origin of the Copepoda, the Mystacocarida, and the Maxillopoda in general, has been regarded as an end-point in the debate on their phylogeny. Neoteny or progenesis is merely an obvious and convenient mechanism to implicate when reduction in numbers of somites and numbers of appendages has occurred, or when a biramous mandibular palp is retained in the adult. What are the functional implications of reducing the numbers of somites or appendages? What is the significance of retaining a biramous mandibular palp in the adult? It is necessary to answer *these* questions in order to understand the evolutionary pathways followed by the maxillopodan groups.

## 2 THE MAXILLOPODAN GROUPS

The largest class included in the Maxillopoda is the Copepoda. The classification of the copepods has recently been placed on a sound basis with the new phylogenetic scheme proposed by Kabata (1979). Although some of the interordinal relationships within the podoplean line are not yet defined by synapomorphies and a new order, the Mormonilloida, Boxshall (1979), has yet to be incorporated, this scheme is already being used by most copepod workers. Kabata's scheme included only a brief consideration of the ancestral copepod. The concept of an ancestral copepod was considered in detail at a discussion session at the First International Conference on Copepoda. The basic uniformity of the Copepoda is such that the characters of the ancestral copepod (Fig.1) were determined largely by consensus of opinion (Boxshall et al. 1983). When copepods are discussed below, it is this organism that is used as the basis for comparison.

The second largest maxillopodan group is the Cirripedia. Newman et al. (1969) considered the presumed characters of the ancestral cirripede stock and arrived at an organism about at the ascothoracican level of organization but possessing compound eyes and generalized copepod-like mouthparts. This organism was probably epizoic on arborescent life forms. This concept may have to be refined when the group represented by the parasitic genera *Basipodella* (see Becker 1975) and *Deoterthron* (see Bradford & Hewitt 1980) is better known, as these carapace-less forms may be related to the Cirripedia, on the

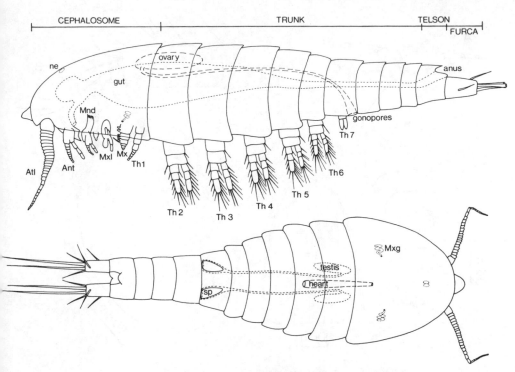

Figure 1. The ancestral copepod: Lateral view of female and dorsal view of male (adapted from Boxshall et al. 1983).

basis of their tagmosis and the presence of a median kentron-like structure in the head. These characters are best displayed in juvenile *Basipodella* (Fig.2) as the adults are rather featureless sacs. These organisms are probably quite common, as the adult figured is one of two taken during a recent RRS *Discovery* cruise off the Azores from a small sample of deep-sea harpacticoids. The ancestral cirripede of Newman et al. (1969) is used herein as the basis for comparison.

The two remaining maxillopodan groups, the Branchiura and Mystacocarida, are both specialized but are relatively uniform in morphology so that the genera *Argulus* and *Derocheilocaris* can be regarded respectively as representative of the ancestral stock of each class.

According to Dahl (1956) the diagnostic characters of the Maxillopoda are six thoracic somites, a well-developed mandibular palp in adults, well-developed maxillules and maxillae adapted to filter feeding purposes in filter feeders, and the absence of gnathobases on the thoracic appendages. The basic body plan of five cephalic, six thoracic and five abdominal somites is found in the Copepoda and the Cirripedia Ascothoracica. Evidence for the unity of the Maxillopoda has also been claimed from comparative spermatology (Grygier 1981) and from the structure of the nauplius eye complex (Elofsson 1966). Among the characters that make the Maxillopoda so heterogeneous are body tagmosis and the presence/absence of a carapace and compound eyes. All of these characters should be considered in any discussion of the phylogeny of the Copepoda and Maxillopoda.

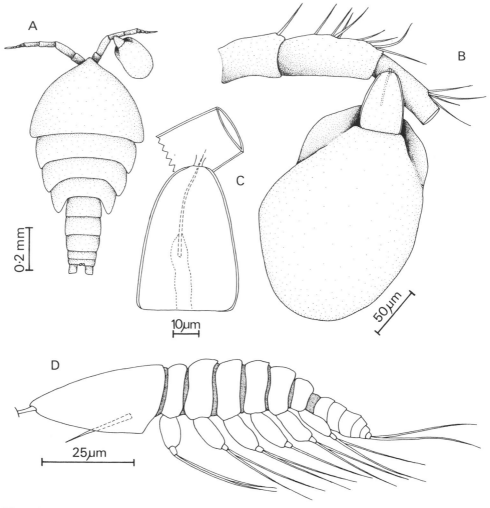

Figure 2. A. Adult *Basipodella* sp. on a deep-sea harpacticoid copepod caught at *Discovery* Stn.10379 (35°N 33°W) in the North Atlantic Ocean. B. The same specimen, dorsal view. C. The same specimen, ventral view of head. D. Juvenile stage of *Basipodella harpacticola* (redrawn from Becker 1975).

## 3 TAGMOSIS

Tagmosis in crustaceans is the combined product of both cephalization and abdominalization. Cephalization comprises two distinct phases which may or may not be synchronous. One is the formation, by fusion of the anteriormost somites, of a cephalon covered by a dorsal shield derived from fused tergites. The other involves the progressive specialization of the cephalic and anteriormost trunk limbs proceeding posteriad from the maxillules. The basic crustacean cephalon comprises five somites but in the Copepoda and Remipedia cephalization has continued producing in both, a cephalosome of six somites (incorporating the maxilliped-bearing somite), or, in some copepods, a cephalothorax of seven somites.

Specialization of the maxillae and of up to three pairs of trunk limbs as maxillipeds has almost certainly occurred independently in a number of crustacean classes.

Abdominalization is the differentiation of an ancestral continuous series of limb-bearing trunk somites into two tagmata, thorax and abdomen, typically by the loss or modification of the limbs of the more posterior somites. The abdomen may be characterized by possession of pleopods, as in malacostracans, or by the absence of limbs. The number of abdominal somites in crustaceans ranges from none in the Remipedia and Conchostraca through to a maximum of 11 in the Cephalocarida (Table 1). Abdominalization in some classes has been accompanied by a reduction in the number of trunk somites. Assuming that the urcrustacean possessed at least 31/32 trunk somites, then all classes except possibly the Remipedia, show some reduction in numbers of somites and this reaches extreme proportions in the Ostracoda. This general trend towards reduced numbers of somites continues within many of the classes. The orders Cladocera and Thoracica, for example, both have further reduced trunks compared to their more plesiomorphic counterparts within the Branchiopoda (the Conchostraca) and Cirripedia (the Ascothoracica) respectively. Reduction in numbers of body somites is clearly a widespread phenomenon within the Crustacea.

Abdominalization of posterior trunk somites and loss of trunk somites are essentially independent processes, so that high numbers of somites may be combined with no abdominalization (Remipedia) or with abdominalization of more than half the trunk somites (Cephalocarida). Similarly low numbers of somites may be combined with abdominalization of only one somite (Ostracoda) or up to half of the trunk somites (Mystacocarida). Both processes may occur progressively from the posteriormost trunk somite anteriad. Abdominalization in cephalocarids, for example, may well have occurred at least in part by this route as indicated by the reduction or loss of the posteriormost pair or pairs of trunk limbs in some genera within the group. Reduction in somite numbers has also occurred by this route as, for example, in the loss of the seventh abdominal somite in most malacostracans.

Progressive loss from the hind end forwards is not the only way the two processes can occur. Paedomorphosis provides a mechanism for both because crustacean development (Fig.3) is essentially the more-or-less regular addition of body somites through a continuous series of molts, followed by the more-or-less regular addition of limbs to these somites typically after a delay of one or more molts. The basic copepod body plan can be produced if the addition of limbs is stopped at a relatively early developmental stage, and if subsequently the posterior somites are lost (pathway A, Fig.3). The result would be the same if addition of body somites ceases at a relatively early stage, and if the posteriormost limbs are subsequently lost (pathway B, Fig.3). If one simultaneously stops the addition of both somites and limbs at the correct stage relative to each other, one has an instant and direct evolutionary mechanism by which the limbless abdomen and reduced number of somites of the basic copepod plan can be produced (pathway C, Fig.3). All three pathways can perhaps be viewed as paedomorphic because, in effect, a larval condition is being exhibited in the adult descendant.

On the basis of comparative anatomy alone it is not possible to decide which pathway is the most likely. The presence of a poorly developed pair of limbs on the last limb-bearing trunk somite can be interpreted as evidence both of progressive loss (if the limbs are vestigial) and paedomorphosis (if the limbs are rudimentary). There are reports of paired structures representing limbs on some postgenital somites of some harpacticoid copepods (see discussion in Boxshall et al. 1983) but these results need confirmation and critical reappraisal. However, it is not of prime importance to determine by which pathway or combination

Table 1. Somatic organizations of the nine extant crustacean classes. The class Branchiopoda is represented by the Conchostraca and Anostraca (with its two distinct body plans), which are tabulated separately.

| | Remi-pedia | Concho-straca | Ano-straca | Cephalo-carida | Malaco-straca | Cope-poda | Cirri-pedia | Mystaco-carida | Bran-chiura | Ostra-coda |
|---|---|---|---|---|---|---|---|---|---|---|
| 1 | Atl | Atl | Atl | Atl | Atl | Atl | Atl | Atl | Atl | Atl |
| 2 | Ant | Ant | Ant | Ant | Ant | Ant | Ant | Ant | Ant | Ant |
| 3 | Mnd | Mnd | Mnd | Mnd | Mnd | Mnd | Mnd | Mnd | Mnd | Mnd |
| 4 | Mxl | Mxl | Mxl | Mxl | Mxl | Mxl | Mxl | Mxl | Mxl | Mxl |
| 5 | Mx ] | Mx | Mx | Mx | Mx | Mx ] | Mx | Mx | Mx | Mx |
| 6 | Mxp ] | P | P-P | P | P | Mxp ] | P♀ | Mxp | P | P |
| 7 | P | P | P-P | P | P | P | P | P | P | P |
| 8 | P | P | P-P | P | P | P | P | P | P | I♂♀ |
| 9 | P | P | P-P | P | P | P | P | P♀♂ | P♀♂ | F |
| 10 | P | P | P-P | P | P | P | P | P | -F | |
| 11 | P | P | P-P | P♀♂ | P♀ | P | P | - | | |
| 12 | P | P | P-P | P | P | P♀♂ | I♂ | - | | |
| 13 | P | P | P-P | P | P♂ | - | - | - | | |
| 14 | P | P | P-P | - | P | - | - | - | | |
| 15 | P | P | P-P | - | P | - | - | - | | |
| 16 | P | P♀♂ | P-P | - | P | T + F | T + F | T + F | | |
| 17 | P | P | P-I♂♀ | - | P | | | | | |
| 18 | P | P | P - | - | P | | | | | |
| 19 | P | P | P - | - | P | | | | | |
| 20 | P | P | P - | - | - | | | | | |
| 21 | P | P | P - | - | T + F | | | | | |
| 22 | P | P | P - | - | | | | | | |
| 23 | P | P | P - | - | | | | | | |
| 24 | P | P | P - | - | | | | | | |
| 25 | P | P | I♂♀–T+F | T + F | | | | | | |
| 26 | P | P | - | | | | | | | |
| 27 | P | P | - | | | | | | | |
| 28 | P | P | - | | | | | | | |
| 29 | P | P | - | | | | | | | |
| 30 | P | P | - | | | | | | | |
| 31 | P | P | - | | | | | | | |
| 32 | P | P | T + F | | | | | | | |
| 33 | P | T + F | | | | | | | | |
| 34 | P | | | | | | | | | |
| 35 | P | | | | | | | | | |
| 36 | P | | | | | | | | | |
| 37 | T + F | | | | | | | | | |

of pathways the evolutionary changes occurred that led to the basic copepod body plan. The real importance should be attached to considering the functional pressures that govern these changes.

Abdominalization results in the differentiation of thorax and abdomen. What are the functions of these tagmata? A limbless abdomen may balance the body in the water, may be used for steering, and may carry fecal or genital products clear of the thoracic limbs (Fig.4). A thorax freed of these functions, and freed of feeding by the retention of a cephalic feeding mechanism, can become specialized for locomotion as it has in copepods,

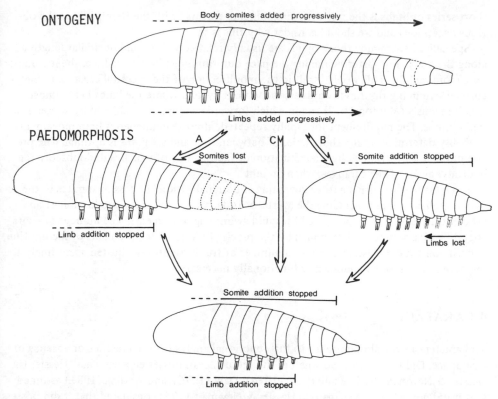

Figure 3. Schematic illustration of crustacean ontogeny and of three possible paedomorphic pathways (A, B and C) by which the basic copepod body plan may be derived.

cirripedes and branchiurans. Similarly the number of thoracic somites is open to functional analysis. As Manton (1977) showed so elegantly, the total number of locomotory trunk limbs in the various arthropodan taxa is correlated in detail with their functional needs, and the same number of limbs may arise repeatedly. In the phylum Uniramia, for example, Manton (1977) provided clear evidence that hexapody evolved independently at least five times. Possession of the same number of legs is not to be regarded as unequivocal evidence of phylogenetic affinity irrespective of functional considerations.

Crustaceans that feed using a primary thoracopod filtering system (Cephalocarida, Branchiopoda and Phyllocarida) tend to have a long series of similar multipurpose thoracopods. Those that retain a cephalic feeding mechanism and have trunk limbs specialized for swimming tend to have a short series of between four and six pairs of legs. Copepods have five pairs of swimming thoracopods, branchiurans have four and cirripedes have six. The phyllocarid *Nebalia* uses four pairs of pleopods for long distance swimming, amphipods use three or four pairs, and mysids use five pairs. Outside the Crustacea, *Limulus* uses five pairs of specialized opisthosomal limbs beating almost simultaneously for providing swimming thrust. Finally, Manton (1977) when considering the limbs of the trilobite *Olenoides* stated 'If about 6 successive legs performed the backstroke almost in unison, the animal would leave the bottom, travel forwards and sink down again'. Empirically it would appear that a short series of limbs is the rule for swimming movements (with the Remipedia as the obvious exception) and we should consider why this is so.

Specialized swimming thoracopods are usually relatively short and of similar length all along the body. For faster swimming the pace duration must be reduced, i.e. the legs must beat more rapidly. This would result in the modification of the ancestral crustacean metachronal swimming rhythm towards a copepod-like system where the limbs beat almost simultaneously (Manton 1977) and in which forward swimming is achieved by successive rapid jumps. The requirement for rapidly repeated, almost simultaneous limb movements probably determines that a short series of between four and six limbs is optimal. The precise number is determined by a combination of factors, one of which is undoubtedly the hydrodynamics of the carapace, when present.

I suggest that the driving force governing abdominalization and loss of somites in the maxillopodan groups (excluding the Mystacocarida, see below) is the production of an efficient thoracic tagma specialized for rapid swimming movements. The number of thoracopod-bearing somites is determined by the precise functional requirements of locomotion in each constituent class, and the total number of trunk somites is adjusted accordingly to preserve as much of an abdomen as functionally necessary.

## 4 CARAPACE

A character of central importance to this functional analysis is the presence or absence of a carapace. Opinions vary as to whether the urcrustacean possessed a carapace. Hessler (in Hessler & Newman 1975), Lauterbach (1974), Dahl (1976), and Watling (1981) assumed that it did not, whereas Newman (in Hessler & Newman 1975) considered that it did possess a well-developed carapace. Newman pointed to the presence of a carapace adductor muscle in the maxillulary/maxillary region, posteroventral to the esophagus in the Phyllocarida, Conchostraca, Cirripedia, and Ostracoda as evidence of the homology of the carapace. Lauterbach (1974) went further and proposed a new taxon, the Palliata, for all carapace-bearing Crustacea. Hessler's (1964) study of the skeletomusculature of *Hutchinsoniella* indicates that the adductor muscle is just one of a series of segmentally arranged horizontal muscles. As it is clear that any adductor muscle should be sited posteriorly within the cephalon in order to gain the necessary mechanical advantage to flex a large posterior outgrowth it is possible that the same muscle within an ancestral series could have been modified as an adductor on more than one occasion.

Dahl (1976) stated that 'it would seem surprising if this simple extension of the posterior edge of the cephalic shield had not been formed more than once to meet various functional demands'. Evidence corroborating this statement can be found in the Copepoda. Members of the order Misophrioida all possess a posteriorly directed carapace-like structure entirely surrounding the next thoracic somite both laterally and dorsally (Boxshall 1982). This structure is derived as a simple extension of the cephalic shield, as predicted by Dahl, but as copepods possess a cephalosome of six somites, it is an extension of the maxilliped-bearing somite not the maxillary somite. In the Maxillopoda the absence of a true carapace in the Copepoda and Mystacocarida probably represents the ancestral crustacean condition and the carapace in both Cirripedia and Branchiura, the derived condition. The presence of branched lateral midgut caeca extending into the carapaces of these two groups is here interpreted as evidence that their common possession of a carapace is a true synapomorphy.

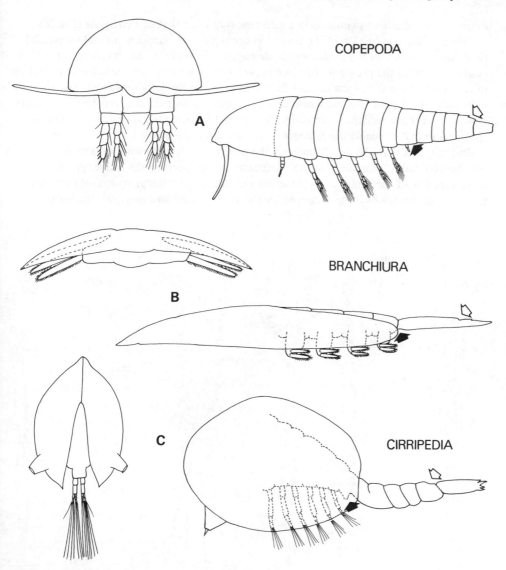

Figure 4. Anterior views and basic functional plans of representative types. A. Copepoda; B. Branchiura; C. Cirripedia. Solid arrows show position of gonopore, hollow arrows the anus.

However, the lateral caeca may prove to be a convergent adaptation to an epizoic or ecto-parasitic mode of life, and thus necessitate reconsideration of the presumed homology of the carapace in these two groups.

The carapace may have a protective, respiratory, trophic, or even reproductive role. It also has a hydrodynamic significance which has been left largely unexplored. The laterally directed carapace of branchiurans (together with the dorso-ventrally flattened body and laterally directed legs) is primarily an adaptation to ectoparasitism on a motile host, but simply by examining an anterior view (Fig.4b) it is possible to infer that the carapace of a branchiuran will affect swimming in a different way from that of a cirripede (Fig.4c).

When extreme reduction of the trunk has occurred (as in cladocerans and ostracods) there are obvious selective advantages in shortening the body so that it can be entirely contained within the protective bivalved carapace. In the Cirripedia the extreme reduction of the abdomen in the cypris larva of thoracicans can be similarly explained. The abdominalization of the posterior trunk somites of ascothoracican cirripedes represents an intermediate condition, as demonstrated by the way the 5-segmented abdomen of *Synagoga* species is folded to enable the carapace to close around the whole body.

In both the Cirripedia and Branchiura the number of thoracic somites is reduced so that the thoracopods are contained entirely beneath the carapace. In the former group, the primary function of the carapace is protective whereas in the latter it is probably streamlining, to lessen the chance of being dislodged by the water currents passing over the host.

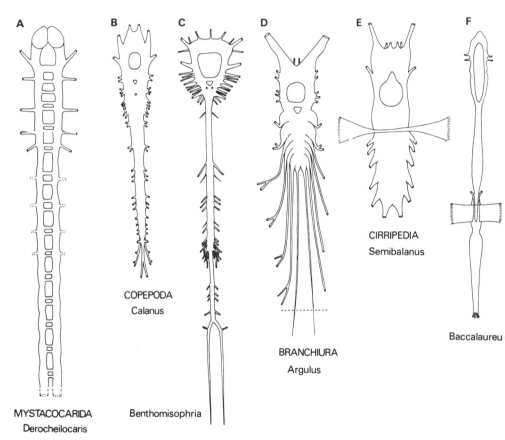

Figure 5. Dorsal view of the central nervous system of A. *Derocheilocaris,* a mystacocarid (based on data from Dahl 1952); B. *Calanus,* a copepod with a nauplius eye (redrawn from Lowe 1935); C. *Benthomisophrià,* a copepod without a nauplius eye (redrawn from Boxshall 1982); D. *Argulus,* a branchiuran (redrawn from Wilson 1902); E. the cypris of *Semibalanus,* a cirripede with compound eyes (redrawn from Stubbings 1975); and F. *Baccalaureus,* an ascothoracican cirripede without compound eyes (redrawn from Pyefinch 1937).

## 5 CENTRAL NERVOUS SYSTEM

The presence of a carapace can profoundly affect the organization of the trunk. In both branchiurans and cirripedes the condensation and anterior displacement of the thoracic ganglia of the ventral nerve cords (Fig.5d,e,f) are closely correlated with the concentration of the thoracopods beneath the carapace and with trunk reduction. In copepods (Fig.5b, c) the paired ventral nerve cords are fused within the thorax but, as in the Cirripedia and Branchiura, this is correlated with the dominance of the thorax as a swimming center. As in many other groups segmental ganglia have been lost in the limbless abdomen but the nervous system of copepods exhibits no trace of ever having been modified in response to the presence of a carapace.

The nervous system of mystacocarids is very different. Limbs are absent from trunk somites six to ten and reduced on somites two to five, but the organization of the central nervous system is uniform throughout the trunk (Dahl 1952). Its ladderlike construction is extremely plesiomorphic and shows no sign of the condensation of the Cirripedia and Branchiura, or of the progressive concentration of the Copepoda. The lack of specialization in the nervous system along the trunk indicates that swimming using a small number of rapidly beating, specialized thoracopods at the anterior end of the trunk has never been a feature of the mystacocarid lineage. This is evidence of the fundamental lack of affinity between the Mystacocarida and the other maxillopodans, in which this type of locomotion has been a key feature of their evolution and of their evolutionary success.

## 6 EYES

In a summary of his work on crustacean nauplius eyes and other photoreceptive frontal organs Elofsson (1966) distinguished four different types of these frontal eyes which he regarded as non-homologous. This has been interpreted as evidence that there are four higher groupings within the Crustacea, namely the Malacostraca, Anostraca, Phyllopoda, and Maxillopoda/Ostracoda. The mystacocarids were excluded from consideration as they lack eyes of any sort. The type of frontal eye found in the Maxillopoda/Ostracoda is distinguished from all others by the presence of tapetal cells. Tapetal cells produce layers of reflecting crystals which form a concave mirror, the function of which varies according to the presence or absence of lens cells or cuticular lenses (Andersson 1979). Cuticular lenses evolved independently within the Ostracoda and Copepoda, and it is probable that lens cells evolved independently in the Copepoda, Branchiura, and Ostracoda. The functional importance of the various components of frontal eyes is often poorly understood but convergent adaptation between photoreceptor organs appears to be a common phenomenon. Phylogenetic inferences based on frontal eye structure should therefore be made with caution.

Despite the discovery of the Remipedia, which lack compound eyes, it is now generally accepted that the ancestral crustacean possessed them and the recent description of vestigial compound eyes in cephalocarids (Burnett 1981) has tended to confirm this. The absence of compound eyes in the remipedian *Speleonectes* can be regarded as secondary, as in many other troglobitic forms. The loss of eyes in the Mystacocarida can similarly be related to its subsurface habitat. Their loss in copepods, which are presumed to have descended from an ancestor living on or near, but not in, the substratum (Fryer 1957, Kabata 1979), is much more difficult to explain. Presumably some selective advantage must have outweighed the reduced visual awareness concomittant with this loss. In those crustaceans that possess compound eyes they do not appear and become functional until the later post-nauplius stages and it is possible that copepods share some functional attributes with the early ontogenetic stages which lack compound eyes. For example, the presence of relatively well-developed and powerful extrinsic antennulary and antennary muscles originating dorsolaterally on the anterior part of the cephalic shield may be antagonistic in some way to good photoreception by compound eyes located in the same region of the head. The loss of compound eyes, however, remains for me the single greatest mystery in the evolution of the copepods.

## 7 POSITION OF THE GONOPORES

The gonopores do not have a fixed position in the Crustacea. They may occur on the first trunk somite, on the 20th, or at various positions inbetween (Table 1). The gonopores are typically on the same somite in male and female individuals, although in the Malacostraca and Cirripedia they are on different somites. In the hermaphroditic Cephalocarida the genital products of both sexes are released through a common gonopore. In those crustaceans, other than female cirripedes, with a trunk differentiated into thorax and abdomen the gonopores are typically located near the boundary between these tagmata. However, their position relative to that boundary is so variable that it cannot be used to define either tagma. The anterior displacement of the gonopore in female cirripedes is assumed to be secondary.

The Copepoda, Branchiura, and Cirripedia originally all had their gonopores located behind the last swimming leg (Fig.4). This is in accord with the thesis that the basic functional plan of all three taxa is determined by the adoption of fast swimming using rapidly beating thoracopods. The precise site of the gonopore also varies. In the Branchiura the gonopores open on the sternite of the somite bearing the last swimming legs, whereas in both copepods and cirripedes they open on the following somite. Also in cirripedes the male gonopore is on the seventh trunk limb which is modified as an intromittent organ, whereas in copepods the gonopores are situated on the sternite posterior to the seventh trunk limbs, which are modified as a mechanism to close off the gonopores.

In Crustacea the gonopores are evolutionarily mobile both in terms of which somite they open on and in terms of the precise location, i.e. on the somites or on its appendage. The position of the gonopore on any given somite may be regarded as part of the basic functional plan determined by factors such as the presence or absence of a carapace and the size of the locomotory tagma. The precise site of the gonopore is determined by other factors, such as whether brooding of eggs occurs on modified limbs, in a brood pouch within the carapace, or in an ovisac on the abdomen. One generalization it is possible to make is that those groups which primitively possessed a spermatophore (Malacostraca, Copepoda, and Mystacocarida) have gonopores on the sternite, whereas in those without (Branchiopoda, Cephalocarida, Cirripedia, and Ostracoda) they are located on the limb, which in males is often modified as an intromittent organ. The Branchiura, with their unique reproductive behaviour of cementing eggs to the substratum, do not conform to this generalization.

## 8 COMPARATIVE SPERM MORPHOLOGY

Phylogenetic inferences based on comparative sperm morphology must be made with caution. Sperm morphology is extremely variable and it is reasonable to assume that the presence of a spermatophore, or of internal fertilization, will have had a profound effect. The similarity in sperm morphology between the Cirripedia, Branchiura, and Mystacocarida has been commented upon, most recently by Grygier (1981). These sperm and those of copepods and cephalocarids are compared with a basic aquatic invertebrate type (Fig.6). The plesiomorphic condition is to have acrosome, nucleus, mitochondria, and 9 + 2 flagellum (the axoneme) arranged serially. The characters shared by the maxillopodan groups (excluding the Copepoda) are: (1) the retention of the four basic elements; (2) the anterior position of the acrosome or pseudacrosome; (3) the location of the mitochondrion in parallel to the axoneme; and (4) the location of the nucleus in parallel to the axoneme (some cirripedes only). The sperm of the ascothoracican cirripedes is more plesiomorphic than the

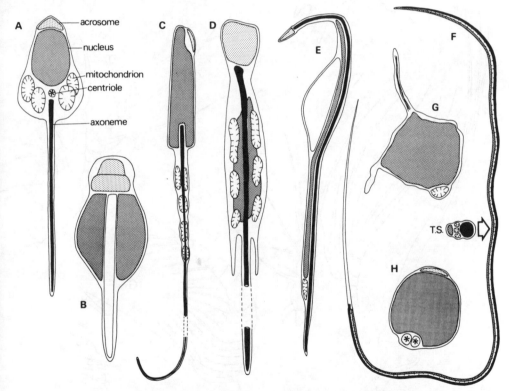

Figure 6. Some crustacean sperms in diagrammatic form, showing the arrangement of the basic organelles. A. A hypothetical ancestral aquatic sperm with the main organelles arranged in series; B. The cephalocarid *Hutchinsoniella* (from micrographs in Brown & Metz 1967); C. An ascothoracican cirripede (from micrographs and data in Grygier 1981); D. The mystacocarid *Derocheilocaris* (redrawn from Brown & Metz 1967); E. The cirripede *Balanus* (redrawn from Munn & Barnes 1970); F. The branchiuran *Argulus* (redrawn from Wingstrand 1972); G. The copepod *Chondracanthus* (from micrographs in Rousset et al. 1978); H. The copepod *Naobranchia* (from micrographs in Manier et al. 1977).

Figure 7. The mandible of A. an adult copepod; B. a naupliar copepod; C. a naupliar cephalocarid; D. a mystacocarid; E. a naupliar cirripede; F. a naupliar branchiuran; G. a phosphatocopine ostracod [B, C and D redrawn from Sanders 1963; E from Barnes & Achituv 1981; F from Wilson 1902; and G from Müller 1979].

others as the nucleus is located anterior to the axoneme. Copepod sperm is aberrant possessing a large nucleus, often irregular in shape, and sometimes possessing centrioles, mitochondria, or processes with a microtubular structure. The cephalocaridan sperm retains some primitive features, but the axoneme is replaced by a posterior projection containing microtubules.

Grygier (1981) correctly stated that if these similarities are accepted as synapomorphies then the Cirripedia would become paraphyletic and the Ascothoracica with their more plesiomorphic sperm should be elevated in taxonomic status. These similarities, however, cannot be synapomorphies because of the large number of true synapomorphies between the Ascothoracica and the other Cirripedia (the positions of the gonopores is one of the more important of these). The ascothoracican sperm is therefore the sperm of the ancestral cirripede stock. The common possession of the four basic elements in cirripedes, branchiurans and mystacocarids is a symplesiomorphy while the rearrangement of these elements from a serial to a parallel position occurred independently in each group. Sperm morphology thus offers no support for the concept of the Maxillopoda as a taxon.

# 9 APPENDAGES

The marked similarities of the cephalic appendages between adult copepods and mystaco-carids have been taken as evidence of close phylogenetic affinity, as have the similarities between the naupliar limbs of copepods and cirripedes. The mandible, for example, has a basic structure of a large coxa bearing a gnathobase medially, and a basis bearing a distal palp comprising a two or three-segmented endopod and an exopod of up to seven segments (Fig.7). This basic plan is common to all maxillopodan groups, other than the branchiurans which have a reduced palp. It is also similar to that of larval cephalocarids (as shown by Sanders 1963) and to that of the Cambrian phosphatocopine ostracods, and it may represent the plan of the mandible of at least the larval urcrustacean and possibly the adult as well. These similarities are real but it is erroneous to interpret them as evidence of close phylogenetic affinity because they are symplesiomorphies and may just represent a common point through which all crustacean classes passed. The basic plan of the antenna and its distribution throughout the Crustacea can be similarly analysed and interpreted.

The antennules and postmandibular limbs are more difficult to interpret. The recent discovery of the Remipedia and Phosphatocopina both with biramous antennules necessitates some updating of the assumption made by Hessler & Newman (1975) that the urcrustacean possessed a uniramous multi-segmented antennule. The loss of one ramus (the endopod?) has probably occurred independently a number of times. The similarities between the uniramous maxillule of larval cephalocarids, branchiurans, and remipedians; the uniramous maxillae of copepods, mystacocarids, branchiurans, and remipedians; and the uniramous maxillipeds of copepods and remipedians probably all represent modifications of the same ramus (the endopod) for a trophic function. It would be unwise, however, to select the superficial similarity between the limbs of any two of these groups as evidence of affinity as many are undoubtedly the product of convergent evolution.

What is the functional basis for assuming that many of the similarities between the more anterior cephalic appendages are symplesiomorphies? Dahl (1976) considered that the urcrustacean possessed a double feeding mechanism; one involving the cephalic limbs and present in the larval stages, and the other involving a long series of metachronally-beating

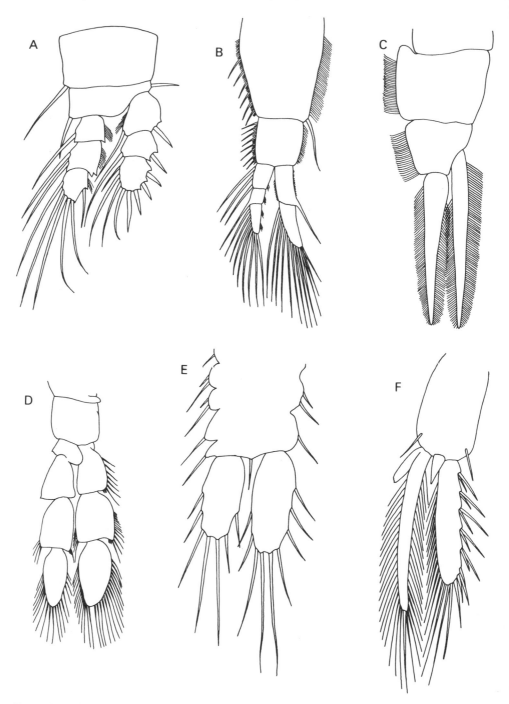

Figure 8. Crustacean trunk appendages specialized for swimming locomotion. A. copepod; B. ascothoracican cirripede; C. branchiuran; D. remipedian (redrawn from Yager 1981); E. lipostracan branchiopod (seventh to eleventh trunk limb, from Scourfield 1926); F. phyllocarid malacostracan pleopod (redrawn from Kensley 1976).

locomotory/trophic trunk limbs present in the adult. As Sanders (1963) points out, both mechanisms persist simultaneously in larval cephalocarids with the cephalic mechanism being lost only in the adult. The retention of the larval cephalic feeding mechanism and the suppression of thoracopod filter-feeding in the adult presents (as Dahl said for the Eumalacostraca) an unlimited field for specialization of the trunk. Similarly it permits specialization of the cephalic feeding appendages and diversification of feeding strategies. In copepods, cirripedes, ostracods, and eumalacostracans this potential has been fully exploited, resulting in considerable adaptive radiation, so that these classes have become highly speciose and are the dominant crustaceans in most habitat types. The relative success of the Cladocera in freshwater habitats may also be attributed to the separation of the locomotory and trophic functions of the ancestral thoracopods, but in this case, the antennae retain the main swimming function and permit the specialization of the thoracopods for feeding. A cephalic feeding mechanism is retained in the Branchiura, Mystacocarida, and Remipedia but adaptive radiation has not occurred. This may be due in the Branchiura to the early specialization of the lineage for an ectoparasitic mode of life. In mystacocarids this failure may be attributed to the reduced functional importance of the trunk limbs insofar as both trophic and locomotory functions are performed by the cephalic appendages. In the Remipedia the lack of serial differentiation within the long trunk may be a more significant factor.

Trunk limbs specialized for swimming are similar in basic morphology in many crustacean groups (Fig.8). Typically, the limb comprises a large flattened protopod and two flattened paddle-like rami. the surface area of which is often increased by marginal fringes of setae. The swimming limb of the remipedian *Speleonectes* (Fig.3d) has a relatively small protopod and the cross-sectional area of the trunk somites is also small. Unless the extrinsic promotor/remotor muscles of any given limb extend into the adjacent somite these muscles will be short and the legs will only beat relatively slowly. The two members of the limb pair in copepods are fused by their medial surfaces to a common intercoxal sclerite. This sclerite (the coupler of Manton 1977) and its associated anatomy was identified by Manton (1977) as the key to the achievement of rapid jumping in the Copepoda.

## 10 SUMMARY

The retention of the ancestral larval cephalic feeding mechanism in the adult stage is the first step in the evolution of all the maxillopodan groups, and of the Eumalacostraca, Ostracoda, and Remipedia. This is a key step in their evolution as it represents a prerequisite for the specialization both of the thorax as a locomotory center and of the head as a trophic center. If we consider the maxillopodan groups only, other important steps are the reduction in numbers of body somites and the differentiation of the trunk into thorax and abdomen. These two steps are independent but are functionally linked. Anatomical evidence suggests that the over-riding functional determinants of body form vary and that, within the Maxillopoda, three independent lineages can be recognised, the Copepoda, the Cirripedia/Branchiura, and the Mystacocarida. In the first two lineages body form is determined by the evolution of fast swimming using rapidly beating, specialized thoracopods. In order to achieve this the legs must beat almost in unison and it is this which determines that a short leg series is optimal. The evolution of this type of locomotion in the Cirripedia/Branchiura lineage is profoundly affected by the presence of a carapace, as can be seen

1- Marrella ; 2- Mimetaster ; 3- Phosphatocopina ; 4- Waptia ; 5- Branchiocaris ;
6- Canadaspis ; 7- Priscansermarinus.

from the condensation and/or anterior displacement of the central nervous system. The independent evolution of the same type of locomotion in copepods has resulted in fusion of the paired ventral nerve cords in the thorax, but the general facies of the central nervous system indicates that this process has not been influenced by the presence of a carapace. It is probable that a carapace has never been a feature of the copepod lineage. In both the Copepoda and the Cirripedia/Branchiura the position of the gonopores is dependent upon the functional differentiation of the trunk into thorax and abdomen, and little phylogenetic significance can be attached to any correspondence in the absolute position of the gonopores on any particular somite.

The evolution of the Mystacocarida has been determined by a completely different set of functional priorities. Swimming using a short series of anteriorly located specialized thoracopods has probably never been a feature of this lineage, as indicated by the extreme plesiomorphy of the central nervous system. After the retention of cephalic feeding and swimming mechanisms the key steps in the origin of the Mystacocarida are the reduction in body size and the reduction of the thoracopods as adaptations to an interstitial mode of life.

In cases where ancestral characters or mechanisms are retained in adult descendants and in cases where numbers of somites or limbs are reduced, it is possible that paedomorphosis is involved. Indeed, paedomorphosis appears to be a widespread evolutionary phenomenon and has been implicated in numerous cladogenetic events both in the Crustacea as a whole and in the Maxillopoda in particular. In the Maxillopoda, however, there is good evidence to suggest that the selective pressures determining the directions of evolutionary change in the Copepoda, the Cirripedia/Branchiura, and the Mystacocarida are different, especially in the latter group. It is probable that reduction in numbers of somites and the differentiation of the trunk into thorax and abdomen has occurred independently in these three lineages and that the Maxillopoda is not a valid taxon. Each of these independent lineages could be derived by a process of paedomorphosis from the ancestors of the Cephalocarida, Malacostraca, and Remipedia with equal ease because the similarities in appendage structure that they share were probably also shared with the larval stages of the ancestors of all these other taxa as well as with the larval urcrustacean itself.

If the taxon Maxillopoda is unacceptable, what should replace it? The studies of Whittington, Briggs and co-workers on the Burgess Shale arthropods indicates that in the late Precambrian and early Palaeozoic there was a large pool of non-trilobite arthropods comprising crustacean-like types and trilobitomorphans. The crustacean types (Fig.9) probably varied considerably both in numbers of somites and in the degree of serial differentiation of the somites and their limbs. Only a few of the basic functional types proved to be successful and I would suggest that these form the crustacean classes extant today. Each class has undergone adaptive radiation to a greater or lesser extent but this involves primarily modification upon the basic functional plan determined when the group first originated.

---

Figure 9. A diagrammatic summary of the present state of knowledge of the phylogeny of the crustacean classes. The stippled area represents an ancestral pool of crustacean types. This probably represents only part of a larger pool comprising both crustacean and trilobitomorphan types and the *Marrella-Mimetaster* lineage may represent a possible intermediate type [known fossil taxa are shown as numbered solid squares]. According to Briggs (this volume), *Branchiocaris* is not a crustacean, and *Priscansermarinus* may not be a barnacle.

The Cirripedia and Branchiura I would tentatively group together because of their common possession of a similar type of carapace. All the other classes appear to be characterised by combinations of unique, plesiomorphic and convergent characters, and I can find no reliable synapomorphies upon which to base any higher groupings at present. Therefore, these classes are all shown (Fig.9) as arising independently from a common ancestral type. This common ancestor does not, however, conform to the urcrustacean envisaged by Hessler & Newman (1975) because, in my opinion, the astonishing lack of serial specialization in the Phosphatocopina separates the Ostracoda from all other Crustacea. In the phosphatoco-pines all limbs, except possibly the antennules, conform to the same basic plan, i.e. that of the mandible (Fig.7g). This plan is retained only in the three anteriormost limbs of other crustacean classes, in most of which the postmandibular limbs are more readily derived from the mixopodium of Hessler & Newman's (1975) urcrustacean. As the phosphatocopine condition appears to be the more plesiomorphic then perhaps we should look for precursors to both in a larger pool of ancestors, comprising possibly both crustacean and trilobitomor-phan types. The contiguous nature of this pool is indicated, more as a matter of faith than evidence, by the asymmetry of the ancestral crustacean pool and by the presence of the trilobitomorphans *Marrella* and *Mimetaster.*

## 11 ACKNOWLEDGEMENTS

I am grateful to the many colleagues in the Departments of Zoology and Palaeontology of the BM(NH) who helped in the preparation of this paper. I would particularly like to record my gratitude to Dr Roger Lincoln both for his pertinent comments on the manuscript and for many enjoyable discussions of the ideas expressed here.

## DISCUSSION

CRACRAFT: I'm not a specialist on crustaceans, but I have spent a little time reading the literature on crustaceans because people who work on them spend a great deal of time talking about how important functional morphology is. It seems to me the bottom line is that structures we deem similar are homologies or synapomorphies. Nowhere in your talk did you indicate how functional morphology can let me know if something is a synapo-morphy or not. It seems we have very few alternatives: either characters have a similar function, in which case they are synapomorphic; or they have a dissimilar function and they are synapomorphic; or they have a dissimilar function and they are not synapo-morphic; or they have a similar function and they are not synapomoprhic. I don't know what you had to say except that you did not want to do cladistic analysis because it is straightforward. You have gone to a phylogenetic diagram which has no information con-tent from what we heard before [Grygier] which had a lot of information content. So I don't understand what functional morphology has to contribute to all of this.
BOXSHALL: I think the characters that those cladograms are constructed upon are inter-pretable in a completely different way.
CRACRAFT: You didn't interpret them. You have no structure to your alternative clado-gram.
BOXSHALL: We have *no* evidence of any structure as yet!

NEWMAN: You've ignored the evidence that has been given here by others.

BOXSHALL: What evidence?

NEWMAN: You threw out the sperm?

BOXSHALL: Yes. If you accept the ascothoracican sperm as the basic cirripede sperm, then these movements from serial to parallel orientation are convergent. I'm willing to accept that most of the these groups might be derived from some kind of malacostracan larva or urmalacostracan larval ancestor. But I think they've come about independently. I don't think they've come from some urmaxillopodan. I think the differences between the mystacocarids and the rest are so great you cannot have anything that would have worked as a common ancestor between them.

CRACRAFT: It seems the bottom line of your methodology is that if it is different it cannot be related. Some of these other gentlemen have pointed out there are similarities. You are pointing out there are differences. Who do we believe?

BOXSHALL: It depends on whether you regard these similarities as synapomorphies or convergences. I tend to regard the characters as convergences which they interpret as synapomorphies.

CRACRAFT: Indeed, the only way you can do that is to show non-congruent characters or have another branching diagram.

BOXSHALL: The noncongruence is in, for example, the position of the gonopores. That disrupts two of the previous cladograms we have seen. That is a complex synapomorphy, not just a single one but a double one!

NEWMAN: I was interested in the comment that where there is a carapace, the nervous system is advanced. You used a balonomorph barnacle as an example, and that of course fits ascothoracicans who have a curiously advanced nervous system. But it was noted in the Treatise that the lepadomorphs were more primitive. *Lepas* for example is very primitive for a thoracican barnacle with a ladder-like arrangement. They are even more generalized than your mystacocaridan, yet you derive your mystacocaridan from others.

BOXSHALL: Doesn't it [central nervous system] stop though at the end of the thorax.

NEWMAN: Yes, though that's because they have a reduced abdomen. That's true.

BOXSHALL: In mystacocarids it goes to the end of the body. Lepadomorphs don't have the end of the body, but the ascothoracicans do and yet in these the nervous system stops.

NEWMAN: In the ascothoracicans the nervous system is condensed into the thorax. You don't have paired ganglia throughout the body, in *Lepas* you do. Because the abdomen is reduced you don't have the pairs there, but you do throughout the thorax.

BOXSHALL: It depends on what nervous system you would put in your cirripede.

NEWMAN: Look at *Lepas*!

BOXSHALL: But in ascothoracicans where they do have an abdomen, it still stops in the thorax.

GRYGIER: Nonetheless, the thorax in the urcirripede in your conception would have had the ladder-like nervous system. Ascothoracids no longer have it but these are supposed to be intermediate in terms of some of the other characters you are discussing, it is just that they are not so in regards the nervous system.

SCHRAM: I feel we must terminate this interesting argument and go on to our next speaker, Dr Robert Hessler of Scripps Institution of Oceanography.

HESSLER: This room has started to heat up, so I think I will remove my coat. It is going to get a lot worse before it gets better.

# REFERENCES

Andersson, A. 1979. *Cerebral Sensory Organs in Ostracodes (Crustacea).* Unpublished PhD Thesis, Lund Univ. 144pp.

Barnard, H.M. 1892. *The Apodidae, a Morphological Study.* London: MacMillan.

Barnes, M. & Y.Achituv 1981. The nauplius stages of the cirripede *Tetraclita squamosa rufocincta* Pilsbury. *J. Exp. Mar. Biol. Ecol.* 54:149-165.

Becker, K.-E. 1975. *Basipodella harpacticola* n.gen., n.sp. (Crustacea, Copepoda). *Helgoländer wiss. Meeresunters* 27:96-100.

Beurlen, K. 1930. Vergleichende Stammesgeschichte. *Fortschr. Geol. Palaeont.* 8:317-586.

Boxshall, G.A. 1979. The planktonic copepods of the northeastern Atlantic Ocean: Harpacticoida, Siphonostomatoida, and Mormonilloida. *Bull. Br. Mus. Nat. Hist. Zool.* 35:201-264.

Boxshall, G.A. 1982. On the anatomy of the misophrioid copepods, with special reference to *Benthomisophria palliata* Sars. *Phil. Trans. Roy. Soc. London* (B), 297:125-181.

Boxshall, G.A., F.Ferrari & H.Tiemann 1983. The ancestral copepod: towards a consensus of opinion at the First International Conference on Copepoda. *Crustaceana* suppl., in press.

Bradford, J.M. & G.C.Hewitt 1980. A new maxillopodan crustacean parasitic on a myodocopid ostracod. *Crustaceana* 38:67-72.

Brown, G.G. & C.B.Metz 1967. Ultrastructural studies on the spermatozoa of two primitive crustaceans, *Hutchinsoniella macracantha* and *Derocheilocaris typicus. Z. Zellforsch. mikrosk. Anat.* 80:78-92.

Burnett, B.R. 1981. Compound eyes in the cephalocarid crustacean *Hutchinsoniella macracantha. J. Crust. Biol.* 1:11-15.

Calman, W.T. 1909. Crustacea. In: E.R.Lankester (ed.), *A Treatise on Zoology* 7:1-346. London: Adam & Charles Black.

Claus, C. 1876. *Untersuchungen zur Erforschung der genealogischen Grundlage des Crustaceen-Systems.* Wien.

Dahl, E. 1952. Reports of the Lund University Chile Expedition 1948-49. 7, Mystacocarida. *Lunds Univ. Arsskr.* 48(6):1-41.

Dahl, E. 1956. Some crustacean relationships. In: K.G.Wingstrand (ed.), *Bertil Hanström: Zoological Papers in Honour of his Sixty-fifth Birthday, November 20th, 1956:*138-147. Lund: Zool. Inst.

Dahl, E. 1963. Main evolutionary lines among recent Crustacea. In: H.B.Whittington & W.D.I.Rolfe (eds), *Phylogeny and Evolution of Crustacea:*1-15. Cambridge: Mus. Comp. Zool.

Dahl, E. 1976. Structural plans as functional models exemplified by the Crustacea Malacostraca. *Zool. Scr.* 5:163-166.

Elofsson, R. 1966. *The Nauplius Eye and Frontal Organs in Crustacea. Morphology, Ontogeny and Fine Structure.* Lund: Zool. Inst.

Fryer, G. 1957. The feeding mechanism of some freshwater cyclopoid copepods. *Proc. Zool. Soc. Lond.* 129:1-25.

Gall, J.C. & L.Grauvogel 1964. Un arthropode peu connu. Le genre *Euthycarcinus* Handlirsch. *Ann. Paléont. Inv.* 50:1-18.

Grygier, M.J. 1981. Sperm of the ascothoracican parasite *Dendrogaster,* the most primitive found in Crustacea. *Int. J. Inv. Repro.* 3:65-73.

Gurney, R. 1942. *Larvae of Decapod Crustacea.* London: Ray Soc.

Hessler, R.R. 1964. The Cephalocarida: Comparative skeleto-musculature. *Mem. Conn. Acad. Arts Sci.* 16:1-97.

Hessler, R.R. & W.A.Newman 1975. A trilobitomorph origin for the Crustacea. *Fossils and Strata* 4:437-459.

Kabata, Z. 1979. *Parasitic Copepoda of British Fishes.* London: Ray Soc.

Kensley, B. 1976. The genus *Nebalia* in South and South West Africa (Crustacea, Leptostraca). *Cimbebasia* 8:156-162.

Lauterbach, K.-E. 1974. Über die Herkunft des Carapax der Crustaceen. *Zool. Beitr.* 20:273-327.

Lowe, E. 1935. On the anatomy of a marine copepod, *Calanus finmarchicus* (Gunnerus). *Trans. Roy. Soc. Edin.* 58:561-603.

Manier, J.-F., A.Raibaut, V.Rousset & F.Coste 1977. L'appareil génital mâle et al spermiogenèse du copépode parasite *Naobranchia cygniformis* Hesse 1863. *Annls Sci. nat.* 19:439-458.

Manton, S.M. 1977. *The Arthropoda. Habits, Functional Morphology, and Evolution.* Oxford: Clarendon Press.

Moore, R.C. 1969. Euthycarcinoidea. In: R.C.Moore (ed.), *Treatise on Invertebrate Paleontology. Part R, Arthropoda 4:* R196-199. Lawrence: Geol. Soc. Am. & Univ. Kansas Press.

Müller, K.J. 1979. Phosphatocopine ostracodes with preserved appendages from the Upper Cambrian of Sweden. *Lethaia* 12:1-27.

Munn, E.A. & H.Barnes 1970. The fine structure of the spermatozoa of some cirripedes. *J. Exp. Mar. Biol. Ecol.* 4:261-286.

Newman, W.A., V.A.Zullo & T.H.Withers 1969. Cirripedia. In: R.C.Moore (ed.), *Treatise on Invertebrate Paleontology. Part R, Arthropoda 4:* R206-295. Lawrence: Geol. Soc. Am. & Univ. Kansas Press.

Pyefinch, K.A. 1937. The anatomy of *Baccalaureus torrensis* sp.n. (Cirripedia Ascothoracica). *J. Linn. Soc. (Zool.)* 40:347-371.

Rousset, V., A.Raibaut, J.-F.Manier & F.Coste 1978. Reproduction et sexualité des copépodes parasites de poissons 1. L'appareil reproducteur de *Chondracanthus angustatus* Heller 1865: anatomie, histologie et spermiogenèse. *Z. parasitenk.* 55:73-89.

Sanders, H.L. 1963. The Cephalocarida: functional morphology, larval development, comparative external anatomy. *Mem. Conn. Acad. Arts Sci.* 15:1-80.

Schram, F.R. 1971. A strange arthropod from the Mazon Creek of Illinois and the trans-Permo-Triassic Merostomatoidea (Trilobitoidea). *Fieldiana, Geol.* 10:85-102.

Schram, F.R., W.D.I.Rolfe 1982. New euthycarcinoid arthropods from the Upper Pennsylvanian of France and Illinois. *J. Paleont.* 56:1434-1450.

Scourfield, D.J. 1926. On a new type of crustacean from the Old Red Sandstone (Rhynie Chert Bed, Aberdeenshire) – *Lepidocaris rhyniensis*, gen. et sp.nov. *Phil. Trans. Roy. Soc. London* (B)214:153-187.

Siewing, R. 1960. Neure Ergebnisse der Verwandschaftsforschung bei den Crustaceen. *Wiss. Z. Univ. Rostock* 9:343-358.

Stubbings, H.G. 1975. *Balanus balanoides. L.M.B.C. Memoirs on Typical British Marine Plants and Animals* 37:1-175.

Tiegs, O.W. & S.M.Manton 1958. The evolution of the Arthropoda. *Biol. Rev.* 33:255-337.

Watling, L. 1981. An alternative phylogeny of peracarid crustaceans. *J. Crust. Biol.* 1:201-210.

Wilson, C.B. 1902. North American parasitic copepods of the family Argulidae, with a bibliography of the group and a systematic review of all known species. *Proc. US Nat. Mus.* 25:635-742.

Wingstrand, K.G. 1972. Comparative spermatology of a pentastomid *Raillietiella hemidactyli*, and a branchiuran crustacean, *Argulus foliaceus*, with a discussion of pentastomid relationships. *K. dansk. Vidensk. Selsk. Skr.* 19:5-72.

Yager, J. 1981. Remipedia. A new class of Crustacea from a marine cave in the Bahamas. *J. Crust. Biol.* 1:328-333.

ROBERT R. HESSLER
*Scripps Institution of Oceanography, La Jolla, California, USA*

# A DEFENSE OF THE CARIDOID FACIES; WHEREIN THE EARLY EVOLUTION OF THE EUMALACOSTRACA IS DISCUSSED

*'The reports of my death are greatly exaggerated'.* S.L.Clemens

## ABSTRACT

The caridoid facies is a suite of features that has long been regarded monophyletic and central to eumalacostracan phylogeny. The present defense of this position considers several recent objections to the idea. Much of the caridoid facies is plesiomorphic and cannot be used to argue monophyly. The caridoid apomorphies are found in all eumalacostracans and occur with the first appearance of this taxon in the fossil record. Imperfectly developed abdominal musculature of hoplocarids reflects the early appearance of this taxon in eumalacostracan evolution. Arguments that hoplocarids evolved independently of other eumalacostracans are rejected. The claim that the carapace is polyphyletic is also considered unsubstantiated. In total, the distribution of caridoid features among taxa and in the fossil record strongly suggests the facies evolved once, concurrent with the advent of the Eumalacostraca. The caridoid facies was only part of the cause for eumalacostracan success; the loss of primitive thoracopodan feeding with the appearance of the thoracic stenopodium is likely to have been a more significant event in the genesis of the Eumalacostraca, but the adaptive forces that stimulated the evolution of the two systems may well have intertwined.

## 1 INTRODUCTION

In the study of malacostracan evolution during the last three-quarters of a century, the concept of the caridoid facies (Calman 1904) has played a dominant role. Its importance was recognized even earlier, for it is embodied in the concept of the Schizopoda (Claus 1885). As seen by its proponents, the caridoid facies is considered a primitive, central morphology from which radiated all the major branches of the Eumalacostraca. This concept and the classification with which it is associated (Hansen 1893, Calman 1904) have had a stabilizing influence on malacostracan systematics from the moment of their inception.

In the last 25 years, however, vociferous discontent with this system has emerged. Some investigators think it probable that the caridoid facies has evolved more than once to serve a pelagic life style (Tiegs & Manton 1958, Dahl 1976). Some regard one of its attributes, the carapace, to be a product of convergence (Dahl, in press, Watling 1981, Kunze 1981, Schram 1981, 1982). Schram (1981:2), in considering the caridoid facies, points out, '. . . the difficulty with facies theories built around "archetypes" is that they run the risk of freezing concepts based on the limited understanding and prejudices of the times in which they were originally formulated'.

So far, these attacks have not been answered. Is it possible that these new perceptions are so obviously correct that no adequate answer can be formulated? Is it time to abandon the caridoid facies and the basic Calman classification? I think not. These criticisms seem to be more speculations than substantial falsifications. A central, monophyletic caridoid facies still seems to be the simplest, the most parsimonious interpretation of the facts. This review of the caridoid facies will document why I hold this opinion.

## 2 NATURE OF THE CARIDOID FACIES

The term 'caridoid facies' was coined by Calman (1904:147) for a series of features shared by the Euphausiacea, Mysidacea, lower Decapoda and 'for the most part also with the Stomatopoda and Leptostraca'. His further consideration in 1909 (p.144) sets the leptostracans aside and includes the Syncarida, which was not mentioned with respect to the caridoid facies in 1904. Calman (1909) diagnosed the facies as follows: (1) carapace enveloping thoracic region; (2) movably stalked eyes; (3) biramous first antenna; (4) scale-like exopod on the second antenna; (5) natatory exopods on the thoracic limbs; (6) elongate, ven

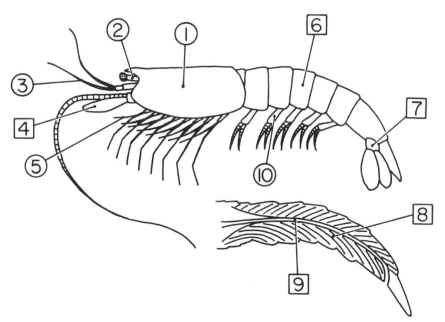

Figure 1. The caridoid facies on a diagrammatic eumalacostracan. Numbers explained in diagnosis on p.146-7. Circled numbers are plesiomorphies; squares indicate caridoid apomorphies.

trally flexible abdomen; (7) tail fan formed by the lamellar rami of the uropods on either side of the telson (Fig.1).

Hessler (1982) extended this list with (8) complex and massive abdominal trunk musculature serving strong ventral flexion; (9) internal organs mainly excluded from the abdomen; (10) pleopods I-V alike, biramous, natatory (Fig.1). These three features are compatible with the general habitus of the caridoid facies which Dahl (1976:165) summarized as 'generally prawn-like and provided with a large carapax'. Hessler also said that the thoracopodal exopods and pleopodal rami are flagelliform, but considering the foliaceous form of these rami in some paleocarid syncarids (Brooks 1969), this seems unduly specific.

The painstaking work of Daniel (1933 and earlier) best documents the necessity of including the nature of the abdominal trunk musculature in the list of features. The same general pattern of complexly intertwining muscles is repeated in the Anaspidacea, Euphausiacea, Mysidacea, and reptantian Decapoda (Daniel 1932). This muscle system is the power source for the caridoid escape reaction, wherein rapid flexion of the abdomen brings the expanded tail fan forward, thus suddenly propelling the animal backward.

Because a large mass of muscle is required, it occupies most of the abdominal cavity. Dahl (1963) noted that the abdomen of caridoids contained a minimum of viscera, even in higher forms where muscles no longer filled the cavity.

Some features attributed to the caridoid facies are actually plesiomorphic to the Eumalacostraca (Fig.1). That is, they are likely to have characterized the urmalacostracan or even the urcrustacean (Hessler & Newman 1975). These include (1) the carapace (to be considered in detail below); (2) movably stalked eyes; (3) biramous first antenna; (10) first five pleopods alike, biramous, natatory. These cannot be caridoid or eumalacostracan apomorphies because they are also found in the Leptostraca and where preserved, other phyllocarids. The natatory thoracopodal exopod (5) is probably also plesiomorphic, since the paddling motion of the exopod is an attribute of the urcrustacean (Hessler & Newman 1975). While these features participate in the caridoid appearance, they cannot be used to argue caridoid monophyly.

The remaining features can. Interestingly, nearly all relate to the caridoid escape reaction. The exceptions are the scale-like antennal exopod, which is used in controlling direction while swimming, and the tail fan, to the extent its initial function was also directional control.

## 3 THE POSITION OF CONVERGENCE IN PHYLOGENETIC THEORY

A common argument against all or parts of a central, ancestral caridoid facies is, 'How do we know it didn't evolve more than once?' Obviously, the answer is that we don't. However, this is not acceptable cause for rejection of monophyly. There is no case where one can rule out absolutely the possibility of convergence.

Monophyly is unprovable; it can only be falsified (Hessler 1982). For this reason, similarities between taxa occupy a special position in phylogenetic theory. They must be accepted as valid indicators of affinity, i.e. synapomorphies, *until compelling reasons force us* to conclude these properties reflect convergence. If we follow any other path, the decision as to whether a feature should be regarded as convergent becomes a matter of personal preference. We will have thrown away the primary objective tool of phylogenetic methodology — similarity.

Thus, it is convergence which must be proven; only then should we prefer it to monophyletic hypotheses. It is for this reason that none of us should embrace arguments of polyphyly except as a goad to hunt for evidence that will falsify the hypothesis of monophyly. Until such evidence surfaces, monophyly remains the most parsimonious paradigm.

## 4 MONOPHYLY OF THE CARIDOID APOMORPHIES

It seems simplest to first consider the phylogeny of the caridoid apomorphies (Fig.1). The plesiomorphic portion of the facies will not be discussed except for the carapace, which will be treated subsequently; the monophyly of the other plesiomorphic features (stalked eyes, for example) within the Malacostraca as a whole has never been criticized.

Schram (1981:3) states that the appearance of crustaceans in the fossil record gives 'difficulty in reconciling the fossils with the caridoid speculations based largely on living forms'. Disregarding the danger of using negative evidence in the fossil record, the statement is certainly not correct when applied to the first appearance of the caridoid morphology versus other eumalacostracan morphologies, such as in tanaids or isopods (Table 1). The earliest eumalacostracans that have been identified from Devonian beds, the eocarid *Devonocaris* and the decapod *Palaeopalaemon* (Brooks 1969, Schram 1969a, 1978, 1982) express the full suite of external caridoid apomorphies (internal characters cannot be considered). Thus, if one wishes to use it, the fossil record supports the appearance of the caridoid facies at the base of the eumalacostracan radiation. However, this first appearance should not be given undue weight. The presence of advanced taxa such as tanaids in the Lower Carboniferous testifies to the rapidity of the eumalacostracan radiation, so that the incompleteness of the fossil record could easily mask important chronological details.

It is important to note that Calman specifically included anaspidacean Syncarida among the taxa whose morphology is dominated by the caridoid facies, even though they lack a carapace (Fig.2). This is completely reasonable. Anaspidaceans express the diagnostic fea-

Table 1. First appearance of malacostracans in the fossil record. All taxa have a carapace *except* those in brackets. The first eumalacostracans, appearing in the Devonian have the full suite of caridoid apomorphies except for the Hoplocarida, where the abdominal musculature has not fully achieved the caridoid morphology.

|  | Peracarids | Other Malacostracans |
|---|---|---|
| Cenozoic | [Amphipoda] | Pancarida |
| Mesozoic |  |  |
| Permian | Cumacea |  |
| Upper Carboniferous | [Isopoda] |  |
| Lower Carboniferous | Spelaeogriphacea | [Syncarida] |
|  | Tanaidacea |  |
|  | Mysidacea |  |
| Devonian |  | Decapoda |
|  |  | Eocarida |
|  |  | Hoplocarida |
| Silurian |  |  |
| Ordovician |  |  |
| Cambrian |  | Phyllocarida |

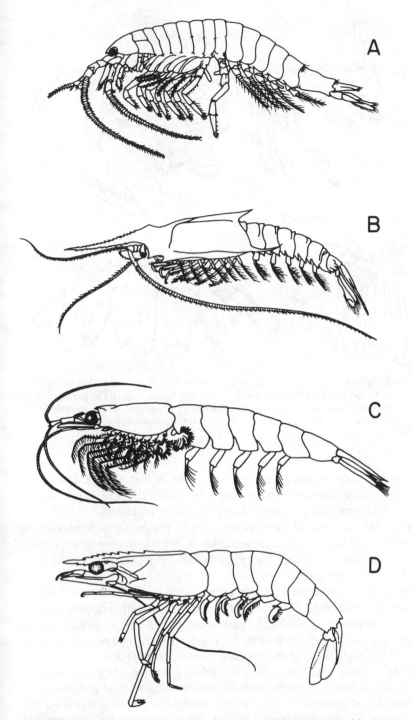

Figure 2. Four major taxa with well-developed caridoid facies. A. anaspidacean syncarid; B. lophogastrid mysidacean; C. euphausiacean; D. peneid decapod. The syncarid deviates in lacking a carapace, the decapod in having vestigial thoracic exopods (after Hessler 1982).

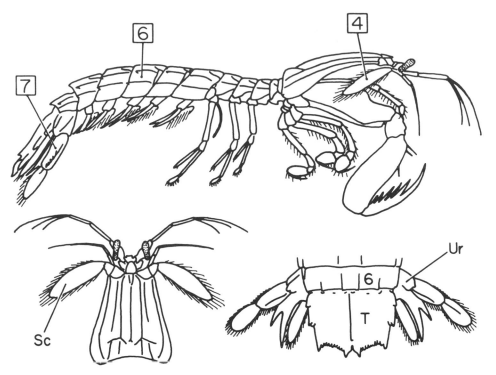

Figure 3. A stomatopod hoplocarid, showing the full external expression of the caridoid apomorphies. Numbers in squares refer to apomorphies listed on p.146-7.(also, see legend, Fig.1).The large protopodal spine separating uropodal exopod and endopod is a stomatopodan apomorphy not seen on early hoplocarids. Symbols: Sc − antennal scale; Ur − uropod; T − telson (after Calman 1909).

tures in every other way, including the escape reaction (Silvey & Wilson 1979, Kunze & Hessler, personal observation). *Anaspides* is caridoid in form even though it is primarily a benthic organism, usually entering the water column only when frightened.

Among the Peracarida, the caridoid apomorphies are fully present in the Mysidacea, and can even be observed in the Mysida, which is specialized in important other ways compared to the Lophogastrida. In the Eucarida, all the caridoid apomorphies can be found in the Euphausiacea and natantian Decapoda.

Thus, all of the caridoid apomorphies occur together on some members of the Syncarida, Peracarida, and Eucarida (Fig.2). This set of features is sufficiently complex that unless contrary evidence were forthcoming, it constitutes a strong synapomorphic suite. In none of the papers criticizing the caridoid facies does such contrary evidence appear. Indeed, Watling's (1981) urmalacostracan incorporates many of these features.

Some divisions do not display the caridoid apomorphies perfectly. The Pancarida is clearly a derived group, as indicated by the absence of an antennal exopod and loss or reduction of pereopods and pleopods. It has a weakly caridoid abdomen as seen from its uropods. The abdominal trunk musculature is modest, but its pattern is basically caridoid, strongly suggesting a caridoid ancestry (Hessler 1964).

The Hoplocarida possess many of the caridoid apomorphies: antennal scale, ventrally flexible abdomen and a tail fan (Fig.3). Their abdominal trunk musculature is not fully caridoid, but shows all the basic caridoid elements except for the transverse muscles (Hessler 1964). Nor do these abdominal trunk muscles fill the abdomen. Instead some of this space is occupied by the gonads (Dahl 1963). These two deviations from the caridoid facies are what one would expect of the evolving caridoid condition (Hessler, in comments following Dahl 1963, 1964). This interpretation is consistent with the belief that hoplocarids are an early eumalacostracan offshoot of the line leading to the full caridoids (Calman 1909, Siewing 1963, Burnett & Hessler 1973, Hessler 1982) (Fig.4). It is compatible with the early appearance (Devonian) of hoplocarids. In this view, the many remarkable features that characterize the Hoplocarida, including those of the digestive system (Kunze 1981) are apomorphies (Table 2). Contrary to Schram's (1973, fig.9) early claim that the hoplostracan phyllocarids are the independent ancestors of the hoplocarids, current thinking does not relate the two taxa (Rolfe 1981, Schram 1982). Thus, as yet there is no concrete evidence to show that the unique hoplocaridan features did not arise in a eumalacostracan that already had evolved a portion of the caridoid facies. That is, the hoplocaridan apomorphies do not disprove caridoid monophyly.

Kunze (1981: this volume) and Schram (1982 and earlier) strongly oppose this point of view. The reader is advised to consult their papers directly; I cannot do justice to their arguments. Only a few aspects of the debate will be aired here. Schram (1982) rightly criticizes Burnett & Hessler (1973) for neglecting the Aeschronectida when they claimed there is no evidence for the coxal origin of pleopodal gills in hoplocarids. However, these epipodal gills do not demonstrate the independent origin of the hoplocarids, for they are best interpreted as extreme plesiomorphies of the urcrustacean ancestor (Siewing 1963, Hessler 1982); they are moot in the present context.

The hoplocaridan uropod has been described as 'blade like' and the telson 'styloid' (Schram 1969b) but, as seen from Schram's figures, this difference from the caridoid tail fan is not fundamental (see in particular his Figure 131b). Many of the early hoplocarids had an abdomen that looks truly caridoid, and a well-developed antennal scale. The latter is uniquely two-segmented (Hessler 1982), a difference whose importance cannot yet be resolved; reduction of podomere number is a common change in crustacean limb evolution.

Schram (1982) emphasizes the lack of a full caridoid escape reaction in stomatopods. However, this is no more than what one would expect in a precaridoid. Even then, the stomatopod flexure has much of the caridoid behavior; it is similar to that of syncarids, as seen in *Anaspides* (Silvey & Wilson 1979). The stomatopod's erichthus larva utilizes the full caridoid escape reaction and the ventral flexion of the abdomen in adults can be so powerful as to launch a disturbed captive animal backward out of its holding tank (Newman personal communication).

Much of the case for independent origin of hoplocarids is built upon the belief that they achieved six pretelsonic abdominal segments by fusion of two of its anterior somites, rather than posterior ones as in other eumalacostracans (Manton 1928a,b). Depending on the theory, different anterior segments are involved. Komai & Tung (1983) and Siewing (1963) thought details of the circulatory system in *Squilla* bespoke fusion of the first and second abdominal somites. However, Burnett (1973) did not find this condition in *Hemisquilla*. Reaka (1975) saw in the pattern of stomatopod molting sutures evidence that the first abdominal somite had fused to the last thoracic.

Two lines of evidence argue in favor of fusion of the sixth and seventh pretelsonic somites rather than anteriorly (Fig.6). First, Shiino (1942) found ontogenetic evidence of it.

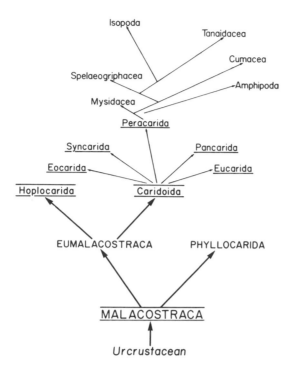

Figure 4. Phylogeny of the Malacostraca. Only in the Peracarida is this carried to order; here, the shortness of the line after its branch point for mysidaceans (lophogastrids) and spelaeogriphaceans reflects their relative lack of major external apomorphies.

Second, the seventh abdominal segment is the only one not known to have an appendage in any phyllocarid. The presence of an appendage on all six pretelsonic abdominal somites in hoplocarids is thus compatible with fusion of the sixth and seventh somites (Hessler 1982). If anterior somites had fused, a limb would have been lost. There is not the slightest evidence such a loss ever took place. These arguments seem more substantial than that of molting sutures, where functional constraints are more likely to dictate pattern than would lingering historical imprint.

Kunze (1981) points out that similarities between phyllocarids and hoplocarids argue against eumalacostracan monophyly. She mentions the cephalic kinesis, cephalic musculature, abdominal viscera and abdominal skeletomusculature. If the last two features are precaridoid, as already suggested, her objection is answered. This may also explain the first two features. Grobben (1917) has found remnants of the cephalic kinesis and musculature in decapods.

The discussion here has dwelt on the question of hoplocaridan affinities because if this taxon evolved independently with respect to the other eumalacostracans, then those aspects of the caridoid facies which hoplocarids do possess must have evolved independently as well. The present analysis does not find sufficient substance in the independent origin argument to require us to accept such convergence (Fig.4).

In summary, the fossil record and the distribution of features among taxa are consistent with the hypothesis that the caridoid apomorphies (antennal scale and abdomen) are monophyletic and must occupy a central position in eumalacostracan evolution.

## 5 THE CARAPACE

The carapace has been the focal point for most of the disagreement with caridoid monophyly. Dahl (1976) questions the necessity of providing the urmalacostracan with a carapace. Watling (1981) uses its absence as a starting point. Schram (1981) regards the possibility of its monophyly unknowable. In the present discussion of these criticisms the issue
will be limited to the Malacostraca; although the monophyly of the carapace between crustacean classes is also subject of much debate, it is irrelevant in the present context.

The distribution of a carapace among taxa is of paramount importance (Table 1). It is
found in all phyllocarids, living and fossil; this taxon is judged more primitive than the
Eumalacostraca on the basis of independent criteria (Hessler 1982). Among the Eumalacostraca, it is present in all groups, living and fossil, except the division Syncarida and
orders Isopoda and Amphipoda. Malacostracans with a carapace appear in the Cambrian
(Rolfe 1969), long before the first appearance of a taxon without one (Carboniferous)
(Schram 1969a). Contrary to Dahl's (1976) claim that there is no obvious reason to endow
the urmalacostracan with a carapace, these lines of evidence lead compellingly to that conclusion.

The arguments against a monophyletic carapace are few. Dahl (1976) sees problems in
deriving syncarids, isopods, and amphipods from an ancestor with a carapace. Watling
(1981) regards the cumaceans and tanaids as too specialized to be satisfactory intermediates between a caridoid with full carapace and the isopods. Possible other objections are
too diffusely expressed to be identified.

One might argue that even if the urmalacostracan possessed a carapace, the ancestral
eumalacostracan lacked one. This is a major requirement of Watling's hypothesis, which
derives this animal from the carapaceless version of the urcrustacean (Hessler & Newman
1975). A difficulty with this senario is the presence of a carapace on all known phyllocarids, thus making necessary a carapaceless ancestor for which there is absolutely no evidence.
Dahl (1976), with Watling's (1981) concurrence, notes the lack of pleura on thoracomeres
covered by a carapace. Therefore, the presence of thoracic pleura in amphipods and isopods
would imply a carapaceless ancestry. However, these two orders do not have thoracic
pleura at all; the lateral outgrowths are epimeral plates of coxae which are broadly attached
to the body.

Amphipods and isopods are specialized in many different ways: unstalked, sessile eyes;
one of the antennular flagella reduced or absent; highly modified maxilliped; no thoracic
exopods; no thoracic epipods (isopods only); little evidence of caridoid abdomen (amphipods only). Therefore, there is no independent reason for regarding their lack of a carapace
as primitive. This is not true with syncarids. Anaspidaceans and paleocaridaceans are won-
derfully primitive eumalacostracans in many ways. One might well question whether their
lack of a carapace is also primitive. However, they are not substantially more primitive than
lophogastrid mysids (Fig.2) or some of the eocarids (whatever they might prove to be),
which have a full carapace. In view of the wide distribution of a carapace among malacostracans, it does not make good sense to give more weight to the lack of a carapace in syncarids than the presence of a carapace in lophogastrids.

The various grades of evolution represented by mysidaceans, spelaeogriphaceans, cumaceans, and tanaidaceans give adequate clues how the carapace could be lost in isopods (Fig.
5). Unfortunately, such clues are totally missing for syncarids and amphipods. A similar
weakness plagues the opposite argument as well; there is no evidence concerning the origins
of these taxa.

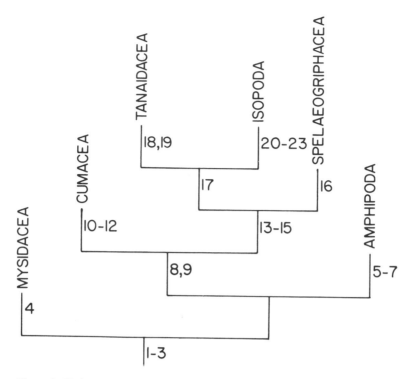

Figure 5. Cladogram of the Peracarida, based on major external features of the thoracic region. The cladogram stems from a presumed ancestor with a full carapace that is not fused to any thoracic segment, thoracopods I-VIII alike, exopods and epipods on all thoracopods, and epipodal respiration. Numbers adjacent to a clade refer to its apomorphies, as listed below. The amphipods could also have been put as a branch from the Mysidacea without loss of parsimony. Apomorphies: (1) thoracomere 1 (T1) fused to cephalon; (2) thoracopod I (T1) a maxilliped; (3) TI epipod a bailer; (4) TI epipod nonrespiratory; (5) carapace lost; (6) TI-VIII exopods lost; (7) TI epipod lost; (8) carapace reduced, not covering T4-8 or T5-8; (9) TII–VIII epipod lost; (10) T2-3 or T2-4 fused to cephalon; (11) TII-V are maxillipeds or involved in trophic functions; (12) TII, VIII exopod always lost, TIII-VII sometimes lost; (13) carapace further reduced, not covering T3-8; (14) carapace with respiratory surface; (15) TI exopod lost; (16) TV-VIII exopods respiratory; (17) TIV-VIII exopod always lost, TII, III exopods vestigial; (18) T2 fused to cephalon; (19) TII involved in nonlocomotory functions; (20) carapace lost; (21) TI epipod nonrespiratory, nonbailer; (22) TII-TIII exopod lost; (23) pleopods respiratory.

Watling's (1981) phylogeny requires the evolution of a carapace four times in the Malacostraca: (1) Phyllocarida and Hoplocarida, (2) Eucarida, (3) non-isopodan mancoids, and (4) Mysidacea and Pancarida. The more traditional scheme espoused here (Figs.4, 5) begins with a carapace and requires no independent derivation. It does require reduction and even loss several times: (1) Hoplocarida, (2) Syncarida, (3) Pancarida, (4) Mysida, (5) Amphipoda, and (6) mancoids (Fig.4). However, it is far easier to envision the multiple reduction or loss of a complex structure than its multiple genesis, as for example in the case of eyes.

Dahl (in press) is concerned with the problem of how animals with many freely articulated thoracomeres, as in the carapaceless isopods or amphipods, could have evolved from an ancestor with thoracomeres fused to the cephalon, as in forms with a carapace like cumaceans or the Mysida. Even disregarding the possibility that fusion is not necessarily evolutionarily irreversible, the problem is not as serious as indicated. Only in the Peracarida and Pancarida is it of any concern; in eucarids, fusion is always complete, and the carapace is never reduced. For all higher peracarids, the lophogastrid Mysidacea give the model for an ancestor with a carapace, and here, only the first thoracomere is fused (Fig.5). The Spelaeogriphacea is the most reasonable representative of a generally intermediate condition with respect to isopod evolution (see below). It too has only one fused thoracomere, as do isopods. Tanaids have two, and cumaceans have three to four fused thoracomeres, rarely five to six. The increased degree of fusion shows a general correlation with the number of thoracopods modified as maxillipeds or primarily devoted to food acquisition and non-locomotory functions. In the taxa with one fused thoracomere, there is only a single maxilliped. The greater number of fused somites in cumaceans and tanaids relates to their five and two pairs of trophic limbs, respectively. Since the chelate second thoracopod in tanaids and trophically specialized thoracopods two to five in cumaceans are surely apomorphies of these respective orders, it seems likely that the greater degree of carapace fusion is also apomorphic (Fig.5). Thus, the peracarids give no examples where degree of fusion needed to be reduced during evolution; the degree of fusion in isopods traces smoothly backward to a lophogastrid-like ancestry.

Watling (1981) takes support for a carapaceless origin of the mancoid line (Cumacea, Tanaidacea, Spelaeogriphacea, Isopoda) from the observation that the more traditional point of view as expressed by Siewing (1963) is inconsistent with the fossil record and cannot explain patterns in the respiratory and circulatory system. His willingness to take literally the order of appearance of peracaridan taxa in the fossil record seems naive, in view of the patchiness of that record.

In the traditional peracaridan phylogeny, the evolution of the respiratory system is not so complicated as Watling suggests. Among peracarids as a whole, only lophogastrid mysidaceans and amphipods rely totally on thoracic epipodal respiration (Fig.5). The primitiveness of this system is documented by its general presence in nonperacarids. Within the Peracarida, the maxillipedal epipod acts as a bailer in mysidaceans, cumaceans, tanaidaceans and spelaeogriphaceans. In the cumaceans, tanaidaceans and spelaeogriphaceans it also acts as a gill. Tanaidaceans and spelaeogriphaceans additionally use the inner surface of the carapace for gas exchange, and spelaeogriphaceans have branchial exopods. Of these respiratory devices, the nonmaxillipedal ones must be considered apomorphies. Isopods are unique in having pleopodal instead of thoracic epipodal respiration. There is no evidence for regarding the isopodan condition as ancestral; its uniqueness labels it an apomorphy whose origin is totally hidden. It must be emphasized that the respiratory surfaces in isopods are the exopod and endopod, and are therefore in no way related to the epipodal abdominal gills of hoplocarids.

Watling (1981) brings the pancarids into the argument because Siewing placed them among the mancoids, but there is no reason to relate them to the mancoid lineage, or perhaps even to the peracarids (Hessler 1982). Leaving them aside (Fig.4), a simple scheme of respiratory evolution begins with the lophogastrids, fits the three non-isopodan mancoids in the middle, and ends with the unique isopods. Here the maxillipedal epipodal bailer gives a bridge between the Mysidacea and the lower mancoids. One strength of this scheme is that it is concordant with a general reduction of the carapace and disappearance of exopods as well as epipods within the mancoid line; the isopods are most advanced in all respects, as opposed to their condition in Watling's senario, which makes them primitive in

lacking a carapace, but advanced in lacking thoracopodal exites. This scenario for respiratory evolution is parsimonious in endowing monophyletic origin to all situations.

In regarding the cumacean and tanaidacean carapace, exites and epipodal respiratory
system more primitive than in isopods, one need not regard those orders more primitive in
all respects. They are obviously uniquely specialized in many ways, but in the characters
under consideration here, they reflect a primitive grade of evolution. The Spelaeographacea
come far closer to what the isopodan ancestor might have been like (Hessler 1982), with a
single fused thoracomere, with reduced carapace, thoracic exites, antennal scale and eye
lobes, but with large swimming pleopods, uropodal tail fan and abdomen. Here, the abdomen has lost the caridoid flexure, and its musculature is modest.

What about the amphipods? Their presence in the Peracarida is based largely on the
presence of a lacinia mobilis and thoracopodal oöstegites. They are different from other
peracarids in many important ways (Siewing 1963). We now know the lacinia mobilis is
not diagnostic of the Peracarida (Dahl & Hessler 1982), leaving thoracic oöstegites as
the primary unifying feature. The especially great isolation of amphipods among peracarids
causes investigators to wonder if they even belong in that division (Watling 1981, Dahl &
Hessler 1982, Hessler 1982). Coupled with the complete lack of intermediate forms,
living or fossil, to guide us, this phylogenetic distance from other taxa has important implications. If we decide amphipods are not related to other peracarids, there is nothing
which compels us to conclude their *immediate* ancestor had a carapace (Dahl 1977); it
might have been more like an anaspidacean than a lophogastrid (Hessler 1982). However,
this does not invalidate any argument that the *ultimate* ancestor had one. The affinities of
the Amphipoda are today the most perplexing issue in malacostracan phylogeny.

The purpose of pursuing these circuitous arguments was to search for reasons that
require the rejection of the monophyletic carapace. Objections of sufficient substance are
not forthcoming. The most parsimonious senario begins with a carapace which is reduced
or even lost in independent lines.

## 6 EARLY MALACOSTRACAN EVOLUTION

The major conclusion of this inquiry is that the caridoid facies, including a carapace, was
present early in eumalacostracan evolution. We know from the Hoplocarida that it was not
yet fully developed when the prime diagnostic eumalacostracan apomorphy, six pretelsonic
abdominal segments, appeared (Table 2). The complex abdominal trunk musculature that
powers the caridoid escape reaction had not achieved its full form, but otherwise the caridoid abdomen was well-developed. At this point, two questions suggest themselves: (1)
What was the original purpose of the caridoid abdomen? (2) Is there any functional significance to the loss of the seventh abdominal segment?

The condition of the abdomen in hoplocarids suggests that the caridoid escape reaction
may not have been the sole selective advantage to the caridoid abdomen. Here, the uropodal tail fan is highly developed in advance of the perfection of the flexing system. Perhaps
originally the tail fan functioned primarily in steering the swimming animal vertically. This
is one function of the tail fan today, in hoplocarids and full caridoids. Secondarily, it is
probably its only function in spelaeographaceans and cirolanid isopods. It would seem to
be relatively simple for the caridoid escape reaction to evolve gradually from this beginning.

The loss through fusion of the seventh abdominal segment probably occurred in concert

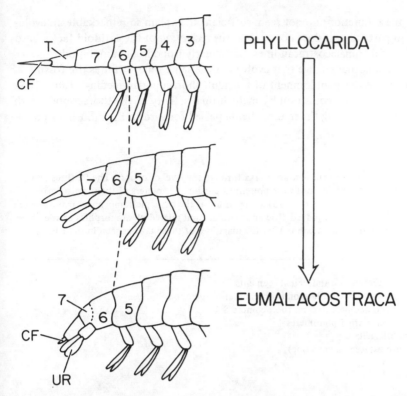

PHYLLOCARIDA

EUMALACOSTRACA

Figure 6. Evolution of the caridoid (eumalacostracan) abdomen from that of a phyllocarid. Symbols: CF – caudal furca; UR – uropod; T – telson; numbers are of abdominal segments.

with the evolution of the tail fan (Fig.6). The uropods are specialized pleopods stemming from the sixth abdominal segment. As already mentioned, no known phyllocarid had a pleopod on the seventh segment. I suggest that the sixth pleopods evolved to uropods because they were the most posterior appendages. Part of their change involved their reorientation to a horizontal, posteriorly directed position. It would seem reasonable that the limbless seventh segment became reduced and fused with the sixth to facilitate the developing tail fan. (According to Manton (1928a,b), fusion allows the uropods to migrate to the posterior end of the seventh segment.) If so, the loss of the seventh segment was not a functionally trivial event, but part of the development of a major eumalacostracan feature.

This hypothesis does not explain why the caudal rami did not play more than a transitory role in tail fan evolution. Nor does it explain why the telson did not fuse at the same time. These do not falsify the hypothesis; they only indicate it is incomplete.

We must consider whether the features of the caridoid facies constitute the driving force for eumalacostracan success. The fossil record leads us to believe most of the major eumalacostracan types were the product of an explosive radiation that occurred in the Devonian-Lower Carboniferous, essentially as soon as the caridoid apomorphies first appeared. Thus, it is reasonable to suspect the caridoid apomorphies played an active role. Functionally, they conferred two capabilities. The antennal scale and tail fan added maneuverability while swimming. The caridoid abdomen provided rapid escape.

Nevertheless, these functions do not seem sufficient to explain so remarkable an evolutionary success. In particular, it does not explain the *reduction* of the caridoid facies in so many of the successful eumalacostracan lines, particularly isopods and amphipods.

The answer lies in another change that evolved at the same time, as far as the fossil record tells us. This was the abandonment of the adult thoracopodan feeding (Table 2), i.e. the primitive system of food acquisition by metachronal activity of the thoracopods, with the food that accumulates along the midline being passed forward to the mouth by proto-

Table 2. External apomorphies of malacostracan taxa through the level of cohort. The features considered here are limited to those relevant to the present discussion. Characters of the urcrustacean model formulated by Hessler & Newman (1975) are regarded as the basic plesiomorphies against which these lists of apomorphies should be judged. The apomorphies and related plesiomorphies of the cohorts Hoplocarida and Caridoida (n.cohort) are the diagnoses of those taxa as seen in their most primitive members.

Class Malacostraca
1. 8 thoracic and 7 pretelsonic abdominal segments
2. Abdominal segments 1-6 with homomorphic, natatory pleopods
3. ♀ gonopore on thoracomere 6; ♂ on thoracomere 8
4. Thoracic endopods with 5 podomeres
5. First antenna biflagellate
6. Carapace (if urcrustacean lacked one)
7. Movable rostrum?
8. Cephalic kinesis?

Subclass Phyllocarida
No meaningful apomorphies, due in part to poor understanding of fossil forms

Subclass Eumalacostraca
1. Second antennal exopod scale-like (for planing)
2. 6th pleopod with fan-like uropod, to help form tail fan
3. Loss of 7th abdominal segment through fusion with 6th
4. Abdomen capable of strong flexion, made possible by coiling of lontigudinal trunk muscles
5. Loss of endites on thoracopods 2-8 along with abandonment of primitive thoracopodan feeding
6. Thoracic limb becomes a stenopodium
   a. protopodal complex small
   b. endopod large, slender, ambulatory
   c. epipods purely branchial
   d. exopods purely natatory

Cohort Hoplocarida
1. Abdomen enlarged
2. First antenna triflagellate
3. Antennal scale 2-segmented ? (not known on early fossils)
4. Thoracopods with 3-segmented protopod, 4-segmented endopod
5. Thoracic epipods 6-8 lost? (not known on fossils)

Cohort Caridoida, new cohort
1. Full caridoid abdominal musculature
2. Pleopodal epipods lost
3. Cephalic kinesis lost
4. Movable rostrum lost
5. Antennal scale 1-segmented
6. Mandible with row of bristles between incisor and molar processes

podal endites (Sanders 1963, Hessler & Newman 1975, Hessler 1982). It was replaced by reliance on the maxillae and the first thoracopod, now a maxilliped (Dahl 1976). This change was either driven by or caused the evolution of the eumalacostracan stenopodium (Hessler 1981). Freedom from the compromises inherent in participation in thoracopodan feeding allowed the epipods to specialize as gills, exopods as swimming organs, and endopods as truly ambulatory structures or anteriorly to be secondarily specialized for new forms of food acquisition or other behavior.

The resulting generalized animal, which must have been like an anaspidacean with a carapace or a lophogastrid that walked, possessed extraordinary evolutionary potential. This was not only a function of the thoracic stenopodium, but also natatory pleopods and the caridoid abdomen. In the major flowering of the Eumalacostraca, different aspects of this generalized form were emphasized in different lineages, as is most vividly revealed in locomotion (Hessler 1981). For example, in the Mysida, thoracic exopods are used for swimming, and the pleopods are reduced or absent; in sergestid decapods, the opposite pertains.

Thus, it would seem that a factor outside of the caridoid facies was the primary key to the eumalacostracan radiation. However, the loss of primitive thoracopodan feeding and the appearance of the stenopodium did appear at the same time as the caridoid apomorphies. The questions remains, was this coincidental, or is there a functional linkage that caused these factors to evolve together?

In concluding, it should be noted that in view of the appearance of most of the caridoid apomorphies and the thoracic stenopodium in the earliest eumalacostracans, the diagnosis of the Eumalacostraca should include this complex of features just as much as it does the presence of six pretelsonic thoracic segments (Table 2). In the past, too much importance has been devoted to abdominal segment count, the apparent trivialness of which is part of the reason why eumalacostracan monophyly has been viewed with such suspicion. An appreciation of the full suite of features that first appeared with the Eumalacostraca, in both the hoplocarids and full caridoids, places the unity of that taxon on much firmer footing.

This does not deny the closer affinity of the full caridoids (Eocarida, Syncarida, Peracarida, Eucarida, and Pancarida) with each other than to the Hoplocarida. This relationship is sufficiently important to justify taxonomic recognition. Therefore, I propose that the full caridoids mentioned above be combined under the name Caridoida, which is a sister group of the Hoplocarida within the Eumalacostraca. The diagnostic differences between these taxa are given as a list of apomorphies in Table 2.

Much speculation has been devoted to the life style of the primitive eumalacostracan. Most commonly, the ureumalacostracan is thought to be primarily, but not exclusively benthic (Manton 1953, Dahl 1976, Watling 1981). For this reason, the observation that the caridoid facies appears best developed in primarily pelagic forms made Dahl (1976) suspicious of its primitiveness. However, this impression is not strictly accurate. While euphausiids, and some lophogastrids and natantian decapods are holopelagic, anaspidaceans, other lophogastrids, mysids and most natantian decapods are benthopelagic, as one would expect in animals outfitted with both ambulatory (thoracic endopods) and natatory (thoracic exopods, pleopods) appendages. The caridoid apomorphies well serve the needs of an animal swimming over the bottom or requiring rapid retreat from some benthic threat. Thus, the combination of the caridoid facies with the primitive thoracic stenopodium is superbly suited for the benthopelagic existence postulated for the eumalacostracan ancestor.

## 7 SUMMARY

Adherence to the principle that synapomorphy is the fundamental hypothesis of phylogeny frees the systematist from the danger of capricious, subjective utilization of convergence in evolutionary constructs. In the present essay, I have considered the major reasons for dissatisfaction with the caridoid facies as a central, primitive eumalacostracan morphology, to see if they constitute valid falsification of the monophyletic hypothesis. None of them qualified. This is not to say that some day a new perception might not reveal a fatal weakness. Monophyletic schemes can never be free of this possibility.

Until then, the caridoid facies must be regarded as an attribute that appeared with the advent of the Eumalacostraca, at the same time as the appearance of the stenopodium. The details of the origin of the eumalacostracan body plan are almost completely unknown; the fossil record is not helpful, and hoplocarids give only the slightest clue. Once evolved, it was undeniably successful. However, its success should not be equated with static perfection. Rather it lies in having provided a remarkable variety of adaptive opportunities, as testified by the multitude of eumalacostracan taxa today.

## 8 ACKNOWLEDGEMENTS

I take pleasure in giving my thanks to the many colleagues who stimulated the writing of this essay. Erik Dahl, Janet Kunze, William Newman, Frederick Schram, Jalle Strömberg, and George Wilson deserve my special appreciation. This study was supported by Grant DEB 77-24614.

## DISCUSSION

SIEG: I have some difficulty with your explanation for the uropods coming from the sixth segment. We have some observations in tanaids, originally made by Claus, where we see the sixth and seventh ganglia in the abdomen in very early stages of development. These ganglia are partially fused, but the nerve for the uropods is coming from the seventh and not the sixth ganglion. Do you have any information that the uropods do in fact develop from the sixth?

HESSLER: The only information I can offer is the work of Manton on *Hemimysis*. This indicated that the limbs migrated to their posterior position from the sixth segment. Your data are very interesting.

SIEG: Could it have evolved twice? In some groups the uropods may have evolved as you've indicated, and in others they may have come from the seventh.

HESSLER: I would say that if both these data sets are incontrovertable, that is the kind of falsification of the scheme presented here I have been talking about.

SIEG: You said that isopods have an epipodite, and that is a common statement in isopod literature. I have looked at some aberrant mysids and done some comparative studies, and I think that what is generally referred to as an epipodite in isopods is an exopodite.

HESSLER: The structure comes off the coxa, which by position makes it an epipod. If you have reasons based on ontogeny that indicate it did not originally come from that spot, then you have good data. If your observations are correct, it would simply make the isopods even more derived as far as I can see.

KUNZE: Do you have any idea as to the functional processes that would have likely led to the loss of the carapace in the course of syncarid evolution, and would these processes have been in operation in amphipods and isopods?

HESSLER: No, I'm sorry, I don't. I cannot explain why syncarids lost their carapace. I feel a little more comfortable with isopods. I feel that in isopods it is the result of gradual progression, as seen in other taxa. However, amphipods again are a problem. I have not the faintest idea where amphipods came from, and therefore, I do not know why they lost the carapace.

WATLING: I will defend myself more readily this afternoon, but I want to offer one quick correction. I do not think I have ever claimed the carapace has evolved four or five times. I think you'll see it's twice.

HESSLER: No, [by having an urmalacostracan and urperacaridan which lack a carapace] you derive it separately in the phyllocarids, the mancoids, the mysids, the eucarids, and the hoplocarids [Watling 1981:202, 204].

WATLING: I disagree. But I want to comment on uses of the fossil record, which I think we are all aware has to be used very carefully. Why do you suppose that all the crustaceans that have shown up as fossils, at least since the Devonian, are all large robust forms? Is there some coincidence with that and the fact they do generally have large carapaces? None of the delicate non-carapace forms show up as fossils.

HESSLER: Syncarids have no carapace and they show up as fossils. *Acadiocaris* is a relatively small and delicate animal and it shows up as a fossil.

WATLING: Those things are very rare. The most abundant fossils are all really large, robust, heavily calcified forms.

HESSLER: So?

WATLING: I would suggest that the fossil record is only of limited usefulness in that the only information you can get out of it is the presence of things. You cannot use it to demonstrate the absence.

HESSLER: The only reason I have used the fossil record at all is that two people [Schram 1981:2, 3, Watling 1981:206-208] have said that the fossil record tends to show the caridoid facies is polyphyletic.

NEWMAN: I am interested in this possibility that uropods may have evolved several times. This brings us to crustaceans as a whole, because the same ennervation problem exists at the other end of the animal in the head. The second antennae are postoral in their origin and then come to migrate anteriorly. We might be able to follow the same model with the uropods, that is, they may have migrated with their ennervation posteriorly. It is just the other end of the animal.

ABELE: I would like to make two comments. One, I do not know what the problem is with the carapace evolving more than once, given the diversity of the carapace and the complete lack of information on its embryological origin, which is only known for a few species. And two, most of the things you've listed are all one thing: the uropods, the flexion, the large musculature in the abdomens. So you have one character, not several.

HESSLER: I agree that all those abdominal characters are very much related.

KUNZE: The antennal scales are rather different between hoplocarids and your caridoids. It has two segments in the hoplocarids, one in the caridoids.

HESSLER: Yes, but who is to say that it is different because it has fewer segments [consider the varying segmentation in decapod pereopods].

WATLING: Another point of contention is the use of the caridoid facies. I would argue that though mysids, euphausids, and shrimp-like forms are definitely caridoid in a Calman sense; thermosbaenaceans, cumaceans, tanaids and spelaegriphaceans are not.

HESSLER: Absolutely; they are very derived but lophagastrids are not.

WATLING: Sure, they are an early offshoot of mysids.

HESSLER: There is nothing wrong with evolving a carapace more than once, but I demand that there be evidence that proves it. Until there is such evidence it is most parsimonious to say it evolved only once. That is all I'm saying. I'm waiting for some falsification of my contention.

GRYGIER: Concerning this lack of limbs on the seventh segment, there is at least one pygocephalomorph [the genus *Pygocephalus*] that has two pairs of furca. I am wondering if at least one of those are the rami of the seventh limb.

HESSLER: I don't know what to do with that; Bill Newman also mentioned this to me. It is possible that is the case, but it is also possible that it is more akin to a situation like that seen in euphausids. which have six spines coming off the telson which can dwindle down to two. We decided those two are spines and not furca, because they are simply two among six.

SCHRAM: I feel compelled to make some comments since I have been an object of attack. Whether you want to call syncarids, pancarids, peracarids, and eucarids Caridoida, as Bob would prefer, or leave the term Eumalacostraca for these as I have suggested is a matter of preference. Our positions are not that far apart. I would point out the apomorphies Bob has used to define the caridoids have different expressions in the hoplocarids, namely, the abdominal musculature and antennal scale. Either these can both be considered apomorphic with regard to some precursor; or as Bob has indicated today for the muscles one is intermediate to the other, that is, one is more primitive and the other advanced. This could indicate separate and coequal groups. These groups are related, since they do have other features that seem to draw them together, but on a much lower level than indicated here.

HESSLER: May I say it seems simpler to keep the term Eumalacostraca for that to which it has always applied, and leave its diagnosis as it always was — a group of animals having lost their seventh segment.

SCHRAM: Until we can sort out what is going on in the phyllocarids and their relationship to these different groups, I prefer to leave it alone. We should simply have three groups in the Malacostraca and not attempt to draw any sister groups because we don't really understand what's going on.

HESSLER: In the traditional view, 'leaving it alone' would mean staying with two groups of Malacostraca.

## REFERENCES

Brooks, H.K. 1969. Eocarida, Syncarida. In: R.C.Moore (ed.), *Treatise on Invertebrate Paleontology, Part R, Arthropoda 4:*R332-359. Lawrence: Geol. Soc. Am. & Univ. Kansas Press.
Burnett, B.R. 1973. Notes on the lateral arteries of two stomatopods. *Crustaceana* 23:303-305.
Burnett, B.R. & R.R.Hessler 1973. Thoracic epipodites in the Stomatopoda (Crustacea): a phylogenetic consideration. *J. Zool. Lond.* 169:381-392.
Calman, W.T. 1904. On the classification of the Crustacea Malacostraca. *Ann. Mag. Nat. Hist.* (7)13:144-158.
Calman, W.T. 1909. Crustacea. In: R.Lankester (ed.), *A Treatise on Zoology, Part 7.* London: Adam & Charles Black.
Claus, C. 1885. Neue Beiträge zur Morphologie der Crustaceen. *Arb. zool. Inst. Univ. Wien* 6:vi,1-108.

Dahl, E. 1963. Main evolutionary lines among recent Crustacea. In: H.B.Whittington & W.D.I.Rolfe (eds), *Phylogeny and Evolution of Crustacea:* 1-15. Cambridge: Mus. Comp. Zool.

Dahl, E. 1976. Structural plans as functional models exemplified by the Crustacea Malacostraca. *Zool. Scripta* 5:163-166.

Dahl, E. 1977. The amphipod functional model and its bearing upon systematics and phylogeny. *Zool. Scripta* 6:221-228.

Dahl, E. (in press). Alternatives in malacostracan evolution. *Rec. Aust. Mus.*

Dahl, E. & R.R.Hessler 1982. The lacinia mobilis: a reconsideration of its origin, function and phylogenetic implications. *Zool. J. Linn. Soc. Lond.* 74:133-146.

Daniel, R.J. 1932. Comparative study of the abdominal musculature in Malacostraca. Part II. The superficial and main ventral muscles, dorsal muscles and lateral muscles, and their continuations into the thorax. *Proc. Trans. Liverpool Biol. Soc.* 46:46-107.

Daniel, R.J. 1933. Comparative study of the abdominal musculature in Malacostraca. Part III. The abdominal muscular systems of *Lophogaster typicus* M.Sars, and *Gnathophausia zoea* Suhm, and their relationships with the musculatures of other Malacostraca. *Proc. Trans. Liverpool Biol. Soc.* 47:71-133.

Grobben, K. 1917. Der Schallenschliessmuskel der dekapoden Crustaceen, zugleich ein Beitrag zur Kenntnis ihrer Kopfmuskulatur. *S.B. Kais, Akad. Wiss. Wien* 126:473-494.

Hansen, H.J. 1893. Contribution to the morphology of the limbs and mouthparts of crustaceans and insects. *Ann. Mag. Nat. Hist.* 12:417-434.

Hessler, R.R. 1964. The Cephalocarida – comparative skeletomusculature. *Mem. Conn. Acad. Arts Sci.* 16:1-97.

Hessler, R.R. 1981. Evolution of arthropod locomotion: A crustacean model. In: C.F.Herreid & C.R. Fourtner (eds), *Locomotion and Exercise in Arthropods:* 9-30. New York: Plenum Press.

Hessler, R.R. 1982. Evolution within the Crustacea. In: L.G.Abele (ed.), *The Biology of Crustacea Vol.I:* 149-185. New York: Academic Press.

Hessler, R.R. & W.A.Newman 1975. A trilobitomorph origin for the Crustacea. *Fossils and Strata* 4: 437-459.

Komai, T. & Y.M.Tung 1931. On some points of the internal structure of *Squilla oratoria. Mem. Coll. Sci. Koyoto Univ.* 6:1-15.

Kunze, J.C. 1981. The functional morphology of stomatopod Crustacea. *Phil. Trans. Roy. Soc. London* (B)292:255-328.

Manton, S.M. 1928a. On the embryology of the mysid crustacean *Hemimysis lamornae. Phil. Trans. Roy. Soc. London* (B)216:363-463.

Manton, S.M. 1928b. On some points in the anatomy and habits of the lophogastrid Crustacea. *Trans. Roy. Soc. Edin.* 56:103-119.

Manton, S.M. 1953. Locomotory habits and the evolution of the larger arthropodan groups. *Symp. Soc. Exp. Biol., VII, Evol:* 339-376.

Reaka, M.L. 1975. Molting in stomatopod crustaceans. 1. Stages of the molt cycle, setagenesis, and morphology. *J. Morph.* 146:55-80.

Rolfe, W.D.I. 1969. Phyllocarida. In: R.C.Moore (ed.), *Treatise on Invertebrate Paleontology, Part R, Arthropoda 4:* R296-331. Lawrence: Geol. Soc. Am. & Univ. Kansas Press.

Rolfe, W.D.I. 1981. Phylocarida and the origin of the Malacostraca. *Geobios* 14:17-27.

Sanders, H.L. 1963. The Cephalocarida: functional morphology, larval development, comparative external anatomy. *Mem. Conn. Acad. Arts Sci.* 15:1-80.

Schram, F.R. 1969a. Polyphyly in the Eumalacostraca? *Crustaceana* 16:243-250.

Schram, F.R. 1969b. Some Middle Pennsylvanian Hoplocarida (Crustacea) and their phylogenetic significance. *Fieldiana: Geol.* 12:235-289.

Schram, F.R. 1973. On some phyllocarids and the origin of the Hoplocarida. *Fieldiana: Geol.* 26:77-94.

Schram, F.R. 1981. On the classification of the Eumalacostraca. *J. Crust. Biol.* 1:1-10.

Schram, F.R. 1982. The fossil record and evolution of Crustacea. In: L.G.Abele (ed.), *The Biology of Crustacea Vol.I:* 93-147. New York: Academic Press.

Schram, F.R., R.M.Feldmann & M.J.Copeland 1978. The Late Devonian Palaeopalaemonidae and the earliest decapod crustaceans. *J. Paleont.* 52:1375-1387.

Shiino, S.M. 1942. Studies on the embryology of *Squilla oratoria* de Haan. *Mem. Coll. Sci. Kyoto Univ.* (B)17:77-173.

Siewing, R. 1963. Studies in malacostracan morphology: results and problems. In: H.B.Whittington & W.D.I.Rolfe (eds), *Phylogeny and Evolution of the Crustacea:*85-103. Cambridge: Mus. Comp. Zool.

Silvey, G.E. & I.S.Wilson 1979. Structure and function of the lateral giant neurone of the primitive crustacean *Anaspides tasmaniae. J. Exp. Biol.* 78:121-136.

Tiegs, O.W. & S.M.Manton 1958. The evolution of the Arthropoda. *Biol. Rev.* 33:255-337.

Watling, L. 1981. An alternative phylogeny of peracarid crustaceans. *J. Crust. Biol.* 1:201-210.

JANET C. KUNZE
*Scripps Institution of Oceanography, La Jolla, California, USA*

# STOMATOPODA AND THE EVOLUTION OF THE HOPLOCARIDA

## ABSTRACT

Recent investigations of the functional morphology of Stomatopoda indicate that the evolution of the predatory stomatopod feeding mechanism was associated with unique mandibular and gastric modifications. These morphological differences, considered in addition to other hoplocaridan features, lend credence to the existence of a hoploid morphotype which evolved from a benthopelagic or benthic detritivore/scavenger. Arguments for and against a polyphyletic origin of the Hoplocarida and Eumalacostraca are discussed in relation to recent morphological and paleontological evidence. These focus on characters diagnostic of the Eumalacostraca and an array of features which describe a discrete hoploid morphotype. The existence of a precaridoid eumalacostracan ancestor for the Hoplocarida is not supported. It is proposed that the Hoplocarida and Eumalacostraca evolved independently from separate 'phyllocarid-like' ancestors.

## 1 INTRODUCTION

The phylogenetic status of the Hoplocarida and, more specifically, the extant Stomatopoda, has been a speculative issue from the time the group was defined by Latreille (1817). Many of these theories (including those of Huxley 1867, Boas 1883, Claus 1885, Grobben 1892, 1919, Haeckel 1896, Giesbrecht 1913, Giesbrecht & Balss 1933, Balss 1938, Siewing 1956, 1963, Glaessner 1957, Brooks 1962, and Secretan 1967) were briefly reviewed by Schram (1969b).

In this chapter the more recent information will be reviewed with particular emphasis on functional morphology to provide an insight into the possible evolutionary lineages that gave rise to the Hoplocarida and modern Stomatopoda. Much of this information is based on comparative studies of the feeding biology and digestive systems of several species of stomatopods with particular emphasis on mandibular and gastric mechanisms (Kunze 1981).

Stomatopods, in contrast with other major malacostracan taxa, display remarkable morphological homogeneity. The group comprises over 350 species which are grouped into 66 genera and 12 families (Manning 1980), primarily on the basis of differences in body armature and ornamentation. These differences, although taxonomically significant, provide little insight into the evolution and adaptive radiation of the group. All stomatopods are carnivores that employ their enlarged second maxillipeds (thoracopods) as powerful implements for prey capture. These raptorial appendages may be used in either of two modes

165

(some employing both), 'spearing' or 'smashing', depending on whether the dactylus is extended or folded during the strike response (Caldwell & Dingle 1975, 1976, Dingle & Caldwell 1978). Such extreme specialization as predators is not paralleled in any of the major extant malacostracan groups. The Decapoda, for example, includes many carnivorous forms, but most augment their food supply by scavenging or carrion feeding. Such augmentative food resources are less significant components of the stomatopod diet.

The morphological and functional features of the feeding and digestive systems are not shared by any malacostracan group. Similarly, locomotion in stomatopods is unique in relation to pereopod morphology and differences in escape behaviour (Schram 1982). In many respects the Stomatopoda exemplify an evolutionary *cul de sac:* they are highly specialized, displaying little morphological divergence since the Jurassic (Holthuis & Manning 1969) and a limited potential for adaptive radiation.

The features that characterize the Stomatopoda are an assortment of plesiomorphies and autapomorphies. They constitute the major issues in the phylogenetic debate as follows.

## 2 CEPHALIC KINESIS — movable ophthalmic and antennular segments and an articulated rostrum

Calman (1909) regarded these features as specializations. Siewing (1963) stated that they were original and linked Stomatopoda with Leptostraca. This was later supported by Schram (1969a,b, 1973) who provided substantial paleontological evidence indicating cephalic kinesis in the earliest hoplocarids. Although it may be argued that rostral articulation is a specialization because of its absence in larval forms, there is no evidence for the secondary acquisition of kinesis between the first and second antennae.

## 3 TRIFLAGELLATE FIRST ANTENNAE

The presence of triflagellate first antennae in extinct aeschronectid hoplocarids indicates that this feature appeared very early in hoplocarid evolution. There is, however, no evidence of their presence in phyllocarids or any other malacostracan group other than palaemonid decapods, suggesting that triflagellate antennules are a convergent feature and apomorphic.

## 4 THORACOPODS

The thoracopods in Stomatopoda comprise five pairs of subchelate maxillipeds and three pairs of biramous pereopods. The maxillipeds lack exopods, having reduced but functional epipods (Burnett & Hessler 1973) and comprise six segments instead of the usual malacostracan complement of seven. Holthuis & Manning (1969) and Schram (1969b) termed the segments distal to the protopod ischiomerus (commonly merus), carpus, propodus, and dactylus. The pereopods consist of a three-segmented protopod, an inner branch of one or two segments and an other branch of two segments. Holthuis & Manning (1969) termed these rami endopod and exopod, respectively. Claus' (1871) work on stomatopod development suggested the outer branch was the endopod and the inner one the exopod, thus invoking a reversal in position during ontogenesis. According to Schram (1969b) the early hoplocarids possessed a one-segmented outer branch and a four-segmented inner branch (comprising ischiomerus, carpus, propodus, and dactylus). From this evidence it is not clear whether the ramus used for walking is the endopod or the exopod or whether these terms can be validly applied to hoplocarids. Holthuis & Manning (1969) provide no justification for their designation other than spatial position. Based on the fossil evidence, a mophological shift of the two rami, such that the primitive segmented inner branch was retained as a locomotory appendage, would be a more parsimonious alternative than the respecialization of a lamellate outer branch as a walking limb with a secondary simplification of the inner branch. Schram (1982) stated that the hoplocarid thoracopod and also the triflagellate antennae were probably 'stochastic derivatives randomly arrived at'. Burnett & Hessler (1973) proposed a derivation from the main precaridoid line by reduction from five limb segments to four. The problem that needs to be resolved is whether the rami used in walking are homologous to the caridoid endopod or exopod and how this reconciles with the fossil evidence without necessitating functionally improbable intermediates.

# 5 ABDOMINAL GILLS

The question as to whether or not abdominal gills are original or secondary is closely tied with the presence of an elongate heart with segmental vessels and ostia. Siewing (1963) and Schram (1969b) concluded that abdominal gills were original, whilst Burnett & Hessler (1973) argued that since pleopodal gills have not been demonstrated in phyllocaridans their acquisition in the Hoplocarida must have been secondary. They also addressed Schram's (1969b) contention that the pleopodal gills in hoplocarids are epipodal using similar logic. Schram (1982) further argued that the presence of pleopodal gills arising from the protopod in aeschronectids and palaeostomatopods was strong justification that the abdominal respiratory structures were original. Hessler (1982) suggested an alternative viewpoint: abdominal gills, if primary, 'may not be an apomorphy of the hoplocaridan line, but rather an extreme plesiomorphy reflecting the state that preceded urmalacostracan subdivision of trunk appendages into pereopods and pleopods, that is, the stage when all limbs had epipods'. The palaeontological evidence and primitive circulatory system indicates that this may be a plesiomorphy. The crucial fossil evidence is, however, lacking.

# 6 CIRCULATORY SYSTEM AND FUSION OF ABDOMINAL SEGMENTS

The circulatory system of stomatopods comprises an elongate tubular heart extending almost the entire body length with segmentally arranged ostia and arteries. This arrangement is not shared by any eumalacostracan. Some controversy arose over the supposed existence of a seventh pair of abdominal arteries (Komai & Tung 1931) which Siewing (1963) interpreted as evidence for the fusion of the first two abdominal segments even though embryological evidence (Shiino 1942) indicated fusion of the posterior segments. Burnett (1972) later refuted Komai & Tung's evidence; however, Reaka (1975) revived the controversy stating that the pattern of molt sutures indicated that the last thoracic and first abdominal segments had fused. Whether or not the molting pattern is a consistent and conservative character indicating a plesiomorphy requires further investigation, such as whether the pattern persists during larval development. The particular molting pattern in stomatopods may also be related to the physical constraints of the body morphology and, therefore, apomorphic. Embryological evidence and details of the circulatory system are, in general, more conservative features than skeletal characteristics, and because of this, it is more valid to assume that fusion occurred between the posterior abdominal segments (as in eumalacostracans) until further evidence indicates otherwise.

## 7 ABDOMINAL MUSCULATURE

Hessler (1964) thought that the stomatopod abdominal musculature was intermediate between the leptostracan and caridoid conditions. The major difference between stomatopods and eumalacostracans is the absence of transverse muscles in stomatopods. Apart from the basic pattern of coiling of muscles, the skeletomusculature is similar in many respects to that of leptostracans. It was suggested by this author (Kunze 1981) that coiling of muscles is probably the only functionally feasible solution to such locomotory requirements as abdominal flexion and it is possible that this pattern may have arisen independently in the Hoplocarida and Eumalacostraca.

## 8 EXTENSION OF THE VISCERA, ESPECIALLY THE GONADS AND DIGESTIVE GLAND, THROUGHOUT THE BODY

This is another plesiomorphic feature shared only by the Phyllocarida. Its persistence in stomatopods suggests an alternative functional model to the caridoid model in which the internal organs and respiratory structures have shifted towards the thoracic region. Dahl (1963) correlated this with the development of powerful abdominal musculature. Stomatopods, on the other hand, display a mode of life which emphasizes tagmatization and functional diversity of the abdomen.

## 9 MANDIBLES AND MANDIBULAR MECHANISM

The mandibles in stomatopods and particularly their functional mechanism contrast those of Syncarida, Peracarida, and Decapoda. The masticatory surfaces of the mandible form an inverted L-shape and lie behind the labrum with the molar process extending into the lumen of the proventriculus (Fig.1). The retention of well-developed molar processes in a large-food feeding malacostracan does not conform to the pattern described by Manton (1964) for *Ligia, Astacus,* and *Carcinus* in which the molar process is reduced associated with the evolution of a transverse biting mechanism. This apparently primitive mandibular arrangement in stomatopods is associated with the dual use of both transverse biting and rolling promotor-remotor movements of the molar processes. The mandibular musculature of stomatopods shares features common to the mandibles in Branchiopoda, Leptostraca, Syncarida, and Peracarida (previously described by Manton) and, to a lesser degree Cephalocarida (Hessler 1964), indicating a general crustacean pattern. As illustrated in Figure 2, the mandibles of cephalocarids (a), branchiopods (b), syncarids (c), mysids (d) and stomatopods (e, f) share a common complement of muscles 2, 3, 4, 5a, 5b, 5c and 6. A functional comparison of these muscles is given in Table 1.

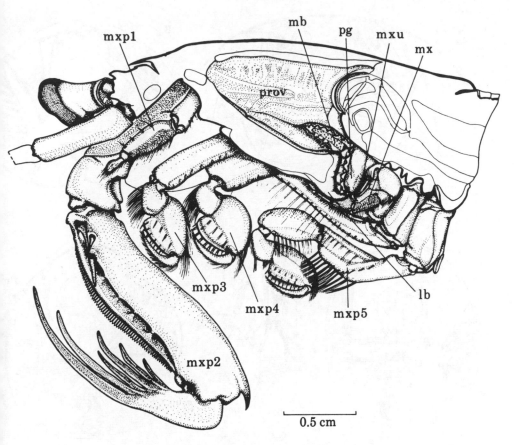

Figure 1. *Alima laevis:* median view of the mouthparts and proventriculus exposed by sagittal section (from Kunze 1981).

A point of explanation should be noted: *Chirocephalus* displays certain modifications which separate it from the Malacostraca and Cephalocarida, these are the absence of muscle 2 and the extensive development of the transverse mandibular tendon. In addition, the muscle 5 group is extensively developed (Fig.2b). The mandibular arrangement of *Chirocephalus* is relatively specialized in comparison with cephalocarids or mysids although, like the latter, it is a fine particle feeder. This contrasts Manton's conclusion that *Chirocephalus* exemplified a primitive construction. *Hutchinsoniella,* on the other hand, provides an excellent primitive template, illustrating not only the origins of the mandibular muscles from unspecialized cephalic limbs (such as the naupliar mandibles), but also the inherent conservatism of crustacean mandibular evolution. The common expression of muscles 2 to 6 (with minor variation) in very distantly related crustacean taxa such as cephalocarids, copepods (Hessler 1964), branchiopods, and malacostracans is a crustacean symplesiomorphy and does not distinguish malacostracan taxa from non-malacostracan taxa.

Superimposed on these common crustacean traits are several modifications of stomatopod mandibles that are associated with the use of molar processes as grinding and chewing structures inside the proventriculus.

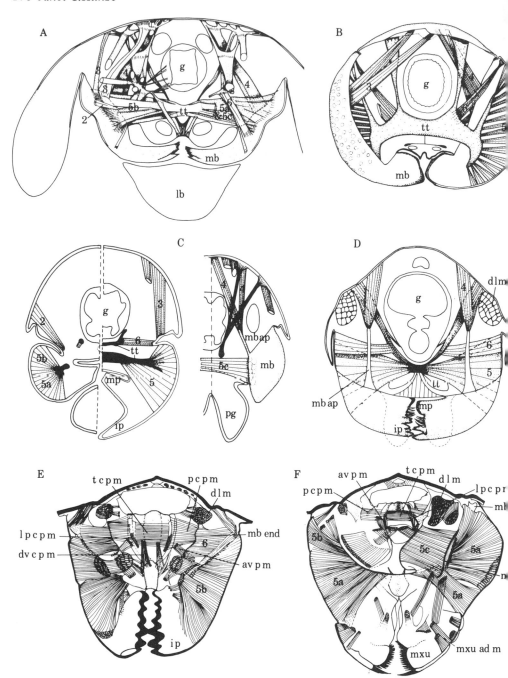

Figure 2. Transverse sections of the mandibles: A. *Hutchinsoniella macracantha,* Cephalocarida, anterior iview (after Hessler 1964); B. *Chirocephalus diaphanus,* Branchiopoda, anterior view (after Manton 1964); C. *Paranaspides lacustris,* Syncarida, montage of sequential sections, left is anterior (after Manton 1964); D. *Hemimysis lamornae,* Mysidacea, anterior view (after Cannon & Manton 1927); E and F. *Alima laevis,* Stomatopoda, posterior views of sequential sections, E. is anterior to F corresponding to levels A and B in figure 4 (after Kunze 1981).

Table 1. Functional comparisons of mandibular muscles in Cephalocarida, Branchiopoda, Syncarida, Mysidacea and Stomatopoda. (Abbreviations: i.p. – incisor process; m.p. – molar process; Branch. – Branchiopoda; Ceph. – Cephalocarida; Mys. – Mysidacea; Stom. – Stomatopoda; Syn. – Syncarida)

| Muscle | Function | Crustacean groups |
|---|---|---|
| Anterior mandibular promotor (m3) | Promotor roll<br>Abduction of i.p.; slight medial rotation of m.p. | Ceph., Branch., Syn., Mys.<br>Stom. |
| Mandibular promotor (m5b) | Promotor roll<br>Abduction of i.p.; and separation of m.p.<br>Medial swing of m.p. | Ceph., Branch.<br>Syn., Mys.<br>Stom. |
| Mandibular promotor (m6) | Promotes movements at right angles to main mandibular roll<br>Promotor roll, abduction of i.p.<br>Medial swing of m.p. | Ceph?, Branch., Mys.<br><br>Syn.<br>Stom. |
| Anterior mandibular remotor (m2) | Absent<br>Promotor roll?<br>Increases strength of remotor roll<br>Remotor roll of mandible body | Branch.<br>Ceph.<br>Syn., Mys.<br>Stom. |
| Mandibular remotor (m4) | Remotor roll causing grinding by m.p. and biting by i.p.<br>Adduction of i.p. | Ceph., Branch., Syn., Mys.<br><br>Stom. |
| Mandibular remotor (m5a) | Remotor roll sim. to m4<br>Adduction of i.p. and lateral swing of m.p. | Ceph., Branch., Syn., Mys.<br>Stom. |
| Transverse mandibular remotor (m5c) | Remotor roll sim. to m4 and m5a<br>Adduction of i.p. | Ceph., Branch., Syn., Mys.<br>Stom. |

Figure 3 shows the orientation and attachment of the mandible to the surrounding en-dophragmal elements. There are two main articulations, a dorsal, ball and socket type of articulation and an anteroventral articulation of the rounded protuberance on the lower anterior margin of the mandible body with the posterior ventral tubercle of the mandibular endophragm (*pvt*, Fig.3). The remainder of the mandible body is attached by flexible cuticle. These two pivots give the mandible an almost vertical axis of movement (*x*, Fig.4). Vertical axes have been shown in *Hutchinsoniella* (Hessler 1964), *Chirocephalus,* and *Nebalia* (Manton 1964), all fine particle feeders. Such an arrangement is, according to Man-ton (1964), usually correlated with grinding and squeezing actions and not suited to dealing with larger sized food. This general conclusion is still valid if one considers the *modus ope-randi* of the stomatopod mandibles and the general features of the mandibular musculature.

A detailed comparative analysis of the mandibular musculature in stomatopods was given by Kunze (1981). The main features of the musculature are illustrated in Figures 2e,f, 3 and 4 and can be divided into four functional groups: muscles that cause a promotor roll (muscle 3, Fig.4); muscles that cause a promotor roll but effect only a medial swing of the molar processes (muscles 5b, 6, Figs.2e,f, 4); muscles that cause a remotor roll (muscle 2, Fig.4); and muscles that cause a remotor roll but secondarily effect adduction of the incisor processes (muscle 4, 5a,c, Figs.2e,f, 4). A comparison of the functions of these muscles in other crustaceans is given in Table 1. The terminology follows Manton (1964).

The functional differences in stomatopods are associated with differences in muscle size,

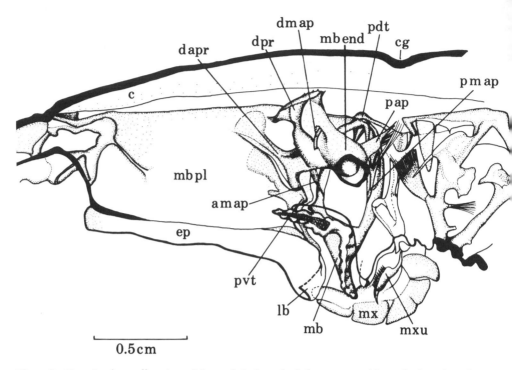

Figure 3. *Alima laevis:* median view of the cephalothoracic skeleton exposed by sagittal section, showing endophragmal structures for muscle attachment (from Kunze 1981).

(e.g. enlargement of muscles 4, 5b and 6) and position of the muscles with respect to the axis of rotation. The other main feature is the replacement of the transverse mandibular tendon (found in *Hutchinsoniella, Chirocephalus, Nebalia,* and anaspidaceans) with an extensive endophragmal bridge (mb.end., Fig.3). This latter modification allows for the enlargement of muscle 6, thus strengthening the grinding and crushing ability of the molar processes. The transverse mandibular remotor 5c is also strengthened by attachment to the intervening endophragmal plate instead of traversing the space between the posterior mandibular margins as in branchiopods, leptostracans, and syncarids.

Stomatopod mandibles share nothing in common with the specialized skeletomuscular arrangements of large-food feeding malacostracans such as *Ligia, Astacus* and *Carcinus* (Manton 1964) which employ transverse biting. Although isopods and decapods have achieved this independently, common to both is the loss of the transverse mandibular tendon and muscles 5b and c, thus allowing a wide mandibular gape. The other major changes in axis of rotation and mandibular articulation differ in the two groups.

The means by which stomatopods have overcome the obstacle of a narrow mandibular gape is by using the incisor processes primarily to grip food whilst it is torn apart by the maxillipeds. Most of the mastication is done by the molar processes inside the proventriculus, aided to some extent by digestive secretions from the digestive gland (see below).

In summary, stomatopods represent an alternative and unique functional model for a large-food feeding malacostracan. They employ a primitive mandibular musculature, elements of which are common to non-malacostracans as well as leptostracans, syncarids, and

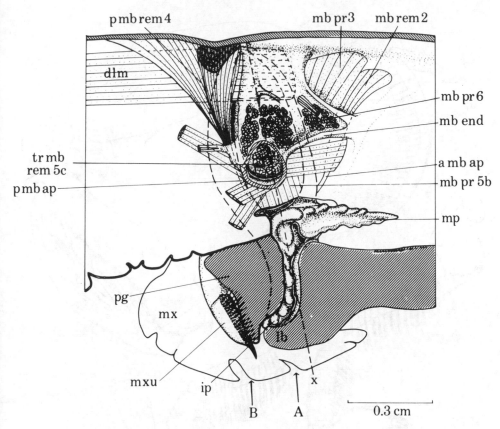

Figure 4. Median sagittal view of the mandibular musculature of *Alima laevis,* right is anterior, A and B arrows indicate planes of section corresponding to those in Figure 2e, f (from Kunze 1981).

mysidaceans. Extensive specialization of the cephalothoracic endophragma, fusion of the labrum to the epistome, absence of an esophagus, and gastric specialization (see below) coupled with intragastric mastication by the molar processes has usurped the role of the incisor processes as the principal masticatory structures. Food is passed into the proventriculus in large pieces. A gastric mill analogous to that found in many peracarids and decapods is absent; instead, the mandibular molar processes provide the only direct mechanical action.

## 10 MOUTHPARTS, FEEDING MECHANISM, AND THE EVOLUTION OF PREDATORY FEEDING IN STOMATOPODS

### 10.1 *Feeding in adult stomatopods*

There has been a general acceptance that early hoplocarids (palaeostomatopods) evolved from marine filter feeders that used their thoracopods as food strainers (Holthuis & Manning 1969, Schram 1969a,b, Caldwell & Dingle 1975). This issue was recently challenged

174 *Janet C.Kunze*

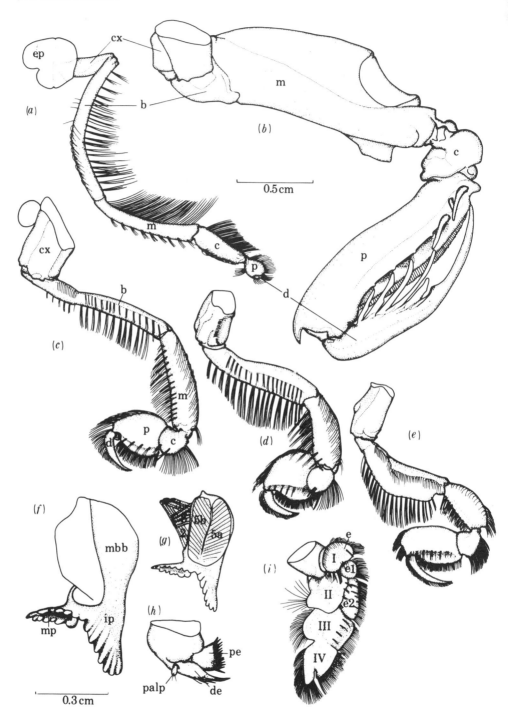

Figure 5. Mouthparts of *Alima laevis:* A. first maxilliped; B. second maxilliped; C. third maxilliped; D. fourth maxilliped; E. fifth maxilliped; F. mandible; G. mandible showing mandibular muscles; H. maxillule; I. maxilla (from Kunze 1981).

on the basis of evidence from the feeding action of living adult and larval stomatopods in addition to existing palaeontological information and comparisons with other filter-feeding malacostracans (Kunze 1981).

The mouthparts of stomatopods comprise the mandibles, maxillules, maxillae, and five pairs of subchelate maxillipeds (Figs.1, 5). The maxillules (Fig.5h) bear spinose endites on the coxa and basis. They are small and function in pushing food between the mandibles.

The maxillae (Fig.5i) consist of four segments supported by calcified plates. There is no clear indication of the origin of this segmentation or whether the endites are homologous with maxillary endites in other malacostracans. The maxillae are ringed with simple setae and act primarily as a screen over the more proximal mouthparts and labrum, preventing the escape of food.

The maxillipeds (Figs.5a-e) are generally similar, comprising six segments and bearing subchelae. The first pair (Fig.5a) are elongate, setose cleaning appendages. The second pair are greatly enlarged, heavily calcified, and armed with teeth on the propodus and dactylus. The remaining third, fourth and fifth pairs of maxillipeds (Figs.5c-e) are similarly shaped, with strong dactylar hooks, propodal spines and setae on the medial and ventral margins forming an interlocking mesh when the limbs are retracted (Fig.6). The limb bases of the maxillipeds are compressed such that they form a row lateromedially (Fig.6) allowing the raptorial appendages to enclose the other maxillipeds when the propodus and dactylus are folded underneath the merus.

The major variations in mouthpart morphology within the Stomatopoda are related to differences in prey and prey-capture mechanisms. Stomatopods feed on a variety of food items including crustaceans, fish, molluscs, polychaetes, and echinoderms. Cannibalism is also well-documented. Examples of dietary specialization in several species of stomatopods (after Kunze 1981) are given in Table 2.

Based on comparative studies of six species of stomatopods (Kunze 1981), several morphological differences in the mouthparts can be identified which characterize the squilloid and gonodactyloid types. The grinding surfaces of gonodactyloid mandibles and the setation of the maxillules are more robust than in squilloids; the maxillary setation is correspondingly reduced; the raptorial limbs of gonodactyloids are specialized as smashing implements with enlargement of the distal propodus and proximal part of the dactylus. Also, the dactylar spines of the raptorial limbs in gonodactyloids are reduced (absent in some) and the basis-merus articulation is specialized to allow for wider lateral extension and greater lateromedial flexibility. The third maxillipeds are less robust, with reduced setation and compressed orally compared with squilloids.

The main difference in feeding between squilloids and gonodactyloids is in the method of prey capture. In squilloids the movement involves an anteroventral swing of the propodus around the carpus, accompanied by a ventral swing of the dactylus (Fig.7a). Prey is either speared with the dactylar spines or grasped between the toothed margins of the propodus and dactylus. The gonodactyloid mechanism does not involve dactylar extension. The propodus is swung anteroventrally from a horizontal position and prey is struck with the 'elbow' on the lower proximal region of the dactylus (Fig.7b).

Subsequent manipulation of prey by the maxillipeds involves rapid, unco-ordinated movements during which food is hooked by the dactylus, positioned underneath the maxillae and pushed between the maxillules for ingestion. The actions of the third, fourth, and fifth maxillipeds are essentially similar to those of the raptorial appendages, except they are not as rapid or powerful. The arrangement and articulations of the limb segments

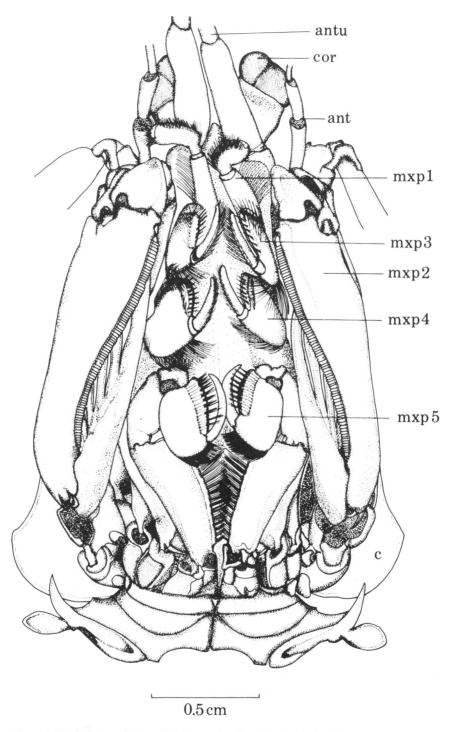

Figure 6. Ventral view of the cephalothorax in *Alima laevis*, showing the arrangement of the maxillipeds, second maxillipeds laterally displayed to expose the other maxillipeds (from Kunze 1981).

Table 2. Summary of the diets of squilloid and gonodactyloid stomatopods based on analysis of gut contents.

| Species | Family | Feeding mechanism | Major prey items | Minor prey items |
|---------|--------|-------------------|------------------|------------------|
| *Alima laevis* | Squillidae | Spearing | Palaemonid crustaceans | Polychaete worms |
| *Anchisquilla fasciata* | Squillidae | Spearing | Cephalopod molluscs | Decapod crustaceans |
| *Oratosquilla nepa* | Squillidae | Spearing | Palaemonid crustaceans | Cephalopod molluscs |
| *Harpiosquilla stephensoni* | Harpiosquillidae | Spearing | Fish | – |
| *Odontodactylus cultrifer* | Odontodactylidae | Smashing | Gastropod and bivalve molluscs | Crustaceans |
| *Gonodactylus graphurus* | Gonodactylidae | Smashing | Gastropod and bivalve molluscs | Crabs |

are similar. The maxillae are not used in feeding but in the expulsion of material from the pre-oral cavity.

No aspect of the feeding repertoire of stomatopods resembles that of the filter-feeding malacostracan. Filter feeding generally requires specialization of the limbs as filtratory structures, a high degree of co-ordination and importantly, a mechanism to direct the suspended food to the filtratory appendages (e.g. production of water currents) and transfer it from the filter to the mouth. The presence of setiferous thoracopods in extinct aeschronectid hoplocarids (Schram 1969a) is not sufficient evidence that these animals were natant filter feeders.

Raptorial feeding on large and/or hard food can often be the end result of an evolutionary trend stemming from a simple raptatory habit (Manton 1977). Manton suggested that ancestral malacostracans were bottom-living in habit and may have been detritus feeders and scavengers. The absence in larval and adult hoplocarids of structures similar to those present in true filter-feeding malacostracans indicates a more parsimonious derivation from a simple raptatory detritivore/scavenger.

Hessler & Newman (1975) argued that the urcrustacean limb was a mixopodium with a flattened protopod, foliaceous epipod and exopod, and stenopodial endopod. Food was concentrated along the ventral midline by the metachronal activity of the postoral limbs and passed to the mouth by protopodal endites. Hessler (1982) states 'the Leptostraca show that the early malacostracan limb was a mixopodium although not necessarily of the precise cephalocarid form, and . . . show what the urmalacostracan must have been like'. Hessler also acknowledges that the Cannon & Manton (1927) view of a primitively biramous limb is exhibited in the remipedian *Speleonectes* (Yager 1981). *Speleonectes,* because of its numerous, serially homologous, enditeless, paddle-like biramous trunk appendages, and absence of tagmosis into thorax and abdomen, may well qualify as the most primitive crustacean. If this is the case, then the mixopodium, by corollary, may be an apomorphy related to a natatory, filter-feeding habit.

## 10.2 *Feeding in larval stomatopods*

Larval stomatopods hatch as antizoeae or pseudozoeae, pass through several pelagic stages, and metamorphose to a postlarva resembling the adult (Giesbrecht 1910). During the first

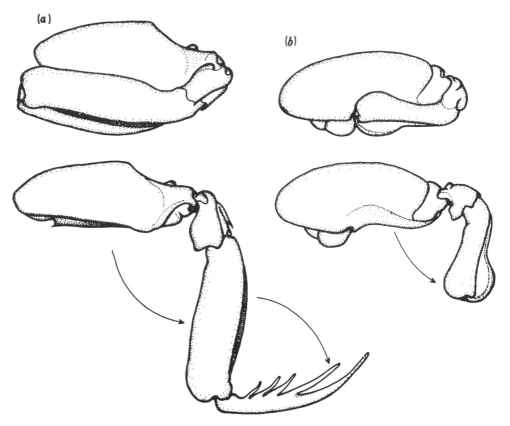

Figure 7. Right raptorial limbs: A. *Anchisquilla fasciata* (Squillidae); B. *Gonodactylus graphurus* (Gonodactylidae); illustrating the differences in the prey-capture strike of squillid and gonodactylid stomatopods, respectively. Arrows indicate direction of movement (from Kunze 1981).

propelagic stages the larvae rely on yolk reserves; no feeding occurs since the mandibles and two pairs of maxillae are not functional and the dactyli of the maxillipeds are not reflexed (Gurney 1937, Manning & Provenzano 1963, Pyne 1972). Feeding only occurs during the pelagic stages when the maxillipeds are subchelate and the feeding mechanism is similar to that of the adult.

### 10.3 *Mouthparts of raptatory feeders vs filter feeders – morphological evidence*

Predatory malacostracans generally have mandibles with strong distal cusps, few or no grinding surfaces, strong articulations with the head and a powerful musculature. Stomatopod mandibles are atypical in having well-developed cusped, molar processes and lacking the capability of a wide mandibular gape. Early stage larval stomatopods lack the well-developed molar processes of the adults and employ only the heavily serrated incisor processes. When present, the molar processes are used as cutting and chewing structures within the gut, not as grinding plates in the pre-oral cavity. Although the mandibular musculature is primitive in terms of its components and general arrangement, these primitive features are shared by raptatory detritivores such as syncarids indicating that the type of post-mandibular limb specialization is an equally important consideration when discussing filter feeding in the Malacostraca.

Stomatopods lack any specialized endites, exopods, or filter plates such as are found in *Nebalia, Euphausia,* and mysids (Cannon 1927, Mauchline 1967, Manton 1977). The maxillules are small, toothed structures; the maxillae are passive during feeding and apparently unspecialized other than to aid in the expulsion of material that is regurgitated from the proventriculus.

There is good fossil evidence that the primitive hoplocarid maxilliped was not subchelate but similar to the other thoracopods, that is, an unspecialized biramous limb (Schram 1969b). Subsequent specialization as rapacious carnivores occurred very early, by late Devonian or early Mississippian and this trend 'continued without interruption from the primitive palaeostomatopods to the Late Carboniferous archaeostomatopodans and tyrannophontids into the Mesozoic, where essentially modern stomatopods are first encountered' (Schram 1979).

It would seem unnecessary to adhere to a filter-feeding hoplocaridan archetype on the basis of the scanty morphological and functional evidence (such as Cannon's 1927 study of *Nebalia*). It is reasonable to envisage a scenario in which the early malacostracans comprised a diversity of forms, some employing biramous limbs to obtain food, others displaying the mixopodial filter-feeding habit. The possibility of a cephalocarid-like urcrustacean or a remipedian-like urcrustacean gives credence to the probability of a premalacostracan radiation in limb structure that was correlated with various feeding modes, including filter feeding and raptatory detritivory.

## 11 GASTRIC STRUCTURE AND FUNCTION

The proventriculus or foregut in stomatopods is a highly specialized functional complex and unique in the Malacostraca. The cardiac stomach (Fig.8) lacks any of the masticatory structures typical of the gastric mills of eucarids and peracarids. It is a thinwalled cuticular sac supported by a few narrow ossicles. The pyloric stomach (Fig.8) comprises complex filtratory structures or ampullae which resemble the ampullary structures of decapods, i.e. they have a ventral system of channels overlain by closely spaced setae and a dorsal 'filter press' which is appressed to the ventral ampullary ossicle (Fig.9). The function of the ampullae is, however, different in the two groups. In decapods ampullae function as a filter during the passage of macerated food into the digestive gland, whereas in stomatopods the ampullae function mainly as a pump to transfer digestive fluids from the digestive gland into the cardiac stomach. A dorsal pyloric stomach connecting with the mid- and hindgut, characteristic of all other malacostracans is vestigial and non-functional in stomatopods. Connecting the cardiac and pyloric regions is a complex system of channels and setose screens, the posterior cardiac plate (Figs.8 and 9). This structure allows bidirectional movement of fluids to and from the ampullae. No coarse particles can pass directly into the midgut, consequently indigestible fragments and particles which cannot pass through the setal screens of the posterior cardiac plate must be regurgitated from the proventriculus through the mouth.

Characterizing the gastric mechanism is the phasic circulation of digested food and diges-

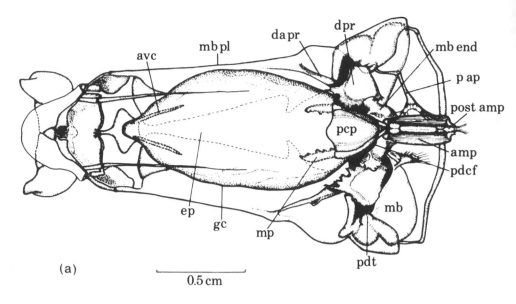

(a)

0.5 cm

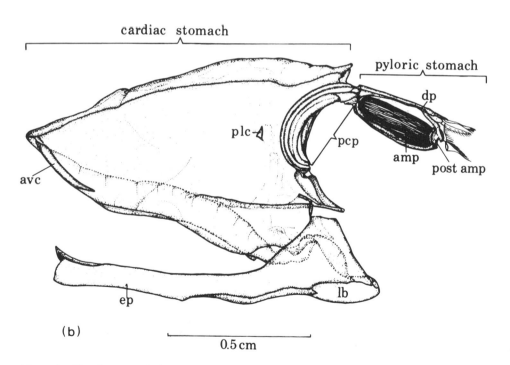

(b)

0.5 cm

Figure 8. *Alima laevis:* A. dorsal view of the proventriculus and surrounding structures after removal of the carapace, left is anterior (underlying structures are superimposed on the transparent gastric cuticle); B. lateral view of the proventriculus after removal of the musculature and cephalothoracic skeleton, left is anterior (from Kunze 1981).

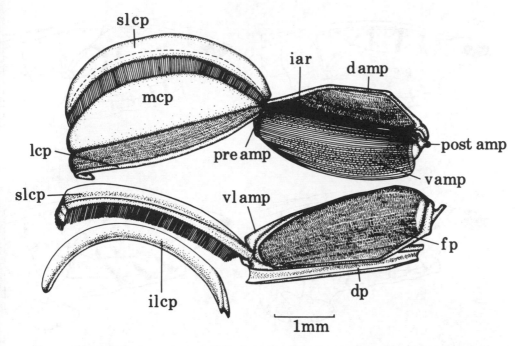

Figure 9. *Alima laevis:* dorsolateral view of the posterior cardiac plate and ampullae. The ossicles of the left side have been separated to expose the ventral ossicles, left is anterior (from Kunze 1981).

tive fluids, respectively, to and from the digestive gland. The major features of the digestive cycle are illustrated in Figures 10 and 11 and are summarized below:

Phase 1. At the onset of ingestion, digestive juices are pumped by the ampullae from the digestive gland into the cardiac stomach (Fig.11a).

Phase 2. Food is macerated by the combined actions of the mandibles, muscular contractions of the gastric wall (Figs.10a-c and 11b), and chemical action of the digestive juices.

Phase 3. Macerated food is circulated into the digestive gland through the posterior cardiac plate and ampullae (Figs.10d, 11b,c).

Phase 4. Non-assimilated food is cycled from the digestive gland into the midgut (Figs.11c, d).

Phase 5. The midgut sphincter (at the junction of the post-ampullary chamber, *pac,* and midgut, *mg,* Fig.11a) is closed and another cycle of pyloric pumping of digestive juices into the cardiac stomach commences (Fig.11e).

Phase 6. Macerated material from the cardiac stomach is again circulated into the digestive gland whilst material is further compacted and moved along the midgut (Fig.11f).

Phase 7. Another cycle of accumulation in the anterior lobe occurs with subsequent shunting of material into the midgut.

These cycles continue until the cardiac stomach is empty.

The histology of the digestive gland differs from that of decapods in that F-cells which have been regarded as precursors to absorptive cells (R-cells) are absent. Secretions appear to be holocrine, a mechanism well suited to sporadic carnivorous feeding strategies (Kunze 1981).

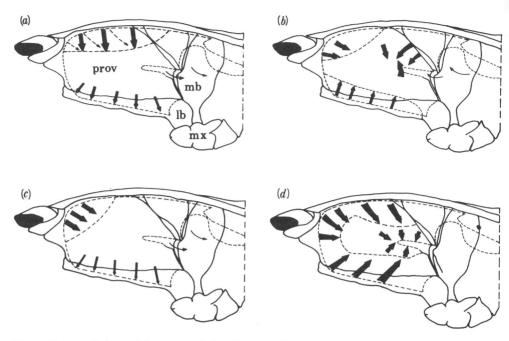

Figure 10. Lateral views of the proventriculus (dashed outline) drawn schematically to illustrate the movements of proventriculus during A-C, maceration of food and D, pumping of food into the pyloric stomach. A. The dorsal cardiac gastric wall moved ventrally and ventral cardiac gastric floor expanded; B. Floor of cardiac stomach raised, lateral cardiac folds brought medially, the rostral part of proventriculus moved posteriorly and molar processes of mandibles occluded; C. Gastric flood lowered and food pressed ventrally by posteroventral movement of rostral part of proventriculus, molar process of the mandibles swung laterally; D. Floor of the cardiac stomach raised, rostral part of proventriculus moved posteriorly, roof of cardiac stomach moved ventrally, and lateral cardiac folds brought medially (from Kunze 1981).

## 12 CONCLUSIONS

The foregoing discussion highlights the major and definitive characters of the Stomatopoda and Hoplocarida, in summary:

1. Cephalic kinesis — plesiomorphic;
2. Triflagellate first antennae — apomorphic;
3. Five pairs of six-segmented subchelate maxillipeds — autapomorphic;
4. Thoracopods with a three-segmented protopod, etiology and homology of the two rami in modern stomatopods unknown (autapomorphic?), primitive condition believed to be an inner ramus of four segments and an outer ramus one-segmented — plesiomorphic;
5. Abdominal gills — plesiomorphic?
6. Tubular heart extending the body length with segmental ostia and arteries — plesiomorphic;
7. Six-segmented abdomen probably the result of fusion of the posterior abdominal segments although it has been suggested (Reaka 1975) that fusion occurred between the last thoracic and anterior abdominal segments;

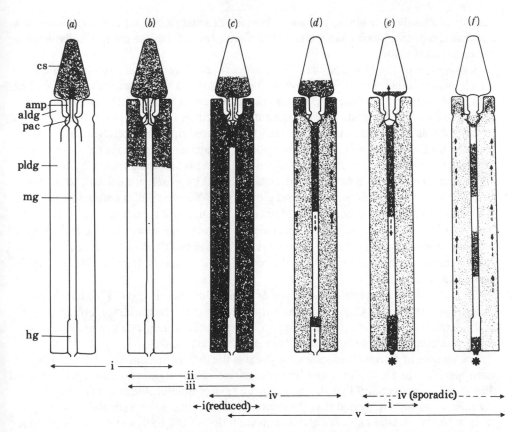

Figure 11. Schematic dorsal views of the alimentary tract, summarizing the major features of the digestive cycle: A. 0-15 min post-ingestion; B. 30-60 min post-ingestion; C. 1-2 h post-ingestion; D. 4 h post-ingestion; E. 8 h post-ingestion; F. 1-2 days post-ingestion. Solid arrows indicate movements of digestive fluids; dashed arrows indicate movements of food particles; stippling indicates distribution and proportion of food particles in different regions of digestive tract; asterisks signify defecation. Phases of the digestive cycle are: (i) secretion and circulation of digestive juices into cardiac stomach; (ii) maceration of food in cardiac stomach; (iii) passage of food particles into digestive gland; (iv) circulation of particles in digestive gland; (v) circulation of non-absorbed particles into midgut, including compaction in anterior lobe of digestive gland and secretion of peritrophic membrane (from Kunze 1981).

8. Absence of transverse abdominal muscles — plesiomorphic;

9. Extension of the viscera throughout the body — plesiomorphic;

10. Mandibular skeletomusculature and mechanism characteristized by a near vertical axis of swing, transverse and rotational movements, replacement of the transverse mandibular tendon with an endophragmal bridge for muscle support, and the use of the molar processes as intragastric masticatory structures — autapomorphic;

11. Gastric mill absent, mastication by mandibular action, dorsal pyloric stomach vestigial, ampullae used as a pump for the anterior transport of digestive fluids — autapomorphic.

The question of polyphyly in the Eumalacostraca (*sensu* Moore 1969) cannot be adequately addressed by arguments pertaining to the characters that define the taxon, i.e.

absence of a bivalved or hinged carapace, absence of caudal furca, and absence of a seventh abdominal segment. These characters, with the exception of the last, are not really diagnostic, as discussed below.

Siewing (1956, 1963) argued that the leptostracan bivalved carapace and its adductor muscles were apomorphic. It can, therefore, be argued that Leptostraca are an independent offshoot of the early Phyllocarida. The important feature of the leptostracan carapace is its lack of fusion to the post-cephalic segments. The extinct Eocarida, purportedly ancestral to the Eucarida and Peracarida (Brooks 1969), also shared this character. Schram (1981) regarded the Eocarida as an unnatural assemblage: 'a catchall category for fossil forms with incomplete information about them'. Schram (1981) in his proposed taxonomy for the Malacostraca erected the cohort Mysoida, characterized by a carapace unfused to the thorax, which included two formerly eocarid groups, the Waterstonellidea and Belotelsonidea. His inclusion within this cohort of the Mysidacea, which have between one and three of the post-cephalic segments fused with the carapace is, however, questionable. The leptostracan carapace, in view of its specialized biramous construction and the fossil evidence of eumalacostracans with carapaces unfused to the post-cephalic segments, is not a reliable or diagnostic character.

Caudal furca are not consistent features of the Phyllocarida. Briggs (1978) has shown them to be absent in the Canadaspididae but present in the Perspicarididae, both grouped in the extinct order Canadaspidida. On the other hand, furca are present in some eumalacostracans such as bathynellacean and stygocaridacean syncarids.

The absence of a seventh abdominal segment is the only consistent and definitive feature separating the Phyllocarida and 'Eumalacostraca'. Some arguments have been presented that suggest the possibility that this may either be symplesiomorphic or convergent. For example, it has been suggested that the transverse groove across the sixth abdominal segment in *Gnathophausia* may indicate incomplete fusion of the sixth and seventh segments. This is supported by the presence of a pair of seventh abdominal somites in some mysidacean embryos (Manton 1928). In defense of the independent ancestry of hoplocaridans, Schram (1973, 1982) suggested that Palaeozoic phyllocarids such as the sairocaridid hoplostracans shared features in common with hoplocarids although were not direct ancestors. In a recent review Schram (1982) expressed some support for Reaka's (1975) suggestion that the hoplocarid abdominal segmentation was a result of anterior segmental fusion. Reaka's evidence is weak and does not adequately demonstrate that the six-segmented abdominal construction of eumalacostracans and hoplocaridans is the result of convergence.

In defense of a polyphyletic origin of Eumalacostraca and Hoplocarida, one must doubt the validity of a taxon based on one reliable character. The crucial issue is whether there are morphological characters and functional processes in hoplocarids and eumalacostracans which have evolved independently and whether a common ancestor (albeit hypothetical) could have given rise to these independently derived forms. Adding further to the problem of establishing the most probable phylogenetic relationships of the Hoplocarida is the heterogeneity of the Phyllocarida, itself an assemblage of distantly related taxa (Briggs 1978) whose affinities are, at best, speculative.

Burnett & Hessler (1973) defended a monophyletic origin of the Eumalacostraca and Hoplocarida and proposed a precaridoid ureumalacostracan ancestor with the following features: (1) sixth and seventh abdominal segments fused; (2) a tailfan composed of uropods and telson; (3) caudal furca reduced; (4) abdomen more flexible than in phyllocarids;

(5) precaridoid abdominal musculature; (6) thoracic epipodal gills; (7) a scale-like exopod on the second antenna. These features are addressed below.

1. It was concluded above that this is the strongest character arguing against polyphyly of eumalacostracans and hoplocarids.

2. There is no reason to believe the pre-hoplocarid ancestor possessed a tailfan *per se*, considering the morphology of the so-called uropods in palaeostomatopods such as *Perimecturus* (Schram & Horner 1978). These structures are no more than articulated spines.

3. The loss of furca must have occurred early during hoplocarid evolution since the group is purportedly 'at least as old as the eumalacostracans' (Schram 1977). Yet, fully developed furca are retained in belotelsonid and waterstonellid eocarids (= mysoids, *sensu* Schram 1981) and persists in some modern syncarids. It is probable that some early Eumalacostraca retained furca while others lost it. There is no evidence for reduced furca in the ureumalacostracan.

4 and 5. Abdominal flexibility and the related precaridoid abdominal musculature could have evolved within the Phyllocarida rather than in the ureumalacostracan. Stomatopods are the only group that displays a 'precaridoid' or intermediate abdominal musculature. The only information on phyllocarids is based on *Nebalia* (Hessler 1964) and the major feature of its skeletomusculature which separates it from stomatopods is the absence of coiled muscles, in other respects the pattern is similar.

6. Thoracic epipods are present in Leptostraca (Calman 1909) and these probably function in respiration. This is not a valid ureumalacostracan character in the sense that it is a plesiomorphy, not an apomorphy.

7. The antennal scale is also not a consistent eumalacostracan feature. It is present in Eucarida, Hoplocarida, and Syncarida; but is absent in most Peracarida (except Mysidacea and most isopods). Schram (1973) indicated an antennal scale in the sairocaridid Hoplostraca which suggested that it arose within the Phyllocarida. Rolfe (1981) disputed Schram's evidence and concluded that the scale was in fact part of an anterior raptorial limb. An important difference between the eumalacostracan antennal scale and that of stomatopods is that in eumalacostracans it arises as a single segment from the protopod, whereas in stomatopods the exopod has two segments: a basal segment and an articulating terminal scale. The presence of an antennal scale on the first antenna in *Nebalia* is perplexing and may reflect morphological plasticity in the expression of this character.

In the light of the previous discussion, the precaridoid eumalacostracan ancestor proposed by Burnett & Hessler can only be validly defined on the basis of its six abdominal segments and possibly the presence of coiled abdominal muscles. Arguing against a monophyletic origin of the Eumalacostraca that includes the Hoplocarida are a suite of symplesiomorphies shared only by the Phyllocarida and autapomorphies. These characters define a discrete hoplocaridan morphotype (Schram calls these the 'hoploid facies') that can be derived with fewer unknown intermediates if an independent phyllocarid-like ancestor is sought than if we constrain the hypothetical ancestor to one characterized by one or two features and assume a divergence of two different functional models from such an ancestor.

In conclusion, the independent origin of the Hoplocarida from an early malacostracan ancestral stock is supported. The existence of a 'hoploid' morphotype has been established. The derivation of both hoploid and caridoid morphotypes from a common ureumalacostracan is a possible alternative, but must be seriously questioned in the light of the available evidence.

# 13 ACKNOWLEDGEMENTS

I would like to express my sincerest thanks to D.T.Anderson (University of Sydney) and R.R.Hessler (Scripps Institution of Oceanography) for many hours of fruitful discussion. R.R.Hessler, F.R.Schram and J.-O.Stromberg kindly offered their valuable criticisms of the manuscript. This work was undertaken while under the tenure of a CSIRO post-doctoral studentship.

Abbreviations used in figures

| | | | |
|---|---|---|---|
| aldg | anterior lobe of digestive gland | mb pl | mandibular pleurite |
| a mb ap | anterior mandibular apodeme | mcp | median cardiac plate ossicle |
| amp | ampullae | mg | midgut |
| ant | antennae | mp | molar process of mandible |
| antu | antennule | mx | maxilla |
| avc | anteroventral cardiac ossicle | mxp1 | first maxilliped |
| avpm | anteroventral pyloric muscle | mxp2 | second maxilliped |
| b | basis | mxp3 | third maxilliped |
| c | carapace | mxp4 | fourth maxilliped |
| cg | cervical groove | mxp5 | fifth maxilliped |
| cor | cornea | mxu | maxillule |
| cp | carpus | mxu add m | adductor muscle of maxillule |
| cs | cardiac stomach | p | propodus |
| cx | coxa | pac | post-ampullary chamber |
| d | dactylus | pap | posterior apodeme of the mandibular endophragm |
| dapr | dorsal anterior process of the mandibular endophragm | | |
| de | distal endite | pcp | posterior cardiac plate |
| dlm | dorsal longitudinal muscle | pcpm | posterior cardiac plate muscles |
| dmap | dorsomedian apodeme of the mandibular endophragm | pdcf | posterior dorsal cardiac fold |
| | | pdt | posterior dorsal tubercle |
| | | pe | proximal endite |
| dp | dorsal pyloric ossicle | pg | paragnath |
| dpr | dorsal process of the mandibular endophragm | plc | posterolateral cardiac ossicle |
| | | pldg | posterior lobe of the digestive gland |
| dvcpm | dorsoventral cardiac plate muscle | p mb ap | posterior mandibular apodeme |
| e | endite | post amp | posterior ampullary ossicle |
| e1, e2 | first and second endites of the second maxillary segment | pr amp | preampullary ossicle |
| | | prov | proventriculus |
| ep | epistome | pvt | posterior ventral tubercle |
| epp | epipod | slcp | superior lateral cardiac plate ossicle |
| fp | filter press | tcpm | transverse cardiac plate muscle |
| g | gut | tt | transverse mandibular tendon |
| gc | gastric cuticle | v amp | ventral ampullary ossicle |
| hg | hindgut | vpr | ventral process of the mandibular endophragm |
| ilcp | inferior lateral cardiac plate ossicle | | |
| ip | incisor process of mandible | x | axis of swing |
| lb | labrum | I, II, III, IV | first, second, third and fourth maxillary segments, respectively |
| lcp | lateral cardiac plate ossicle | | |
| lpcpm | lateral posterior cardiac plate muscles | 2 | anterior mandibular remotor 2 |
| m | merus | 3 | mandibular promotor 3 |
| mb | mandible | 4 | posterior mandibular remotor 4 |
| mb ap | mandibular apodeme | 5a | mandibular remotor 5a |
| mbb | mandible body | 5b | mandibular promotor 5b |
| mbc | articulation of mandibular condyle | 5c | transverse mandibular remotor 5c |
| mb end | mandibular endophragm | 6 | mandibular promotor 6 |
| mbp | posterior margin of mandible | | |

# REFERENCES

Balss, H. 1938. Stomatopoda. In: *H.G.Bronn's Klassen und Ordnungen des Tierreichs* 5(1), 6(2):1-173. Leipzig: Academische Verlags.

Boas, J.E.V. 1883. Studien über die Verwandtschaftsbeziehungen der Malakostraken. *Morphol. Jb.* 8: 485-579.

Briggs, D.E.G. 1978. The morphology, mode of life and affinities of *Canadaspis perfecta* (Crustacea; Phyllocarida) Middle Cambrian, Burgess Shale, British Columbia. *Phil. Trans. Roy. Soc. London* (B) 281:439-487.

Brooks, H.K. 1962. The Paleozoic Eumalacostraca of North America. *Bull. Am. Paleont.* 44:163-338.

Brooks, H.K. 1969. Eocarida. In: R.C.Moore (ed.), *Treatise on Invertebrate Paleontology, Part R, Arthropoda 4:*332-345. Lawrence: Geol. Soc. Am. & Univ. Kansas Press.

Burnett, B.R. 1972. Notes on the lateral arteries of two stomatopods. *Crustaceana* 23:303-305.

Burnett, B.R. & R.R.Hessler 1973. Thoracic epipodites in the Stomatopoda (Crustacea): A phylogenetic consideration. *J. Zool.* 169:381-392.

Caldwell, R.L. & H.Dingle 1975. Ecology and evolution of agonistic behaviour in stomatopods. *Naturwiss.* 62:214-222.

Caldwell, R.L. & H.Dingle 1976. Stomatopods. *Sci. Amer.* 234:80-89.

Calman, W.T. 1909. Crustacea. In: E.R.Lankester (ed.), *A Treatise on Zoology, Part VII:*1-346. London: A. and C.Black.

Cannon, H.G. 1927. On the feeding mechanism of *Nebalia bipes. Trans. Roy. Soc. Edinb.* 55:355-370.

Cannon, H.G. & S.M.Manton 1927. On the feeding mechanism of a mysid crustacean, *Hemimysis lamornae. Trans. Roy. Soc. Edinb.* 55:219-253.

Claus, C. 1871. Die Metamorphose der Squilliden. *Abh. kgl. Ges. Wiss. Göttingen* 16:111-163.

Claus, C. 1885. Neue Beiträge zur Morphologie der Crustaceen. *Arb. zool. Inst. Univ. Wien* 6:1-105.

Dahl, E. 1963. Main evolutionary lines among recent Crustacea. In: H.B.Whittington & W.D.I.Rolfe (eds), *Phylogeny and evolution of Crustacea:*1-15, 17-26. Cambridge: Mus. Comp. Zool.

Dingle, H. & R.L. Caldwell 1978. Ecology and morphology of feeding and agonistic behaviour in mudflat stomatopods (Squillidae). *Biol. Bull. Mar. Biol. Lab.* 155:134-149.

Giesbrecht, W. 1910. Stomatopoden. *Fauna Flora Golf. Neapel* 33:1-240.

Giesbrecht, W. 1913. Crustacea. In: A.Lang (ed.), *Handbuch der Morphologie der Wirbellosen Tiere. Bd.4, Arthropoden:*9-252. Jena: Fischer.

Giesbrecht, W. & H.Balss 1933. Crustacea. In: *Handwörterbuch der Naturwissenschaften* 2:800-840.

Glaessner, M.F. 1957. Evolutionary trends in Crustacea (Malacostraca). *Evolution* 11:178-184.

Grobben, K. 1892. Zur Kentniss des Stammbaumes und der Systems der Crustaceen. *S.B. Akad. Wiss. Wien* 101:237-274.

Grobben, K. 1919. Über die Musculatur des Vorderkopfes der Stomatopoden und die systematische Stellung dieser Malakostrakengruppe. *S.B. Akad. Wiss. Wien* 128:185-214.

Gurney, R. 1937. Notes on some decapod and stomatopod Crustacea from the Red Sea. III-V. *Proc. Zool. Soc. Lond.* 107:319-336.

Haeckel, E. 1896. In: *Systematische Phylogenie. 2. Der Wirbellose Tiere:*645-662. Berlin: Reimer.

Hessler, R.R. 1964. The Cephalocarida: Comparative skeletomusculature. *Mem. Conn. Acad. Arts Sci.* 16:1-97.

Hessler, R.R. 1982. Evolution within the Crustacea. In: L.G.Abele (ed.), *The Biology of Crustacea, Vol.I:*149-185. New York: Academic Press.

Hessler, R.R. & W.A.Newman 1975. A trilobitomorph origin for the Crustacea. *Fossils & Strata* 4:437-459.

Holthius, L.B. & R.B.Manning 1969. Stomatopoda. In: R.C.Moore (ed.), *Treatise on Invertebrate Paleontology, Part R, Arthropoda 4:*R535-552. Lawrence: Geol. Soc. Am. & Univ. Kansas Press.

Huxley, T.H. 1867. *A Manual of the Anatomy of Invertebrated Animals.* London: Churchill.

Komai, T. & Y.M. Tung 1931. On some points of the internal structure of *Squilla oratoria. Mem. Coll. Sci. Engng. Kyoto Imp. Univ.* 6:1-15.

Kunze, J.C. 1981. The functional morphology of stomatopod Crustacea. *Phil. Trans. Roy. Soc. London* (B)292:255-328.

Manning, R.B. 1980. The superfamilies, families, and genera of recent stomatopod Crustacea, with diagnoses of six new families. *Proc. Biol. Soc. Wash.* 93:362-372.

Manning, R.B. & A.J. Provenzano, jr 1963. Studies on development of stomatopod Crustacea. I. Early larval stages of *Gonodactylus oerstedii* Hansen. *Bull. Mar. Sci. Gulf Caribb.* 13:467-487.

Manton, S.M. 1928. On the embryology of the mysid crustacean, *Hemimysis lamornae. Phil. Trans. Roy. Soc. London* (B)216:363-463.

Manton, S.M. 1964. Mandibular mechanisms and the evolution of arthropods. *Phil. Trans. Roy. Soc. London* (B)247:1-183.

Manton, S.M. 1977. *The Arthropoda: Habits, Functional Morphology and Evolution.* Oxford: Clarendon Press.

Mauchline, J. 1967. The feeding appendages of the Euphausiacea (Crustacea). *J. Zool.* 153:1-43.

Moore, R.C. 1969. Eumalacostraca. In: R.C.Moore (ed.), *Treatise on Invertebrate Paleontology, Part R, Arthropoda 4:*332. Lawrence: Geol. Soc. Am. & Univ. Kansas Press.

Pyne, R.R. 1972. Larval development and behaviour of the mantis shrimp *Squilla armata* Milne Edwards (Crustacea: Stomatopoda). *J. Roy. Soc. N.Z.* 2:121-146.

Reaka, M.L. 1975. Molting in stomatopod crustaceans: I. Stages of the molt cycle, setagenesis, and morphology. *J. Morph.* 146:55-80.

Rolfe, W.D.I. 1981. Phyllocarida and the origin of the Malacostraca. *Geobios* 14:17-27.

Schram, F.R. 1969a. Polyphyly in the Eumalacostraca? *Crustaceana* 16:243-250.

Schram, F.R. 1969b. Some middle Pennsylvanian Hoplocarida (Crustacea) and their phylogenetic significance. *Fieldiana:Geol.* 12:235-289.

Schram, F.R. 1973. On some phyllocarids and the origin of the Hoplocarida. *Fieldiana:Geol.* 26:77-94.

Schram, F.R. 1977. Paleozoogeography of late Paleozoic and Triassic Malacostraca. *Syst. Zool.* 26:367-379.

Schram, F.R. 1979. The genus Archaeocaris, and a general review of the Palaeostomatopoda (Hoplocarida: Malacostraca). *Trans. San Diego Soc. Nat. Hist.* 19:57-66.

Schram, F.R. 1981. On the classification of Eumalacostraca. *J. Crust. Biol.* 1:1-10.

Schram, F.R. 1982. The fossil record and evolution of Crustacea. In: L.G.Abele (ed.), *The Biology of Crustacea, Vol.I:*93-148. New York: Academic Press.

Schram, F.R. & J.Horner 1978. Crustacea of the Mississippian Bear Gulch limestone of central Montana. *J. Paleont.* 52:394-406.

Secretan, S. 1967. Proposition d'un nouvelle compréhension et d'un nouvelle subdivision des Archaeostraca. *Ann. Paléo.* 53:153-188.

Shiino, S.M. 1942. Studies on the embryology of *Squilla oratoria* de Haan. *Mem. Coll. Sci. Engng. Kyoto Imp. Univ.* 17:77-174.

Siewing, R. 1956. Untersuchungen zur Morphologie der Malacostraca (Crustacea). *Zool. Jb. (Anat. Ont. Tiere)* 75:39-176.

Siewing, R. 1963. Studies in malacostracan morphology: Results and problems. In: H.B.Whittington & W.D.I.Rolfe (eds), *Phylogeny and Evolution of Crustacea:*85-103. Cambridge: Mus. Comp. Zool.

Yager, J. 1981. Remipedia, a new class of Crustacea from a marine cave in the Bahamas. *J. Crust. Biol.* 1:328-333.

ERIK DAHL
*Department of Zoology, Lund, Sweden*

# MALACOSTRACAN PHYLOGENY AND EVOLUTION

## ABSTRACT

Malacostracan ancestors were benthic-epibenthic. Evolution of ambulatory stenopodia probably preceded specialization of natatory pleopods. This division of labor was the prerequisite for the evolution of trunk tagmosis. It is concluded that the ancestral malacostracan was probably more of a pre-eumalacostracan than a phyllocarid type, and that the phyllocarids constitute an early branch adapted for benthic life. The same is the case with the hoplocarids, here regarded as a separate subclass with eumalacostracan rather than phyllocarid affinities, although that question remains open. The systematic concept Eumalacostraca is here reserved for the subclass comprising the three caridoid superorders, Syncarida, Eucarida and Peracarida. The central caridoid apomorphy, proving the unity of caridoids, is the jumping escape reflex system, manifested in various aspects of caridoid morphology. The morphological evidence indicates that the Syncarida are close to a caridoid stem-group, and that the Eumalacostraca *sensu stricto* were derived from pre-syncarid ancestors. Advanced hemipelagic-pelagic caridoids probably evolved independently within the Eucarida and Peracarida.

## 1 INTRODUCTION

The differentiation of the major crustacean taxa must have taken place very early, probably at least partly in the Precambrian (cf. Bergström 1980). From the Cambrian we have proof of the presence of branchiopods (Simonetta & Delle Cave 1980, Briggs 1976), ostracods (Müller 1979, 1981), and cephalocarids or forms resembling them (Müller 1981). It has also been more or less generally accepted that phyllocarid malacostracans were represented in the Cambrian faunas. However, a critical review of modern redescriptions of the best-preserved taxa, *Perspicaris* (Briggs 1977), *Canadaspis* (Briggs 1978) and *Plenocaris* (Whittington 1974), has shown that the morphology at these taxa in various respects is incompatible with the universally accepted definition of the Malacostraca (Dahl, in press 2). These findings make the malacostracan nature of less well-defined taxa, such as *Hymenocaris* and others, still more doubtful. Pending the discovery of new evidence, it would appear we have no proof that Malacostraca existed in the Cambrian even if that may well have been the case. The earliest unequivocal Malacostraca are, as demonstrated by Rolfe (1962) for *Ceratiocaris*, the Archaeostraca, known from the Ordovician onwards.

189

## 2 GENERAL REMARKS ON THE MALACOSTRACA

The Malacostraca constitute a well-defined taxon, the main diagnostic features being the strictly uniform tagmosis and the presence of paired appendages on all segments except, in the Phyllocarida, the seventh abdominal segment.

It deserves recalling in this context that while the cephalon is well-defined, even if thoracic segments may become secondarily incorporated, the terms thorax and abdomen, when applied for example to the branchiopods, are used much less strictly than in the case of the Malacostraca (Dahl 1963) and that the segmental composition of these tagmata varies from group to group. The Malacostraca stand out as a very isolated crustacean taxon, and we have no definite indication of a close relationship to any of the others. The similarities with the Cephalocarida, although providing possible indications concerning evolutionary trends within the Crustacea (Sanders 1963, Hessler 1964, 1982, Anderson 1973), are rather symplesiomorphies than anything else.

I wish to call attention here to two organ complexes which underline the high degree of malacostracan isolation, mainly because their importance has been neglected or misunderstood. These organ complexes are the frontal eyes (sensu Elofsson 1966c) and the compound eyes. With respect to the frontal eye complex Elofsson (1965, 1966a,b,c) demonstrated the presence of four distinct patterns, one of which is found within the Malacostraca, the other three in the non-malacostracan groups. The frontal eyes of the Malacostraca differ from those of the other taxa in having everse receptor cells resembling those of the compound eye retina.

The differences between malacostracan and non-malacostracan compound eyes are even greater and more fundamental (Elofsson & Dahl 1970). Neither type could be derived from the other without a complete breakdown and rebuilding of the whole neuronal connection pattern in the lamina ganglionaris in the medulla externa-medulla interna region. Hessler & Newman (1975) failed to see the essential problem, and referred it to the separation of the medulla externa from the medulla interna, only partial in *Nebalia,* which is a point of minor importance. Paulus (1978) in an attempt to explain away existing differences between compound eyes by means of sweeping generalizations and deliberate over-simplifications, sought to minimize this piece of adverse evidence by means of a phrase concerning ontogenetical processes, which could not have been less relevant.

In 1963 at the Harvard Conference on Crustacea it was stated that we have no evidence enabling us to derive any crustacean taxon at what is now regarded as the class level from another such taxon (Dahl 1963). This, unfortunately, is still the case, the only possible exception being the taxa forming the core of the Maxillipoda (Hessler 1982). In the case of the Malacostraca from conchostracan ancestors, as proposed by Lauterbach (1975) appears very unlikely on account of various fundamental differences (Dahl 1976, Hessler 1982) not least in the structure of the protocerebral sense organs.

On the other hand, there exists a considerable degree of agreement concerning fundamental traits of the morphology of the ur-crustacean as an animal with numerous appendage-bearing segments and a subterminal mouth, with an atrium oris opening posteriad behind a labrum and receiving food from behind by means of a ventral transport mechanism (Calman 1909, Dahl 1956, Hessler & Newman 1975, Lauterbach 1980, Hessler 1982). We do not know whether the early Malacostraca, like the Cephalocarida, had two modes of feeding, or only a cephalic filter feeding mechanism (Cannon & Manton 1927, Manton 1930, Attramadal 1981). Both methods use appendages behind the mouth (Sanders 1963,

Dahl 1956, 176, Hessler 1982). This question will be discussed below in section 5.

It is important that the Malacostraca, in contrast to all other more or less multi-seg-mented Crustacea, have retained a full set of appendages; and that a crucial event in the fixa-tion of the malacostracan functional model system was the establishment of a rigid tagmo-sis with a thorax of eight segments and an abdomen bearing six pairs of appendages (Cal-man 1909, Lauterbach 1975, 1980, Dahl 1976, and in press 2). It was in fact only due to the organization with a specialized locomotory pleon and a thorax with appendages free to differentiate in a great variety of directions that a foundation could be laid for the singular evolutionary success of the Malacostraca. This was denied non-malacostracan groups such as Copepoda and Cirripedia which have a short abdomen devoid of limbs and a limited number of thoracic segments, possibly as a result of neotenic evolution (Gurney 1942, Hessler 1982). It would appear that a limited number of thoracic appendages is not a suf-ficient basis for large scale evolutionary flexibility.

Another important point was the fixation of the gonopores on the sixth (female) and eighth (male) thoracic segments. The position of the female gonopore was a prerequisite for the participation of the thoracopods in brood care, an arrangement peculiar to certain malacostracan taxa (Phyllocarida and Peracarida) and with bearings upon the evolution of these taxa.

## 3 EARLY MALACOSTRACAN ADAPTATIONS

The ancestral malacostracans were certainly benthic-epibenthic, a conclusion that appears now to be widely accepted. Manton (1953) showed that an adaptation of the anterior trunk-limbs for walking is advantageous for arthropods in the process of adapting ambula-tory habits. She concluded that in the case of the Malacostraca tagmatization took place at a stage when benthic locomotion was predominant, pleopod adaptation for swimming came later in connection with an increasing tendency to move above the bottom. A tagmatiza-tion so profoundly influencing the total structural plan would probably not have occurred if predominantly natatory habits had preceded ambulatory ones. These conclusions gain strength from the retention of ambulatory thoracic endopods in all pelagic eumalacostra-cans.

If this argument holds, and it appears to have a sound foundation in observed facts and less of a need for hypothetical intermediary forms than possible alternatives, it would imply that the stenopodium is a very old attribute, i.e. an adaptation to ambulatory habits in ancestral malacostracans.

There also exists ontogenetic evidence that bears upon the evolution of the malacostra-can structural plan which has been largely overlooked in the current discussion of malaco-stracan evolution and phylogeny. As pointed out by Anderson (1973) the ontogeny of the postnaupliar segments and their appendages in primitive malacostracans follows a pattern in principle identical with that found in the Cephalocarida (Sanders 1963) with a serial for-mation of segments and later of appendages in an anteroposterior direction. This has been demonstrated for Leptostraca (Manton 1934). Syncarida (Hickman 1937) and for Peraca-rida (Manton 1928) in mysids, in amphipods (Weygoldt 1958), in tanaids (Scolle 1963), and in isopods (Strömberg 1965, 1967, 1971). In these taxa no evidence of embryoniza-tion of larvae later than the nauplius has been found (Anderson 1973). Only in the Euca-rida and, probably, the Hoplocarida 'did an evolution of post-naupliar larval specializations precede the embryonization of larval stages in association with increased yolk' (Anderson 1973). In view of this ontogenetic retention of embryonized larval specializations in the Eucarida Anderson concluded 'that the malacostracan ancestors of the Leptostraca, Syncarida, and Peracarida did not have larval stages other than the nauplius'.

## 4 SYNCARIDA, EUCARIDA, AND PERACARIDA

### 4.1 *The caridoid concept*

The caridoid facies (see Hessler, this volume, fig.1) constitutes an important manifestation of eumalacostracan morphology, considered as ancestral among the eumalacostracans by many authors including Calman (1909), Siewing (1963), and Hessler (1982). In a previous paper, however, (Dahl, in press 1), I noted the need for a fresh evaluation of the caridoid morphological type with respect to its significance in eumalacostracan evolution and phylogeny. This will be one of the main subjects of the present chapter.

Calman (1909) listed the characteristics of the caridoid facies in its most advanced form. Hessler (1982) amended this list to comprise the following features: (1) a carapace enclosing the thorax, (2) movable stalked eyes, (3) biramous antennules, (4) exopod of the antenna scale-like, (5) all thoracopods with flagelliform exopods, (6) abdomen well-developed with complex and massive musculature, designed for flexing the tail fan, (7) uropods and telson together forming a tail fan, (8) pleopods 1-5 alike with two flagelliform rami, (9) internal organs mainly excluded from the abdomen. These various features will be discussed in some detail below.

Advanced caridoids, possessing all the features listed by Hessler (1982) are found among the Peracarida (order Mysidacea) and the Eucarida (orders Decapoda and Euphausiacea). The third eucarid order, the Amphionidacea, although caridoid, is too imperfectly known to be included in this discussion. Hessler (1982) after defining the caridoid facies, added the following important sentence, which deserves being quoted in full: 'The close adherence to caridoid morphology of the benthic *Anaspides* demonstrates that a 'shrimplike' habitus *sensu stricto* (as in euphausiids and peneids) is not a necessary quality of the caridoid facies *in its most meaningful sense*'. (The last italics are mine.) Hessler then went on to say that my own reluctance to accept 'the unity of caridoids' (Dahl 1976, and in press 1) might be due to my using a too restrictive concept. However, as early as at the Harvard conference on Crustacea in 1963 (Dahl 1963) I discussed the caridoid concept in its widest possible sense calling attention also to caridoid parallels in non-malacostracan crustaceans. I have found no reason to abandon this broad concept, which I share with Hessler. It is just when seen in this very broad perspective that the question of the unity of the caridoids becomes particularly interesting. Various aspects of this question will be dealt with below.

Looking somewhat more closely upon the caridoid features listed by Hessler, one finds that they are not all of the same diagnostic significance. Hessler (1982) pointed out that in his opinion the large carapace (1) the stalked eyes, (2), and the biramous antennule (3) were to be regarded as malacostracan plesiomorphies rather than diagnostic caridoid features. This is undoubtedly so with (2) and (3), while (1), the question of the carapace/ cephalothorax, will be discussed in more detail in section 4.3. In addition to the remarks made by Hessler (1982), it may be added that the biramous pleopod (8) is also a malacostracan plesiomorphy and that the exclusion of the viscera from the abdomen (9) is a direct effect of the strong development of the abdominal musculature (6). Concerning (4), the antennal scale, (6), the abdominal musculature, and (7), the tail fan, some further remarks are necessary. The antennal scale is, in its typical form, a unique caridoid feature, present in all eucarid and peracarid caridoids and in paleocaridacean and anaspidacean syncarids. Its function has not been understood, and tentative suggestions have been made concerning aid in swimming (Calman 1909) and stabilizing effects (Tattersall & Tattersall 1951). Recently, however, Mr Y.Attramadal (in preparation) has found experimentally in mysids that it plays an important part in giving effect to the caridoid escape reflex. This explains why it is lacking in fast-swimming isopods and amphipods which have uniramous antennae and do not jump. Within the Peracarida it is found only in the caridoid Mysidacea and, in a very vestigial form, in *Spelaeogriphus.*

It should be noted in this context that the generally ellipsoid expansion of the second, not the first, article of the stomatopod antennae is said to have a definite rudder function (Balss 1938). It is thus probably not homologous with the caridoid antennal scale.

It then becomes evident that in caridoids the strong abdominal musculature, the antennal scale, and the tail fan as well as the appropriate receptors of external stimuli and the co-ordinating nervous mechanisms are all components of one single integrated functional system, designed to produce the jumping escape reaction. This system constitutes, in fact, in its entirety the only truly diagnostic feature of the caridoid facies. As such it is of the greatest importance for the understanding of the interrelationships of the advanced eumalacostracan superorders, to which we shall revert in section 4.4.

However, before leaving the morphological aspects of the caridoid facies, something should be said about the carapace. The carapace is certainly not a diagnostic feature of the caridoids, for it is lacking in the syncarids and present in various non-caridoid peracarid orders. However, in the form in which it is present in euphausiids, natantian decapods, lophogastrids, and certain mysids it can be said to be at least characteristic of the most advanced exponents of the caridoid facies.

The advanced caridoid has a very large carapace/cephalothorax provided with a rostrum. This rostrum is long and laterally compressed in most natantian decapods and some euphausiids and lophogastrids (*Gnathophausia*), short but still laterally compressed in certain natantians (e.g. Pasiphaeidae), most euphausiids, and certain lophogastrids. The wide distribution of the carapace and rostrum in the most advanced caridoids, indicates they possess advantageous hydrodynamic properties; however, such have never been properly investigated. In the case of the euphausiids this argument gains strength from the fact that the carapace is not directly involved in the respiration/ventilation system (see section 4.2). Therefore, it is difficult to imagine any function other than a hydrodynamic one. The lateral compression suggests a function of cleaving the water like the bow of a ship. In this context, it is suggestive that certain caridoid members of the taxa in question which have become partly benthic have no or only a small rostrum (e.g. the caridean genera *Crangon* and *Pontophilus*) or a horizontally flattened rostral plate (e.g. the lophogastrid genus *Lophogaster*).

### 4.2 *Manifestations of functional systems*

Although the basic structural plan of the Eumalacostraca can be readily recognized in all its members, variations within the frame of this plan are nevertheless considerable. The molding forces, which call forth morphological adaptations, have both exogenous and endogenous components. The exogenous forces are clearly connected with the habitat and mode of life: pelagic, epibenthic, benthic, fossorial, tubicolous. The endogenous forces are based upon the requirements of the various functional systems in forming their external manifestations. The habitus always reflects both these kinds of influence, but their relative importance is not always immediately evident. Thus in the advanced caridoids discussed in the previous section the effects of exogenous forces shaping the taxa into a pattern suitable for a pelagic or hemipelagic life tend to obscure deep-lying differences between the same taxa. The present section will deal with external expressions of functional systems which are at least partly due to the effect of endogenous forces.

The functional systems exerting the greatest influence upon the structural patterns of the three superorders under discussion are those involved in respiration and ventilation, locomotion, alimentation, and reproduction. Some remarks on ontogenetic patterns will also be included.

The primary respiratory organs are the thoracic epipods. Originally, there appear to have been at least two on each thoracopod (this is the number still found in the syncarids), while there appears to be three epipods in the Decapoda (Calman 1909). Not more than one epi-

Table 1. Respiratory organs of caridoids

| Taxon | Thoracic epipod | | | | Accessory respiratory organs |
| | Morphology | Exposure | Occurrence, Series | Ventilation | |
| --- | --- | --- | --- | --- | --- |
| **Syncarida** | | | | | |
| *(Anaspides)* | Simple | Exposed | P1-P7 | Autochthonous + thoracic exopods and pleopods 1-2 | Thoracic exopods |
| *(Bathynella)* | | | T1-T6 | | |
| **Eucarida** | | | | | |
| Euphausiacea | Complex | Exposed | T1-T8 | Thoracic exopods | – |
| Decapoda *(Peneus)* | Complex | Enclosed by branchiostegal folds | P1-P7 P2-P7 P3-P7 | Maxillary exopod (= scaphognathite) | Inner wall of branchiostegal folds |
| **Peracarida** | | | | | |
| Proximal, epipod series partly oöstegites in female, lost in male | | | | | |
| Lophogastrida | Complex | Enclosed by branchiostegal folds | P2-P7 | P1 (mxp) epipod plus thoracic exopods | – |
| Mysida *(Boreomysis)* | Lost except on P1, non-respiratory | ” | – | P1 (mxp) epipod | Inner wall of branchiostegal and carapace folds |
| Tanaidacea *(Apseudes)* | Lost except on P1, complex | ” | P1 | P1 (mxp) epipod | ” |
| Cumacea *(Diastylis)* | Lost except on P1, complex | ” | P1. | P1 (mxp) epipod | ? |
| Amphipoda | Simple | Exposed | P2-P6(P7) | Pleopods | ? |
| Isopoda | Lost | – | – | – | Pleopods |

pod is retained in the Peracarida, but there are good reasons to presume that in the female the proximal epipod on certain thoracic legs has been transformed into an oöstegite, the whole proximal epipod series having been lost in the male (Claus 1885, Calman 1909, Siewing 1956, Dahl 1977). In *Gammarus pulex* L. transmission electron microscope examination revealed a close similarity in the general structural patterns of epipod and oöstegite, the main difference consisting in the much richer vascularization of the epipod (Dahl, unpublished). Also the mutual positions of oöstegite and epipod in *Gammarus* are identical with those of the two epipods in *Anaspides* (Dahl 1977). It seems safe to conclude that the oöstegite in the *Gammarus* female, and by inference those of other Peracarida, are transformed epipods of the proximal series.

In decapods and certain peracarids the inner wall of the carapace may serve as a respiratory organ, and the same is the case with the thoracic exopods of the syncarids. This may also apply to certain pleopods. The simplest respiratory system is found in the Anaspidacea where the two flat and uncomplicated epipods, present on all thoracopods except the last one, are ventilated, partly by means of autochthonous vibration, partly by the beating of the thoracopod exopods and the two anterior pleopods. Only the amphipods have a system built on the same principle but with a more sophisticated ventilatory system (Dahl 1977). Exposed respiratory epipods are also found in the Euphausiacea; however, there the epipods themselves are highly complicated. One series of epipods has been lost, possibly in connection with the proliferation of the other series. Ventilation is provided by the thoracic exopods which are exclusively ventilatory and do not take part in locomotion (Mauchline 1980).

In comparison, all other taxa show further complications. The actual situation encountered in those taxa where conditions are reasonably well-known are summarized in Table 1, where the progressive degrees of complication in comparison with the simple syncarid system have been noted. These various degrees of complication must not, however, be understood directly to represent evolutionary lines. Evolutionary implications are there, but they will be discussed in a later connection.

Some further brief comments may be of interest. It is obvious that the enclosure of gills into branchiostegal chambers leads to a demand for specialized ventilatory mechanisms and that the answer is different in the Decapoda and the Peracarida. It is also obvious that an evolution away from simple and generalized epipod respiration has proceeded much farther in peracarids than in eucarids. In this respect the Isopoda with their exclusive reliance upon pleopod respiration stand apart from all other Peracarida.

Concerning locomotory adaptations Manton (1953) concluded that ambulatory habits in malacostracans preceded natatory ones (see section 3), and that this was a basic reason for tagmatization. In the Malacostraca in general, including the Leptostraca and apparently also the Archaeostraca, tagmatization is complete as far as locomotion is concerned, with a complete segregation between thoracopod and pleopod activity.

The Syncarida, however, constitute a remarkable exception (Manton 1930). In the Anaspidacea, and also in some Palaeocaridacea (e.g. *Squillites*), the endopods of pleopods 1-5 are reduced. The exopod is used both for walking and for swimming. When walking the pleopods 'move in a series with the thoracic endopodites' (Manton 1930). The thoracic exopods in *Anaspides* are constantly in motion, swinging antero-posteriorly and 'when stationary the first one or two pairs of pleopods beat gently in series with the thoracic exopodites' (Manton 1930). When swimming the thoracic exopods and the pleopods also beat metachronally in series, although more rapidly. In *Paranaspides* Manton did not find the same degree of co-ordination, but occasionally the thoracic and pleopod exopods beat in series.

The fact that 'both in swimming and walking *Anaspides* uses the thorax and abdomen as a single functional region' appeared to Manton to indicate a persistence of a functional continuity taken over from early Malacostraca before a functional tagmatization was complete. Manton also considered the alternative possibility that it might be a result of the 'crawling habits' of *Anaspides*. Nothing similar, however, is known to occur in any other malacostracan with comparable habits. Therefore the first alternative proposed by Manton appears more likely.

Some comments should be made upon the fact that the co-ordination of the pleopodal exopod is with the thoracic endopods when walking, and with the thoracic exopods when ventilating and swimming. This, however, is only logical for in all Malacostraca a locomotory pleopod always acts as one single unit. Therefore one has to expect that either type of co-ordination with the thoracopods should be with the whole pleopod. The loss of the endopod in the Anaspidacea and certain Palaeocaridacea may be connected with the participation of the pleopods in walking, for a biramous pleopod acting as a whole may cause difficulties, especially on rough ground.

Whereas the recent freshwater anaspidids have a reduced pleopod endopod the situation was more diversified in the Carboniferous Palaeocaridacea living in fresh water, brackish water, and marine habitats (Brooks 1962c, Schram & Schram 1974, Schram 1981a). The genera *Acanthotelson* and *Palaeocaris* had biramous pleopods with laminate rami, while *Praeanaspides* had biramous annulate pleopods, and *Squillites* possessed a uniramous annulate pleopod with the endopod being reduced. It should be noted here that among the peracarids multiarticulate pleopodal rami are only found in Mysidacea and Amphipoda. In the Eucarida and other Peracarida they are uni- or biarticulate.

There is a trend towards reduction of pleopods in interstitial eumalacostracans and also in some other peracarid taxa, viz. the Thermosbaenacea and females of Mysida and Cumacea. In these cases natatory functions are performed by the thoracopod exopods. On the other hand there exists an opposite trend, certainly evolved independently, towards more or less complete reduction of thoracopodal exopods and an emphasis on pleopod swimming in natatory decapods, amphipods, and isopods. Outwardly thorax-abdomen tagmosis is much less evident in the Syncarida than in the Peracarida and Eucarida.

In comparison with the functional systems dealt with above, alternative alimentary adaptations exert a profound influence upon the topography of the cephalon (Dahl 1956) but less so on the general structural plan. There exists very striking differences in the cephalon between malacostracans which use a maxillary filtering mechanism and convey the filtered food into the atrium oris from behind, and those where the emphasis lies upon mandibular browsing. In the former, found in its most typical form in mysids, the cephalon is opisthognathous and the opening of the atrium oris is directed posteriad. In the latter, found, for example, in most isopods and in talitrid amphipods, the cephalon is prognathous and the atrium oris opens obliquely anteriad. Every intergradation occurs and the connection with the feeding method is generally quite clear. A prognathous condition, which has certainly evolved independently in isopods and talitrids, leads to a 'tipping over' of the anterior part of the cephalon. The tip of the labrum becomes the anterior end of the body, the antennules become more or less dorsal and the compound eyes lie behind them. These topographical relationships are different from the normal crustacean plan, and recall the situation found in many insects and myriapods. The incorporation of one or more of the anterior thoracopods into the alimentary apparatus appears to have taken place repeatedly and at least partly independently.

In the reproductive system, taken in its widest sense, the formation of a marsupium, apparently by transformation of certain female epipods into oöstegites distinguishes the Peracarida from the other two superorders. This feature forms a somewhat tenous link between the various peracaridan orders (Fryer 1964, Dahl & Hessler 1982), since the question of whether the formation of oöstegites has taken place more than once remains open.

Transformation of pleopods into male copulatory organs occurs among Syncarida and Eucarida but never in the Peracarida.

There exists no ontogenetic pattern common to all the three superorders. The marine Eucarida have normally a more or less complicated ontogeny with series of free-swimming larval stages, the Peracarida a direct lecitotrophic one. The Syncarida also have direct development. Within the Eucarida the patterns of larval development are similar enough to leave no doubt about the comparatively close relationship between euphausiids and decapods. Nevertheless, the euphausiid larvae are easily recognized and in their development, in contrast to that of the Decapoda, there is no well-marked metamorphosis (Gurney 1942). Within the Decapoda only the Dendrobranchiata hatch as nauplii, but throughout the order there exists an embryonic nauplius stage. The embryos of the Syncarida and the Peracarida, despite their direct development, pass through a distinct nauplius stage but show little indication of a previous existence of later specialized larval stages (Anderson 1973).

Thus a nauplius stage, though actually only hatching in euphausiids and peneids, is recognizable in the ontogeny of all three superorders. Free-swimming larval stages constitute a means of early ontogenetic exploitation of pelagic resources. The acquisition of different types of larval development in eucarids (and also in hoplocarids) appears to be apomorphic and may have evolved independently in euphausiids and decapods.

## 4.3 *The carapace/cephalothorax*

The genetic potential for forming dorsal and lateral folds or shields in the cephalic-thoracic regions appears to be inherent in the basic crustacean organization. Traditionally a structure of this kind is referred to as a carapace. It is not certain that the general application of this term is correct but that it is a matter for future research. In the present connection the problems possibly involved need not be considered, for the carapace/cephalothorax of the Malacostraca always appears to be formed in the same manner. Within the nine classes of Crustacea now recognized carapace structures in the widest sense are found in five, viz. the Branchiopoda, Ostracoda, Cirripedia, Branchiura, and Malacostraca; but are typically lacking in the Cephalocarida, Copepoda, Mystacocarida, and Remipedia. Out of the first-mentioned five classes two, the Branchiopoda and the Malacostraca, include taxa at the ordinal or higher levels which do not possess a carapace.

This in no way contradicts the statement made above concerning a probable basic potential for carapace formation. However, a structure such as the carapace, generally voluminous and demanding for its formation and maintenance a very considerable expenditure of energy, is not just there. If present, it will have a function. In all carapace-bearing Malacostraca one or more such functions can be recognized.

In the Malacostraca a free carapace fold is comparatively rare. More often the dorsum of the thorax, or part of it, is covered by a vaulted, unsegmented shield, continuous with the cephalon and referred to as the cephalothorax. The free carapace fold, when present,

forms a posterior continuation of this shield. In the three superorders now under consideration a well-developed carapace/cephalothorax occurs in all members of the Eucarida, in the
order Mysidacea of the Peracarida, and not at all in the Syncarida. Of the six remaining
peracaridan orders, four have some kind of carapace/cephalothorax, viz. the Tanaidacea,
Cumacea, Spelaeogriphacea, and Thermosbaenacea, while the Isopoda and Amphipoda
have none. However, especially in many amphipods, the dorsum of thoracic and abdominal
segments form folds which may project posteriad over the adjoining segment. Similar structures are found in abdominal segments of decapod larvae. It seems not improbable that the
formation of such structures is related to that of other dorsal folds.

According to the classical concept the carapace is a fold growing out from the maxillary
segment, and if segments behind the maxillary segment are included in a cephalothorax
this is supposed to be due to a fusion of the carapace fold to the dorsal integument. This is
the view still taken by Hessler in his latest very important paper on the evolution of Crustacea, and it plays a considerable part in his derivation of the Eumalacostraca from advanced
caridoid ancestors (Hessler 1982). In a paper now in press (Dahl, in press 1) and quoted by
Hessler (1982) I tried to point out the difficulties confronting the classical carapace concept
in the malacostracan context, but without throwing any doubt upon the general potential
for carapace formation mentioned above. Some further comments upon this problem
appear to be necessary.

It has in fact been known for a very long time that the malacostracan carapace/cephalo-
thorax is not basically an outgrowth from the maxillary segment. This was shown by Manton in *Nebalia* as early as 1934 (see Manton 1934, Figs.3, 4, 5, 7 and 8). The carapace in
*Nebalia* begins to form as lateral folds on either side of the body. These folds extend backwards to include the third thoracic segment and then curve dorsad so that a continuous fold
is formed. By the dorsal fold growing in from behind it is, figuratively speaking, lifted up
from the dorsum. This 'lifting' by means of a fold formation proceeds forwards so that the
third and second segments become disengaged. It is not known whether this is effected
during one single molt. The formation of the anterior part of the free carapace fold
actually then proceeds from behind forwards instead of the other way round as demanded
by the classical theory. Simultaneously, the posterior portion of the fold continues to grow
posteriad, to cover the back and sides of the thorax.

In all other Malacostraca for which observations are available the carapace/cephalothorax
formation follows the same general patterns as in *Nebalia,* although a large free carapace
fold is formed only in the Mysidacea and the hoplocarid larvae. In the Mysidacea the free
carapace fold is attached to the second to fourth thoracic segment. Otherwise the process
of formation is essentially the same as in *Nebalia.* This is definitely the case in *Praunus*
(Dahl, unpublished, see also Manton 1928 on *Hemimysis,* fig.15).

In *Meganyctiphanes* (Euphausiacea) I have studied the formation of the cephalothorax
from the first calyptopis through furcilia stages up to the adult. Branchiostegal folds are
formed on the lateral sides of the whole thoracic area growing backwards at the same rate
as differentiation proceeds. Dorsad of the fold a vaulted cephalothoracic shield is formed,
very thin in the early stages. At the posterior margin of the thorax it is continuous with
the abdominal integument. In this area a slight folding of the posterior margin is indicated.
In *Euphausia* this fold is more distinct and produced (Gurney 1942), but in principle there
is no difference between the two genera.

In decapods, information on cephalothorax formation in the peneid genus *Gennadas* is
found in Gurney (1942). In the protozoea no thoracic segment is included in the cephalo-

thorax and the border between the maxillary segment and the thorax is actually indicated by a small fold (Gurney 1942, Fig.52) which can be interpreted as a carapace fold. However, from the molt of the third protozoea to the first zoea the whole thorax has been incorporated into the cephalothorax. In some other peneid genera the process includes more steps (Gurney 1942). It is obvious, however, that it proceeds along the same lines as in other malacostracans, i.e. by means of progressive branchiostegal fold formation and the production of a continuous cephalothoracic shield. The continuity of the dorsal thoracic integument with the abdominal integument could be verified by the writer in larvae of the natantians *Crangon* and *Leander*. Evidence of branchiostegal fold formation in the crab *Pilumnus* is found in Anderson (1973, fig.124).

The larvae of the Hoplocarida may constitute an exception from the general rule concerning cephalothorax formation in the malacostracans, for they have a large carapace fold attached to the maxillary segment (Calman 1909). Details concerning the mode of formation of this carapace are not available. If it grows out from the maxillary segment it would mean another indication of the isolated position of the Hoplocarida within the malacostracans.

According to Schram (1979) the pelagic Carboniferous genus *Waterstonella* had a large and thin carapace, enveloping the thorax and the proximal parts of the thoracopods and not attached to any of the thoracic segments. *Waterstonella* was provisionally placed among the Eocarida, i.e. 'unassignable schizopodous caridoids' (Schram 1979). For obvious reasons, it is impossible to evaluate its significance in the present context, for it is the mode of formation of the carapace rather than the final product which is of primary relevance to the present discussion.

The second fundamental aspect of the classical carapace theory is the postulate that in those malacostracans which possess a more or less complete cephalothorax this has been formed by a complete or partial fusion of a free maxillary segment carapace fold to the dorsum of the thorax (see above). As far as can be seen from the literature this has been assumed an axiom and no one seems to have produced any evidence to support it. In order therefore to verify or falsify this fusion hypothesis I have examined histologically a large number of malacostracans, where possible both larvae and adults, of the orders Decapoda, Euphausiacea, and Mysidacea (including the genera *Lophogaster* and *Eucopia*) together with a few species of Cumacea and Tanaidacea, all from the very extensive collections of sectioned Crustacea of the Department of Zoology in the University of Lund. If a carapace fold grew out from the maxillary segment and at some ontogenetic stage become fused to the dorsum it should, in the crudest case, show up as the two superficial layers of a three-layered dorsal integument. There should at the very least exist some indication that at some ontogenetic stage such a fusion had taken place. No such indication was found in any of the species investigated by me. Nor was any such indication to be expected for, as shown above, the cephalothoracic shield, in all cases where its development could be traced, is formed in a different manner.

When the branchiostegal folds grow outward, dorsal segmental borders between cephalon and thorax or between thoracic segments are not formed in the areas involved in the process. The simple dorsal integument comes to form a continuous dorsal and lateral shield, often more or less heavily calcified. In the Eucarida, it is always continuous with the abdominal integument, in the Cumacea and Tanaidacea with the segmented thoracic integument directly behind the shield. Apart from the sometimes heavy calcification referred to above, there is nothing which distinguishes the integument of the cephalothoracic area from that of adjacent areas of cephalon or body. Thus, it would simply be an act of realism to accept the fallacy of the hypothesis of the fusion of a free carapace fold to the thoracic dorsum as part of the formation of a cephalothoracic shield.

In summary, it can be stated that carapace/cephalothorax formation in *Nebalia* and those Eumalacostraca which have been examined starts with the formation of branchiostegal folds and leads to the formation of a histologically simple dorsal and lateral cephalothoracic shield from the posterior margin of which a free carapace fold in some cases grows out. The hoplocarid larvae may constitute an exception which merits further research. The normal process in the Malacostraca differs from that found in the Notostraca where a free carapace fold grows out from the posterior margin of the maxillary segment (Dahl, in press 1). It is the application of this process in the Notostraca to all Crustacea which has produced confusion, and has led to erroneous conclusions. A closer study of the formation of mantle folds in cirripeds, shell valves in ostracods, and the dorsal shield in the Branchiura might lead to clarification of the carapace concept in regard to all Crustacea.

### 4.4 *Evolution and interrelationships of the caridoid superorders*

Morphologically and ecologically the fully developed caridoid facies can be seen as a functional system adapted to marine pelagic existence, including the transition between pelagic and benthic habits.

More or less perfected caridoids, partly belonging to recent taxa, partly of uncertain relationships, were present in the Carboniferous and Devonian (Brooks 1962b, Schram et al 1978). These and other records, particularly in papers by Schram (1969, 1970, 1974a,b, 1978, 1979, 1981) indicate a Devonian radiation of the Eumalacostraca, including caridoid types. At least some of these genera had a strong caridoid abdomen, a tail fan and antennal scales, indicating the possibility, even the probability, that they possessed the caridoid escape reaction. This applies to the Devonian genera *Devonocaris* (Brooks 1962b) and *Palaeopalaemon* (Schram et al. 1978), and to the Carboniferous genera *Belotelson* and probably *Peachocaris* (Schram 1974). At least some of the more heavily-built Carboniferous pygocephalomorphs, now considered to be peracarids related to the lophogastrids, also had the combination of antennal scales and a well-developed tail fan, among them *Tealliocaris* and *Pygocephalus* (Schram 1979). The evolution of the caridoid escape reaction indicates an adaptation to predation by agile predators; and the coincidence in time between taxa with the external attributes associated with this escape reflex and the Devonian radiation of marine crossopterygians may be more than a coincidence.

The synapomorphies connected with the escape reflex alone provide evidence that the three caridoid superorders are more closely related to each other than any of them are to the Hoplocarida or the Phyllocarida. As recorded by Hessler (1982) and above (section 4.1), it is among the mysidaceans, euphausiids, and natant decapods that we find the habitus typical of advanced caridoids. In some of them, notably the lophogastrids, euphausiids, and peneids, we find a large number of plesiomorphies retained. It is perhaps not surprising therefore that Calman (1909) and Siewing (1963) in their figures of the 'ur-malacostracan' presented a caridoid of a plesiomorphic type and that this interpretation found a wide acceptance. Their habitual similarity notwithstanding, these advanced caridoids are, however, unmistakably representative of their respective orders: Euphausiacea, Decapoda, and Mysidacea.

The crucial question is whether the advanced caridoid facies was inherited from the ancestral eumalacostracan! If so, the non-caridoid taxa within the three superorders (and also the Hoplocarida) have evolved as a result of the disintegration of an ancestral caridoid pattern. This is the line taken by Hessler (1982, this volume). An alternative is that the final perfection of the advanced caridoid form proceeded independently within the Eucarida and the Peracarida.

It is implicit in the hypothesis of the advanced caridoid ancestor that the evolution of the caridoid superorders took place under pelagic or semipelagic conditions. The absence of a carapace in the Syncarida is then merely an instance of the disintegration mentioned above, and a logical consequence of the stand taken by Hessler (1982). It appears, however, that the significance of the Syncarida has been underrated in discussions of malacostracan phylogeny and evolution. The syncarids are the only eumalacostracans in which the tagmosis is less advanced morphologically *and* functionally than in the other members of the subclass. In the Palaeocaridacea and the Bathynellacea, the first thoracic segment is not incorporated into the cephalon. Furthermore, the Syncarida are the only Malacostraca which do not have a complete functional thorax-abdomen tagmosis. Instead the two tagmata, as stated above, are co-ordinated 'as a single functional region' both in swimming, walking, and ventilation of epipodal branchiae (Manton 1930).

The respiratory system of the Anaspidacea, with its double rows of simple epipods ventilated by thoracic exopod and pleopod beating, is the least complicated one found in any eumalacostracan taxon. Ventilation by non-locomotory thoracic exopods occurs only in two other plesiomorphic taxa, the lophogastrids and the euphausiids, but in the former group it is supplementary to the beating of maxillipedal epipods. The euphausiids are the only caridoids, besides the syncarids, which have exposed respiratory epipods, in all others they are enclosed in branchiostegal folds and have apomorphic ventilatory mechanisms (scaphognathites or specialized maxillipedal epipods). Even in the euphausiids one series of respiratory epipods has disappeared and those of the remaining series have become enlarged and complicated.

With respect to sensory systems Hanström (1934, 1937, 1947) found that the optic ganglia and the optic nerve in the compound eye of *Anaspides, Paranaspides,* and *Micraspides* are less complex than those of decapods and stomatopods, and that the protocerebral associative centers and the deutocerebrum are less complicated. Elofsson & Dahl (1970) demonstrated the transformations in the topographical relationships of the optic ganglia during malacostracan ontogeny. In the embryos the medulla externa length axis forms a 90° angle with those of lamina ganglionaris (which is the case also in adult branchiopods) and the medulla interna. Later a 90° rotation of the medulla externa takes place so that all three length axes become parallel. In the Anaspidacea, however, there is only a partial rotation. A recent control of the conditions in *Anaspides* and *Paranaspides* showed that the angle between the length axes of lamina and medulla externa in the adults was 45 to 60°. Also, in the Euphausiacea the embryonic topographical relationships are partly retained. There, however, the arrangement of the perikarya of the lamina ganglionaris and medulla externa have become arranged in a manner which is unique among the Malacostraca and certainly apomorphic.

The anatomy of the jumping reflex system in regards the tail fan and the antennal plate are well-developed not only in the Anaspidacea but also in those Palaeocaridacea where the state of preservation permits comparisons (Brooks 1962c, Schram 1979). Concerning the trunk musculature, the abdominal part of which supplies the motor element in the caridoid reflex system, it has been shown that the most primitive caridoid type is to be found in the Syncarida (*Paranaspides*) which lack specialized caridoid elaborations (Dahl 1931a,b, 1932, 1933) and therefore should form the basis of comparisons between advanced peracarid and eucarid caridoids (Hessler 1964).

Finally, as shown by Anderson (1973) there are no indications of larval stages later than the nauplius in Phyllocarida, Syncarida, and Peracarida. This indicates that pelagic larvae are apomorphic in Eucarids.

As is apparent from this survey, the Anaspidacea and, as far as we can judge, the Palaeo-caridacea in system after system are found to be more primitive than any other caridoids, and with respect to the thorax-abdomen functional tagmosis even more primitive than the phyllocarids.

When Hessler (1982) writes that 'Anaspidaceans are no more primitive than . . . *Waterstonella* and the euphausiaceans, whose first thoracopod is also unmodified as a maxilliped', this does not reveal a full recognition of salient facts concerning the Syncarida. Moreover, the statement is hardly adequate, for as indicated above the euphausiids with their complete tagmosis, their strict division of labor between thoracopods and pleopods, their more derived respiratory system, their more complex musculature (Daniel 1929), and their pelagic larvae are certainly less primitive than the Syncarida. Concerning *Waterstonella* we simply have not enough information to make a qualified statement concerning its degree of primitivity.

Siewing (1956, 1963) on the basis of a comprehensive comparative anatomical investigation concluded that the Peracarida and the Eucarida were derived from ancestors of the Syncarida. Daniel (1933) drew the same conclusions from his musculature studies. Fryer (1964) on the other hand saw the evolution of the three superorders as a radiation from common ancestors among pygocephalomorphs and eocarids; however, this presumed common foundation has since been undermined. The pygocephalomorphs are now regarded as probable mysidaceans; and the eocarids, as originally understood, have been shown to include a conglomerate of species belonging to established taxa or which are too imperfectly known to permit conclusions concerning their true affinities (Schram 1979). The account of the functional systems given above and in sections 4.1 and 4.2 strongly supports the conclusions drawn by Siewing & Daniel that the caridoid superorders were derived from pre-syncarid ancestors.

The caridoid escape reaction with all its morphological attributes is unique and constitutes a set of interdependent synapomorphies clearly illustrating the genetic relationship between the three superorders containing caridoid forms. In comparison cephalic and/or thoracic shields and valves, 'carapaces' in a sense so wide that a single relevant definition is difficult to formulate, fulfill a variety of functions separately elaborated within the taxa possessing them. Where carapaces can serve no functions, they are not formed (section 4.3). Within the three superorders a fully developed 'caridoid' carapace/cephalothorax is found only in the Eucarida and the order Mysidacea. In all of these, except the Euphausiacea, and also in some peracarid orders with a less developed carapace/cephalothorax it is always involved in the respiratory system, which, as pointed out above, includes apomorphic adaptations not present in the Syncarida. In all advanced caridoids it is also highly probable that the carapace/cephalothorax fulfils hydrodynamic functions. Finally, in benthic decapods calcification of the exoskeleton and particularly of the cephalothorax provides protection against predation. Consdering the presence of a cephalothorax in the euphausiids, where it has no respiratory function, one might speculate that the possible order of which these three functions evolved was (1) hydrodynamic, (2) respiratory, (3) protective. It is almost certain that the sequence 2-3 is correct. The sequence 1-2 is more doubtful, however, for in the non-caridoid peracarid orders the function of the cephalothorax as a respiratory system appears to be the only one, and we have no means to tell whether this is a secondary or a primary situation.

It was presumed by Hessler (1982) that fully caridoid ancestors of the Syncarida later lost their carapace. There exists no evidence, however, that syncarids were anything but benthic-epibenthic, and secondarily interstitial and subterranean. If so, a hydrodynamically advantageous carapace/cephalothorax was hardly required. The plesiomorphic arrangement of the respiratory epipods on the thoracic legs (which were used by Hessler to exemplify an unmodified stenopodium) and the ventilatory system connected with them do not require a carapace, and if they had ever been included in a branchiostegal respiratory system they would most probably have been modified and had a different ventilatory system. Only if we accept a carapace/cephalothorax, even without a function, as a malacostracan prerequisite (section 4.3) do we have a reason to presume its existence in a syncarid ancestor. The general plesiomorphic nature of the syncarid organization, however, does not support such a presumption.

If on the other hand the advanced caridoid facies was independently perfected within the Eucarida and the Mysidacea, then it is less difficult to see how a carapace/cephalothorax with the respiratory apomorphies connected with it came to be superimposed upon the different basic epibenthic organizational patterns of eucarids and peracarids.

Presuming then that all three caridoid superorders were derived from pre-syncarid ancestors we have to consider the mutual relationships between the three taxa. The Syncarida have generally been considered to be more closely related to the Eucarida than to the Peracarida. Synapomorphies shared with the former group are antennular statocysts, receptaculum seminis in the female, and sexual dimorphism in the pleopods (Calman 1909). Further evidence along the same lines was recently produced by Schminke (1978) who demonstrated remarkable similarities between peneid larvae and bathynellacean syncarids, and by Hessler (in press) who showed that the coxal-body articulation was similar in syncarids and eucarids. In both these cases symplesiomorphies are probably involved.

It appears probable that the peracarids deviated earlier than the Eucarida from common-pre-syncarid ancestors. It is not certain that the formation of oöstegites was a unique event. However, presuming this was the case, evolution of peracarids from pre-syncarid ancestors may have proceeded along separate benthic, epibenthic, and hemiphelagic-pelagic lines. A detailed discussion of peracarid evolution and interrelationships falls outside the scope of the present chapter. Moreover, these questions are under consideration by others. Consequently, only a few remarks will be made here on the two carapace-less orders. The amphipods with their undifferentiated epipods and the curvature of their embryos appear to share syncarid symplesiomorphies. Moreover, their respiratory-ventilatory system, like that of the syncarids, provides no evidence of having ever been enclosed by branchiostegal folds. The last can also be said about isopods, which are certainly not closely related to amphipods (Siewing 1951). Isopods have always been considered to be derived from mysids by a series of reductions. The argumentation has always appeared to be a circular one in that, being (axiomatically) derived from mysids, isopods show a reduction of mysid features. No tangible supporting evidence has been produced for this proposition. It appears that, like the amphipods, the isopods represent a separate line which had neither a caridoid habitus nor a caridoid respiratory and ventilatory system. A derivation of more advanced caridoids from pre-syncarid ancestors removes the difficulties involved in deriving amphipods and isopods from advanced caridoids.

Looking upon the advanced caridoids we find, as already noted above, that some of them, notably lophogastrids, euphausiids, and peneids, have features plesiomorphic within their respective taxa or within the Eumalacostraca in general. It appears most likely that these undoubtedly ancient forms independently and at different times deviated from the main current of epibenthic hemipelagic evolution in order, figuratively speaking, to go up into pelagic habitats in order to exploit its resources. The marine and above all the open ocean pelagic system is one of the most stable of habitats, even the most dramatic changes in the history of the earth having been buffered by its enormous water masses. Once adapted to it a taxon stands a good chance of long survival with comparatively little change.

In summary, it appears most probable that the caridoid orders Syncarida, Eucarida, and Peracarida evolved from pre-syncarid ancestors without a carapace/cephalothorax but otherwise of an incipient caridoid type with a caridoid escape reflex. Perfection of the advanced caridoid facies adapted to a hemipelagic-pelagic life and with a well-developed carapace/cephalothorax took place independently within the Eucarida and Peracarida, in the latter only in the order Mysidacea.

The writer is in full agreement with Hessler (1982, and this volume) about the unity of the caridoids, but finds the explanation of their origin given above the more plausible one. A derivation of syncarids from advanced caridoids by means of a disintegration of caridoid features appears improbable: (1) because of the profound primitive nature of the syncarids, and (2) because it would demand a derivation of eumalacostracans from hemipelagic-pelagic ancestors, which, for reasons given in sections 3 and 4, appears unlikely.


## 5 MALACOSTRACAN ANCESTORS

Because of their long fossil record and primitive features in their structural plan, the Phyllocarida have been more or less universally regarded as ancestors of the Eumalacostraca. Calman (1909), however, interpreted them as an early branch of the main malacostracan evolutionary line, and Manton (1953) appears to have favored a similar interpretation. Dahl (1976, and in press) found it difficult to derive a eumalacostracan functional model from a phyllocarid ancestor. Rolfe (1981) indicated that Dahl (1976) had paid too little attention to the Archaeostraca, and this criticism is partly justified even if information on archaeostracan functional systems is poor.

The phyllocarids retain many plesiomorphic features, but that is the case also in other taxa and particularly so in the syncarids. Indeed, with respect to the imperfect thoracoabdominal tagmosis, the recent Anaspidacea are more plesiomorphic than the Phyllocarida, and the Palaeocaridacea and the Bathynellacea share with the Leptostraca a cephalon in which no thoracic segments have become incorporated. Some syncarid taxa also possess furca. In other respects, such as the retention of the seventh abdominal segment, the Phyllocarida are more plesiomorphic. [It should be noted that the statements by Siewing (1956, 1969) about the presence of a seventh segment in the palaeocaridacean genus *Uronectes* (= *Gampsonyx*) is erroneous. Unfortunately, Brooks (1962c) in his graphic reconstruction of *Uronectes* drew seven well-developed abdominal segments. It is to be surmised that this is a mistake.]

Feeding in the Leptostraca takes place by means of ventral food transport effected by thoracopod beating, and by final sorting and ingestion with the aid of maxillulae and maxillae. The presence of numerous juvenile stages in certain Archaeostraca does not in itself provide arguments in favour of a different mode of feeding, although differences in the size of particles handled by various size classes appear likely. The big mandibles found in certain large Archaeostraca (Rolfe 1969, 1981) could possibly be used for picking food items directly from the bottom. However, even if such feeding took place it would not be a cephalic filter feeding of the type referred to above, and we have no evidence that such a possible mode of feeding replaced the ventral feeding transport. The large mandibles are more likely to have been an adaptation to large food particles, e.g. shelled animals, transported by the ventral mechanism of a large archaeostracan. Archaeostracan general morphology makes it likely that in a way similar to that found in the Leptostraca they plowed in mud and used a similar feeding mechanism. The sediment gut filling found in certain Archaeostraca might point that way, even if Rolfe (1981) is of the opposite opinion. The living Crustacea best suited for a functional comparison, i.e. the Notostraca, plow in mud in a way comparable to that employed by the recent Phyllocarida and use a morphologically different but functionally similar ventral food transport mechanism. They have very strong mandibles, are mainly carnivorous (Lundblad 1920) and ingest considerable quantities of sediment from which they obtain prey organisms which are crushed by their strong mandibles. An investigation of gut contents in sectioned specimens of two species of *Triops* (Dahl, unpublished) revealed the presence of crushed remains of ostracods and probably other arthropods, organic debris, and inorganic particles including large sandgrains. It is, however, by no means surprising that the intestine of non-filtering carnivores sometimes contain sediment. I have frequently seen this in raptorial amphipods where the histological methods employed made it clear that the sediment was often contained in the intestines of prey animals.

Manton (1953), as discussed in more detail in section 3.2, maintained that the evolution of a stenopodium with an ambulatory endopod was a prerequisite for the formation of a malacostracan thoraco-abdominal tagmosis and a division of labor between thoracopods and pleopods. It was maintained by Sanders (1963) and Hessler & Newman (1975) that cephalocarids were related and perhaps ancestral to malacostracan ancestors. Powerful beating of cephalocarid legs produces locomotion (Sanders 1963) but they are not ambulatory. If a cephalocarid ventral food filtering and transport mechanism had been retained in malacostracan ancestors the origin of a functional demand for tagmatization is difficult to understand. Malacostracan stenopodia are not suited for ventral food filtration and transport and the beating of pleopods would interfere with a filtration current system.

Phyllocarida have efficient natatory pleopods, but they are inactive when the ventral feeding mechanism is in function. Phyllocarid thoracopods, insofar as we understand them and in any case in the Leptostraca, are non-ambulatory. This might imply that phyllocarid ancestors in the course of a specialization for benthic life either completely or partly lost the locomotory function of the thoracopods and became secondarily adapted for thoracopod filtering and food transport. This might explain the vicariating thoracopod and pleopod function mentioned above.

The argument that ventral food transport is secondary gains force from the fact that the phyllocarid trunk limb, like malacostracan stenopodia in general, is devoid of enditic structures. In those taxa where ventral food transport is primary, the Branchiopoda and the Cephalocarida, a variety of enditic structures and their armature play important parts in the transport mechanism. The Leptostraca for the same purpose are restricted to the use of the setal armature of the smooth medial borders of the thoracopods (Cannon 1927).

And it is worth noting that the leptostracan limb, particularly that of the genus *Paranebalia* (generally considered most plesiomorphic), has more in common with a primitive stenopodium than with a cephalocarid limb.

An aspect which has not been discussed in the present context is the phyllocarid brood care and its role in the functional system. As pointed out above (section 2) the position of the female malacostracan gonopore on the medial side of the base of the sixth thoracopod facilitates the brood care prevalent in some of the taxa. In decapods it allows the female to cement fertilized eggs to the pleopods by bending the abdomen below the thorax. In phyllocarids and peracarids the thoracopods become directly involved in the process by forming brood chambers. In the Leptostraca the whole series of thoracopods of the ovigerous female grow long setae on the distal part of the endopod, and thus form a large brood chamber covering the whole ventral side of the thorax. This has far-reaching functional consequences, as observed by Claus (1888) in his studies of living *Nebalia*. Claus found that in ovigerous females the thoracopod movement which produces the feeding current and the transport of food was reduced to small swinging movements, barely sufficient to ventilate the respiratory surfaces and the eggs in the brood chamber, while food intake was more or less completely suspended. Checking this observation on sectioned ovigerous *Nebalia* females I found very little matter in the intestine.

Although this type of brood care works well enough in the mud-living Leptostraca, it affords no sound basis for evolutionary radiation, adding, as it does, one more function to impede thoracopod differentiation. Peracarids have overcome this problem by means of the transformation of the proximal epipods of some thoracopods into an oöstegite. Together these oöstegites form a brood chamber which leaves the endopods free for other tasks.

Nothing definite is known about brood care in the Archaeostraca. Figures of *Nahecaris* (Broili 1928) show rounded thoracopod tips, possibly with setae. In *Ceratiocaris* Rolfe (1962) recorded the presence of very slender distal parts of thoracopod endopods. In both genera thoracopod structure would appear to be compatible with a brood care function similar to that found in the Leptostraca. Rolfe (1981), however, referred to a possible occurrence of pelagic larvae in other archaeostracan species and, if confirmed, that might open fresh alternatives.

As far as we know, the structure and function of the phyllocarid abdomen tends to bear out the isolated position of the taxon. Both in the Leptostraca and the Archaeostraca (cf. figures of *Nahecaris* in Broili, 1928 and of *Ceratiocaris* in Rolfe 1969 and 1981) there exists a tendency to reduction of the posterior pleopods, whereas the four (in *Ceratiocaris* possibly the five) anterior pairs form a series more or less distinctly decreasing in length and volume posteriad. This arrangement may be involved in mechanical aspects of pleopod swinging. It does not form a good starting point for the evolution of a pleon of the eumalacostracan type. Rolfe (1981) realized the problems inherent in the structure of the phyllocarid abdomen and indicated that possible phyllocarid ancestors of the Malacostraca may have to be found among still undiscovered Phyllocarida with a full set of pleopods.

Despite the plesiomorphies retained by the Phyllocarida, the balance seems to be at least slightly in favor of their being derived from benthic-epibenthic ancestors of a basically eumalacostracan type rather than being themselves ancestral to the eumalacostracans.

## 6 THE POSITION OF THE HOPLOCARIDA

It is obvious that the hoplocarids stand well apart from the rather close-knit assemblage of caridoid superorders. The evidence, i.e. the roomy abdomen, indicates a separation at a stage when advanced natatory habits had not yet been adopted by eumalacostracan ancestors. Later some Hoplocarida, notably the genus *Kallidecthes,* within the extinct order Aeschronectida appears to have been natatory, but the general morphology of the Hoplocarida including the extinct forms indicates a predominantly benthic mode of life throughout the history of the taxon.

Recent discussions of hoplocarid-eumalacostracan affinities largely center around a derivation of the hoplocarids from caridoid ancestors (Hessler 1982, and this volume), and quite logically so considering the hypothesis maintained by Hessler (1964, 1975, and 1982) and others that the ancestral eumalacostracan was a more or less advanced caridoid. But if, as maintained in section 4, the caridoids were more probably derived from ancestors of a pre-syncarid type, a separate hoplocarid derivation from benthic-epibenthic pre-caridoid ancestors becomes more plausible.

The most important apomorphies within the hoplocarid structural plan are the folding raptorial thoracopods, well-developed in the Carboniferous palaeostomatopods (Schram 1969b), the highly diagnostic proventriculus (Kunze 1981), and the specific larval types. The diagnostic value of other features discussed in this context (cf. Siewing 1956, Burnett 1973, Burnett & Hessler 1973, Hessler 1982) appears more doubtful (see Kunze, this volume). The fossil record of the hoplocarids extends to the Devonian, and the considerable degree of differentiation then attained indicates a comparatively long previous history at the hoplocarid level.

A direct derivation of the hoplocarids from phyllocarids instead of from eumalacostracans was advocated by Schram (1969c, 1973) and Kunze (1981) and to some extent supported by Rolfe (1981). Even if the sum of the evidence hitherto presented can hardly be said to be convincing the possibility cannot be written off. On the other hand, the close and well-substantiated relationships between the three caridoid superorders and the distance between each one of them and the hoplocarids makes a subclass Eumalacostraca comprising these four superorders of equal rank somewhat incongruous. Recognizing the value of existing apomorphies, the above-mentioned incongruity, and the still doubtful relationships between the Hoplocarida and other malacostracans it appears justified at least provisionally to remove the superorder Hoplocarida from the subclass Eumalacostraca and to elevate it to subclass rank.

# 7 THE HIGHER SYSTEMATICS OF THE MALACOSTRACA

In recent years, the debate about malacostracan phylogeny and evolution has gained impetus at least partly because new arguments have been introduced and concepts of long standing questioned. This has led to some divergences of opinion concerning the interpretation of malacostracan relationships.

Schram (1981b) recognizing the difficulties involved, tried to eliminate confusion and apparent contradictions by means of an analysis of the Malacostraca based on a random association of characters, deliberately reducing the demand of phylogenetic coherence. For obvious reasons the malacostracan system resulting from this approach differs considerably from a natural system also in those aspects where we have fairly definite information about actual affinities. Even if some aspects of malacostracan systematics remain obscure, the approach chosen by Schram appears over-pessimistic. The situation is not quite so bad.

Rolfe (1981) pointed out that a Hennigian analysis of malacostracan systematics has never been performed. It is doubtful, however, whether such an analysis would fulfil a purpose. The history of malacostracan evolution is certainly much longer than its fossil record, which starts at least in the Ordovician for the phyllocarids and in the Devonian for the hoplocaridans and eumalacostracans. But this certainly tells far less than enough about the actual age of the taxon Malacostraca. The absence of cephalocarids from the Cambrian to the Recent (Müller 1981) and of phyllocarids from the Permian to the Recent (Rolfe 1969) shows how erratic this record may be (Brooks 1969, Bergström 1980). Dealing with taxa, the early history of which is so obscure, will unavoidably lead to problems in discerning what is plesiomorphic and what is apomorphic within the respective units. When the necessary choice has been made in this respect, however, the logical rigidity (which is the strength of the cladistic method) makes the final result a foregone conclusion, irrespective of whether it is correct or not (see Schram, this volume). However suitable for low-level taxonomic analysis, where fundamentals are secure, the Hennigian method is hardly meaningful in a situation where the possibility of discerning basic facts is in doubt, for it does not in itself recognize controversy. This is probably the reason why it has not been applied in the present context.

The present survey is based on a review of comparative morphology of functional systems, their presumed origins, and integrations and segregations. It need hardly be pointed out that progressive integration or segregation, where discernible, gives important indications with respect to apomorphies. It appears as if the present approach has been fruitful, especially in the case of the caridoid superorders where the sum effects on the structural plans of the caridoid escape reaction, in conjunction with the respiratory, ventilatory and locomotory systems, has led to what may be an improved understanding of the taxa and their interrelationships. It has also introduced some new arguments into the discussion of the early evolution of the Malacostraca, and appears to have shed some new light upon the relative merits of the hypotheses of a phyllocarid versus a eumalacostracan type of ancestor.

It is concluded that the malacostracan ancestor was possibly of an incipient eumalacostracan rather than of a phyllocarid type, and that the phyllocarids, mainly as a result of progressive benthic adaptation of an ancestor of the former type, came to represent the earliest known lateral branch in malacostracan evolution. The hoplocarids are also presumed to be an early benthic side branch stemming from a basically eumalacostracan parentage. The caridoid superorders constitute the typical Eumalacostraca and are certainly closely related. They share in the caridoid escape reaction and its comprehensive synapomorphic system.

The general primitive nature of the Syncarida, including also the muscular part of the caridoid reflex system, led to the conclusion that syncarids and their ancestors rather than advanced caridoids were the parents of the unique caridoid constellation of superorders. This would imply that a final perfection of the caridoid facies in its most advanced form took place independently in eucarids and mysidacean peracarids. It may also explain why the orders Amphipoda and Isopoda among the Peracarida appear never to have had a carapace. The widespread opinion that the Eucarida are more closely related to the syncarids than the Peracarida gains further support.

The higher classification of the Malacostraca with one single deviation from that of Calman (1909) appears as follows. All definitions are as formulated by Calman (1909) with the exception of the reference to the Stomatopoda in the diagnosis of the Eumalacostraca, which should be deleted.

Class Malacostraca
  Subclass Phyllocarida
  Subclass Hoplocarida
  Subclass Eumalacostraca
    Superorder Syncarida
    Superorder Eucarida
    Superorder Peracarida

# 8 ACKNOWLEDGEMENTS

The present investigation was supported by grants from the Swedish Natural Science Research Council and by the University of Lund, granting me research facilities also after my retirement. Technical assistance from Miss Ylwa Andersson, Mrs Gunilla Bergh, Mrs Gunnel Behm, Mrs Ingrid Lukell and Miss Inger Norling is gratefully acknowledged. For discussions on problems dealt with in this paper, I am greatly indebted to many colleagues, among them Professor D.T.Anderson, Dr J.Bergström, Professor R.Elofsson, Dr F.R.Schram, my collaborators, Mr Y.G.Attramadal and S.B.Johnson, and particularly Professor R.R.Hessler who has patiently listened to and commented upon my numerous heresies.

# REFERENCES

Anderson, D.T. 1973. *Embryology and phylogeny in annelids and arthropods.* Oxford: Pergamon Press.
Attramadal, Y.G. 1981. On a non-existent ventral filtration current in *Hemimysis lamornae* (Couch) and *Praunus flexuosus* (Müller) (Crustacea: Mysidacea). *Sarsia* 66:283-286.
Balss, H. 1938. Stomatopoda. In: *Bronns Klassen u.Ordn.d. Tierreichs. V.1,* 6(2):1-173. Leipzig: Fisher Verlag.
Bergström, J. 1980. Morphology and systematics of early arthropods. *Abh. naturwiss. Ver. Hamburg (NF)* 23:7-42.
Briggs, D.E.G. 1976. The arthropod *Branchiocaris* n.gen., Middle Cambrian, Burgess Shale, British Columbia. *Bull. Geol. Surv. Canada* 264:1-29.
Briggs, D.E.G. 1977. Bivalved arthropods from the Cambrian Burgess Shale of British Columbia. *Palaeontol.* 20:595-621.
Briggs, D.E.G. 1978. The morphology, mode of life, and affinities of *Canadaspis perfecta* (Crustacea: Phyllocarida), Middle Cambrian, Burgess Shale, British Columbia. *Phil. Trans. Roy. Soc. London* (B)281:439-487.
Broili, F. 1928. Beobachtungen an *Nahecaris. Sitz.-ber. Bayer. Akad. Wiss. math.-naturw.* 1928:1-19.
Brooks, H.K. 1962a. On the fossil Anaspidacea, with a revision of the classification of the Syncarida. *Crustaceana* 4:229-242.
Brooks, H.K. 1952b. Devonian Eumalacostraca. *Ark. Zool.* (2)15:307-315.
Brooks, H.K. 1962c. The Paleozoic Eumalacostraca of North America. *Bull. Amer. Paleontol.* 44:160-338.
Brooks, H.K. 1969. Syncarida. In: R.C.Moore (ed.), *Treatise on Invertebrate Paleontology. Part R. Arthropoda 4:*R345-359. Lawrence: Geol. Soc. Am. & Univ. Kansas Press.
Burnett, B.R. 1973. Notes on the lateral arteries of two stomatopods. *Crustaceana* 23:303-305.
Burnett, B.R. & R.R.Hessler 1973. Thoracic epipodites in the Stomatopoda (Crustacea): a phylogenetic consideration. *J. Zool. Lond.* 169:381-392.
Calman, W.T. 1909. Crustacea. In: Sir Ray Lankester (ed.), *A treatise on Zoology Vol.7.* London: Adam & Charles Black.
Cannon, H.G. 1927. On the feeding mechanism of *Nebalia bipes. Trans. Roy. Soc. Edinb.* 55:355-369.
Cannon, H.G. & S.M.Manton 1927. On the feeding mechanism of a mysid crustacean, *Hemimysis lamornae. Trans. Roy. Soc. Edinb.* 55:219-253.

Claus, C. 1885. Neue Beiträge zur Morphologie der Crustaceen. *Arb. zool. Inst. Wien* 6:1-108.

Claus, C. 1888. Ueber den Organismus der Nebaliiden und die systematische Stellung der Leptostraken. *Arb. Zool. Inst. Wien* 8:1-147.

Dahl, E. 1956. On the differentiation of the topography of the Crustacean head. *Acta Zool.* 37:123-192.

Dahl, E. 1963. Main evolutionary lines among Recent Crustacea. In: H.B.Whittington & W.D.I.Rolfe (eds), *Phylogeny and Evolution of Crustacea:* 1-14. Cambridge: Mus. Comp. Zool.

Dahl, E. 1976. Structural plants as functional models exemplified by the Crustacea Malacostraca. *Zool. Scr.* 5:163-166.

Dahl, E. 1977. The amphipod functional model and its bearing upon systematics and phylogeny. *Zool. Scr.* 6:221-228.

Dahl, E. (in press). Alternatives in malacostracan evolution. *Rec. Aust. Mus.*

Dahl, E. (in press). The subclass Phyllocarida (Crustacea) and the status of some early fossils. *Vidensk. Medd. Dansk. Naturh. Foren.*

Dahl, E. & R.R.Hessler 1982. The crustacean lacinia mobilis: a reconsideration of its origin, function and phylogenetic implications. *Zool. J. Linn. Soc.* 74:133-146.

Daniel, R.J. 1929. The abdominal musculature system of *Meganyctiphanes norvegica* (M.Sars). *Proc. Trans. Liverpool Biol. Soc.* 43:149-180.

Daniel, R.J. 1931a. The abdominal muscular systems of *Paranaspides lacustris* (Smith). *Proc. Trans. Liverpool Biol. Soc.* 46:26-45.

Daniel, R.J. 1931b. Comparative study of the abdominal musculature in Malacostraca. Part I. The main ventral muscles of the typical abdominal segments. *Proc. Trans. Liverpool Biol. Soc.* 45:57-71.

Daniel, R.J. 1932. Comparative study of the abdominal musculature in Malacostraca. Part II. The superficial and main ventral muscles, dorsal muscles and lateral muscles, and their continuations into the thorax. *Proc. Trans. Liverpool Biol. Soc.* 46:46-107.

Daniel, R.J. 1933. Comparative study of the abdominal musculature in Malacostraca. Part III. The abdominal muscular systems of *Lophogaster typicus* M.Sars and *Gnathophausia Zoëa* Suhm, and their relationships with the musculatures of other Malacostraca. *Proc. Trans. Liverpool Biol. Soc.* 47:71-133.

Elofsson, R. 1965. The nauplius eye and frontal organs in Malacostraca (Crustacea). *Sarsia* 19:1-54.

Elofsson, R. 1966a. The nauplius eye and frontal organs of the non-Malacostraca (Crustacea). *Sarsia* 25:1-128.

Elofsson, R. 1966b. Notes on the development of the nauplius eye and frontal organs of decapod crustaceans. *Acta Univ. Lund. II, 1966* 27:1-23.

Elofsson, R. 1966c. The nauplius eye and frontal organs in Crustacea morphology, ontogeny and fine structure. *Diss. inaug. Lund:* 1-12.

Elofsson, R. & E.Dahl 1970. The optic neuropiles and chiasmata of Crustacea. *Z. Zellforsch.* 107:343-360.

Fryer, G. 1964. Studies on the functional morphology and feeding mechanism of *Monodella argentarii* Stella (Crustacea: Thermosbaenacea). *Trans. Roy. Soc. Edinb.* 66:49-90.

Gurney, R. 1942. *Larvae of Decapod Crustacea.* London: Ray Soc.

Hanström, B. 1934. Neue Untersuchungen über Sinnesorgane und Nervensystem der Crustaceen. IV. *Ark. Zool.* 26A(24):1-66.

Hanström, B. 1937. Neue Untersuchungen über Sinnesorgane und Nervensystem der Crustaceen. *V.K. Fysiogr. Sällsk. i Lund Förh.* 5(16):1-14.

Hanström, B. 1947. The brain, the sense organs, and the incretory organs of the head of the Crustacea Malacostraca. *Acta Univ. Lund.* (2), 43(9):1-45.

Hessler, R.R. 1964. The Cephalocarida comparative skeletomusculature. *Mem. Conn. Acad. Arts Sci.* 16:1-97.

Hessler, R.R. 1982. Evolution within the Crustacea. In: L.G.Abele (ed.), *Biology of the Crustacea, vol.1:* 149-185. New York: Academic Press.

Hessler, R.R. 1982. The structural morphology of walking mechanisms in malacostracan crustaceans. *Phil. Trans. Roy. Soc. London.* (B)296:245-298.

Hessler, R.R. & W.A.Newman 1975. A trilobitomorph origin for the Crustacea. *Fossils and Strata* 4:437-459.

Hickman, V.V. 1937. The embryology of the syncarid crustacean, *Anaspides tasmaniae. Pap. Proc. Roy. Soc. Tasmania* 1936:1-35.

Kunze, J.C. 1981. The functional morphology of stomatopod Crustacea. *Phil. Trans. Roy. Soc. London* (B)292:255-328.

Lauterbach, K.-E. 1975. Ueber die Herkunft der Malacostraca (Crustacea). *Zool. Anz.* 194:165-179.

Lauterbach, K.-E. 1980. Schlüsselereignisse in der Evolution des Grundplans der Mandibulata (Arthropoda). *Abh. naturwiss. Ver. Hamburg (NF)* 23:105-161.

Lundblad, O. 1920. Vergleichenede Studien über die Nahrungsaufnahme einiger schwedischen Phyllopoden. *Ark. Zool.* 13(1):1-32.

Manton, S.M. 1928. On the embryology of a mysid crustacean, *Hemimysis lamornae. Phil. Trans. Roy. Soc. London* (B)438:363-463.

Manton, S.M. 1930. Notes on the habits and feeding mechanisms of *Anaspides* and *Paranaspides* (Crustacea, Syncarida). *Proc. Zool. Soc. London* 1930:791-800.

Manton, S.M. 1934. On the embryology of the Crustacean *Nebalia bipes. Phil. Trans. Roy. Soc. London* (B)498:163-204.

Manton, S.M. 1953. Locomotory habits and the evolution of the larger arthropodan groups. *Symp. Soc. Exp. Biol. VII Evolution:*339-376.

Mauchline, U. 1980. The biology of mysids and euphausiids. *Mar. Biol.* 18:1-677.

Müller, K.J. 1979. Phosphatocopine ostracodes with preserved appendages from the Upper Cambrian of Sweden. *Lethaia* 12:1-27.

Müller, K.J. 1981. Arthropods with phosphatized soft parts from the Upper Cambrian 'Orsten' of Sweden. *US Geol. Surv. Open-File Report* 81-743:147-151.

Paulus, H.F. 1978. Eye structure and the monophyly of the Arthropoda. In: A.P.Gupta (ed.), *Arthropod phylogeny:*299-383. New York: Van Nostrand Reinhold.

Rolfe, W.D.I. 1962. Grosser morphology of the Scottish Silurian phyllocarid crustacean *Ceratiocaris papilio* Salter in Murchison. *J. Palaeontol.* 36:912-932.

Rolfe, W.D.I. 1969. Phyllocarida. In: R.C.Moore (ed.), *Treatise on Invertebrate Paleontology. Part R. Arthropoda 4:*R296-331. Lawrence: Geol. Soc. Am. & Univ. Kansas Press.

Rolfe, W.D.I. 1981. Phyllocarida and the origin of the Malacostraca. *Geobios* 14:17-24.

Sanders, H.L. 1963. The Cephalocarida. Functional morphology, larval development, comparative external anatomy. *Mem. Conn. Acad. Arts Sci.* 15:1-80.

Schminke, H.K. 1981. Adaptation of Bathynellacea to life in the interstitial ('Zoea Theory'). *Int. Rev. ges. Hydrobiol.* 66:575-637.

Scholl, G. 1963. Embryologische Untersuchungen an Tanaidaceen (*Heterotanais oerstedi* 'Kröyer) ). *Zool. Jb. Anat.* 80:500-554.

Schram, F.R. 1969a. Some Middle Pennsylvanian Hoplocarida (Crustacea) and their phylogenetic significance. *Fieldiana Geol.* 12:235-289.

Schram, F.R. 1969b. Polyphyly in the Eumalacostraca? *Crustaceana* 16:243-350.

Schram, F.R. 1970. Isopod from the Pennsylvanian of Illinois. *Science* 169:854-855.

Schram, F.R. 1973. On some phyllocarids and the origin of the Hoplocarida. *Fieldiana Geol.* 26:77-93.

Schram, F.R. 1974a. The Mazon Creek caridoid Crustacea. *Fieldiana Geol.* 30:9-65.

Schram, F.R. 1974b. Palaeozoic Peracarida of North America. *Fieldiana Geol.* 33:95-124.

Schram, F.R. 1979. British Carboniferous Malacostraca. *Fieldiana Geol.* 40:1-129.

Schram, F.R. 1981a. Late Paleozoic crustacean communities. *J. Palaeontol.* 55:126-137.

Schram, F.R. 1981b. On the classification of Eumalacostraca. *J. Crust. Biol.* 1:1-10.

Schram, F.R., R.M.Feldmann & M.J.Copeland 1978. The late Devonian Palaeopalaemonidae and the earliest decapod crustaceans. *J. Palaeontol.* 52:1375-1387.

Schram, J.M. & F.R.Schram 1974. *Squillites spinosus* Scott 1938 (Syncarida, Malacostraca) from the Mississippian Heath Shale of Central Montana. *J. Palaeontol.* 48:95-104.

Siewing, R. 1956. Untersuchungen zur Morphologie der Malacostraca (Crustacea). *Zool. Jb. Anat.* 75: 39-176.

Siewing, R. 1963. Studies in malacostracan morphology: results and problems. In: H.B.Whittington & W.D.I.Rolfe (eds), *Phylogeny and evolution in Crustacea:*85-103. Cambridge: Mus. Comp. Zool.

Simonetta, A. & R.Delle Cave 1980. The phylogeny of the Palaeozoic arthropods. *Boll. Zool.* 47(suppl.): 1-19.

Strömberg, J.-O. 1965. On the embryology of the isopod *Idothea. Ark. Zool.* (2)17:421-473.

Strömberg, J.-O. 1967. Segmentation and organogenesis in *Limnoria lignorum* (Rathke) (Isopoda). *Ark. Zool.* (2)20:91-139.

Strömberg, J.-O. 1971. Contribution to the embryology of bopyrid isopods. *Sarsia* 47:1-46.

Tattersall, W.M. & O.S.Tattersall 1951. *The British Mysidacea.* London: Roy. Soc.

Weygoldt, P. Die Embryonalentwicklung des Amphipoden *Gammarus pulex pulex* (L.). *Zool. Jb. Anat.* 77:51-110.

Whittington, H.B. 1974. *Yohoia* Walcott and *Plenocaris* n.gen., arthropods from the Burgess Shale, Middle Cambrian, British Columbia. *Bull. Geol. Surv. Canada* 231:1-21.

LES WATLING

*Program in Oceanography, and Zoology Department, University of Maine, Walpole, USA*

# PERACARIDAN DISUNITY AND ITS BEARING ON EUMALACOSTRACAN PHYLOGENY WITH A REDEFINITION OF EUMALACOSTRACAN SUPERORDERS

## ABSTRACT

As a follow-up to an earlier paper (Watling 1981) all eumalacostracan orders (except Decapoda and Amphionidacea) are examined with respect to their carapace, compound eyes, mandible, maxilliped, blood system, and developmental patterns. On the basis of these features, the eumalacostracan orders are grouped into the following Superorders: Syncarida, Brachycarida, Isopoda, Amphipoda and Eucarida. Syncarida and Eucarida represent one phylogenetic lineage, Isopoda and Brachycarida a second, and Amphipoda a third. No basis is found for maintaining the Peracarida as a Superorder; the Mysidacea is considered to be a separate line of the Eucarida, the Isopoda is elevated as a distinct Superorder, and the remaining orders are grouped as the Brachycarida.

## 1 INTRODUCTION

The Superorder Peracarida was established as a Division within the Subclass Malacostraca by Calman in 1904. Two features, the presence of a lacinia mobilis on the mandible and oöstegites in egg-bearing females, served to define the group. Associated with the latter, peracarids have been further characterized by direct development, with the young more or less resembling the adult upon hatching. Watling (1981) felt additional features were of equal importance such as the form of the mandible, teloblastic development of most postnaupliar somites, and the incorporation of the first thoracomere into the head. None of these characters, with the exception of oöstegites, occur exclusively in the Peracarida as presently constituted, and most do not occur in some species that are otherwise apparently 'peracaridan'. It can readily be seen for example, that if the thermosbaenaceans are not included, peracarids are definable on the basis of only a single feature, the brood pouch, and if thermosbaenaceans are included a single but different feature, the lacinia mobilis, defines the group. It seems likely, therefore, that the Superorder Peracarida is a polyphyletic assemblage of orders, as tentatively suggested in a previous paper (Watling 1981).

In this paper the phyletic affinities of the peracaridan orders will be re-examined in the context of the Eumalacostraca as a whole. Several features of the Euphausiacea and Syncarida, in addition to the peracarid orders, will be considered. The Decapoda and Amphionidacea will not be dealt with in much detail in this account. The major features to be considered include: carapace, compound eyes, mandible, maxilliped, blood system, and developmental patterns. These were chosen for a variety of reasons. They are relatively conser-

213

vative, and have been considered previously in a generally superficial manner. For example, the only attribute of the mandible thought to have any phylogenetic importance was the presence or absence of a lacinia mobilis. In the following account, it will be seen that this is probably the mandible's least phylogenetically important characteristic. Each feature is reviewed in detail and discussed in the following section. In each case, the major problem becomes the separation of those characters that are especially related to life-habit or habitat from those which contain relevant phylogenetic information. Often, this distinction cannot be satisfactorily made since a particular feature has been examined in only a single, or a few, species within each order.

## 2 REVIEW OF PHYLOGENETICALLY IMPORTANT FEATURES

### 2.1 *Carapace*

#### 2.1.1 *Characteristics*
2.1.1.1 Anaspidacea: absent.
2.1.1.2 Bathynellacea: absent.
2.1.1.3 Thermosbaenacea: usually covers two pedigerous thoracic somites in male and immature females; forms an enlarged brood chamber in mature females, sometimes extending to posterior end of thorax. A 'branchial' epipod from thoracopod (= maxilliped) extends posterodorsally into a chamber formed by the carapace inner wall and lateral walls of head.
2.1.1.4 Spelaeogriphacea: covers second pedigerous somite laterally; thoracopod one (= maxilliped) epipod extends into branchial cavity formed by inner carapace walls and lateral walls of head.
2.1.1.5 Tanaidacea: covers second pedigerous somite; thoracopod one (= maxilliped) branchial, extends into cavity formed by inner carapace wall and lateral walls of head.
2.1.1.6 Cumacea: usually covers three, but occasionally as many as six, pedigerous somites; epipod of thoracopod one (= maxilliped one), elaborately plicated and extends backwards into enlarged cavity formed by lateral walls of head and thorax and inner carapace walls.
2.1.1.7 Isopoda: absent.
2.1.1.8 Amphipoda: absent.
2.1.1.9 Lophogastrida: covers all thoracic somites both dorsally and laterally, but is not fused to any dorsally.
2.1.1.10 Mysida: covers all thoracic somites laterally but usually not the last dorsally; may be fused dorsally to three anterior thoracic somites.
2.1.1.11 Euphausiacea: covers all thoracic somites dorsally and laterally; fused to all thoracic somites dorsally.

2.1.2 *Discussion.* Dahl (in press) reviewed the embryological development of the carapace in malacostracans and concluded that a true maxillary fold carapace is not present in any of the orders. Rather, the carapace develops from branchiostegal folds that grow laterad and ventrad, and sometimes also posteriad, from the cephalon and thorax of the embryo. This results in a cephalothoracic shield type of carapace developed to varying degrees. In four of the above Orders (Anaspidacea, Bathynellacea, Isopoda, Amphipoda) the carapace is absent. Since in those orders where a carapace is present, thoracic segmentation is lost, and since it is probably difficult to restore lost segments (Siewing 1958), Dahl (1976, in press) suggested that the absence of a carapace was the basic condition. Watling (1981) postulated the ancestral form of the malacostracan carapace to be short, as seen in male thermosbaenaceans. Aside from the unconventional use of the carapace as a dorsal brood space in mature female thermosbaenaceans the short carapace has been retained as a respiratory structure in the Spelaeogriphacea, Tanaidacea, and Cumacea. On the other hand, the Lophogastrida and Mysida have combined hydrodynamic as well as respiratory functions into an elongate carapace, while the euphausiids appear to use their elongate carapace solely for hydrodynamic purposes. The development of the carapace thus appears to have begun with branchiostegal folds of the head and first thoracic somites, a feature retained by Lophogastrida, Thermosbaenacea, and Spelaeogriphacea. The Tanaidacea have added a branchiostegal fold on the second thoracic somite and the Cumacea on thoracic somite three (possibly to thoracic somite six). Of those orders with an elongate carapace, the Mysida have added branchiostegal folds on thoracic somites two and three while such folds appear to be present on all thoracic somites in the euphausiids.

## 2.2 *Compound eyes*

2.2.1 *Characteristics.* Syncarids and peracarids exclusive of Mysidacea use apposition optics, whereas the mysids and euphausiids have refracting superposition facets (Land 1980, Fincham 1980). Sessile compound eyes are found in cumaceans, tanaids (although the eyes are on projections of the head), isopods, amphipods, and the anaspidacean family Koonungidae. Anaspididae, lophogastrids, mysidans, and euphausiaceans have stalked compound eyes. Non-functional eye-lobes are found in spelaeogriphaceans while bathynellaceans and thermosbaenaceans are blind.

2.2.2 *Discussion.* Because of their association with many other primitive features, stalked compound eyes using apposition optics appears to be the plesiomorphic condition for this group of malacostracan orders. There is a trend to sessile eyes and finally a loss of compound eyes in the syncarids associated with the reduction of the body as life habits become interstitial. The ground-water living thermosbaenaceans and cave-dwelling spelaeogriphaceans are also blind; however, in the latter, eye lobes are still present. Tanaids have eye lobes with few ommatidia, whereas the few remaining ommatidia in the cumacean eye are mid-dorsal and sessile. The presence of refracting superposition facets in mysid and euphausiacean eyes seems to indicate a common ancestry or phylogenetic linkage for these groups although it is possible that this is a convergent feature.

## 2.3 *Mandible*

2.3.1 *Characteristics.* The structure, mode of operation, and evaluation of the basic malacostracan mandible has been summarized in Manton (1977). In the following account the treatment of the mandible structure and its attachment to the head will be standardized using terminology of Fryer (1965) and Manton (1977).

2.3.1.1 Anaspidacea (Fig.1a,b): rolling dual-purpose type with large boat-like basal portion; hinge-line (= axis of swing) oblique; distal portion consists of low rounded molar and broad coarsely-toothed incisor process, between which is a row of small immovable teeth; palp of three articles or absent (Thompson 1894, Cannon & Manton 1929, Manton 1977).

2.3.1.2 Bathynellacea (Fig.1c): rolling dual-purpose type with boat-like basal portion; hinge-line oblique in Bathynellidae; molar spinous, incisor narrow, appears to be designed

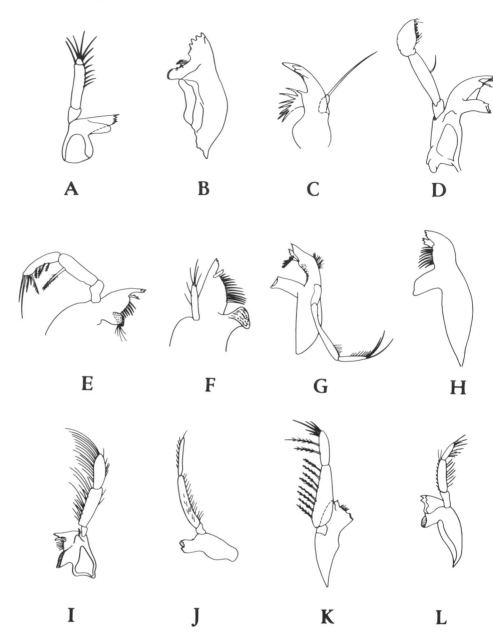

Figure 1. Diagrammatic representation of mandibles as reviewed in the text. (a) *Koonunga cursor* Sayce, 1907 (from Sayce 1908); (b) *Parastygocaris andina* Noodt, 1962 (from Noodt 1962); (c) *Notobathynella chiltoni* Schminke, 1973 (from Schminke 1973); (d) *Disconectes phalangium* (G.O.Sars) 1864 (from Wilson & Hessler 1981); (e) *Monodella sanctaecrucis* Stock, 1976 (from Stock 1976); (f) *Spelaeogriphus lepidops* Gordon, 1957 (from Gordon 1957); (g) *Carpoapseudes serratospinosus* Lang, 1968 (from Lang 1968); (h) *Diastylis rathkei* (Kroyer) 1841 (from G.O.Sars 1900); (i) *Gammaropsis nitida* (Stimpson) 1853 (from G.O.Sars 1894); (j) *Lophogaster typicus* M.Sars, 1862 (from M. Sars 1862); (k) *Euchaetomera typica* G.O.Sars, 1883 (from G.O.Sars 1885); (l) *Euphausia pellucida* Dana, 1852 (from G.O.Sars 1885).

for cutting hard particles during adduction, spine row absent; palp of three articles in Bathynellidae but reduced in Parabathynellidae (Schminke 1973, Serban 1980).

2.3.1.3 Thermosbaenacea (Fig.1e): rolling, dual-purpose type with large boat-like basal portion; hinge-line oblique; distal portion consists of columnar molar, and narrow finely-toothed incisor process with lacinia mobilis, row of lifting spines present; palp of three articles (Fryer 1965, Stock 1976).

2.3.1.4 Spelaeogriphacea (Fig.1f): rolling, dual-purpose type with large boat-like basal portion (not figured by Gordon 1957); hinge-line oblique (personal observation); distal portion consists of low molar, and weak narrow incisor process with lacinia mobilis, row of lifting spines present; palp uniarticulate.

2.3.1.5 Tanaidacea (Fig.1g): rolling, dual-purpose type with large boat-like basal portion; hinge-line oblique; distal portion consists of tall columnar molar process, widely separated from weak narrow finely-toothed incisor process; with lacinia mobilis and small group of lifting spines situated distally; palp of three articles, often reduced or absent (Lang 1968, Gardiner 1975).

2.3.1.6 Cumacea (Fig.1h): rolling, dual-purpose type with large boat-like basal portion (except in Leuconidae where basal portion has become shortened); hinge-line oblique; distal portion consists of widely separated molar and incisor processes; molar is columnar; incisor process narrow and finely-toothed; lacinia mobilis and lifting spines present in most forms; palp absent (G.Sars 1900).

2.3.1.7 Isopoda (Fig.1d): transverse-biting type with short basal portion; hinge-line horizontal; molar process various but often large, columnar; incisor usually narrow, strongly-toothed; lacinia mobilis and lifting spines often present; palp of three articles, occasionally reduced or absent (G.Sars 1899, Manton 1977).

2.3.1.8 Amphipoda (Fig.1i): transverse-biting type with short basal portion; hinge-line horizontal; molar process various but often large, columnar; incisor usually narrow but may be broadened, strongly-toothed; lacinia mobilis and lifting spines typically present; palp of three articles, occasionally reduced or absent (G.Sars 1894).

2.3.1.9 Lophogastrida (Fig.1j): rolling dual-purpose type with large boat-like basal portion; hinge-line oblique; distal portion with low, rounded molar process in close proximity to elongated, coarsely-toothed incisor process; lacinia mobilis present only on left mandible; spine row absent; palp of two articles (M.Sars 1862, Manton 1928).

2.3.1.10 Mysida (Fig.1k): rolling dual-purpose type with large boat-like basal portion; hinge-line oblique; distal portion with low, rounded molar and narrow incisor processes, slightly separated by short spine row; lacinia mobilis present; palp of three articles (Cannon & Manton 1927, Mauchline 1980).

2.3.1.11 Euphausiacea (Fig.1l): rolling dual-purpose type with large boat-like basal portion; hinge-line oblique; distal portion with low, rounded molar in close proximity to elongated, coarsely-toothed incisor process; lacinia mobilis and spine row present only in some larval stages (Knight 1975, 1978, Weigmann-Haass 1977, Mauchline 1980).

2.3.2 *Discussion.* Three basic types of mandibles can be discerned in the above account. The first type (I), a rolling, dual-purpose mandible with oblique hinge-line and molar and incisor processes in close proximity is found in the syncarids, lophogastrids, mysids, and euphausiids. Within this group a lacinia mobilis is not found in the syncarids, but is seen in lophogastrids, mysids, and larval euphausiids. Further, Manton (1928, 1977) noted the extensive similarities in the form of the lophogastrid and euphausiid mandibles. These differed

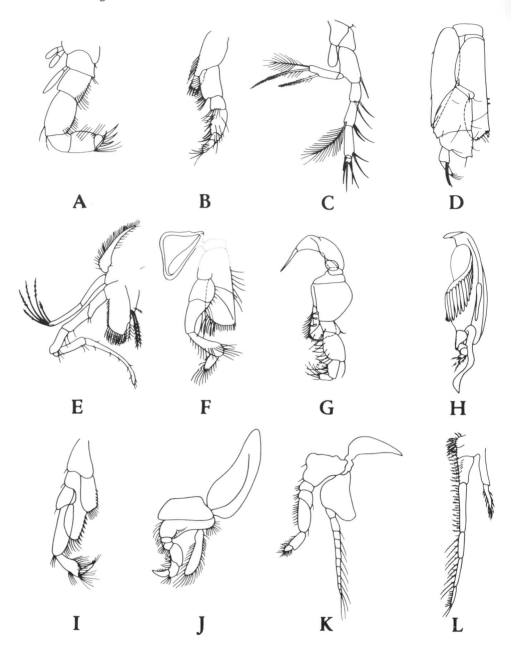

Figure 2. Diagrammatic representation of thoracopod 1 as reviewed in the text. (a) *Koonunga cursor* Sayce, 1907 (from Sayce 1908); (b) *Parastygocaris andina* Noodt, 1962 (from Noodt 1962); (c) *Notobathynella chiltoni* Schminke, 1973 (from Schminke 1973); (d) *Disconectes phalangium* (G.O.Sars) 1864 (from Wilson & Hessler 1981); (e) *Monodella sanctaecrucis* Stock, 1976 (from Stock 1976); (f) *Spelaeogriphus lepidops* Gordon, 1957 (from Gordon 1957); (g) *Apseudes diversus* Lang, 1968 (from Lang 1968); (h) *Diastylis rathkei* (Kroyer) 1841 (from G.O.Sars 1900); (i) *Gammaropsis nitida* (Stimpson) 1853 (from G.O.Sars 1894); (j) *Lophogaster typicus* M.Sars, 1862 (from M.Sars 1862); (k) *Siriella thompsoni* (H.Milne-Edwards) 1837 (from G.O.Sars 1885); (l) generalized euphausiid (from Mauchline 1967).

principally in the presence of the lacinia mobilis in lophogastrids at a position where only a fixed cusp exists in euphausiids. The second type (II), a rolling, dual-purpose mandible with oblique hinge-line and widely separated incisor and molar processes is found in thermosbaenaceans, spelaeogriphaceans, tanaids and cumaceans. In this group, the basic rolling movement of the mandible, primarily controlled by massive promotor and remotor muscles connected to a transverse mandibular tendon, is retained. However, there is a tendency towards elongation of the distal portion of the mandible, which when coupled with the oblique hinge-line, effectively places the incisor near the front margin of the head. A series of lifting spines are often present between the incisor and molar processes, presumably to aid in the movement of food posterodorsally to the mouth. The third type (III), a transverse-biting mandible with horizontal hinge line and, consequently, much shortened basal portion, is found only in the Amphipoda and Isopoda. In this type the primary movement of the mandible is controlled by an adductor-abductor set of muscles connected to upwardly-directed apodemes arising from the mesial and lateral margins, respectively. A transverse mandibular tendon is absent. Because there is no rolling action, this type of mandible is designed primarily for cutting. The molar processes may be present, but function as anti-slip surfaces to keep the food in place near the mouth. In both amphipods and isopods there is a tendency for the incisor processes to become elongate, anteriorly-directed structures with concomitant reduction of the molar.

## 2.4 *Maxilliped*

### 2.4.1 *Characteristics*

2.4.1.1 Anaspidacea (Fig.2a,b): with two short, fleshy, outwardly-directed epipods (absent in the Psammapsidae); exopod short and fleshy, bi-articulate; endites not present in microphagous forms such as *Koonunga* and *Psammapsides,* but present on coxa in macrophagous *Anaspides,* project mesially; endopod 'leg-like', similar to thoracopod two, consists of five articles (Thomson 1894, Sayce 1908, Noodt 1963, Schminke 1974).

2.4.1.2 Bathynellacea (Fig.2c): epipod occasionally present (e.g. in some species of *Notobathynella* and *Hexabathynella*); endites not developed, exopod of one to three articles; endopod of four articles; appendage not 'typically leg-like', but similar to thoracopod two (Schminke 1973).

2.4.1.3 Thermosbaenacea (Fig.2e); epipod a flattened lobe, backwardly-directed; exopod of two articles or absent; coxal endite shorter and mesial to large, anteriorly-directed basal endite; endopod slender, elongate, five articles; structure of endopod and exopod similar to that of thoracopod two (Stock 1976).

2.4.1.4 Spelaeogriphacea (Fig.2f): fleshy, flattened epipod directed dorsally into branchial cavity; exopod absent; coxal endite absent, basal endite enlarged, wider than endopod proximal articles, projects anteriorly; endopod slender, of five articles; appendage not similar in structure to thoracopod two (Gordon 1957).

2.4.1.5 Tanaidacea (Fig.2g): in *Apseudes,* flattened epipod directed dorsally into branchial cavity; exopod absent; coxal endite absent, basal endite projects anteriorly, slightly wider than endopod (= palp) proximal articles; endopod occasionally with five, but usually four, articles, shortened, not similar to thoracopod two which is modified as a cheliped (Lang 1968, Gardiner 1975).

2.4.1.6 Cumacea (Fig.2h): epipod highly modified for respiration; exopod absent; basal endite narrower than endopod (= palp) proximal articles, projects anteriorly; endopod occasionally with five, but usually four, articles, shortened, not similar to thoracopod two which is modified and reduced as a second maxilliped (G.Sars 1900).

2.4.1.7 Isopoda (Fig.2d): flap-like epipod forms lateral cover to mouthpart bundle; exopod absent; basal endite wider than endopod (= palp) proximal articles, projects anteriorly; endopod of five articles, often reduced, not similar in structure to thoracopod two (G.Sars 1899).

2.4.1.8 Amphipoda (Fig.2i): epipod and exopod absent; coxae and proximal portion of bases coalesced medially; anteriorly projecting endites usually on basis and ischium (generally referred to as inner and outer plates, respectively); remaining four articles of endopod form palp, not similar to structure of thoracopod two (G.Sars 1894).

2.4.1.9 Lophogastrida (Fig.2j): in *Lophogaster,* flattened epipod extends posterodorsally into branchial cavity; exopod slender, two or three articles; coxal and basal endites small, directed mesially; endopod of four to five articles, somewhat 'leg-like', but shorter than thoracopod two (M.Sars 1862, G.Sars 1885).

2.4.1.10 Mysida (Fig.2k): epipod usually flattened, directed backwards into branchial cavity; exopod of two articles, the second elongate and annulated; endites often absent but occasionally project mesially from basis, ischium and merus; endopod shortened, five articles, occasionally similar in structure to thoracopod two (G.Sars 1885).

2.4.1.11 Euphausiacea (Fig.2l): epipod a shortened lobe; exopod of two articles, the second elongate and annulated; mesially-directed endites weak; endopod slender, 'leg-like', similar to thoracopod two (Mauchline 1967, Mauchline & Fisher 1969).

2.4.2 *Discussion.* The variability seen in the structure of this appendage undoubtedly reflects the influence of both the phylogenetic history and the life habits of the species in the orders. Still, some patterns emerge. In the syncarids (Type I) the endopod is generally quite similar in structure to thoracopod two, the epipod(s) and exopod, when present, are lobate and lateral, and endites develop mesially. Reductions of epipods and exopods within this group can be quite clearly related to the interstitial life habits of the stygocaridans and bathynellaceans. Lophogastrids, mysidans, and euphausiids (Type II) also have more or less leg-like maxillipedal endopods, the branchial epipod usually occurs with a two article natatory exopod, and coxal and basal endites, if developed, project mesially. A third type (III), with strongly modified endopod, no exopod (except in thermosbaenaceans), a fleshy epipod which is directed posteriorly into a branchial chamber formed by the inner carapace wall and the lateral walls of the head, and an anteriorly projecting basal endite, is found in thermosbaenaceans, spelaeogriphaceans, tanaidaceans and cumaceans. The isopodan type (IV) differs from the previous group only in the epipod being used as a lateral cover for the mouthpart bundle. Finally, the amphipodan maxilliped (Type V) differs considerably from all the others. It lacks epipod and exopod, has mesially fused coxae, and basal and ischial endites.

2.5 *Blood system*

2.5.1 *Characteristics.* In this section, we will be concerned only with the size and location of the heart and major arteries. Even so, for many orders only rudimentary information is available or only single species have been examined.

2.5.1.1 Anaspidacea (*Anaspides tasmaniae*) (Fig.3): heart extends from thoracic somite 2 to abdominal somite 4; one pair ostia; five pairs lateral arteries posteriorly, the last subdivided; posterior four pairs of lateral arteries to bases of pleopods and uropods; descending

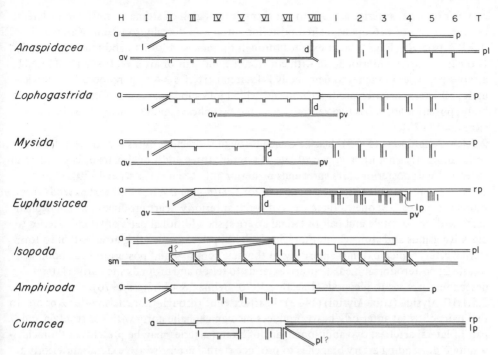

Figure 3. Diagrammatic representation of the blood system for several eumalacostracan orders. All lateral and posterolateral arteries are paired, but for clarity are drawn singly; others are unpaired and are located medially. The widened tube represents the dorsally-located heart, all narrow tubes represent blood vessels. Abbreviations: H – head; I-VIII – thoracic somites; 1-6 – abdominal somites; T – telson; a – anterior aorta; av – anterior branch of ventral artery; d – descending artery; 1 – lateral artery; lp – left posterior aorta; p – posterior aorta; pl – posterolateral artery; pv – posterior branch of ventral artery; rp – right posterior aorta; sn – subneural artery. Sources of drawings: Anaspidacea, from Siewing (1963); Lophogastrida, from Belman & Childress (1976); Mysida, from Mauchline (1980); Euphausiacea, from Mauchline & Fisher (1969); Isopoda, from Kaestner (1970); Amphipoda, from Kaestner (1970); Cumacea, from Kaestner (1970).

artery in thoracic somite 8 to all thoracopods and posterior mouthparts; anterior aorta to brain and eyestalks (Siewing 1963).

2.5.1.2  Bathynellacea: heart of variable length, greatest length is from thoracic somite 2 to 4; no ostia; no arteries except for anterior and posterior aortae (Kaestner 1970).

2.5.1.3  Thermosbaenacea (*Thermosbaena mirabilis*): heart short, in thoracic somite 1; one pair of ostia; posterior aorta short, anterior aorta with divisions to mouthparts and front of head (Siewing 1963, Kaestner 1970).

2.5.1.4  Spelaeogriphacea: heart extends the length of thorax (personal observation); no other details known.

2.5.1.5  Tanaidacea (*Apseudes*): heart extends through thoracic somites 3-8; two pairs of ostia; anterior aorta widens to form cor frontale and supplies anterior part of head; four pairs lateral arteries open near gut caeca; blood then travels to post-mandibular appendages via longitudinal lacuna (Kaestner 1970).

2.5.1.6  Cumacea (*Diastylis rathkei*) (Fig.3): heart extends through thoracic somites 3-6;

one pair of ostia; anterior aorta with cor frontale in region of stomach; four pair of lateral arteries extend to pereopods; paired posterior aortae run dorsally to telson (Kaestner 1970).

2.5.1.7  Isopoda (Fig.3): heart extends through thoracic somite 7 to abdominal somite 3; two pairs of ostia; anterior aorta with cor frontale and pair of arteries; first pair of lateral arteries produces branches to pereopods 1-4; lateral arteries 2-4 to pereopods 5-7 individually; all lateral arteries produce branches which join subneural artery running length of body; posterior aorta has five ventrally directed branches opening near pleopod bases (Kaestner 1970).

2.5.1.8  Amphipoda (Fig.3): heart extends through thoracic somites 2-7; three pairs of ostia; anteriorly an aorta and pair of lateral arteries; three additional pairs of lateral arteries extend to gut; posterior aorta surrounds posterior gut caecum (Kaestner 1970).

2.5.1.9  Lophogastrida (Fig.3): description for *Gnathophausia ingens;* heart extends through thoracic somites 2-8; one to three pairs of ostia; anteriorly heart produces median cephalic aorta with cor frontale and pair of lateral arteries; six additional pairs of lateral arteries to digestive glands and stomach; in thoracic somite 8, paired descending arteries join to form sternal artery which extends anteriorly to thoracic somite 4 and posteriorly to abdominal somite 2; posterodorsal median artery extends to telson and uropods and carries lateral descending arteries to pleopods (Kaestner 1970, Belman & Childress 1976).

2.5.1.10  Mysida (tribe Mysini) (Fig.3): heart extends through thoracic somites 2-6; one to two pairs of ostia; anteriorly heart produces median cephalic aorta with cor frontale and pair of lateral arteries; two additional pairs of lateral arteries may be present; in thoracic somite 6 descending artery branches to produce sternal artery which extends anteriorly to the mouthparts and posteriorly to thoracic somite 8; posterodorsal median artery extends to telson and uropod, with lateral arteries in each abdominal somite (Mauchline 1980).

2.5.1.11  Euphausiacea (*Meganyctiphanes norvegica*) (Fig.3): heart extends through thoracic somites 3-6; two pairs of ostia; anteriorly heart produces median aorta and pair of lateral arteries; other lateral arteries absent; descending artery divides in thoracic somite 6 to produce sternal artery which extends anteriorly to thoracopods and posteriorly to abdominal somite 4; posterodorsal paired arteries bear lateral arteries in each abdominal somite; right posterodorsal artery extends to telson (Mauchline & Fisher 1969).

2.5.2  *Discussion.* A blood system consisting of an elongate dorsal blood vessel, part of which is expanded and muscularized to form a heart, posterior ventrally-directed arteries, and descending artery with branches of variable length anteriorly and posteriorly (Type I), is found in Anaspidacea, Lophogastrida, Mysida, and Euphausiacea. With the exception of that seen in isopods, all other blood systems considered show reductions of this pattern. Amphipods (Type III) as well as cumaceans, tanaids, thermosbaenaceans, and probably spelaeogriphaceans (Type II) do not have a descending artery with ventral branches. The isopodan system (Type IV), on the other hand, is very elaborate and differs from all the others in having paired anterior aortae, a subneural artery running the length of the body, and lateral arteries leading directly to the pereopods. Yet, it is possible to see in Figure 3 that the isopodan system is similar to the anaspidacean system in possessing a branched posterolateral artery and an anterior lateral artery that, by position, may be homologous to the descending artery (marked by d?), and possibly can be derived from it. While the cumacean system is highly reduced it maintains certain affinities with that of isopods. Also clear is the fact that the blood vessel pattern, rather than relative length and position of the heart, contains the greatest amount of phylogenetic information.

## 2.6 *Developmental patterns*

2.6.1 Anaspidacea: eggs are attached to substratum; when they hatch they resemble adults, but eyes are not stalked, and they lack a rostrum and pleopod endopods.

2.6.2 Bathynellacea: eggs deposited in substratum; hatches as larva with antennae, mouthparts, first pair of thoracopods and ten somites; body somites, limbs and limb buds added with successive moults; seven to eight juvenile stages in *Bathynella natans*.

2.6.3 Thermosbaenacea: eggs incubated dorsally under enlarged carapace; in *Monodella*, hatch as manca without last two pairs of pereopods or pleopods.

2.6.4 Spelaeogriphacea: unknown.

2.6.5 Tanaidacea: eggs incubated in a ventral brood pouch composed of oöstegites from pereopods 1-4 (= thoracopods 3-6); oöstegites shed between broods, hatch as manca without last pair of pereopods or pleopods; two manca stages before juvenile stage.

2.6.6 Cumacea: eggs incubated in ventral brood pouch composed of oöstegites from thoracopods 3-6; oöstegites shed between broods; hatch as manca without last pair of pereopods, but all limbs non-functional and abdomen incomplete; moults three times before leaving brood pouch as functional manca.

2.6.7 Isopoda: eggs incubated in a ventral brood pouch composed of oöstegites from pereopods 1-5, though occasionally one, four or seven oöstegites may occur; oöstegites shed between broods; hatch as manca without last pair of pereopods.

2.6.8 Amphipoda: eggs incubated in ventral brood pouch composed of oöstegites from pereopods 2-5 or 4 and 5; oöstegites not shed between broods; hatch as miniature adults, not lacking any appendages.

2.6.9 Lophogastrida: eggs incubated in ventral brood pouch composed of oöstegites on all pereopods; oöstegites not shed between broods; hatch as miniature adults, not lacking any appendages.

2.6.10 Mysida: eggs incubated in ventral brood pouch composed of oöstegites on last two or three pairs of thoracic legs, rarely on all pereopods; oöstegites not shed between broods; hatch as miniature adults, not lacking any appendages.

2.6.11 Euphausiacea: eggs shed freely into water, occasionally carried in mass loose among the thoracic legs or even attached to the sternal surface of the thorax near the bases of the posterior thoracic legs.

# 3 CURRENT PHYLOGENETIC ARRANGEMENTS

Until quite recently (Watling 1981, Schram 1981), the only view of peracarid phylogeny was that proposed by Siewing (1956, 1963) and subsequently modified by Fryer (1965). Problems with the Siewing scheme were reviewed and a radically different arrangement proposed (Watling 1981). In that paper, I suggested that the ancestral peracarid was very syncarid-like and gave rise to two lines: the mancoids, leading from the isopods through the spelaeogriphaceans and tanaids to cumaceans; and a second, heterogeneous line leading from amphipods to mysids and thermosbaenaceans. This scheme also had its difficulties, primarily in the amphipod, mysid and thermosbaenacean line. First among these is the fact that thermosbaenaceans have a manca-like stage, and further, they have a short-carapace that, at the very least, anticipates the respiratory function seen in other short-carapace forms. In addition, only the possession of an antennal gland and the hatching of larvae with all extremities unite the amphipods and mysids, two features which are not unique to these orders within the Eumalacostraca. Thus, even this phylogenetic proposal is incomplete and is in need of further revision.

## 4 A NEW PHYLOGENY OF THE EUMALACOSTRACA

A new view of eumalacostracan phylogeny is given in Figure 4. It is based primarily on the patterns seen in the review of features in Section 2. Three basic lineages are seen as having arisen in some unknown way from the hypothetical eumalacostracan ancestor proposed by Dahl (1976) and discussed by Watling (1981).

First is the syncarid/eucarid line which is characterized by the retention of the rolling, dual-purpose mandible with molar and incisor processes in close proximity and with a lacinia mobilis in at least some larval stages, maxilliped with more or less leg-like endopod and mesially-directed endites (when present), and blood system with a thoracic descending artery. The development of a long respiratory-hydrodynamic carapace distinguishes the eucarids from the syncarids. The caridoid forms show two different trends in embryonization of the larvae. There is a nauplius larva in primitive decapods (Order Dendrobranchiata) and euphausiids; however, a thoracic brood pouch and direct development is found in pygocephalomorphs, lophogastrids, and mysids while in higher decapods eggs are attached to abdominal appendages and the developmental sequence becomes increasingly abbreviated. Besides the characteristics of the mandible, maxilliped, and blood system held in common, euphausiids and mysids are unique in possessing refracting superposition facets in the compound eyes.

A second line begins with the Isopoda, which has no carapace and a strongly modified maxilliped endopod with anteriorly projecting basal endite. This latter feature, and the manca condition of the juvenile, provide some indication of a common ancestry for Isopoda and the short carapace-bearing Thermosbaenacea, Spelaeogriphacea, Tanaidacea, and Cumacea. Most isopods differ from the Brachycarida in having developed a transversely-

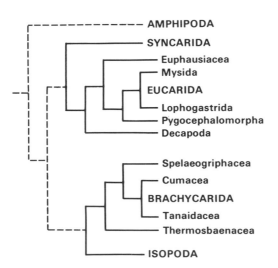

Figure 4. Phylogenetic arrangement of the Eumalacostraca showing monophyletic origin of Superorders Amphipoda, Syncarida, Schizopoda, Eucarida, Brachycarida, and Isopoda. Dashed line indicates uncertainty of Superorder relationships but represents current 'best guess'.

biting mandible with horizontal hinge-line, and in their retention of a blood system with an elaborate set of lateral arteries. The synapomorphies of Brachycarida include a rolling, dual-purpose mandible with widely separated incisor and molar processes, row of lifting spines, and oblique hinge line; and a blood system with a reduced set of lateral arteries.

The third line involves only the Amphipoda. They differ from both of the other lines in the maxilliped form, transversely-biting mandible, reduced blood system, modification of pleopods 4 and 5 as structures similar to the uropods, and a host of other features. The true relationships of this line in reference to the others is difficult to ascertain since amphipods have some features in common with each of the other lines.

## 5 A REVISED CLASSIFICATION OF THE EUMALACOSTRACA

From the patterns of characteristics associated with the structures considered in the foregoing account, the orders dealt with can be grouped as follows: Group 1, Anaspidacea, Bathynellacea; Group 2, Thermosbaenacea, Spelaeogriphacea, Tanaidacea, Cumacea; Group 3, Isopoda; Group 4, Amphipoda; Group 5, Lophogastrida, Mysida, Euphausiacea. It is proposed here that each of these groups, the last including the decapods, represent monophyletic units at the level of superorder. Their diagnostic features and component orders are summarized below.

Superorder Syncarida Packard, 1885
   Characteristics: no carapace; compound eyes primitively stalked, apposition optics; mandible of rolling, dual-purpose type with oblique hinge-line and molar and incisor processes in close proximity (Type I); maxilliped primitively with endopod similar in structure to thoracopod 2, epipods and exopod (when present) lobate and lateral, and endites mesially-directed (Type I); blood system with elongate dorsal blood vessel, posterior ventrally-directed arteries, and thoracic descending artery with ventral branches (Type I); excretory organs as maxillary glands.
   Component orders: Anaspidacea Calman, 1904; Bathynellacea Chappuis, 1915; Paleocaridacea Brooks, 1962.

Superorder Brachycarida Schram, 1981
   Characteristics: carapace short, forming branchial chamber; compound eyes on short lobes or sessile, apposition optics; mandible of rolling, dual-purpose type with oblique hinge-line and widely separated incisor and molar processes (Type II); maxilliped with endopod strongly modified, exopod absent (except in Thermosbaenacea). epipod fleshy and directed posteriorly into branchial chamber, and anteriorly projecting basal endite (Type III); blood system with lateral arteries but without thoracic descending artery with ventral branches (Type II); excretory organs as maxillary glands.
   Component orders: Thermosbaenacea Monod, 1972; Spelaeogriphacea Gordon, 1957; Tanaidacea Dana, 1853; Cumacea Kroyer, 1846.

Superorder Isopoda Latrielle, 1817
   Characteristics: carapace absent; compound eyes sessile, apposition optics; mandible of transverse-biting type with horizontal hinge-line (Type III); maxilliped with endopod strongly modified, exopod absent, epipod directed anteriorly along mouthpart bundle and

anteriorly projecting basal endite (Type IV); blood system with lateral arteries leading directly to pereopods and sternal artery but without thoracic descending artery (Type IV); excretory organs as maxillary glands.

Component orders: Valvifera Sars, 1882; Anthurida Leach, 1814; Flabellifera Sars, 1882; Microcerberida Chappuis and Deboutteville, 1960; Asellota Latreille, 1803; Phreatoicida Stebbing, 1893; Gnathiida Leach, 1814; Oniscoida Latreille, 1803; Epicarida Latreille, 1831.

Superorder Amphipoda Latreille, 1816.

Characteristics: carapace absent; compound eyes sessile, apposition optics; mandible Type III; maxilliped lacks epipods and exopods, has medially fused coxae and anteriorly projecting basal and ischial endites (Type V); blood system with few lateral arteries, no thoracic descending artery, posterior aorta goes to posterior gut cecum (Type III); excretory organ as antennal gland.

Component orders: Gammarida Latreille, 1803; Ingolfiellida Hansen, 1903; Caprellida Leach, 1814; Hyperiida Latreille, 1831.

Superorder Eucarida Calman, 1904

Characteristics: carapace long, combining respiratory and hydrodynamic functions, formed from branchiostegal folds on some or all thoracic somites; compound eyes stalked, with refracting or reflecting superposition facets, occasionally using apposition optics; mandible Type I; maxilliped primitively with more or less leg-like endopod, branchial epipod, 2-articulate (often article 2 annulated) natatory exopod, and weak, if developed, mesially projecting coxal and basal endites (Type II); blood system Type I; excretory organs as antennal glands, but reduced maxillary glands also exist in one order.

Component orders: Pygocephalomorpha Beurlen 1930; Lophogastrida Boas, 1883; Mysida Boas, 1883; Euphausiacea Dana, 1852; Dendrobranchiata Bate, 1888; Pleocyemata Burkenroad, 1963.

# 6 ACKNOWLEDGEMENTS

I would like to thank especially Dr F.R.Schram for his continued interest in my ideas and for repeatedly encouraging me to go beyond my classical crustacean training. Numerous individuals provided help in various ways. Professor Erik Dahl kindly sent a copy of his in press manuscript containing his ideas about crustacean carapaces. Specimens of groups not found in North America were obtained from Drs R.Y.George and J.Dearborn, or collected with the aid of Dr J.Grindley. Drs F.R.Schram and J.C.Kunze read an earlier version and Ms Patrice Rossi drafted the figures.

# DISCUSSION

SIEG: You classed mandibles into several types, some were rolling, others monocondylic, others dicondylic. That is no reason to separate the isopods because in the tanaids you have both types — rolling types in so-called 'monokonophorans' and dicondylic in 'dikonophorans'. This would show that there would be parallel evolution of mandibles in isopods and tanaids.

WATLING: I am not claiming that all transverse biting types are synapomorphic. The arrangement of musculature in those has not been properly examined. But I think one can say that the rolling dual purpose type is synapomorphic with relation to the *Chirocephalus* type. The muscle arrangement and transverse mandibular tendon on all these are all the same. What I would like to know is in the tanaids if in the transition from that to the transverse biting types what are the modifications in the muscles? I would not expect a similar apodemal arrangement as in isopods.

SIEG: They are very similar. There have been comparative studies of both tanaids and isopods done in Germany [Lauterback on tanaids, Scheleoske on *Asellus,* Schmalfuss on *Tylos*] that indicate they are very similar.

KUNZE: I would just like to point out that in some isopods such as cirolanids as well as asellotes have a rolling mandible.

# REFERENCES

Belman, B.W. & J.J.Childress 1976. Circulatory adaptations to the oxygen minimum layer in the bathypelagic mysid *Gnathophausia ingens. Biol. Bull.* 150:15-37.

Calman, W.T. 1904. On the classification of the Crustacea Malacostraca. *Ann. Mag. Nat. Hist.* (7)13: 144-158.

Cannon, H.G. & S.M.Manton 1927. On the feeding mechanism of a mysid crustacean, *Hemimysis lamornae. Trans. Roy. Soc. Edin.* 55:219-252.

Cannon, H.G. & S.M.Manton 1929. On the feeding mechanism of the syncarid Crustacea. *Trans. Roy. Soc. Edin.* 56:175-189.

Dahl, E. 1976. Structural plans as functional models exemplified by the Crustacea Malacostraca. *Zool. Scripta* 5:163-166.

Dahl, E. (in press). Alternatives in malacostracan evolution. *Rec. Aust. Mus.*

Fincham, A.A. 1980. Eyes and classification of malacostracan crustaceans. *Nature* 287:729-731.

Fryer, G. 1965. Studies on the functional morphology and feeding mechanism of *Monodella argentarii* (Crustacea: Thermosbaenacea). *Trans. Roy. Soc. Edin.* 66:49-90:

Gardiner, L.F. 1975. The systematics, post marsupial development, and ecology of the deep-sea Family Neotanaidae (Crustacea: Tanaidacea). *Smith. Contr. Zool.* 170:1-265.

Gordon, I. 1957. On *Spelaeogriphus,* a new cavernicolous crustacean from South Africa. *Bull. Brit. Mus. (Nat. Hist.) Zool.* 5:31-47.

Kaestner, A. 1970. *Crustacea. Invertebrate Zoology, Vol.3.* New York: Interscience Publishers.

Knight, M.D. 1975. The larval development of Pacific *Euphausia gibboides* (Euphausiacea). *Fish. Bull.* 73:145-168.

Knight, M.D. 1978. Larval development of *Euphausia fallax* Hansen (Crustacea: Euphausiacea) with a comparison of larval morphology within the *E.gibboides* species group. *Bull. Mar. Sci.* 28:255-281.

Land, M.F. 1980. Compound eyes: old and new optical mechanisms. *Nature* 287:681-686.

Lang, K. 1968. Deep-sea Tanaidacea. *Galathea Rept.* 9:23-209.

Manton, S.M. 1928. On some points in the anatomy and habits of the lophogastrid Crustacea. *Trans. Roy. Soc. Edinb.* 56:103-119.

Manton, S.M. 1977. The Arthropoda, Habits, Functional Morphology, and Evolution. Oxford: Clarendon Press.

Mauchline, J. 1967. Feeding appendages of the Euphausiacea (Crustacea). *J. Zool., London* 153:1-43.

Mauchline, J. 1980. The biology of mysids and euphausiids. *Adv. Mar. Biol.* 18:1-677.

Mauchline, J. & Fisher 1969. The biology of euphausiids. *Adv. Mar. Biol.* 7:1-454.

Noodt, W. 1963. Anaspidacea (Crustacea, Syncarida) in der sudlichen Neotropis. *Verh. Deutsch. zool. Ges. Wien* 1962:568-578.

Sars, G.O. 1885. Report on the Schizopoda collected by HMS 'Challenger' during the years 1873 to 1876. *Challenger Rept. Zool.* 13:1-228.

Sars, G.O. 1894. Amphipoda. *Crustacea of Norway* 1:1-711.

228 *Les Watling*

Sars, G.O. 1899. Isopoda. *Crustacea of Norway* 2:1-270.

Sars, G.O. 1900. Cumacea. *Crustacea of Norway* 3:1-115.

Sars, M. 1862. Beskrivelse over *Lophogaster typicus.* Universitetsprogram for Andet Halvaar, Christiania, pp.1-37.

Sayce, O.A. 1908. On *Koonunga cursor,* a remarkable new type of malacostracous crustacean. *Trans. Linn. Soc.* (2)11:1-16.

Schminke, H.K. 1973. Evolution, System und Verbreitungsgeschichte der Familie Parabathynellidae (Bathynellacea, Malacostraca). *Akad. Wiss. Lit. Mainz, math.-nat. Kl., Mikrofauna Meersbodens* 24: 1-92.

Schminke, H.K. 1974. *Psammaspides williamsi* gen.n., sp.n., ein vertreter einer neuen Familie mesopsammaler Anaspidacea (Crustacea, Syncarida). *Zool. Scripta* 3:177-183.

Schram, F.R. 1981. On the classification of the Eumalacostraca. *J. Crust. Biol.* 1:1-10.

Serban, E. 1980. La mandibule et l'individualisation des ensembles évolutifs majeurs dans l'ordre des Bathynellacea (Malacostraca, Podophallocarida). *Bijd. Dierk.* 50:155-189.

Siewing, R. 1956. Untersuchungen zur morphologie der Malacostraca (Crustacea). *Zool. Jb. (Anat.)* 75: 39-176.

Siewing, R. 1958. Anatomie und histologie von *Thermosbaena mirabilis,* ein beitrag zur phylogenie der reihe Pancarida (Thermosbaenacea). *Akad. Wiss. Lit. Mainz, math.-nat., Kl. Abh.* 7:197-270.

Siewing, R. 1963. Studies in malacostracan morphology: results and problems. In: H.B.Whittington & W.D.I.Rolfe (eds), *Phylogeny and Evolution of Crustacea:*85-103. Cambridge: Mus. Comp. Zool.

Stock, J.H. 1976. A new genus and two new species of the crustacean order Thermosbaenacea from the West Indies. *Bijd. Dierk.* 46:47-70.

Thomson, G.M. 1894. On a freshwater schizopod from Tasmania. *Trans. Linn. Soc. London Zool.* (2)6: 285-303.

Watling, L. 1981. An alternative phylogeny of peracarid crustaceans. *J. Crust. Biol.* 1:201-210.

Wiegmann-Haass, R. 1977. Die Calyptopis und Furcilia-stadien von *Euphausia hanseni* (Crustacea: Euphausiacea). *Helgol. wiss. Meeresunter* 29:315-327.

Wilson, G.D. & R.R.Hessler 1981. A revision of the genus *Eurycope* (Isopoda, Asellota) with descriptions of three new genera. *J. Crust. Biol.* 1:401-423.

JÜRGEN SIEG
*Universität Osnabrück, Abt. Vechta, West Germany*

# EVOLUTION OF TANAIDACEA

## ABSTRACT

The systematic position of the orders within Peracarida is discussed. This leads to the reconstruction of an ancestral Tanaidacean type. An overview of the comparative morphology
of the appendages shows that there are four different phylogenetic lines, each represented
by a suborder: Anthracocaridomorpha, Apseudomorpha, Neotanaidomorpha, and Tanaidomorpha. Because our knowledge of morphology and anatomy of the Apseudoidea is rather
poor, only a first approximation of a natural arrangement of this superfamily can be
attempted. There is much more information for the Tanaidomorpha; and, therefore, the
phylogenetic relationships for these are discussed in detail. The established system also
allows an interpretation of several other lines of information such as geographical and vertical distribution as well as phylogenetic development of sexual dimorphism.

## 1 INTRODUCTION

Interest in Tanaidacea, a small peracaridan order closely related to the Isopoda, has increased during the past three decades. There are several reasons for this. More exact field
studies have been undertaken on known forms, extensive research has begun on marine
benthic faunas of abyssal and hadal depths. Numerous investigations in recent years have
indicated that the Tanaidacea are among the numerically abundant invertebrates in marine
habitats (Gage & Coghill 1977, Gardiner 1975, Grassle & Sanders 1973, Jumars 1976,
Livingston 1977, Odum & Heald 1972). Remarkably, abundance increases with depth such
that numbers of tanaids often exceed that of Isopoda (Sanders et al. 1965, Menzies &
George 1967, Belyaev 1968). However, the difficulties of identifying specimens to species
or even family level have hindered rapid research in the biology and distribution of the
group and even obscured a deeper understanding of their evolution.

It is largely due to the efforts of the late Professor Lang that our knowledge has been
established of the main features of the order's systematics. In addition to describing many
new species, Lang also tried to erect a 'natural system'. It was Lang (1956, 1970) who
established the subordinal status of the 'Apseudidae' and 'Tanaidae'. He also was able to
define several additional families in both these suborders by examining all species very carefully and providing excellent drawings even of structures originally thought to be unim-
portant at the time. Lang's work has provided a sound foundation for understanding the
phylogeny of this order.

More recent studies on tanaidaceans, mainly on comparative morphology, forced us to reflect on the different phylogenetic lines pointed out by Lang (1956, 1967, 1970, 1971, 1973) and Sieg (1973). For example, it has now been recognized that the 'dikonophor' is a plesiomorphic character, and thus not appropriate to basing the definition of a suborder. Rather, a consideration of the mouthparts, sexual dimorphism, and structure of the marsupium have played an important role in allowing the 'Dikonophora' to be reorganized and split into the suborders Neotanaidomorpha and Tanaidomorpha (Sieg 1976, 1980b). However, nothing heretofore has been advanced concerning the relationship within the Apseudoidea due to ignorance by other scientists of phylogenetic considerations when new taxa were described within this group.

Extensive studies upon the evolutionary process within the Tanaidacea (Sieg, in press) now allow, for the first time, the proposal of a phylogenetic scheme showing the possible relationships of all known families. It is evident that this scheme is only the beginning of an understanding of the evolution of the Tanaidacea. There is still a great lack of information that can clarify all the questions that have been raised. Some of these will be discussed later. I hope, however, that it now might be easier to recognize phylogenetically important characters in the definition of taxa, and this hopefully will allow the eventual stabilization of the systematics of the Tanaidacea. This in turn will allow ecologists an opportunity to recognize the importance of tanaidaceans within marine habitats and food chains.

## 2 ON THE SYSTEMATIC POSITION OF THE TANAIDACEA

Up to a few years ago there was no debate concerning phylogenetic relationships within the Peracarida. As pointed out in many papers by Siewing (1953-1963) there were several major points which still demand our attention. The first is the heterogeneity of 'Mysidacea'. There is no doubt that Siewing was correct in treating both suborders, Mysida and Lophogastrida, created by Boas (1883) as equivalent to Amphipoda, Cumacea, etc. It is true that

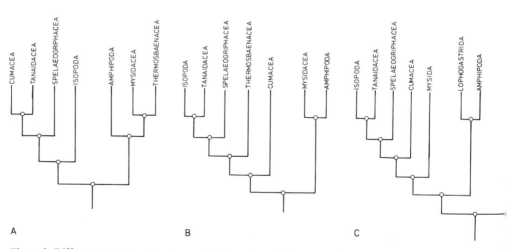

Figure 1. Different phylogenetic schemes for the orders of the superorder Peracarida. A. after Watling 1981; B. after Fryer 1964; C. after Siewing 1953. All schemes are redrawn and modified.

Siewing never formally recognized Mysida and Lophogastrida as orders, but in all his manifold diagrams it is easy to perceive that he treated both groups as equivalent to orders. These two groups are indeed related taxa, but in my opinion the 'Mysidacea' must be split up, as was the 'Schizopoda' a century ago. Most of the characters that unite them together are plesiomorphic ones. Therefore, it is also likely that both Mysida and Lophogastrida are derived from different ancestors.

The second important point is that there is no close relationship between the Isopoda and Amphipoda (Siewing 1951, 1953). The anatomical and morphological facts would appear to argue against uniting both orders in a higher taxonomic category contrary to some Palaeozoic data which might suggest doing so (Schram 1981). It is one of the disastrous obstructions working on Palaeozoic forms that we only can recognize the coarse and not the fine structures of the specimens. This may be explained only by one example. Such a character as articulation of peraeopods are normally not visible on fossil forms, but in our case this is a phylogenetically important point (Hessler 1980). In my opinion, nearly all the characters which are evident on the known fossils of the two orders belong to a functional-morphological complex which has to be accepted as derived convergently in Isopoda and Amphipoda.

Another problem in achieving an understanding of the phylogenetic relationships within the Peracarida results from the incorporation of Thermosbaenacea into the Peracarida (Freyer 1964, Watling 1981, Fig.1b). There exists no important character which would allow such an action. We have to accept that Thermosbaenacea is very closely related to Peracarida. Therefore, we should expect to find many characters in thermosbaenaceans that are typically peracaridan, but these are all plesiomorphies and do not tell us anything about proximity of relationship to the peracarids. The main point is that the Thermosbaenacea show us another way of solving the problems of brooding, which was not as successful as that developed by Peracarida.

Several authors (Tiegs & Manton 1958, Dahl 1963, 1976, 1977) have questioned the caridoid facies of Calman (1909) as a primitive pattern for the malacostracans. If this is accepted, the peracaridan system has to be changed totally, as was done by Watling (1981, Fig.1a). In this scheme, that which have heretofore been considered advanced (apomorphic) would now be placed at the base of the peracarid lineage (becoming plesiomorphic).

Watling's arrangement would also have important effects in determining relationships within peracaridan orders. For example, Watling presumes, among others, parallel evolution of the maxillule palp in Cumacea/Tanaidacea and Mysidacea (*Gnathophausia*)/Thermosbaenacea because the supposedly primitive Isopoda and Amphipoda of Watling's scheme do not possess a maxillulary palp. If this character really is a plesiomorphic feature then we would have to consider the palpless tanaids as more primitive than those possessing one, and consequently demand a palpless ancestor for all Peracarida. This would seem to be unlikely if Eucarida (e.g. Euphausiacea) are to be considered as the closest related group of the Peracarida, because then we have to accept another parallel evolution of this structure. However, to accept the lack of a maxillulary palp as a plesiomorphy presents also great difficulties within the Tanaidacea, since while the Neotanaidae, which lack the palp, may be considered primitive, the Kalliapseudidae, which also lack a palp, are very highly specialized.

Also the ventilatory exopods of thoracopods 2-4 in Spelaeogriphacea are suggested as 'independently acquired apomorphic feature' (Watling 1981:206). This also seems to be unlikely because parallel evolution must have taken place too often. In both peracaridan lines of Watling's scheme the plesiomorphic orders lack exopodites on all peraeopods. This normally led to the assumption that the common stem form of these groups also lacked exopodites, an obscure idea because there is no doubt that the typical crustacean limb possessed an endo- and exopodite. Therefore, exopodial reduction in Isopoda and Amphipoda has to be accepted. This reduction also must be accepted as a parallel process each in Spelaeogriphacea and Tanaidacea because Cumacea do have fully developed exopodites. It seems more likely that we only have to accept a gradual exopodial reduction.

These are only two of the many inconsistencies in Watling's phylogenetic scheme which mitigate against its acceptance. One cannot ignore the fact that there are also some difficulties in understanding the evolution of Peracarida within the context of the scheme postulated by Siewing and others. However, to solve these problems will require extensive scientific work on comparative morphology and anatomy with an aim towards reconstructing stem forms. In this connection, Lauterbach's (1973, 1974) arguments in favor of the caridoid facies has great relevance. According to Lauterbach (1974), the hypothetical stem form of Crustacea was not characterized by the typical carapace but by a lateral expansion surrounding the whole head and by pleural expansions ('Pleurotergit') in all postcephalic segments. A situation, well represented in recent Cephalocarida (*Lightiella, Sandersiella*). From this feature the carapace has developed as caudal outgrowths of the lateral expansion of the head. A process we still can observe during the ontogeny of some conchostracan genera (*Limnadia, Cyzicus, Cyclestheria*). Lauterbach also was able to make sure that the carapace adductor muscle in cephalocarids is a homolog to that occurring in all other Crustacea. So, only a monophyletic origin of the carapace can be accepted, and forces us to demand a caridoid facies both for Eumalacostraca and consequently also for Peracarida.

However, the discussion comes to an end on these different points. We only will be able to clarify peracarid systematics after we have been able to clarify the various peracarid archetypes. Until that time, I prefer to retain Siewing's scheme (Figure 1C) as one which includes as few incompatibilities as possible.

## 3 RECONSTRUCTION OF AN URTANAIDACEAN

Relevant information for the reconstruction of the hypothetical urtanaidacean can be derived from the known fossils as well as from studies of comparative morphology of recent species. The fossils present information concerning primitive body shape and segmentation. The recent species will offer information dealing mainly with the form and function of appendages.

The basal stock of the tanaids is characterized by having the first two thoracic segments fused with the cephalon and a body shape more or less cylindrical. The pleon consisted of six free segments and an elongated telson, as well as a pair of cylindrical, multisegmented, biramous uropods. This type is well represented by the lower Carboniferous genus *Anthracocaris* (Fig.2a). These features present some difficulties, when we consider the upper Carboniferous form *Cryptocaris* (Fig.2b). In this latter species, having real chelipeds (Schram 1974), the uropods are bisegmented and flattened, more akin to those known from *Spelaeogriphus* and which consequently have to be accepted as a plesiomorphic structure. As cylindrical, more or less multisegmented, and mostly biramous uropods are a characteristic feature in recent Tanaidacea, we have to demand this also for the common stem form. Therefore, when erecting a consequent phylogenetic system one might be justified in putting *Cryptocaris* into a separate order of its own.

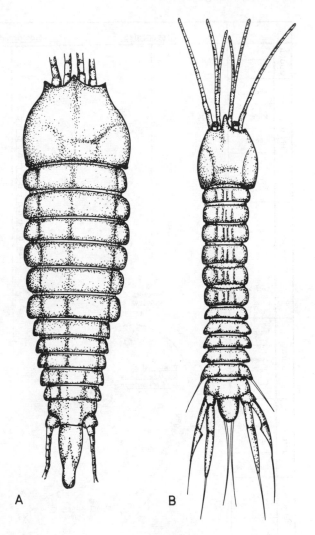

Figure 2. Palaeozoic Tanaidacea. A. *Anthracocaris scotica* (Peach) 1882, dorsal view, (from Sieg 1980b, after Schram 1979); B. *Cryptocaris hootchi* Schram 1974, dorsal view, from Sieg 1980 (after Schram 1974).

A

B

The first antenna was biramous, and consisted of a four-jointed peduncle, a feature some-times discussed for *Nebalia,* and known from different plesiomorphic isopod genera (*Asellus, Synassellus, Jaera, Eurydice, Sphaeroma,* etc.; see Scheloske 1977). Such an antennular peduncle might be a synapomorphy of the isopods and tanaids. Similarities between these two orders are also found in the peduncle of the second antenna, which must have been originally six-jointed. A feature still represented in some plesiomorphic tanaid genera (*Archaeotanais, Arctotanais;* Sieg 1980). In those cases where we find a five-jointed peduncle it can be pointed out that in one case the first joint of the antennal peduncle is reduced and in other instances the third and fourth joint are fused (Sieg, in press).

The mouth-parts of the urtanaidacean must have contained monocondylic mandibles. The right mandible was somewhat reduced as compared with the left, but possessed distally a lacinia mobilis and a row of strong setae. The first maxilla consisted of an inner and outer endite, the latter bore a two-jointed palp with many setae distally. The second maxilla and

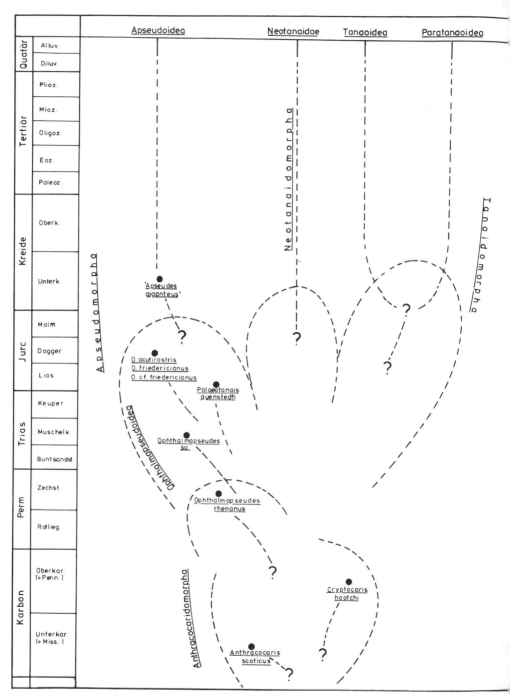

Figure 3. Stratigraphic distribution of known fossil Tanaidacea.

labium (hypopharynx; paragnaths) looked very much like those of recent Apseudoidea. Of this, the distal labial palp seems to be a synapomorphy in tanaids.

The first thoracopod was already modified as a maxilliped, a peracaridan synapomorphic feature, and characterized by an epipodite directed backwards and by the lack of an exopodite. In contrast, in Isopoda we find an exopodite, albeit in a reduced state, and not an epipodite as this structure is usually called falsely. This was long ago pointed out by Tschetwerikoff (1911) when discussing Nusbaum's (1886) embryological data, but this observation has been largely ignored in the literature.

The second thoracopod was already transformed into a cheliped bearing a reduced exo- and epipodite (oöstegite).

The third through eighth pairs of rod-like thoracopods (peraeopods) are subdivided in groups of three anteriorly directed and three posteriorly directed sets. The first peraeopod must have had an exopodite, while it may be only suggested for the second. In the other peraeopod pairs the exopodites are wanting, but the first through fourth peraeopods have coxal epipodites in the female.

Pleopods were found on pleonites one through five, the sixth pleomere developed the styliform uropods.

Concerning the sensory organs we can only deduce something regarding the eyes. It seems that these were lobed but moveable. This is derived from literature on *Ophthalmapseudes* and examination of recent *Langitanais* which have an articulating membrane between eye-lobe and carapace. Lauterbach (1970) has been able to demonstrate in *Tanais cavolinii (= T.dulongii)* a reduced muscle which probably was used for eye-movements.

## 4 MAIN EVOLUTIONARY LINES (SUBORDERS) IN TANAIDACEA

Sieg (1980) divided the Tanaidacea into four suborders. The oldest record of the order is from the Carboniferous, with forms characterized by six free pleonites. This group of species forms the suborder Anthracocaridomorpha and is at the moment represented by *Anthracocaris* and *Cryptocaris*. Both forms are so different that they have to be placed into two distinct families, Anthracocarididae (cf. Anthracocaridae, Sieg 1980) and Cryptocarididae (cf. Cryptocaridae, Sieg 1980). The latter may represent a separate order as indicated before (see above).

From this basal stock there arose three different lines: Apseudomorpha, Neotanaidomorpha and Tanaidomorpha. Of these, so it seems, only the Apseudomorpha are represented by fossils which are all combined at present in the genus *Ophthalmapseudes*. Originally the type species *Ophthalmapseudes rhenanus* from middle Permian strata (Zechstein) was suggested as a link between fossil Apseudomorpha (Ophthalmapseudidae) and Anthracocaridomorpha. But recent investigations seem to indicate that this species probably has to be referred to the Anthracocaridomorpha, since the pleon seems to consist of six free segments (Schram, personal communication, Fig.3). Not all species of *Ophthalmapseudes*, however, are so characterized, and a major revision of the Mesozoic Tanaidacea seems necessary.

Depending on general body-shape and segmentation, the Mesozoic species of *Ophthalmapseudes* shows strong similarities to Apseudoidea (= Monokonophora). Therefore, Ophthalmapseudoidea and Apseudoidea are both referred to as superfamilies to the suborder Apseudomorpha (Fig.3). *Palaeotanais* also has to be included in the Ophthalmapseudidae.

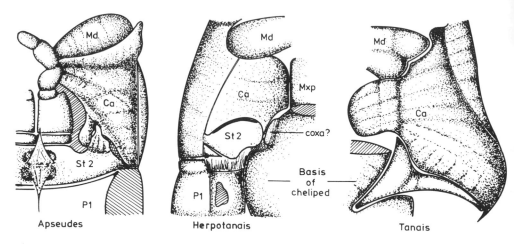

Figure 4. Construction of the carapace region of the suborders of recent Tanaidacea. Ca: wall of carapace, Md: mandible, Mxp: maxilliped, P.1: peraeomere 1, St2: sternite of second thoracic segment.

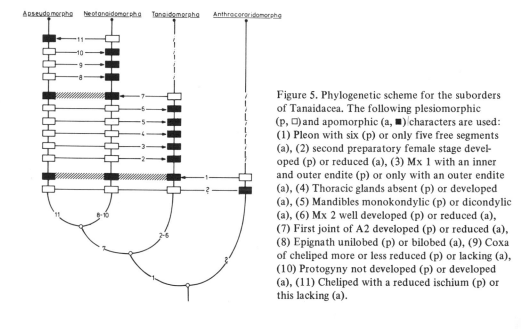

Figure 5. Phylogenetic scheme for the suborders of Tanaidacea. The following plesiomorphic (p, □) and apomorphic (a, ■) characters are used: (1) Pleon with six (p) or only five free segments (a), (2) second preparatory female stage developed (p) or reduced (a), (3) Mx 1 with an inner and outer endite (p) or only with an outer endite (a), (4) Thoracic glands absent (p) or developed (a), (5) Mandibles monokondylic (p) or dicondylic (a), (6) Mx 2 well developed (p) or reduced (a), (7) First joint of A2 developed (p) or reduced (a), (8) Epignath unilobed (p) or bilobed (a), (9) Coxa of cheliped more or less reduced (p) or lacking (a), (10) Protogyny not developed (p) or developed (a), (11) Cheliped with a reduced ischium (p) or this lacking (a).

However, there exists some doubt on the systematic position of '*Apseudes giganteus*', recently described from Cretaceous (Malzahn 1979). Chelae characters (mainly shape of dactylus and fixed finger) and loss of 'stalked-eyes' indicate that this species is more developed than *Ophthalmapseudes,* and may belong to the Apseudoidea, representing a distinct Mesozoic family.

During the Triassic and Cretaceous the Neotanaidomorpha and Tanaidomorpha also developed. In the evolution of the Neotanaidae, a reconstruction of the respiratory chamber (carapace) took place (Fig.4) which led to the distinct shape of the epignath, which is bilobed and the dorsal lobe covered with a bunch of setae. Other neotanaid synapomorphies are the first three pairs of peraeopods flattened, loss of maxillular palp, and sexual-dimorphism of mouthparts.

In this connection, it should be mentioned that 'dikonophoran' cheliped construction is not a characteristic feature for the Tanaidomorpha and Neotanaidomorpha. Pagurapseudidae, Synapseudidae, and Whitelegginae within the Apseudoidea possess a cheliped having a reduced coxa and swollen basis (Fig.6). In these families the extrinsic muscles of the leg insert laterally on the basis. It also is shown, that the so-called 'coxae' of the Neotanidae are in fact rudimentary sternal plates of the second thoracic segment and not true coxae (Fig.4). Consequently that structure formerly interpreted as a coxa in Anarthruridae (Lang 1971) should be called a 'pseudocoxa' (Sieg in press) (Fig.6).

## 5 EVOLUTIONARY LINES WITHIN TANAIDOMORPHA

This suborder of recent Tanaidacea is noted for the poverty of structural features useful in diagnosing the group. In all tanaidomorph groups we find regressive evolutionary changes (degradation) which mark a loss of complexity in these morphological lines, i.e. homonomous elements are reduced in number and that these reduced numbers are fixed (Sieg 1978). On the other hand, this group shows the greatest variability in sexual-dimorphism, and the main evolutionary trends within the suborder are the alterations in the post-marsupial development. This will be discussed below.

Comparison of body appendages provides little information on phylogenetic changes in pleopods, uropods, and peraeopods. All these structures in tanaidomorphs are very homogenous, and only the latter show some phylogenetic variations mainly in the dactylus and terminal spine (Sieg 1973, 1976).

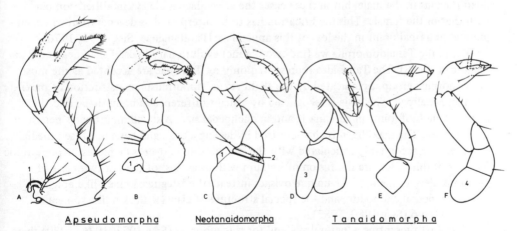

Apseudomorpha    Neotanaidomorpha    Tanaidomorpha

Figure 6. Chelipeds of various tanaidacean taxa. A. *Apseudes setosus* Lang, after Lang 1968; B. *Whiteleggia multicarinata* (Whitelegge), after Lang 1970; C. *Neotanais micromopher* Gardiner, after Gardiner 1975; D. *Leptognathia paraforcifera* Lang, after Lang 1968; E. *Paranarthrura subtilis* Hansen, after Lang 1971; F. *Paragnathotanais typicus* Lang, Lang 1971a. All redrawn and modified. 1 – coxa, 2 – ischium, 3 – side piece, 4 – pseudocoxa.

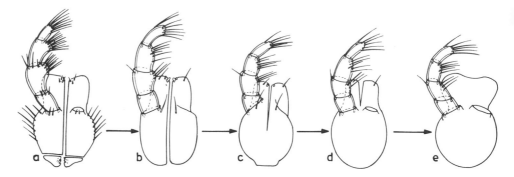

Figure 7. Evolutionary changes in the maxilliped within Tanaidomorpha. A. Tanaidae; B. Pseudozeuxidae/Paratanaidae; C. *Paratanais*/Leptognathiidae/Agathotanaidae; D. Cryptocopinae/Nototanaidae/Anarthruridae; E. Pseudotanainae/Nototanaidae. Redrawn and modified after Sieg 1976.

Some differences are found in the chelipeds. As noted above, the coxa is reduced in this suborder, and the formerly so-called 'coxa' (e.g. Lang 1971, Sieg 1973, 1976) should be named 'Seitenstück' (side-piece). However, in addition, this 'Seitenstück' can be reduced or fused with the carapace and cephalothorax, and instead of the true coxa a secondary segmentation of the basis took place leading to the formation of a pseudocoxa (Sieg in press).

Greater changes take place in the maxillipeds (Fig.7). First the coxae are reduced, then the bases are fused medially, and finally the endites grow together. In this manner, we get the typical plate-like maxilliped (Sieg 1973, 1976, 1980) characteristic of apomorphic Tanaidomorpha (e.g. Pseudotanaidae, etc.).

Of great interest is the loss of mouth-parts in the male phase. In primitive families (Tanaidae, Pseudozeuxidae) there is no structural difference in the mouthparts of the sexes. In apomorphic families (Leptognathiidae, Pseudotanaidae, Nototanaidae) normally the maxilliped remains in the male, but in these cases the appendage is always modified compared with that of the female. This modification has to be interpreted as degeneration which culminates at a final point in the loss of this appendage (Paratanaidae; Sieg 1973, 1976).

Within the Tanaidomorpha we find two distinct evolutionary lines: Tanaoidea and Paratanaoidea (Fig.8). The Tanaoidea represented only by Tanaidae are accepted as the most plesiomorphic group, as indicated by the slight sexual dimorphism, construction of maxilliped and cheliped. The male is recognized by a slight difference in body shape and the relative length of joints in antenna 1, and in chelipedal size. Also, all mouthparts persist in the male sex. The maxilliped is characterized by having a coxa and basis not fused medially. Chelipeds are secondarily articulated with the cephalothorax over 'side-pieces'. Apomorphic features of this group are the fusion of ischium with basis, loss of the last two pairs of pleopods, development of the unique ovisacs instead of oöstegites, kidney-like epignath, loss of uropodial exopodite, and the special structure of claw (with skin-like spines) of peraeopods 4-6 (Sieg 1980, 1982).

When first presenting a 'natural system' for this suborder (Sieg 1973, 1976, 1980) there was a great gap between Tanaoidea and Paratanaoidea. However, this gap was closed recently with the description of a so-called 'connecting link', the Pseudozeuxidae. This family intercedes between Tanaidae and Paratanaidae. The pseudozeuxids possess, not only slight sexual dimorphism in body shape, antenna 1, and chelipeds; but also presence of all mouthparts in the male. On the other hand the maxillipeds are characterized by the loss of the coxae. Further this family also has oöstegites instead of ovisacs, and the ischium is not fused with the basis in the peraeopods (Sieg 1982).

As pointed out above the reduction of the mouthparts in the male (or better: reduction of the male phase, see below!) is the main evolutionary trend within the Paratanaoidea. The first step is the reduction of all mouthparts except maxillipeds, palp of maxilla 1, and the epignath. This can be seen for example in Nototanaidae (*Androtanais, Nototanais*) where we often find rudiments of these structures (Sieg 1980c). The Anarthruridae, Leptognathiidae, Pseudotanaidae, and Agathotanaidae (?, no males are known until now) have only the maxillipeds. Only in Paratanaidae is there a total reduction of maxillipeds (Lang 1973, Sieg 1973, 1976), but the remaining unfused basis of maxillipeds force us to understand the family as the most plesiomorphic after Pseudozeuxidae within the Paratanaoidea.

The lack of sufficient information on the heterogenous Leptognathiidae make it nearly impossible at the moment to give exact phylogenetic relationships between the remaining families, and published schemes (Sieg 1976, 1980) only represent shadows of the truth. There is too much information uncovered by recent investigations which is inconsistent with the present system, and which indicates that a revision is necessary. Some may be mentioned here.

In connection with studies of the New Zealand and Australian tanaidacean fauna, I have seen many samples collected by diving or hand collection, very rich in species composition and/or in amount of specimens, which allowed the recognition of undescribed males of known species in many different genera. For example, there is no doubt any longer that in *Paratanais* the males have only a slightly reduced maxilliped, as published by Kussakin & Tzareva (1972). Each maxilliped in this genus has a broad basis which is fused with the proximal part of the opposite one, and the broad basis shows tendencies towards a fusion of the lateral margins with the carapace. This is typical for males of Leptognathiidae and has been described by G.O.Sars (1896) as an epistome. Also the body shape of males in *Paratanais* is different to that of other Paratanaidae. In this regard, *Bathytanais* is very similar, in that *Bathytanais* is consequently more closely related to *Paratanais* than to *Leptognathia*.

A second point is the heterogeneity of Leptognathiidae. Comparative studies on the peraeopods favors the assumption that the complex genera *Leptognathia* and *Typhlotanais* each represent a different phylogenetic line (probably at the family-level). *Typhlotanais* shows resemblance on one side to *Paratanais* and on the other side to the Pseudotanaidae. In this connection, it might be mentioned that Kudinova-Pasternak (1969) has found in females of *Typhlotanais magnificus* an incomplete set of oöstegites with a large number of eggs. This shows the tendency of reduction of marsupial plates in this group. One line may have led to recent Pseudotanaidae having only one pair of oöstegites on the fourth pair of peraeopods. *Leptognathia* in comparison shows affinities to Agathotanaidae. In many *Leptognathia* species the chelipeds are shifted forward and medially. From this situation the chelipedial type of Agathotanaidae can be easily derived.

The two remaining families Nototanaidae and Anarthruridae show resemblances to each other, based mainly on the morphological structures in the male. Of which the most important is the different body shape of males to that of Leptognathiidae. In this regard, the two families are related to Paratanaidae.

So, if we are to present a phylogenetic scheme of this suborder which might contain more truth than the former ones, it should perhaps look like that of Figure 8.

240 *Jürgen Sieg*

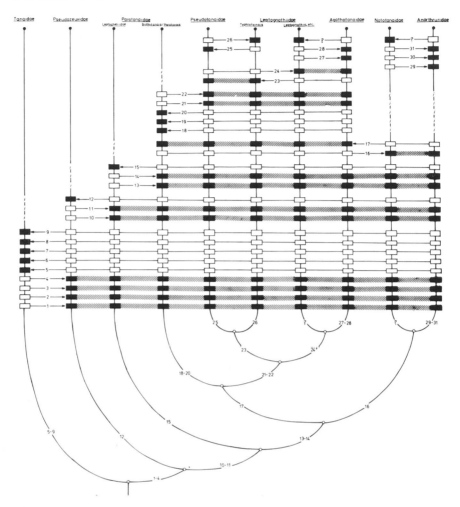

Figure 8. Suggested phylogenetic relationship within Tanaidomorpha. The following plesiomorphic
(p, □) and apomorphic (a, ■) characters are used: (1) Lacinia mobilis of right mandible present (p) or
absent (a), (2) Mandibles with biciliate setae very close to lacinia mobilis (p) or such setae lacking (a),
(3) Maxilliped with coxa (p) or the coxa lacking (a), (4) Epignath kidney-shaped (p) or small (a), (5)
Respiration current running from caudal to rostral (p) or from rostral to caudal (a), (6) Peraeopods 1-6
with (p) or without ischium (a), (7) Marsupium formed by oöstegital sheets (p) or represented by
ovisacs (a), (8) Uropods biramous (p) or uniramous (p), (9) Claw of peraeopods 4-6 without thin spines
(p) or with such spines (a), (10) A1 flagellum equal in both sexes (p) or different (a), (11) Mouth-parts
in the male well-developed (p) or reduced (a), (12) Pleon developed normally (p) or all pleonites
reduced (a), (13) Basis of maxilliped unfused (p) or fused (a), (14) Endopodite of uropods at least 3-
jointed (p) or only 2-jointed (a), (15) Mouthparts in the male except maxilliped more or less reduced
(p) or maxilliped also reduced (a), (16) Basis of maxilliped partly fused (p) or totally fused (a), (17)
Body shape of male vary little from female (p) or extremely different (a), (18) Basis of maxilliped developed normally (p) or enlarged (a), (19) Fourth point of A1 shorter (p) or longer (a) than second and
third joint combined, (20) Second and third joint of A2 cylindrical (p) or triangular (a), (21) Mandibles
and endites of Mx 1 existing in the male (p) or reduced (a), (22) Flagellum of A1 in males at least 6-
jointed (p) or only 4-jointed (a), (23) Dactylus of peraeopods 4-6 with a long terminal spine (p) or both
forming more or less complete claw (a), (24) Chelipeds in lateral position (p) or shifted anteriorly and
medially (a), (25) Marsupium formed by 4 (p) or 1 pair (a) of oöstegites, (26) Spines of merus, carpus,

# 6 EVOLUTIONARY LINES WITHIN RECENT APSEUDOMORPHA (= APSEUDOIDEA)

In the Apseudoidea we find a totally different adaptive radiation than in Tanaidomorpha. Instead of evolutionary changes in mouthparts accompanied by sexual dimorphism, we find changes in the peraeopods. Therefore, sexual dimorphism is less pronounced in the apseudoideans. However, it should be mentioned again, this is only a first attempt at a phylogenetic system, and it is clear that there will be changes in the future.

Originally, the six pairs of peraeopods were subdivided into a group of three anteriorly directed and three posteriorly directed pairs. This primary subdivision is also found in all recent Tanaidacea but is superimposed on secondary subdivisions only in Apseudoidea.

We have to demand as a plesiomorphic feature a rod-like walking leg type for all peraeopods. This is realized, however, only in Tanaidomorpha and Gigantapseudidae in the recent Tanaidacea (Fig.9a). In Neotanaidomorpha and remaining Apseudoidea at least one pair of peraeopods (usually the P.1) is always transformed.

Lang (1971) has shown that these transformations led to different leg types: digging type (fossorial), climbing leg, and the rod-like walking leg. He also demonstrated that there exist various combinations of rod-like types with the fossorial type (*Parapseudes* type, *Apseudes setosus* type, *A.tuberculatus* type; Fig.9e). There seems to be little doubt that the evolution of the fossorial leg is monophyletic. The original rod-like form is increasingly flattened and thereby transformed slowly to the characteristic digging shape. The climbing leg in comparison seems to be at least diphyletic. It will be shown that this type of leg must have derived from the rod-like and also from the fossorial type.

Therefore, in discussing evolutionary changes in peraeopods, we have to reconstruct the different steps which led to a situation in which there are up to six different pairs of legs in one individual.

It would appear that the plesiomorphic state is one in which there are groups of peraeopods (P.1-P.3/P.4-P.6), are all rod-like but only slightly differentiated, though homonomous in each group. Therefore, Gigantapseudidae (Fig.9a) has to be accepted as the most plesiomorphic taxon, because all six pairs of peraeopods are rod-like and the cheliped is of the normal Apseudoidea type.

The first evolutionary change which took place was the transformation of the P.1 to a fossorial leg. That means that we have to subdivide the remaining Apseudoidea in two groups: one having the P.1 as a fossorial type and one having it as a rod-like or climbing type. It is also clear, that other changes took place afterwards, so we have to discuss the evolution within each of these groups separately.

Having a relatively unmodified P.1 is true only for Metapseudidae (Metapseudinae, Synapseudinae, Tanzanapseudinae) and Pagurapseudidae. Therefore, these two families have to be accepted as the sister group of all other Apseudoidea. This is supported by another synapomorphy. In both families, except some Metapseudinae (see below), we find a cheliped characterized by the reduction of the coxa (Fig.6b).

---

and propodus developed normally (p) or changed (a), (27) Cheliped with (p) or without side-piece (a), (28) Outer edge of labium without (p) or with (a) a tip, (29) Flagellum of A1 in the male 2- or 3-jointed (p) or 1-jointed (a), (30) In addition to remnants of mandibles and endites of Mx 1, only slightly reduced maxilliped is found in the male (p) or these structures are lacking, and palpus of maxilliped is reduced (a), (31) Chelipeds without (p) or with pseudocoxa (1).

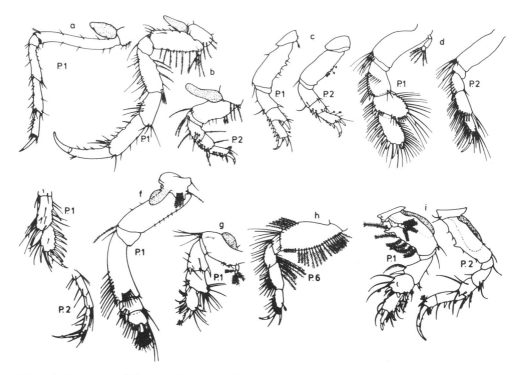

Figure 9. Leg types and their combination within Apseudoidea. A. P.1 of *Gigantapseudes adactylus* Kudinova-Pasternak 1978 (redrawn from Kudinova-Pasternak 1978); B. P1/P2 of *Pagurapseudes laevis* Menzies 1940; C. P1/P2 of *Metapseudes aucklandiae* Stephensen 1927 (redrawn from Lang 1970); D. P1/P2 of *Discapseudes holthuisi* Bacescu & Gutu 1975; E. P1/P2 of *Apseudes tuberculatus* Lang 1968 (redrawn from Lang 1970); F. P1 of *Kalliapseudes bahamensis* Sieg 1982; G. P1 of *Cirratodactylus floridensis* Gardiner 1973 (redrawn from Gardiner 1973); H. P6 of *Discapseudes holthuisi* Becescu & Gutu 1975; I. P1/P2 of *Calozodion wadei* Gardiner 1973 (redrawn from Gardiner 1973). Legs of the same species enlarged identically (drawings from original unless otherwise noted).

The next evolutionary step in the peraeopods led to the unique shape of P.2-P.6 in Pagurapseudidae, which are an interesting analogy to those of Paguridae. Only the P.1 is represented by the plesiomorphic rod-type, all others (Fig.9b) are transformed to legs remarkably developed to fix their possessor to the smooth walls of gastropod shells. Messing (1980) has given extended information on this fact. Contrary to this, in Metapseudidae (except *Cyclopoapseudes* and *Apseudomorpha*) all six pairs of peraeopods are transferred into the climbing type. The legs are all very similar so that even the primary anterior-posterior subdivision is obliterated (Fig.9c).

Metapseudidae consequently have to be redefined. *Cyclopoapseudes* and *Apseudomorpha* (all species?) have to be excluded, and the remaining genera should be grouped into the subfamilies Metapseudinae (*Cryptapseudes, Curtipleon, Metapseudes, Synapseudes*) and Tanzanapseudinae (*Tanzanapseudes*).

Within the second evolutionary line (Fig.10), in which the P.1 is a fossorial limb, several differentiations of the succeeding peraeopods took place.

The next step of differentiation in shape of peraeopods is found in P.6, a fact overlooked up to now. In Kalliapseudidae, Cirratodactylidae, Anuropodidae, several Apseudidae (in part, e.g. *Discapseudes, Halmyrapseudes, Calozodion*) and in *Cyclopoapseudes/Apseudomorpha* we find a P.6 characterized by flatness, tergal as well as sternal border of basis, merus, and carpus covered with long, bicilate setae, and the propodus with a long row of tergal spines (Fig.9h). In addition, the dactylus and terminal spine are well developed and form a characteristic claw.

In contrast to this, we find the rod-like type in P.6 in Sphyrapidae, Tanapseudidae, Apseudellidae, Leviapseudidae and several Apseudidae (e.g. *Apseudes setosus, Apseudes tuberculatus*), which is of course a plesiomorphic feature.

Within the above mentioned families, *Cyclopoapseudes/Apseudomorpha* and *Calozodion* are recognized as foreign elements because these taxa do not have the P.1-P.3 modified as fossorial legs, but rather as climbing legs. This was the reason Lang (1971) included the first two genera in Metapseudidae. In comparing the P.1, however, of *Cyclopoapseudes* and *Metapseudes* (Fig.9c,i), we see structural differences (more flattened, shape of claw) which force us to accept a separate development of these legs. In connection with the knowledge of structure of P.6, it is now clear that in these three genera the climbing type P.1 was derived from the fossorial type and not, as in Metapseudidae, from the rod-like type.

Within the remaining families having a modified P.6 the next evolutionary stage is found in P.2/P.3, and was recognized by Lang (1971) though he was not able to understand the phylogenetic importance of this. The author showed, that there exist genera and species in which the P.2/P.3 is flattened (e.g. *Parapseudes,* Fig.9d) and others in which the P.2/P.3 are rod-like (*Apseudes setosus* and *A.tuberculatus* type, Fig.9e). This situation can only be interpreted as a continuation of a process which had led to a fossorial limb in P.1. Because of the phylogenetic importance of the structure of P.6 we have to accept for now its parallel evolution in, for example, Cirratodactylidae/Kalliapseudidae and Sphyrapidae.

All remaining families having a modified P.6 (Anuropodidae, Apseudidae part., Cirratodactylidae, Kalliapseudidae) have these flattened P.2/P.3. Therefore, they have to be suggested as the sister group of *Cyclopoapseudes/Calozodion*, and which can be subdivided into two groups.

One phylogenetic line is represented by Cirratodactylidae (Fig.9g) and Kalliapseudidae (Fig.9f). These two families are characterized by many synapomorphies of which the most important are the loss of the maxillary palp and the transformation of dactylus with terminal spine into a sensory organ.

In contrast to these, Anuropodidae, and one part of the Apseudidae are plesiomorphic. It should be mentioned here that I think that Anuropodidae at least represent only a subfamily and not more. Because a degressive feature like the loss of an appendage, in this case the uropods, is not convincing enough to establish a family.

Within the other families having no modified P.6, the differentiation of P.2/P.3 allow again a subdivision.

In Apseudellidae, Leviapseudidae, and a second part of Apseudidae have the plesiomorphic rod-like limb in P.2/P.3. In contrast to these, Sphyrapidae, Tanapseudidae, and a third part of Apseudidae have these legs flattened (similar to P.1).

Within these two subgroups, only a little can be stated about the phylogenetic relationship, but it seems that Apseudellidae is the sister group of Leviapseudidae/Apseudidae (in part).

Putting this puzzle together, we get the following phylogeny outlined in Figure 10. But it should be remembered that this is only a first attempt at a 'natural system' of the Apseu-

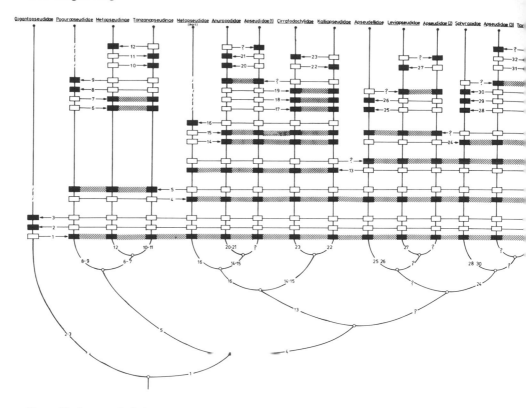

Figure 10. Suggested phylogenetic relationship within Apseudoidea. (1) All peraeopods cylindrical (p) or at least one peraeopod (usually P1) modified (a); (2) Dactylus of P4 developed normally (p) or reduced (a); (3) Inner endite of Mx 1 with strong setae (p) or with fine setae (a); (4) P1 cylindrical or modified as a climbing leg (in these cases P6 identical to P1) (p) or P1 fossorial (sometimes it is transformed secondarily to a climbing leg, but then P6 is different from P1) (a); (5) Cheliped with elongated basis ('Apseudes'-type) (p) or thickened ('Dikonophoran'-like) (a); (6) P1 cylindrical (p) or transformed to climbing-leg (a); (7) All pleonites developed normally (p) or more or less reduced (a); (8) P2-P6 without (p) or with sucker-like structures on merus to propodus (a); (9) Uropods with normal (p) or with strong, spine-like setae (a); (10) Peraeonites without (p) or with strong lateral expansions (a); (11) Pleonites at least partly (p) or totally fused (a) in dorsal view; (12) Exo- and endopods of pleopods bisegmented (p) or only unisegmented (a); (13) P6 similar to P4/P5 (p) or transformed (a); (14) Pleotelson with lateral indenture indicating sixth pleonite (p) or without such an indenture (a); (15) P2/P3 different (p) or similar to P1 (a); (16) P1 fossorial (p) or transformed to a climbing-leg (a); (17) Dactylus and terminal spine of peraeopods developed normally (p) or transformed (a); (18) Mx 1 with (p) or without (a) palp; (19) Sexual dimorphism less (p) or strongly (p) pronounced in chelipeds; (20) Uropods developed (p) or lacking (a); (21) Flagella of A1 and A2 multisegmented (p) or reduced (a); (22) Dactylus and terminal spine of peraeopods with curls (p) or at least P1 with sensory organ (a); (23) Maxilliped with (p) or without (a) coxa; (24) P2/P3 different (p) or similar to P1 (a); (25) Maxillular palp present (p) or absent (a); (26) Epignath large ('Apseudes'-type) (p) or similar to that of Tanaidomorpha (a); (27) Inner distal seta of maxillipedal endite developed normally (p) or transformed (a); (28) First peraeonite more or less coalesced (p) or fused (a) with the cephalothorax; (29) P1 without (p) or with (a) sexual dimorphism; (30) A2 with (p) or without (a) squama; (31) A1 with (p) or without (a) an inner flagellum; (32) Armament of carpus and propodus of P1 developed normally (p) or reduced (a).

doidea, and that it contains certain inconsistencies. One point may be mentioned here. Lang (1971) postulated that the transformation of inner disto-caudal seta of the maxillipe-dal endite is monophyletic. It would seem, however, that we have to express certain doubts concerning this fact. In the scheme presented here, Lang's opinion is accepted and, there-fore, the Whiteleggiinae are thus considered as a subfamily of Leviapseudidae; but White-leggiinae are totally different in respect to shape of cheliped and also of P.1 from Leviap-seudinae. There is a great similarity with the chelipedial type known from Metapseudidae/ Pagurapseudidae. If the transformation of the maxillipedal endite setae is polyphyletic, we only have to accept that a fossorial limb has developed twice. In my opinion, the P.1 repre-sented by *Whiteleggia/Pseudowhiteleggia* is more related to the climbing limb than to the original fossorial type.

Finally it should be noticed that this interpretation of the phylogeny of Apseudoidea makes it necessary to split Apseudidae into three families. However, this will be done in another paper after more detailed studies of *Apseudes* limb types are completed.

# 7 EVOLUTION AND DISTRIBUTION

Distribution of a species depends partly on the necessary conditions which must exist in a particular habitat, and the ability to reach such habitats. Since the oceans and seas normally have similar conditions over long distances, a species which is more or less planktonic or which has planktonic larvae naturally will settle in larger areas than one which is benthic or sessile. Also one must take into consideration that there exists many opportunities for additional distribution through the agency of man, so that often one can only discuss spe-cies distribution in relation to recent ecological factors and not from a historical point of view. One exception to this — besides Bathynellacea and Cephalocarida — is the order Tanaidacea.

Biological researches have shown that these animals are mostly tube-building species that seldom leave their tubes. This is true of all stages, including juvenile animals (sexually im-mature) which brood in the parent tubes. When building a new tube, juveniles bore through the wall of the 'parent tube' and build their own nearby (Bückle-Ramirez 1965). Therefore, the tanaids in an area have spotty distribution with high population densities. Recent inves-tigations in the Mediterranean Sea (Ischia, Gulf of Naples; Valentin 1978 and personal com-munication) also indicate that specimens seldom occur in the night plankton. This agrees with results from other regions (Hobson & Chess 1976, 1978) and show that tanaids have a different behavior to that of say the Cumacea. We have to suspect that it takes a great deal of time to extend the area of distribution of tanaids over greater distances. For this reason, it is possible to discuss the distribution patterns within the Tanaidacea from a his-torical point of view.

Unfortunately, the systematics within Apseudoidea are so unstable at the moment that we cannot give any information of distribution above species level. However, this can only be done within certain limits in Neotanaidomorpha and Tanaidomorpha.

Reviewing the literature, it is easily found that all families of the mentioned suborders are more or less distributed worldwide. This is true, even if in some cases we do not have any record for a particular region at present. Therefore, only at the family level is discus-sion of vertical distribution possible. This is of interest, because it may allow us to evaluate matters referring to deep-sea settlement within Tanaidacea (Fig.11).

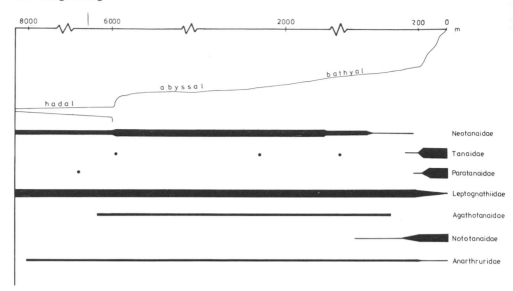

Figure 11. Vertical distribution of different tanaidacean families.

Vertical distribution of the plesiomorphic Neotanaidae, the only family of the suborder Neotanaidomorpha, was recently discussed extensively by Gardiner (1975) in an excellent monograph. The range for the family is 223 to 8 300 m and indicates it as a deep-sea taxon. Even the relatively shallow location of *Neotanais laevispinosus* (677 m), *N. tricarinatus* (587-805 m), *N. armiger* (598 m) and *N. antarcticus* (223-390 m) do not hinder such an interpretation because the mean depth of all the stations from which the family has been collected is 3 110 m. Unfortunately, Gardiner has given no extensive information that would suggest phylogenetic relationships within the family. Species are only combined into species groups. Consequently, we are not able to clarify if a species showing bathyal distribution is plesiomorphic or not. In addition, there exist no criteria which indicate from which part of the world the settlement of the deep-sea took place, e.g. if the plesiomorphic species only occur in the antarctic or subantarctic regions, e.g. as in Tanaidae.

Little has been published about vertical distribution in Tanaidomorpha. Recently it was shown (Sieg 1980, 1980a) that Tanaidae are a typical shallow water family. There exist few records that mention this family from depths greater than 1 000 m. *Synaptotanais abyssorum* has been described from 1 301 m at Indonesia (Nierstrasz 1913), *Protanais birsteini* from 6 090-6 135 m in Kurile trench (Kudinova-Pasternak 1970), and Shiino (1978) mentioned two species of *Langitanais* caught at about 3 000 m (*L. willemoesi* – 2 925 m, *L. magnus* – 3 025 m). However, normally the members occur from the intertidal up to 200 m and are characteristic inhabitants of algal mats, hydroid colonies, and such like.

Paratanaidae and Nototanaidae are also only known from shallow waters. Records for depths greater than 200 m are scarcely found. Mentioned here are the paratanaids *Heterotanoides ornatus* (Japanese trench: 7 370 m; Kudinova-Pasternak 1977), *Paratanais euelpis* (South African coast: 229-230 m; Barnard 1920); and the nototanaids *Nototanais dimorphus* (Kerguelen Islands: 232,4 m; Beddard 1886), as well as *Tanaissus lilljeborgi* (?) (southern Atlantic Ocean: 1 105, 1 660 m; Kudinova-Pasternak 1975). Checking the ecological data we find, in contrast to Tanaidae, that the members of these families inhabit normally sandy (Paratanaidae) or muddy (Nototanaidae) bottoms. If we find specimens in algal mats, these are characterized by a great deal of detritus and/or sand. Tanaidae and Paratanaidae/ Nototanaidae have nearly the same depth range, but settle different ecological niches.

The remaining families Agathotanaidae, Anarthruridae, Leptognathiidae, and Pseudo- tanaidae mainly occur in deeper waters.

This is true for Anarthruridae only within certain limits. The family has been subdivided into two subfamilies (Sieg 1978) of which Nesotanainae occur in very shallow waters (tidal regions), and Anarthrurinae in depths greater than 100 up to 8 015 m (*Anarthruropsis langi;* Kudinova-Pasternak 1977).

None of the other three families has been recorded from tidal regions. Pseudotanaidae are known down to 6 850 m (*Cryptocopoides arctica;* Kudinova-Pasternak 1978), but members of this family are mainly caught in depths between 50-300 m. Agathotanaidae, at the moment, have not been found in depths less than 1 281 m (*Agathotanais ingolfi;* Bar- nard 1920). The greatest known depth for this family is 6 890 m (*Paragathotanais typicus;* Kudinova-Pasternak 1978).

Finally, the Leptoganthiidae are known from about 30 m down to hadal depths (*Lep- tognathia elegans:* 8 006 m; Kudinova-Pasternak 1965). It seems to be a 'typical' deep-sea taxon. As in Nototanaidae, ecological data from shallow water stations indicate that forms like *Leptognathia* and *Typhlotanais* nearly always occur on muddy bottom.

How might these data be interpreted from an evolutionary point of view? I would point out that only a preliminary hypothesis can be advanced at this time, but this may be tested by additional work on phylogenetic relationships.

The known vertical distributions of the different families indicate that the deep-sea has been settled by Tanaidacea several times, and that the abyssal fauna is a composite of ple- siomorphic (archaic) and apomorphic (young) taxa. We do not know anything about Apseu- doida but this superfamily has settled the deep-sea at least two times (Gigantapseudidae,

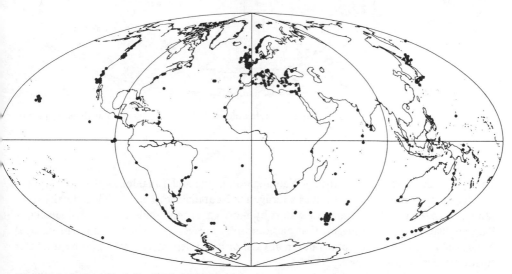

Figure 12. Geographical distribution of Tanaidae.

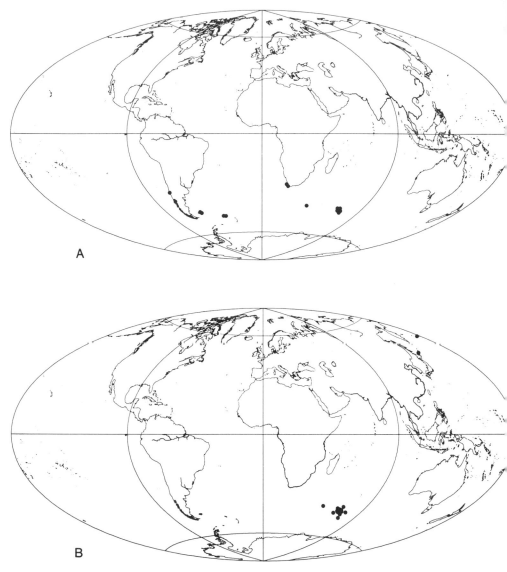

Figure 13. Geographical distribution of archaic subfamilies of Tanaidae. A. Archaeotanainae; B. Langitanainae.

Leviapseudidae). This is also true for Neotanaidomorpha and Tanaidomorpha.

The deep-sea must be suggested as a refuge for Neotanaidae. This phylogenetically old family has been replaced in shallow waters by more advanced families, e.g. Tanaidae and Paratanaidae. This would suggest that plesiomorphic species are to be found in the bathyal or abyssal zone and apomorphic ones in abyssal or hadal depths, something to be tested in future studies.

The last mentioned suggestion is true on the family level within Tanaidomorpha. The

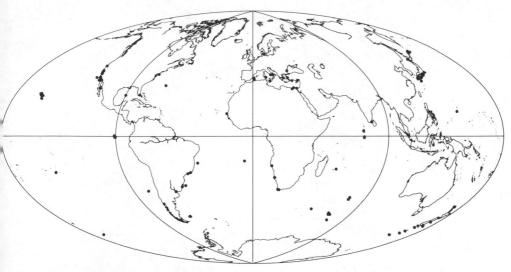

Figure 14. Geographical distribution of Pancolinae.

plesiomorphic families Tanaidae and Paratanaidae occur from the tidal zone down to about 100-200 m, each occupying a different ecological niche.

The apomorphic families Leptognathiidae, Pseudotanaidae, and Agathotanaidae do not occur in the tidal zone, but inhabit muddy bottoms of deeper waters. The preference for muddy bottom must be thought of as a prior condition for deep-sea settlement. It would thus appear that Leptognathiidae are a starting point for this dispersion, followed by Agathotanaidae and Pseudotanaidae.

As mentioned above, geographic distribution only can be discussed within a family when there is sufficient knowledge concerning the phylogenetic relationships of the subordinated taxa. This is applicable only in two families (Pseudotanaidae, Sieg 1977; Tanaidae, Sieg 1980). Of these, horizontal distribution of the latter has been discussed recently (Sieg 1980a). Therefore, a short overview is merely given here. There is no doubt that the Tanaidae is a 'shallow-water' family distributed worldwide between the polar circles, even if there exist few records referring to depths greater than 200-300 m (*Synaptotanais abyssorum, Protanais birsteini, Langitanais magnus, L.willemoesi;* Fig.12).

In lower taxa, e.g. subfamilies, the maps have different appearances. The phylogenetically old ones, Archaeotanainae and Langitanainae, are only recorded from cold water areas. The Archaeotanainae (Fig.13a) occur in antarctic cold waters or in the antarctic transition area. The center of the Langitanainae (Fig.13b) probably is found in the same region. But there is an isolated occurrence in the northern Pacific (Kurile Islands, Kiska Island) of *Arctotanais alascensis* which is, in comparison to *Langitanais,* an apomorphic taxon. This may mean that the ancestors have migrated from the south to the north along the west coast of America. An explanation for the phenomenon may be found in Pancolinae.

At first glance Pancolinae (Fig.14) show a distribution not unusual to that of the family. Even the two attached tribes Pancolini and Anatanaini do not show a different distribution pattern. However, it looks quite different when examining the record at the generic level. The plesiomorphic genus *Zeuxoides* (Fig.15a) again only occurs in antarctic cold waters or

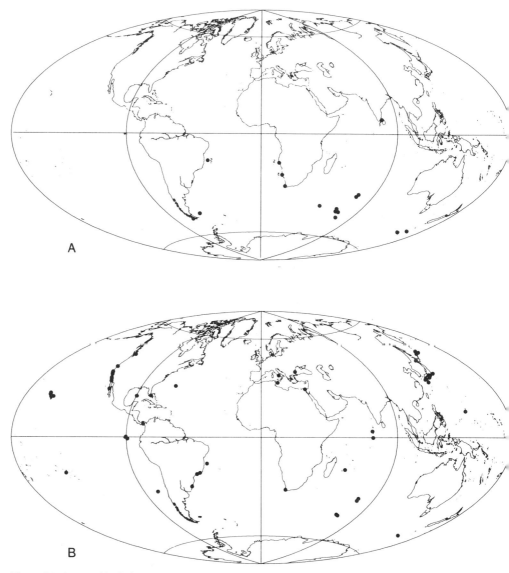

Figure 15. Geographical distribution of subordinate taxa of Pancolinae. A. *Zeuxoides;* B. *Zeuxo.*

in subantarctic regions. In comparison, *Zeuxo* (Fig.15b) can be regarded as the typical taxon of the family in tropical warm waters. The genus contains many species which are unfortunately not easily classified and so at the moment there is little known about the distribution of the different species.

This is possible for *Anatanais* with some caution. This genus contains — at the moment — three species: *A.novaezealandiae, A.lineatus* and *A.pseudonormani.* Evaluation of different characters leads to the opinion that *A.novaezealandiae* is the most plesiomorph species, followed by *A.lineatus* and *A.pseudonormani.* Their relationship is correlated with their

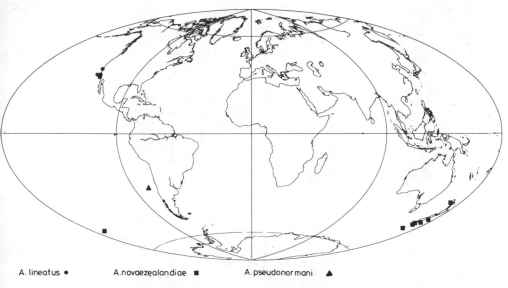

A. lineatus ●       A.novaezealandiae ■       A. pseudonormani ▲

Figure 16. Geographical distribution of species of the genus *Anatanais*.

distribution (Fig.16). The genus is only recorded from the west coast of America (north-wards to southern California) and southern New Zealand. Of this, *A.novaezealandiae* occurs only near the subantarctic islands of New Zealand. *A.lineatus* is known from Masitierra, Juan Fernandez, and *A.pseudonormani* from southern California (Sieg 1980:147-162).

## 8 EVOLUTION AND SEXUAL DIMORPHISM

Sexual dimorphism is something tanaids are especially known for. As in many other cases, this 'typical feature' occurs only in some groups and has developed in various taxa to very different degrees. Sexual dimorphism is correlated with postmarsupial development.

Originally, it appears that there was no sexual dimorphism in tanaids. Such a lack of di-morphism is rare in recent groups, but is encountered frequently in Apseudoidea and some Tanaidae, e.g. *Allotanais* and *Singularitanais* (Sieg 1980).

The first step towards sexual dimorphism was the enlargement of the male cheliped. Even if we do not know the biological reasons, it can be observed in many groups. Because mouthparts have not yet been reduced in this stage, a continuous growth of male cheliped interrupted by moultings can be observed. Messing (1980) was able to show this in an exemplary manner for *Pagurapseudes*.

The second step in evolution of sexual dimorphism (Fig.17a) is represented by the addi-tional elongation of the first antennae. The enlargement of the chelae probably led to pro-longation of these segments. This can be verified in Tanaidae both in different species and in the postmarsupial development of a certain species. In several genera (e.g. *Zeuxo* and *Tanais*) we find mature males with the first antenna little elongated and with relatively small chelipeds; in contrast to several other species, e.g. *Sinelobus stanfordi,* where we find strong dimorphism. In the latter species as well as in several members of the genus *Zeuxo*

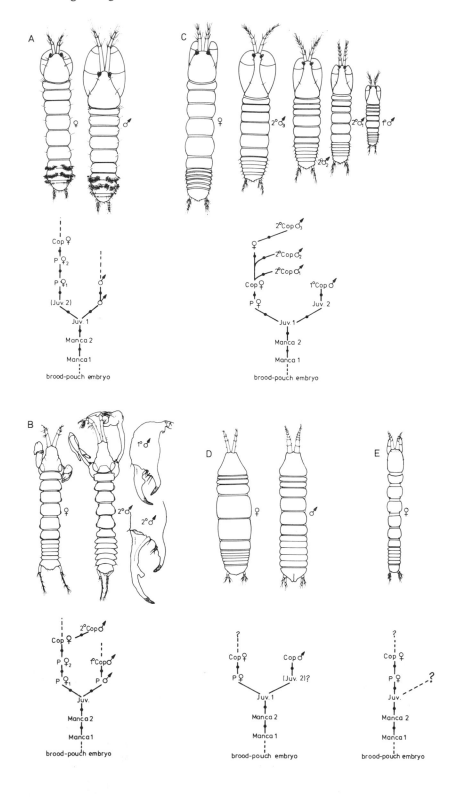

this allometrical growth of first antenna and cheliped can be observed in the postmarsupial development of males. This type of sexual dimorphism is characteristic for Tanaidae. It is scarcely found in Apseudoidea, where normally in apseudoids the first antenna is only little elongated.

In Neotanaidae and other Tanaidomorpha (i.e. excluding Tanaidae) sexual dimorphism is marked by striking differences (Fig.17b,c). Most important is a reduction of mouthparts accompanied by multiplication of terminal segments of first antenna in the male. Additional changes can be noted in the shape of cephalothorax and in the proportions of peraeon segments. Furthermore, the reduction of mouthparts forces shortening of postmarsupial development. Comparative morphology indicates Neotanaidae and Paratanaoidea represent two different phylogenetical lines, so parallel evolution of protogyny must be accepted. Sex is determined by androgenic glands (Charniaux-Cotton et al. 1966), and such fluctuating hormone levels thus easily explains the evolution of protogyny. Reduced mouthparts ensure that the life of mature or primary males is very short compared with that of the females. This led to problems in fertilization. Protogyny was chosen as one solution to this problem. To guarantee a second brood, or fertilization of those females not previously inseminated initially, secondary males are produced by the transformation of females. Initially, secondary males are only derived from copulatory females as in Neotanaidae (Cop ♀ 1, Gardiner 1975). In neotanaids, protogyny led to only two types of males, recognized by different chela forms (Fig.17b).

In Paratanaidae, protogyny is much more complicated and leads to the highest degree of male diversity (Fig.17c). Here species are found which have up to four different males (e.g. *Heterotanais oerstedi, Leptochelia dubia*). Bückle-Ramirez (1965), who has cultured *Heterotanais oerstedi*, was able to show that there exist three ways to produce secondary males (Fig.17c). These are classified as type 1 (from unfertilized females), type 2 (from females, which had a brood), and type 3 (from females, which had a brood and have moulted to a habitual female, which in *Heterotanais oerstedi* is characterized by a slight blue color). Other differences in the first antenna (multisegmented flagellum) allows that males of the same species are often more distinct in regard to each other than females of separate species. Another phenomenon observed is the reduction in the number of preparatory female phases (P ♀ 2, Gardiner 1975) in Tanaidomorpha. Increase of oöstegite sheets at the end of this preparatory female phase may be interpreted as a recapitulation of a time when there were two preparatory female stages.

In more apomorphic families there was a reduction of the male phase. Leptognathiidae and Pseudotanaidae have only one kind of male which is totally different from the female and possibly represents the primary male of protogynous families (Fig.17d). In this case, the male is normally much smaller, the first antenna is multisegmented, mouthparts are reduced (except for the maxilliped), and all proportions of segments are totally different to

Figure 17. Postmarsupial development and sexual dimorphism in Tanaidacea. A. Gonochoristic type as demonstrated in Pagurapseudidae by Messing 1980; typical for Apseudoidea, in Tanaidae the second juvenile stage may be lacking, therefore, it is written in parenthesis. B. Hermaphroditic type of Neotanaidae characterized by producing two different type males (after Gardiner 1975). C. Hermaphroditic type of Paratanaidae characterized by producing four different type males (from Sieg 1978a, after Bückle-Ramirez 1965). D. Gonochoristic type as suggested for Leptognathiidae or Pseudotanaidae. E. Parthegonetic type as suggested for several Leptognathiidae (?) and Agathotanaidae (?).

those of the females. Sometimes males bear pleopods even when females lack them.

Finally, certain Paratanaoidea present a special case. In many species males have never been recorded (Fig.17e), for which it seems justified to accept the loss of the male phase. Consequently, for these species parthenogenesis must be suggested. On the family level, this may be true for Agathotanidae.

## 9 ACKNOWLEDGEMENTS

This work was partially supported by grants from the Deutsche Forschungsgemeinschaft, whose help I gratefully acknowledge. The ideas contained herein are the result of ten years of ongoing study, and has profited by discussions with innumerable colleagues. Special thanks are due to Dr F.R.Schram who critically read the manuscript and offered many helpful suggestions as to content and grammatical expressions.

## REFERENCES

Barnard, K.H. 1920. Contribution to the Crustacean Fauna of South Africa. *Ann. S.A. Mus.* 17:319-438.

Beddard, F.E. 1886. Report on the Isopoda collected by HMS 'Challenger' during the years 1873-1876. *Challenger Rept. (Zool.)* 17:1-78.

Belyaev, G.M. 1968. The ultra-abissal bottom fauna. In: L.A.Zenkevitch (ed.), *The Pacific Ocean. Biology of the Pacific Ocean, Part 2:*217-238. Moscow: Academy of Sciences, USSR.

Boas, J.E.V. 1883. Studien über die Verwandtschaftsbeziehungen der Malakostraken. *Gegenb. Morph. Jb.* 11:112-116.

Bückle-Ramirez, L.F. 1965. Untersuchungen über die Biologie von *Heterotanais oerstedti* (Krøyer). *Z. Morph. Ökol. Tiere* 55:711-782.

Calman, W.T. 1909. The Crustacea. In: R.Lankester (ed.), *A Treatise on Zoology:*190-195.

Charniaux-Cotton, H., C.Zerbib & J.J.Mensy 1966. Monographie de la glande androgène des Crustacés supérieurs. *Crustaceana* 10:113-136.

Dahl, E. 1963. Main evolutionary lines among recent Crustacea. In: H.B.Whittongton & W.D.I.Rolfe (eds), *Phylogeny and Evolution of Crustacea:*1-15. Cambridge: Mus. Comp. Zool.

Dahl, E. 1976. Structural plans as functional models exemplified by the Crustacea Malacostraca. *Zool. Scripta* 5:163-166.

Dahl, E. 1977. The Amphipod functional model and its bearing upon systematics and phylogeny. *Zool. Scripta* 6:221-224.

Dahl, E. 1980. Alternatives in Malacostracan evolution. In: D.J.G.Griffin (ed.), *International Conference. Biology and Evolution of Crustacea.* Sydney: Australian Museum.

Fryer, G. 1964. Studies on the functional morphology and feeding mechanism of *Monodella argentarii* Stella (Crustacea: Thermosbaenacea). *Trans. Roy. Soc. Edinb.* 66:49-90.

Gage, J.D. & G.G.Goghill 1977. Studies on the dispersion patterns of Scottish sea loch benthos from contiguous core transects. In: B.C.Couel (ed.), *Ecology of marine benthos:*319-337. Columbia: Univ. South Carolina Press.

Gardiner, L.F. 1972. A new genus of a new monokonophoran family (Crustacea: Tanaidacea), from southeastern Florida. *J. Zool.* 169:237-253.

Gardiner, L.F. 1973. A new species of the genera Synapseudes and Cycloapseudes with notes on morphological variation, postmarsupial development, and phylogenetic relationships within the family Metapseudidae (Crustacea: Tanaidacea). *Zool. J. Lin. Soc.* 53:25-58.

Gardiner, L.F. 1973a. *Calozodion wadei,* a new genus and species of apseudid tanaidacean (Crustacea) from Jamaica, West Indies. *J. Nat. Hist.* 7:499-507.

Gardiner, L.F. 1975. The systematics, postmarsupial development, and ecology of the deep-sea family Neotanaidae (Crustacea: Tanaidacea). *Smithson. Contr. Zool.* 170:1-264.

Grassle, J.F. & H.L.Sanders 1973. Life histories and the role of disturbance. *Deep-Sea Res.* 20:643-659.

Hessler, R.R. 1980. Structural morphology and evolution of walking mechanisms in the Eumalacostraca. In: D.J.D.Griffin (ed.), *International Conference. Biology and evolution of Crustacea.* Sydney: Australian Museum.

Hobson, E. & J.R.Chess 1976. Trophic interactions among fishes and zooplankters near shore at Santa Catalina Island, California. *Fish. Bull.* 74:567-598.

Hobson, E. 1978. Trophic relationships among fishes and plankton in the lagoon at Enewetak Atoll, Marshall Islands. *Fish. Bull.* 76:133-153.

Jumars, P.A. 1976. Deep-sea species diversity: Does it have a characteristic scale? *J. Mar. Res.* 34:217-246.

Kudinova-Pasternak, R.K. 1965. Deep-sea Tanaidacea from the Bougainville Trench of the Pacific. *Crustaceana* 8:75-91.

Kudinova-Pasternak, R.K. 1969. A case of extramarsupial development of eggs in *Thyphlotanais magnificus* (Crustacea: Tanaidacea) living in the tube. *Zool. Zhur.* 48:1737-1738.

Kudinova-Pasternak, R.K. 1970. Tanaidacea kurilo-kamciatkogo jeloba. *Trudy Inst. Okeanol.* 86:341-380.

Kudinova-Pasternak, R.K. 1975. Tanaidacea (Malacostraca) of the deep-sea trench Romansch and Guinea hollow. *Zool. Zhur.* 54:682-687.

Kudinova-Pasternak, R.K. 1978. Gigantapseudidae fam.n. (Crustacea, Tanaidacea) and composition of the suborder Monokonophora. *Zool. Zhur.* 57:1150-1161.

Kudinova-Pasternak, R.K. 1978a. Tanaidacea (Crustacea, Malacostraca) from the deep-sea trenches of the western part of the Pacific. *Trudy Inst. Okeanol.* 108:115-135.

Kussakin, O.G. & L.A.Tzareva 1972. Tanaidacea from the coastal zones of the middle Kurile Islands. *Crustaceana (Suppl.)* 3:237-245.

Lang, K. 1949. Contribution to the systematics and synonymies of the Tanaidacea. *Ark. Zool.* (1)42:1-14.

Lang, K. 1956. Neotanaidae nov.fam., with some remarks on the phylogeny of the Tanaidacea. *Ark. Zool.* (2)9:469-475.

Lang, K. 1956a. Kalliapseudae, a new family of Tanaidacea. In: K.G.Wingstrand (ed.), *Bertil Hanström: Zool. papers in Honour of his 65th Birthday:* 205-225. Lund: Zool. Inst.

Lang, K. 1958. Protogynie bei zwei Tanaidaceen-Arten. *Ark. Zool.* (2)11:535-540.

Lang, K. 1967. Taxonomische und phylogenetische Untersuchungen über die Tanaidaceen. 3. Der Umfang der Familie Tanaidae Sars, Lang und Paratanaidae Lang nebst Bemerkungen über den taxonomischen Wert der Mandibeln und Maxillulae. Dazu eine taxonomisch-monographische Darstellung der Gattung *Tanaopsis* G.O.Sars. *Ark. Zool.* (2)19:343-368.

Lang, K. 1968. Deep-sea Tanaidacea. *Galathea Rept.* 9:23-209.

Lang, K. 1970. Taxonomische und phylogenetische Untersuchungen über die Tanaidaceen. 4. Aufteilung der Apseudidae in vier Familien nebst Aufstellung von zwei Gattungen und einer neuen Art der Familie Leiopidae. *Ark. Zool.* (2)22:595-626.

Lang, K. 1971. Taxonomische und phylogenetische Untersuchungen über die Tanaidaceen. 6. Revision der Gattung *Paranarthrura* Hansen 1913, und Aufstellung von zwei neuen Familien, vier neue Gattungen und zwei neuen Arten. *Ark. Zool.* (2)23:361-401.

Lang, K. 1973. Taxonomische und phylogenetische Untersuchungen über die Tanaidaceen (Crustacea). 8. Die Gattungen *Leptochelia* Dana, *Paratanais* Dana, *Heterotanais* G.O.Sars und *Nototanais* Richardson. Dazu einige Bemerkungen über die Monokonophora und ein Nachtrag. *Zool. Scripta* 2:197-229.

Lauterbach, K.E. 1970. Der Cephalothorax von *Tanais cavolinii* Milne-Edwards. Ein Beitrag zur vergleichenden Anatomie und Phylogenie der Tanaidacea. *Zool. Jb. (Anat.)* 87:94-204.

Lauterbach, K.E. 1974. Über die Herkunft des Carapax der Crustaceen. *Zool. Beitr. (n.s.)* 20:273-327.

Livingstone, R.J. et al. 1977. The biota of the Apalachicola Bay system: Functional relationships. *Florida Marine Resources, Florida Dept. Nat. Resources Publs.* 26:75-100.

Malzahn, E. 1979. *Apseudes giganteus* nov.spec. – Die erste Scherenassel aus der Kreide. *Ann. naturh. Mus. Wien* 82:67-81.

Menzies, R.J. & R.Y.George 1967. A re-evaluation of the concept of hadal or ultraabyssal fauna. *Deep-Sea Res.* 14:703-723.

Messing, C.G. 1980. *Pagurapseudes (Crustacea: Tanaidacea) in southeastern Florida: Functional mor-*

phology, post-marsupial development, ecology, and shell use. Thesis University of Miami.

Nierstrasz, H.F. 1913. Die Isopoden der Siboga-Expedition I. Isopoda Chelifera. – *Siboga Exp.* 32A:1-56.

Nusbaum, J. 1886. L'embryologie d'*Onisus murarius. Zool. Anz.* 9:454-458.

Odum, W.E. & E.J.Heald 1972. Trophic analysis of an estuarine mangrove community. *Bull. Mar. Sci.* 22:671-738.

Sanders, H.L. et al. 1965. An introduction to the study of deep-sea benthic faunal assemblages along the Gay-Head-Bermuda transect. *Deep-Sea Res.* 12:845-867.

Sars, G.O. 1896. *An account of the Crustacea of Norway 2 (Isopoda):*1-270. Bergen.

Scheloske, H.W. 1977. Skelett und Muskulatur des Cephalothorax von *Asellus aquaticus* (L.) (Asellidae, Isopoda). Ein Beitrag zur vergleichenden Anatomie der Crustacea-Malacostraca. *Zool. Jb. (Anat.)* 97:157-293.

Schram, F.R. 1974. Palaeozoic Peracarida of North America. *Fieldiana Geol.* 33:95-124.

Schram, F.R. 1979. British Carboniferous Malacostraca. *Fieldiana Geol.* 40:1-129.

Schram, F.R. 1981. On the classification of Eumalacostraca. *J. Crust. Biol.* 1:1-10.

Shiino, S.M. 1979. Tanaidacea collected by French scientists on board the survey ship 'Marion-Dufresne' in the region around the Kerguelen Islands and other subantarctic Islands in 1972, 1974, 1975, 1976. *Sci. Rep. Shima Marineland* 5:1-122.

Sieg, J. 1973. *Ein Beitrag zum Natürlichen System der Dikonophora, Lang.* Thesis, Univ. Kiel.

Sieg, J. 1976. Zum Natürlichen System der Dikonophora Lang (Crustacea, Tanaidacea). *Z. zool. Syst. Evolut.-forsch.* 14:177-198.

Sieg, J. 1977. Taxonomische Monographie der Familie Pseudotanaidae (Crustacea, Tanaidacea). *Mitt. zool. Mus. Berlin* 53:3-109.

Sieg. J. 1978. Aufteilung der Anarthruridae Lang in zwei Unterfamilien sowie Neubeschreibung von *Tanais willemoesi* Studer als Typus-Art der Gattung Langitanais Sieg (Tanaidacea). *Crustaceana* 35 (2):119-133.

Sieg, J. 1978a. Bemerkungen zur Möglichkeit der Bestimmung der Weibchen bei den Dikonophora und der Entwicklung der Tanaidaceen. *Zool. Anz.* 200:233-241.

Sieg, J. 1980. Taxonomische Monographie der Tanaidae Dana 1849 (Crustacea: Tanaidacea). *Abh. Senckenberg. naturf. Gesell.* 537:1-267.

Sieg, J. 1980a. Phylogenetic relationship and distribution of the shallow-water family Tanaidae (Crustacea: Tanaidacea). In: D.J.G.Griffin (ed.), *International Conference. Biology and Evolution of Crustacea.* Sydney: Australian Museum.

Sieg, J. 1980b. Sind die Dikonophora eine polyphyletische Gruppe? *Zool. Anz.* 205:401-416.

Sieg, J. 1980c. Revision der Gattung *Nototanais* Richardson, 1906 (Crustacea, Tanaidacea). *Mitt. zool. Mus. Berlin* 56:45-71.

Sieg, J. 1982a. Über ein 'connecting link' in der Phylogenie der Tanaidomorpha (Tanaidacea) *Crustaceana* 43:65-77.

Sieg, J. 1982b. Neuere Erkenntnisse zur Phylogenie der Tanaidacea (in press).

Siewing, R. 1951. Besteht eine engere Verwandtschaft zwischen Isopoden und Amphipoden. *Zool. Anz.* 147:166-180.

Siewing, R. 1953. Morphologische Untersuchungen an Tanaidaceen und Lophogastriden. *Z. wiss. Zool.* 157:333-426.

Siewing, R. 1960. Neue Ergebnisse der Verwandtschaftsforschung bei den Crustaceen. – *Wiss. Z. Univ. Rostock (Math.-nat.)* 9:343-358.

Siewing, R. 1963. Studies in Malacostracan morphology: Results and problems. In: H.B.Whittington & W.D.I. Rolfe (eds), *Phylogeny and Evolution of Crustacea:*85:104. Cambridge: Mus. Comp. Zool.

Tiegs, O.W. & S.M.Manton 1958. The evolution of the Arthropoda. *Biol. Rev.* 33:255-337.

Tschetwerikoff, S. 1911. Beiträge zur Anatomie der Wasserassel (*Asellus aquaticus* L.) *Bull. Soc. imper. natur. Moscou (n.s.)* 24:377-509.

Valentin, C. 1978. Zur Biozoenotik und Substratpraeferenz von Benthos crustaceen: Ein Beitrag zur Biologie und Autökologie der Cumaceen aus dem Küstenbereich der Insel Ischia (Golf von Neapel/ Italien). Thesis, Univ. Kiel.

Watling, L. 1981. An alternative phylogeny of peracarid Crustaceans. *J. Crust. Biol.* 1:201-210.

E. L. BOUSFIELD
*National Museum of Natural Sciences, Ottawa, Canada*

# AN UPDATED PHYLETIC CLASSIFICATION AND PALAEOHISTORY OF THE AMPHIPODA

## ABSTRACT

Based on new materials and a reappraisal of natural classificatory concepts proposed earlier (Bousfield 1977, 1979), an updated classification of amphipod subordinal, superfamily, and family groups is presented. The arrangement of superfamilies is slightly altered, but modifications reflect more closely recent new taxonomic concepts and the overall plesio-apomorphic range of character states within the higher taxa. Notable changes include submergence of superfamily Niphargoidea as a family within the reconstituted superfamily Crangonyctoidea; transfer of superfamily Gammaroidea to the base of the Melphidippoidea and superfamily Bogidielloidea to a position near the Melphidippoidea and Hadzioidea; transfer of Phreatogammaridae Bousfield to Melphidippoidea and Paracrangonyctidae Bousfield to superfamily Liljeborgioidea; creation of *Paraleptamphopus* family group within superfamily Eusiroidea, and *Paracalliope* family group within the Oedicerotoidea; the formal recognition of all these changes await confirmation by colleagues now publishing on those groups. The assessment of plesio-apomorphy of character states, and its use in defining superfamily concepts is described. Within the superfamily Oedicerotoidea, recent studies on the antipodal genera and outgroup families demonstrate a strong ecological basis to the plesio-apomorphic condition and range of character states. A probable palaeohistory of the Amphipoda is reconstructed from (1) the limited fossil record of the amphipods themselves and more extensive fossil records of animals and plants with which they are intimately associated; (2) modern distributions of amphipod subgroups of low vagility and low eurytopicity, in relation to established geochronology of present-day continents, oceans, and lacustrine basins in which the animals are endemic; and (3) comparative morphological relationships of higher amphipod taxa. The most modern and specialized subordinal and superfamily groups evolved during and since Cretaceous times, the more primitive extant groups probably date from the Jurassic, but the ancestral, prototype amphipods may not have appeared before the Triassic or possibly late Palaeozoic.

## 1 INTRODUCTION

The Amphipoda may prove to be the largest single ordinal group of malacostracan crustaceans (Bousfield 1981b, 1982a). Their remarkable diversity of form and color matches their broad ecological diversity. A phyletic classification is very much needed to reflect the evolutionary pathways that resulted in such diversity, and to provide a meaningful basis to

257

subdisciplines of amphipod biology including biogeography, ecology, behaviour, and physiology. Early arrangements of higher gammaridean taxa were semi-phyletic (e.g. Sars 1895, Stebbing 1906). Later, Barnard (1958, 1969) opted for alphabetical arrangements. More recently, however, Bowman & Grüner (1973) proposed a phyletic classification of hyperiidean amphipods that provided a useful model for phyletic systems of gammaridean amphipods developed very recently (Bousfield 1977, 1979, 1982a) and adopted in part by Lincoln (1979).

These recent phyletic proposals have received both positive and negative commentary, the details of which will not be treated here. In summary, however, some support has come from students of surface ultrastructure (Holman 1981, Bousfield & Halcrow, in preparation) including antennal calceoli (Lincoln & Hurley 1981), and from systematists involved in certain freshwater groups, especially the Crangonyctoidea (Holsinger 1977), Liljeborgioidea (Sebidae), and Bogidielloidea (Holsinger & Longley 1980) and in marine groups including the family-monotypic Ampeliscoidea (Dickinson & Wigley 1981) and Corophioidea (Myers 1981), and the terrestrial Talitridae (Friend 1980). Negative critique has come from the studies of Karaman & Barnard (1979), Ruffo (1979), Barnard & Karaman (1980), Holsinger & Longley (1980) and Stock (1981), mainly concerning certain freshwater groups (Gammaroidea, Salentinellidae) and the Hadzioidea. Disagreements tend to be subjective; they reflect philosophical differences in how higher taxa should be defined (e.g. 'rigidly', or by 'best-fit' criteria), in the need for phyletic revisions (e.g. retention of 'good old Gammaridae')

Table 1. Index to high-level taxonomic characters and character states

| Char. no. | Symbol | Character state or condition | | |
|---|---|---|---|---|
| | | Plesiomorphic (0) | Intermediate (1) | Apomorphic (2) |
| 1 | Reprod. Morph. | Pelagic male, I | Pelagic male, II Benthic male III | Benthic male, IV |
| 2 | Gn.1 and 2 | Subsimilar, non-sexually dimorphic | Weakly sexually dimorphic (amplexing) | Strongly dimorphic amplexing in male |
| 3 | Antennal Armature | Calceoli present (most members) | 'Brush setae' (A2) (most members) | Neither calceoli nor 'brush setae' |
| 4 | L.L. | Inner lobes lacking | Inner lobes partly developed | Inner lobes distinct, or fused |
| 5 | Md.Lft. Lacinia | 5-dentate | 4-dentate, 6-dentate | 3-0 dentate, multi-dentate |
| 6 | P5-7 | Homopodous, equal | Homopodous, unequal | Heteropodous |
| 7 | Coxae, P5-7 | Posterolobate | Aequilobate | Anterolobate |
| 8 | U1 and 2 rami | Lanceolate; serially spinose | Lanceolate, bare; linear, serial-spines | Linear, irregular-spinose or bare |
| 9 | U3 rami | Foliaceous, biramous | Weakly setose; unequally biramous | Non-foliaceous; (often uniramous) |
| 10 | Telson | Deeply bilobate; notch and spine apex | Shallow-bilobate; totally bilobate | Lobes totally fused |
| 11 | Gill P7 | Present, large (most members) | Present, small, occ. lacking | Lack (all members) |
| 12 | Brood plate | Broad, setose | Sublinear (convex margins) | Linear, few setae |

vs. implementation of modern superfamily concepts), the need to subdivide families and genera (Amphipoda are heavily oversplit, vs. undersplit), and in operational methodology (few-character vs. multiple-character analysis).

Whatever the degree of support for phyletic classification of the Amphipoda at the present time, I would advise colleagues to: (1) *look at specimens,* not make 'arm-chair' revisions from the literature, which may be incomplete and inaccurate; (2) *look at all body regions and appendages,* since evolution did not take place just in the gnathopods, or just in the mouthparts of amphipods, taxonomic analyses and conceptualizations should encompass broad suites of characters including pleopods, brood plates, coxal gills (e.g. Dickinson 1982, in press), and sexual dimorphism; and (3) *utilize modern numerical taxonomic methods* to enhance the objectivity, reliability, and broader applicability of the results.

Since recent development of new higher taxonomic groupings, and re-analysis of old groupings have required some changes in the original phyletic concepts, my purpose here is to: (1) present an updated and revised classification of the Amphipoda and review some of the major taxonomic characters on which it is based; (2) reaffirm the basis for plesio-apomorphic rating of character states, exemplified by the range of character states within a single superfamily; (3) compare the plesio-apomorphic ratings and relative positions of all gammaridean superfamilies and three other suborders, in developing a phyletic classification; and (4) present a plausible time frame for evolution within the Amphipoda.

## 2 PLESIO-APOMORPHY OF HIGHER TAXONOMIC CHARACTER STATES

Let us first review the plesiomorphic, intermediate, and apomorphic conditions of 12 selected higher taxonomic characters (subordinal, superfamily and family level) (Table 1). Most of these characters were initially recognized and delineated in previous papers (e.g. Bousfield 1979). A more recently discovered character (Fig.1) is the dentition of the left mandibular lacinia mobilis (Bousfield 1981a, 1982c). The 5-dentate (5-toothed) condition is considered plesiomorphic, from which evolved in one direction a proliferation of teeth (6-dentate is intermediate, multi-dentate is apomorphic), and in the other a loss of teeth (4-dentate is intermediate, and uni-dentate is apomorphic). These 12 characters encompass

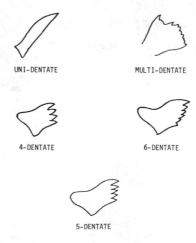

UNI-DENTATE          MULTI-DENTATE

4-DENTATE          6-DENTATE

Figure 1. Mandibular left lacinia mobilis dentition.          5-DENTATE

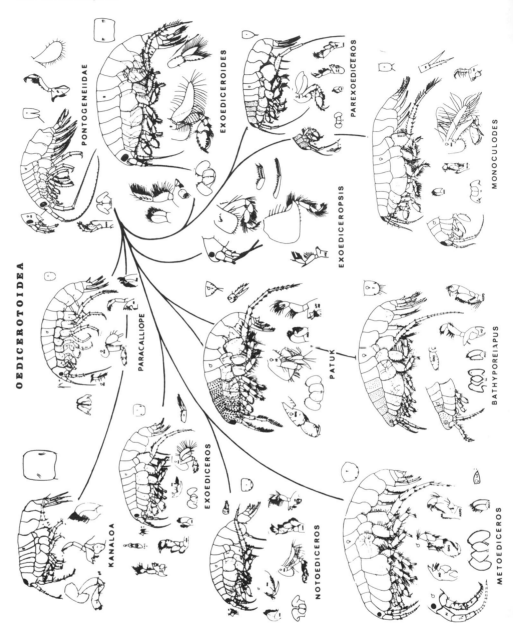

a broad spectrum of body regions and appendages. Some of these characters (e.g. antennae, peraeopods, and uropods) might readily show in fossil impressions. Their plesiomorphic, intermediate and apomorphic states have been derived mainly through outgroup comparisons within the Amphipoda and other Peracarida. For purposes of comparative analyses, these character states have been assigned numerical values of 0, 1 and 2 respectively.

Significantly, lengthier analyses that employ about a dozen additional characters of coxal plates, mouth-parts, accessory flagellum, eyes, anterior head lobes, pleopods, etc. (also mostly diagnosed in Bousfield 1979), do not appreciably change the phyletic relationships of the higher amphipod taxa developed here. In some of these additional characters, bi- or multi-directional apomorphic trends are highly probable. Thus, although the moderately elongate antennal accessory flagellum (~ 6 segments) is believed plesiomorphic, and reduction or total loss of accessory flagellum is apomorphic, an extremely elongate state (20-plus segments) must also be apomorphic; such is well developed in several abyssal gammaroideans of 25-million-year-old Lake Baikal whose presumed littoral ancestors have only a short accessory flagellum.

## 3 PLESIO-APOMORPHY OF CHARACTER STATES IN SUPERFAMILY OEDICEROTOIDEA

Both the derivation and application of characters states and their plesioapomorphic ratings can be illustrated rather clearly within superfamily Oedicerotoidea and closest outgroup family (Fig.2). This superfamily is characterized by (1) a 2, 2, 2, 1 similarity-pairing of thoracic legs or peraeopods (i.e. gnathopods are subsimilar, peraeopods 3 and 4 are subsimilar; peraeopods 5 and 6 are subsimilar, but *peraeopod 7 is unique,* very different in size and form, *invariably elongate,* with elongate dactyl); (2) appendages (especially peraeopods) are usually fossorial in form and armature (i.e. burrowing type); (3) telson lobes are fused, plate-like, weakly armed; and (4) integument is often strongly pitted. Typically also the rostrum is well-developed with eyes fused dorsally at its base (giving the group its name: *oedi-ceros*); the accessory flagellum is minute or lacking; the coxal plates (especially 5 and 6) are deep; gnathopods are often fossorial, with elongate carpal lobes; and the abdomen is large, the pleopods powerful.

Figure 2 indicates four main lines of oedicerotid evolution from a presumed eusiroidean (pontogeniid-calliopiid) ancestral type. Three of the genera are new to science as a result of recent field work in Tasmania: (*Notoediceros* n.g., type species – *N.tasmaniensis* n.sp.; *Exoediceroides* n.g., type species – *E.maximus* n.sp.; and *Parexoediceros* n.g., type species – *P.latimerus* n.sp.). Summary diagnoses of these new taxa are given in the appendix to this paper, and full diagnoses in Bousfield & Escofet (1982, in prep).

The free-swimming epibenthic eusiroidean ancestral type has well-developed calceolate antennae; strong rostrum; lateral eyes; basic mouthparts, with weakly developed inner lobes of the lower lip, and 5-dentate mandibular left lacinia; subsimilar, non-amplexing gnathopods (male); non-fossorial, homopodous, postero-lobate peraeopods; medium coxal plates; lanceolate, serially spinose uropods 1 and 2; foliaceous uropod 3; lobate telson; large coxal gill on peraeopod 7; and broad, setose brood plates (female). All these represent a combination of generally very primitive characters.

Let us now follow the plesio-apomorphic range of these 12 characters within the Oedicerotoidea, based largely on selected anti-boreal genera (of the southern hemisphere). The four main evolutionary lines (each possibly of formal subfamily or family significance) can be distinguished: (1) To the top left is the *Paracalliope* family group that occurs on non-sandy bottoms, in brackish and fresh waters of the 'Gondwanaland' region; *Kanaloa* of Hawaii may be slotted here, pending development of further information on its sexual dimorphism. Eyes are lateral; coxal plates are medium; legs are only slightly fossorial; and urosome segments 2 and 3 are usually fused. (2) To the far lower left is the *Exoediceros-Metoediceros* group that occurs on sand and silty sand, mainly intertidally, in estuaries and brackish waters of southeast Australia, including Tasmania, and southern South America. These are short-rostrate, deep-plated, fossorial, lateral-eyed animals with free urosomal segments. (3) To the centre left is the *Patuki-Bathyporeiapus* group that occurs intertidally on exposed sand beaches of New Zealand and southern South America. These are deflex-rostrate, lateral-eyed, deep-plated, fossorial types with free urosome segments and often deeply pitted integument. *Exoediceropsis* of South America, with long antennae, probably stands near the base of this group. (4) To the lower right is the *Exoediceroides-Monoculodes* group, the 'typical' oedicerotid members, that occurs mainly subtidally to the abyss, on various sediments, and shows high generic diversity in the northern hemisphere. These are strongly rostrate, dorsal-eyed, deep-plated, strongly fossorial animals with free urosome segments. Species diversity within the genus *Monoculodes* is very great, and merits subgeneric or possibly even further generic division.

With respect to the 12 taxonomic characters of Table 1: (1) *The pelagic male form* is variously lost in the three intertidal and estuarine groups to the left. The antennae become short, the pleopods and tail fan less powerful; the pelagic male form is retained in the subtidal groups to the right. (2) *Gnathopods 1 and 2* are sexually dimorphic and strongly amplexing (male) in the two estuarine groups to the left; weakly amplexing in the centre group;

Table 2. Genera of Oedicerotoidea and relative values of character states.

| Genus | Character state | | | | | | | | | | | | Totals | |
|---|---|---|---|---|---|---|---|---|---|---|---|---|---|---|
| | 1 | 2 | 3 | 4 | 5 | 6 | 7 | 8 | 9 | 10 | 11 | 12 | No/24 | % |
| PONTOGENEIIDAE and CALLIOPIIDAE (most) | 0 | 0+ | 0 | 1− | 0 | 0 | 0 | 0 | 0+ | 1+ | 0+ | 0 | 2+ | 8 |
| *Exoediceroides* | 0 | 1 | 0 | 2 | 0 | 2 | 1+ | 0 | 0 | 2 | 0+ | 1 | 9+ | 38 |
| *Exoediceropsis* | 0 | 0 | 0 | 2− | 0? | 2 | 1? | 1− | 2− | 2 | 1? | 1? | 12? | 50 |
| *Parexoediceros* | 0 | 0 | 0 | 2 | 0 | 2 | 0+ | 1+ | 2− | 2 | 2 | 1 | 12+ | 51 |
| *Monoculodes* | 0+ | 0+ | 2 | 2 | 0+ | 2 | 1+ | 1 | 2 | 2 | 2 | 2 | 16+ | 67 |
| *Exoediceros* | 2 | 2 | 0+ | 2 | 0 | 2 | 1 | 1 | 0+ | 2 | 1 | 1 | 14+ | 58 |
| *Notoediceros* | 2 | 2 | 2 | 2 | 0 | 2 | 1 | 2− | 2 | 2 | 2 | 1+ | 20 | 84 |
| *Metoediceros* | 2 | 2 | 1 | 2 | 1 | 2 | 1 | 1+ | 2 | 2 | 2 | 2 | 20+ | 85 |
| *Patuki* | 2 | 1+ | 2 | 2 | 1 | 2 | 1 | 2 | 2 | 2 | 1 | 2 | 20 | 83 |
| *Bathyporeiapus* | 2 | 1− | 1 | 2 | 1 | 2 | 1+ | 2− | 2 | 2 | 1 | 2 | 19 | 79 |
| *Paracalliope* | 2 | 2 | 0+ | 1+ | 1 | 1 | 0 | 0 | 2 | 2 | 2 | 0 | 13+ | 54 |
| *Kanaloa* | 2? | 1? | 2? | 1? | 1? | 1 | 0 | 0+ | 2 | 1+ | 1? | 0? | 12? | 51? |

Character state values: plesiomorphic: 0; intermediate: 1; apomorphic: 2; Maximum total value (all character states apomorphic): 24; +: some members with advanced character states; −: some members with primitive character states; ?: precise information on character state unavailable.

non-amplexing (with strong carpal lobes) in the subtidal group. (3) *Calceoli* are reduced in number, and confined to the flagellum of antenna 2 (males only), or lost entirely in the three intertidal groups; retained in both sexes in subtidal groups, or lost entirely, or replaced by brush setae in some genera (e.g. *Synchelidium*). (4) *The lower lip* retains its tall narrow form but develops distinct inner lobes in the *Paracalliope* and *Patuki* groups, but becomes low and broad with very large inner lobes in the other two groups. (5) *The mandible left lacinia mobilis* remains primitively 5-dentate in the *Exoediceros* and *Monoculodes* groups but becomes 4-dentate or much reduced in the *Paracalliope* and *Patuki* lines. (6) *Peraeopods 5-7* remain nearly homopodous and non-fossorial in the *Paracalliope* group, but are strongly heteropodous and fossorial in the others. (7) *Coxal plates 5-7* remain shallow and primitively posterolobate in *Paracalliope,* but become mainly deep and aequilobate in the other three (fossorial) lines. (8) *Uropods 1 and 2* remain lanceolate and serially spinose to bare in the *Paracalliope* and *Monoculodes* groups, but trend to linear and randomly spinose to bare in the intertidal sand-burrowing groups. (9) *Uropod 3* remains foliaceous only in the most primitive members of the *Exoediceros* and *Monoculodes* groups; becomes linear and reduced or vestigial in the intertidal groups; but remains narrow-lanceolate in subtidal members. (10) *The telson* is plate-like, apex rounded or acute, in all groups, with only a trace of a notch in the most primitive members of the subtidal group. (11) *Coxal gill on P7* ranges from large to lacking in the subtidal group; vestigial in the intertidal sand-burrowers; but is totally lost in the most brackish and estuarine types, especially in *Paracalliope* and *Metoediceros.* (12) *Brood plates* remain broad in *Paracalliope;* medium broad to linear in the sub-tidal group; but narrow to distally clavate in intertidal sand-burrowers.

Similar plesio-apomorphic trends can be detected in other major characters such as male: female size ratio (small in *Patuki* and *Monoculodes* lines, large in the other two); form of mandibular palp (smaller, clavate, or lacking in intertidal sand-burrowers; elongate falcate in the others); and mouthpart armature (generally reduced in more advanced members of all series).

In summary, the plesio-apomorphic state of major taxonomic characters here considered tends to be ecologically correlated. Thus, the primitive condition is more usual in the marine, subtidal, and least fossorial members, and the advanced state more frequent in the estuarine, intertidal, and strongly fossorial members of a phyletic series within the superfamily. Similar trends are indicated in other sand-burrowing amphipod groups (e.g. Pontoporeioidea, Phoxocephaloidea, some Gammaroidea) but detailed confirmation awaits further study.

The relative states of plesio-apomorphy of the genera and generic groupings are also given in Table 2. The percentage values in the right-hand column may be considered the plesio-apomorphic index of the given taxon; low values indicate a plesiomorphic, high values an apomorphic taxon. These values confirm the relationships and trends noted above. Thus, the eusiroidean ancestral group is extremely low (8 %); the *Paracalliope* group intermediate ($\sim$ 50 %); the subtidal northern burrowing group also intermediate (38-67 %); but the southern intertidal groups of both estuarine and sand beach environments are strongly apomorphic in these character states (58-85 %). Thus, southern intertidal oedicerotoids may be primitive in a few characters such as the foliaceous uropod 3 and lateral eye but are actually derived or advanced in most other characters, and conform with the ecological basis (marine to fresh-water to hypogean) previously observed in several other gammaroidean superfamilies (Bousfield 1980).

Table 3. Range of plesio-apomorphy in subordinal and superfamilies of Amphipoda.

Character state graph column: **CHARACTER STATE — Family envelope plesio-apomorphy (%)** (scale 0–25–50–75–100). Character number key — 0: Plesiom.; 1: Interm.; 0: Apomorph.

| Subordinal and superfamily groups | No. of families | 1 | 2 | 3 | 4 | 5 | 6 | 7 | 8 | 9 | 10 | 11 | 12 | Total No.24 | % |
|---|---|---|---|---|---|---|---|---|---|---|---|---|---|---|---|
| 1. TALITROIDEA | (10) | 2 | 2 | 2 | 0 | 0+ | 0+ | 0 | 2 | 2 | 1 | 2 | 0+ | 13+ | 55 |
| 2. LEUCOTHOIDEA | (12) | 2 | 2- | 2 | 1 | 0 | 0+ | 0+ | 1 | 2 | 2 | 2 | 0 | 14 | 58 |
| 3. OEDICEROTOIDEA | (2+) | 1 | 1 | 1 | 2 | 0+ | 2 | 1 | 1 | 1+ | 1+ | 1+ | 1+ | 14+ | 54 |
| 4. EUSIROIDEA | (7) | 0+ | 0+ | 1 | 1 | 0+ | 0 | 0 | 0+ | 0+ | 1 | 0+ | 0 | 3+ | 15 |
| 5. CRANGONYCTOIDEA | (4) | 2 | 1 | 1 | 1 | 1 | 0 | 0 | 2 | 2 | 2 | 1 | 0 | 13 | 54 |
| 6. PHOXOCEPHALOIDEA | (3) | 1 | 0 | 0 | 2 | 1 | 2 | 0 | 0 | 0 | 0 | 1 | 2 | 9 | 38 |
| 7. LYSIANASSOIDEA | (2+) | 0+ | 0+ | 0+ | 0 | 1 | 1- | 0 | 0 | 0 | 1- | 0+ | 2 | 5+ | 21 |
| 8. SYNOPIOIDEA | (2) | 1 | 0 | 1 | 1 | 0 | 0+ | 0 | 0 | 0 | 0+ | 0 | 1+ | 4+ | 17 |
| 9. STEGOCEPHALOIDEA | (4) | 2 | 0 | 1 | 0 | 0+ | 1 | 0 | 0 | 0 | 1 | 1 | 0+ | 6+ | 25 |
| 10. PARDALISCOIDEA | (5) | 1 | 0 | 1 | 2 | 0+ | 1 | 1 | 0 | 0+ | 0 | 0 | 0 | 6+ | 26 |
| 11. HYPERIIDEA | (21) | 1 | 0 | 1 | 2 | 2 | 1 | 1 | 0+ | 1 | 2 | 2 | 0 | 13+ | 54 |
| 12. LILJEBORGIOIDEA | (4) | 2 | 1 | 2 | 2 | 0 | 1 | 1 | 0+ | 1+ | 1 | 2 | 2 | 15+ | 62 |
| 13. DEXAMINOIDEA | (5) | 1+ | 1 | 1 | 1 | 0+ | 1+ | 1+ | 0+ | 1 | 0+ | 0+ | 1 | 8+ | 35 |
| 14. AMPELISCOIDEA | (1) | 1 | 0 | 1 | 2 | 0+ | 2 | 1+ | 0 | 0 | 0+ | 2 | 2 | 11+ | 47 |
| 15. PONTOPOREIOIDEA | (2) | 1 | 1 | 0+ | 2 | 2- | 2 | 1+ | 0+ | 1- | 1 | 2 | 0 | 13+ | 54 |
| 16. GAMMAROIDEA | (10) | 1 | 2- | 1 | 0 | 1 | 2 | 2 | 0+ | 0+ | 1 | 0+ | 0+ | 10+ | 42 |
| 17. MELPHIDIPPOIDEA | (4) | 1 | 1 | 1 | 2- | 1 | 1 | 2 | 1 | 0+ | 0+ | 1 | 2- | 13 | 53 |
| 18. HADZIOIDEA | (3) | 2 | 2- | 2 | 2- | 1 | 2 | 2 | 2 | 1+ | 0+ | 2 | 2 | 20- | 83 |
| 19. BOGIDIELLOIDEA | (2) | 2 | 1- | 2 | 2 | 1 | 2 | 2 | 1+ | 2 | 2 | 2 | 2 | 21 | 86 |
| 20. COROPHIOIDEA | (9) | 2 | 2 | 2 | 2 | 1 | 2 | 2 | 2 | 2 | 2 | 2 | 0+ | 21+ | 88 |
| 21. CAPRELLIDEA | (6) | 2 | 2 | 2 | 2 | 1 | 2 | 2 | 2 | 2 | 2 | 2 | 0 | 21 | 87 |
| 22. INGOLFIELLIDEA | (2) | 2 | 0 | 2 | 2 | 1 | 1+ | 2 | 1 | 0 | 2 | 2 | 2 | 17 | 71 |

# 4 PLESIO-APOMORPHIC CHARACTER STATES WITHIN AMPHIPOD SUBORDERS AND SUPERFAMILIES

We may now evaluate the plesio-apomorphic state of these same 12 taxonomic characters in all 22 subordinal and gammaridean superfamily groups. The results are summarized in Table 3. Values of component families (calculations supplied on request) are contained within the almond-shaped envelopes for each higher taxon. The vertical arrangement of the taxa (left column) is semi-phyletic. Thus, cup-calceolate eusiroideans with short-pedunculate antennae and spinose broad-homopodous peraeopods are uppermost, with presumed most primitive groups somewhere between eusiroideans and lysianassoideans near the middle; and those with primarily non-calceolate, mainly long antennae, and setose narrow-homopodous and heteropodous peraeopods are lowest.

The most noteworthy observations are as follows: (1) Most gammaridean superfamilies (and other suborders) are tightly defined. Component families are included within a range of 30-40 % or less in two-thirds of the superfamilies. Component genera of individual families range even more narrowly, usually within values of 10-15 %. Superfamilies of exceptionally wide range (40-60 %), e.g. Lysianassoidea, Oedicerotoidea, Gammaroidea, Hadzioidea, are probably polyphyletic, or inadequately subdivided taxonomically. (2) The most primitive superfamilies (values of 5-50 %) are typically: (a) marine, especially in the non-calceolate group; (b) encompass many abyssal to hadal species; and (c) have pelagic reproductive (mating) morphology, non-amplexing gnathopods. (3) The most advanced superfamily and subordinal group (values of 50-95 %) are typically: (a) shallow-marine and intertidal; (b) benthic, commensal, parasitoid, tubicolous, or with very specialized feeding requirements; and (c) have non-pelagic mating morphology, more or less sexually dimorphic, amplexing gnathopods (male). (4) Intermediate superfamily groups (values of 25-75 %) are well represented in fresh waters, brackish waters, and on land. Except for some Pontoporeiidae, such ecotypes do not mate pelagically; gnathopods are variously sexually dimorphic and usually amplexing in males. Hypogean members are more apomorphic than epigean members of a given group. (5) From a classification standpoint, the positions of non-gammaridean suborders seem particularly significant. Their character states range from moderately to very apomorphic, but 'envelopes' for component families are smaller than those for most gammaridean superfamilies. On this basis, the case for inclusion of the Hyperiidea and Caprellidea within the Gammaridea at superfamily (or perhaps infraordinal level) is strengthened. Moreover, these two groups possess no unique morphological structures (not derivable by loss or modification of existing gammaridean structures), and ecologically are narrowly stenotopic (exclusively marine and epigean). In the case of the phyletic-relict suborder Ingolfiellidea, however, apomorphic character states are mainly derived, and associated with the hypogean life style of extant members. They tend to mask the more plesiomorphic features of this group such as the subsimilar (usually), non-sexually dimorphic, non-amplexing, carpochelate gnathopods; the elongate peduncular segment 3 of antenna 2; the independently flexing lanceolate uropods 1 and 2; and moveable (separated) ocular lobes (presumed to be vestigial stalked eyes), a most plesiomorphic character within the Amphipoda and unique to the Ingolfiellidea. Although encompassing only one-tenth the number of described species, the extant ingolfiellid group is less stenotopic than either the hyperiid or caprellid groups, and occurs in both marine and freshwater habitats. Moreover, their wide-spread benthic hypogean habitus almost certainly 'required' an ancestry of epigean pelagic or epibenthic types of greater morphological diversity. In effect, the phyletic relationship of the Ingolfiellidea to the other Amphipoda might be considered somewhat analogous to the position of the stalk-eyed Tanaidacea vis-à-vis the sessile-eyed Isopoda, and therefore more fully meriting its present subordinal recognition. (6) Any or all characters can be convergent across or within various groups. Morphological convergence has proven to be a most significant factor in frustrating production of a universally acceptable phyletic classification of the Amphipoda. (7) The present phyletic system has surprising concordance with the semi-phyletic arrangements of Sars (1895) and Stebbing (1906) who had far fewer families and no superfamilies available to them. Thus, 14 families and family groupings of the 25 Norwegian families treated by Sars match 14 of the present 19 gammaridean superfamilies, in somewhat similar order; 17 families and family groupings of 41 world-wide gammaridean families treated by Stebbing match 17 of the superfamilies here, but less well, and in less similar order. On the other hand, Barnard (1974) patterned the relationships of more than 50 gammaridean families into nine major groupings that radiated from a presumed most primitive 'pool of species in Gammaridae' (that also included families Haustoriidae and Phoxocephalidae). Six of these groupings can be correlated more or less with six of the present superfamilies but in rather different phyletic order.

The major weakness of previous systems, however, was an uncritical retention of the concept of family Gammaridae (*sensu* Stock 1980). Thus, the plesio-apomorphic range of 'Gammaridae' (Table 3) is about 75 %, greatly exceeding all individual amphipod subordinal and superfamily values, and exceeding the values of all other recognized families by more than 200 %. Moreover, a credible superfamily and/or family definition should encompass only those taxa that bear a testably close phyletic relationship to the type genus and species, in this case *Gammarus pulex;* most members of 'old family Gammaridae' (~ 1 300 species) do not. These taxonomic inconsistencies render the concept (if it can truly be defined) of 'good old Gammaridae' (e.g. Stock 1980, Barnard & Karaman 1980) unrealistic. The total demise of this outmoded concept would therefore greatly enhance the prospects of finding a phyletic classification of wide acceptance among amphipodologists.

# 5 THE PHYLETIC CLASSIFICATION OF THE AMPHIPODA UPDATED

Based on the considerations (above), the phlogenetic arrangement presented earlier (Bousfield 1979) has been updated and is given in Table 4. Rationale and basis for the new arrangement, and formal and informal treatment of various levels of new taxa are provided in Appendix I.

The most important changes are as follows: (1) Recognition of the co-primitive (plesiomorphic) positions of lysianassid and eusiroidean and presumed immediate descendant superfamily groups (of Bousfield 1979); (2) submergence of superfamily Niphargoidea to family status within the reconstituted superfamily Crangonyctoidea; (3) transfer of superfamily Gammaroidea to the base of the Melphidippoidea, and superfamily Bogidielloidea to a position near the Melphidippoidea and Hadzioidea (see also Bousfield 1980); (4) transfer of Phreatogammaridae Bousfield 1982 to Melphidippoidea; (5) transfer of Paracrangonyctidae Bousfield to superfamily Liljeborgioidea (see also Bousfield 1980); (6) creation of *Paraleptamphopus* family group within superfamily Eusiroidea (see also Bousfield 1980); (7) creation of *Paracalliope* family group within the Oedicerotoidea (Fig.3 above, and Appendix I); (8) addition of family Neomegamphopidae Myers (1981)

Table 4. Suborders, superfamilies, and family groups of Amphipoda. References to original taxonomic literature provided in Stebbing (1906), Barnard (1969), Bowman & Grüner (1973), and Bousfield (1979).
* This encompasses Phoxocephalopsidae Barnard & Clark 1982, Urohaustoriidae Barnard & Clark 1982, and Zobrachoidae Barnard & Drummond 1982.
** Name change from Melitoidea, recommended by Barnard & Karaman (1980). Component family and/ or subfamily units of literature not yet formalized.

---

Superfamily Eusiroidea
  Family Pontogeneiidae Stebbing 1906
    Calliopiidae Sars 1893
    (incl. Gammarellidae Bousf. 1977)
    Eusiridae Stebbing 1888
    Paramphithoidae Stebbing 1906
    Amathillopsidae Pirlot 1934
    Bateidae Stebbing 1906
    *Paraleptamphopus* family group
Superfamily Oedicerotoidea
  Family Oedicerotidae Lilljeborg 1865
    *Paracalliope* family group
Superfamily Leucothoidea
  Family Pleustidae Buchholz 1874
    Amphilochidae Boeck 1872
    Leucothoidae Dana 1852
    Anamixidae Stebbing 1897
    Maxillipiidae Ledoyer 1973
    Colomastigidae Stebbing 1899
    Pagetinidae K.H.Barnard 1932
    Laphystiopsidae Stebbing 1899
    Nihotungidae Barnard 1972
    Cressidae Stebbing 1899
    Stenothoidae Boeck 1871
    Thaumatelsonidae Gurj. 1938
Superfamily Talitroidea
  Family Hyalidae Bulycheva 1957
    Dogielinotidae Gurjanova 1954
    Hyalellidae Bulycheva 1957
    Najnidae Barnard 1972
    Ceinidae Barnard 1972 revised
    Talitridae Costa 1857, emend Bulycheva 1957
    Eophliantidae Sheard 1938, emend Barnard 1969
    Phliantidae Stebbing 1899, emend
    Temnophliantidae Griffiths 1975
    Kuriidae Barnard 1964
Superfamily Crangonyctoidea
  Family Paramelitidae Bousfield 1977
    Neoniphargidae Bousfield 1977
    Niphargidae S.Karaman 1962
    Crangonyctidae Bousfield 1973, emend 1977
Superfamily Phoxocephaloidea
  Family Urothoidae Bousfield 1979*
    Phoxocephalidae Sars 1891
    Platyischnopidae Barnard & Drummond 1979

Superfamily Lysianassoidea
  Family Lysianassidae Dana 1849
    Uristidae Hurley 1963
    (several additional families under review; cf. Barnard 1969)
Superfamily Synopioidea
  Family Synopiidae Dana 1853
  Family Argissidae Walker 1904
Superfamily Stegocephaloidea
  Family Stegocephalidae Dana 1952
    Acanthonotozomatidae Stebbing 1906
    Ochlesidae Stebbing 1910
    Lafystiidae Sars 1893
Superfamily Pardaliscoidea
  Family Pardaliscidae Boeck 1871
    Stilipedidae Holmes 1908
    Hyperiopsidae Bovallius 1886
    Astyridae Pirlot 1934
    Vitjazianidae Birstein & Vinogradov 1955
Superfamily Liljeborgioidea
  Family Liljeborgiidae Stebbing 1888
    Sebidae Walker 1908
    Salentinellidae Bousfield 1977
    Paracrangonyctidae Bousfield 1982
Superfamily Dexaminoidea
  Family Atylidae Lilljeborg 1865
    Anatylidae Bulycheva 1955
    Lepechinellidae Schell. 1926
    Dexaminidae Leach 1814
    Prophliantidae Nicholls 1940, revised J.L. Barnard 1969
Superfamily Ampeliscoidea
  Family Ampeliscidae Bate 1861
Superfamily Pontoporeioidea
  Family Pontoporeiidae Sars 1892
    Haustoriidae Stebbing 1906
Superfamily Gammaroidea
  Family Acanthogammaridae Garjej. 1901
    Anisogammaridae Bousfield 1977
    Gammaroporeiidae Bousfield 1979
    Gammaridae Leach 1814
    Pontogammaridae Bousfield 1977
    Typhlogammaridae Bousfield 1979
    Mesogammaridae Bousfield 1977
    Macrohectopidae Sowinsky 1915
    *Behningiella-Zernovia* family group?
    *Iphiginella-Pachyschesis* family group?
Superfamily Melphidippoidea
  Family Melphidippidae Stebbing 1899

Table 4 (continued).

| | |
|---|---|
| *Hornellia-Cheirocratus* family group | SUBORDER HYPERIIDEA |
| *Megaluropus* family group | Infraorder PHYSOSOMATA |
| Phreatogammaridae Bousfield 1982 | Superfamily Lanceoloidea |
| Superfamily Hadzioidea** |   Family Lanceolidae Bovallius 1887 |
|   Family Hadziidae Karaman 1932 |     Chuneolidae Woltereck 1909 |
|     Melitidae Bousfield 1973 |     Microphasmidae S. & P. 1931 |
|     Carangoliopsidae Bousfield 1977 | Superfamily Scinoidea |
| Superfamily Bogidielloidea |   Family Archaeoscinidae Stebbing 1904 |
|   Family Bogidiellidae Hertzog 1936 |     Scinidae Stebbing 1888 |
|     Artesiidae Holsinger 1980 |     Mimonectidae Bovallius 1885 |
| Superfamily Corophioidea |     Proscinidae Pirlot 1933 |
|   Family Ampithoidae Stebbing 1899 | Infraorder PHYSOCEPHALATA |
|     Biancolinidae Barnard 1972 | Superfamily Vibiloidea |
|     Isaeidae Dana 1853 (Photidae) |   Family Vibiliidae Dana 1852 |
|     Ischyroceridae Stebbing 1899 |     Cystosomatidae Willemoes-Suhm 1875 |
|     Neomegamphopidae Myers 1981 |     Paraphronimidae Bovallius 1887 |
|     Aoridae Stebbing 1899 (revised) | Superfamily Phronimoidea |
|     Cheluridae Allman 1847 |   Family Hyperiidae Dana 1852 |
|     Corophiidae Dana 1849 (revised) |     Dairellidae Bovallius 1887 |
|     Podoceridae Stebbing 1906 |     Phrosinidae Dana 1853 |
| |     Phronimidae Dana 1853 |
| SUBORDER CAPRELLIDEA | Superfamily Lycaeopsoidea |
| Infraorder CAPRELLIDA |   Family Lycaeopsidae Chevreux 1913 |
| Superfamily Phtisicoidea | Superfamily Platysceloidea |
|   Family Paracercopidae K. & V. 1972 |   Family Pronoidae Claus 1879 |
|     Phtisicidae Vassilenko 1968 |     Anapronoidae B. & G. 1973 |
|     Dodecadidae Vassilenko 1968 |     Lycaeidae Claus 1879 |
| Superfamily Caprelloidea |     Oxycephalidae Bate 1861 |
|   Family Caprogammaridae K. & V. 1966 |     Platyscelidae Bate 1862 |
|     Aeginellidae Vassilenko 1968 |     Parascelidae Bovallius 1887 |
|     Caprellidae White 1847 | |
| | SUBORDER INGOLFIELLIDEA |
| Infraorder CYAMIDA |   Family Ingolfiellidae Hansen 1903 |
| Superfamily Cyamoidea |     Metaingolfiellidae Ruffo 1969 |
|   Family Cyamidae White 1847 | |

to superfamily Corophioidea; and (9) addition of family Artesiidae Holsinger & Longley (1980) to superfamily Bogidielloidea.

Within superfamily Gammaroidea, the formally proposed families Acanthogammaridae and Typhlogammaridae, and two other informally proposed family groups (Bousfield 1977, 1979) would undoubtedly benefit from the study of pertinent fresh materials, and from thorough multiple-character analysis. However, the validity of the concepts would be independent of the 'rigid' characterization demanded, unrealistically, by some authors (e.g. Barnard & Karaman 1980).

## 6 PALAEOHISTORY OF THE AMPHIPODA

In concluding this review, I would like to briefly outline the derivation of a plausible time frame for evolution of the Amphipoda (see Bousfield 1982b). Major clues are provided by (1) the fossil record of the animals themselves, or of other animals and plants with which some amphipod groups are intimately associated; (2) the recent continental distributions of amphipod subgroups of low eurytopicity and low vagility, in relation to established chronology of plate tectonics and formation of oceanic and lacustrine basins; and (3) comparative morphological relationships of the higher amphipod taxa.

The fossil record of the Amphipoda is limited to less than two dozen authentic species, in three (possibly four) superfamilies (Crangonyctoidea, Gammaroidea, Corophioidea, possibly Hadzioidea), none older than Upper Eocene, and of little direct value here (Bousfield 1982b). However, the probable minimum ages of selected subordinal, superfamily, and family groups within the Amphipoda, based on continental geochronology, are summarized in Table 5 (from Bousfield 1982b, Tables 1 and 2). Laurasia (North America and Eurasia) separated from Gondwana (southern continents) mainly during late Triassic to mid-Jurassic (180-150 my BP), and the break-up of Gondwana occurred mainly during Cretaceous to early Tertiary (Howarth 1981). These 'drift' chronologies indicate the probable minimum age of the presumed common ancestor to generic and/or family groups of amphipods now widely isolated from each other on the separated continents.

In some instances (e.g. for most Tertiary groups), the probable maximum age of the presumed common ancestor is suggested. Thus, groups of advanced morphology such as Hadziidae and Bogidiellidae, confined mainly to Tethyan Sea margins or epicontinental areas, are probably not older than Middle Cretaceous; highly apomorphic subgroups such as Bateidae and Haustoriidae that occur within narrowly limited Tethyan marine areas (western Atlantic and Caribbean regions) are probably not older than early Tertiary. Species of Acanthogammaridae endemic to Lake Baikal, and of Hyalellidae endemic to Lake Titicaca, are not older than the lake basins (maximum 25 and 10 million years, respectively). The Cyamidae (whale lice) must be more recently evolved than the earliest whales (fossil record to lower Eocene). Furthermore, terrestrial amphipods (Talitridae) of Australia and South Africa are unlikely to be older than their obligatory food and habitat plants (i.e. angiosperm leaf litter, *fide* Friend 1980) which did not evolve until Middle and Upper Cretaceous (Bousfield 1981a).

Despite their problematical nature, the geochronological data of Table 5 indicate that most Recent superfamilies and families of Amphipoda had evolved by Cretaceous times, and that both marine and freshwater groups were present in the Jurassic, about 150 million years ago. Primitive freshwater groups (e.g. Crangonyctoidea) tend to occur on the highest numbers of continents, the more advanced groups (e.g. Gammaroidea) on fewest continents. Plesiomorphic marine groups (e.g. Eusiroidea) are dominant in the Pacific Basin but relatively scarce in Caribbean-Mediterranean (Tethyan) seas.

Comparative morphology of continental groups presumed to date from mid-Jurassic would suggest that more plesiomorphic and possibly ancestral marine superfamily groups such as Eusiroidea, Lysianassoidea, and Synopioidea, should also date from those times or earlier. However, gradualistic extrapolation of a precise period of origin of the Amphipoda from presumed Jurassic morphotypes is not realistic because of probable 'punctualistic' evolution (Stebbings & Ayala 1981) that apparently characterized the evolution of most, if not all, animal life. The 'abrupt', near-simultaneous appearance of most peracaridan ordinal groups in the late Palaeozoic fossil record underscores this phenomenon.

For example, freshwater Astacura (Decapoda), whose modern obligate-continental distribution remarkably parallels that of crangonyctoidean amphipods, occur as fossils from mid-Jurassic to Lower Triassic; though reptant fossil decapods, including extinct groups, extend back to the Devonian. The weak-flying Ephemeroptera (Insecta), whose obligate-

Table 5. Estimated minimum geological ages of higher amphipod taxonomic groups; marine, freshwater, and terrestrial environments.

| Superfamily salinity range | Geological Age | | |
| --- | --- | --- | --- |
| | Jurassic and Lower Cretaceous | Upper Cretaceous | Tertiary |
| Freshwater | CRANGONYCTOIDEA | | |
| Freshwater | EUSIROIDEA (M) | EUSIROIDEA (FW) (*Paraleptamphopus* gp.) | EUSIROIDEA (Bateidae) |
| Brackish and marine | TALITROIDEA (M) (Hyalidae) | TALITROIDEA (FW & T) (Hyalellidae, Talitridae) | GAMMAROIDEA (most groups) |
| | INGOLFIELLIDEA | OEDICEROTOIDEA PONTOPOREIOIDEA (Pontoporeiidae) LILJEBORGIOIDEA HADZIOIDEA (Hadziidae) | PONTOPOREIOIDEA (Haustoriidae) HADZIOIDEA (Melitidae, Carangoliopsidae) |
| | | MELPHIDIPPOIDEA (M) BOGIDIELLOIDEA | MELPHIDIPPOIDEA (Phreatogammaridae) |
| Marine | LYSIANASSOIDEA PHOXOCEPHALOIDEA LEUCOTHOIDEA (Pleustidae) | LEUCOTHOIDEA (commensal groups) STEGOCEPHALOIDEA | HYPERIIDEA |
| | SYNOPIOIDEA | PARDALISCOIDEA DEXAMINOIDEA COROPHIOIDEA (Isaeidae, Ampithoidae, Podoceridae) CAPRELLIDEA (Caprellida) | AMPELISCOIDEA COROPHIOIDEA (apomorphic families) CAPRELLIDEA (Cyamida) |

continental (non-oceanic) freshwater nymphs closely occur with epigean freshwater amphipods, have changed relatively little since their fossil ancestors 'suddenly' appeared in the Permian (Kukalova-Peck 1973).

Within the Peracarida, some Recent suborders, superfamilies, and even families of Mysidacea, Cumacea, Tanaidacea and Isopoda (Phreatoicidea) occur as fossils in the Permo-Carboniferous, and the partly carapaced Spelaeogriphacea (now relict in hypogean fresh waters of South Africa) is related to the long-extinct *Anthracocaris* (Carboniferous). The recent evidence extends the fossil record of all these peracaridan groups (except Cumacea) back to the Carboniferous (Schram 1982).

One might also attempt a reconstruction of the ancestral or prototype amphipod by combining the most primitive features of all extant groups. Primordial amphipods may have evolved in high-energy coastal marine and estuarine environments, and exploited the detrital food resources of stony, gravelly, and rubble bottoms; such environments are ideally suited to slender, flexible, lightly chitinized bodies, strongly ambulatory peraeopods, forward-pushing tail fan, and where the loss of stalked eyes (yet vestigial in the Ingolfiellidea), antennal scale, and bulky carapace would also be advantageous. Exploitation of vascular plant food resources being washed from the land into late Palaeozoic seas may have stimulated the development of an efficiently chewing, compact buccal mass. Ability to nourish the young, *in situ,* by means of lecithotropic development of eggs in the protective thoracic brood pouch, was preadaptive in the peracaridan ancestral type. Lightly chitinized animals living in high energy environments are unlikely to be fossilized, a probability that agrees with the absence of fossil amphipods, or their immediate progenitors, from Mesozoic deposits (especially Jurassic) that contain fossil mysids, cumaceans, tanaids, and isopods, and from Triassic and Permian deposits in which fossils of the more heavily chitinized amphipod-like phreatoicid isopods are known.

Descendants of the architype amphipods, aided by relatively slight specializations of body form and appendages, presumably penetrated the food resources of upper estuaries, fresh waters, and terrestrial environments, both epigean and hypogean, infaunal and neritic, from the shore lines into the deep sea, and geographically from the tropics to polar waters. Hypogean and tube-building modes of life were presumably also facilitated by small size, slender body and lack of carapace (no malacostracans with full carapace are phreatic). More extreme modifications and sclerotization of the body and appendages, and loss of pelagic reproductive life cycle, typify amphipod groups that have become nestlers or intimate associates with sponges, tunicates, or coelenterates; epi-parasites; open ended tube-builders or wood borers; or developed as 'armoured' epizoans, active predators, free burrowers in sand, or live on land.

The Amphipoda possess several primitive malacostracan features, especially of the heart, gut, and mouthparts, and the large, laterally compressed abdomen, that relate them to the Mysidacea and (possibly) Euphausiacea. Some authors, (e.g. Watling 1981) have suggested an independent origin of amphipod crustaceans from syncarid-like ancestral eumalacostracans, and separate superordinal ranking of the Amphipoda. To date, evidence for this is ambivalent, and plesio-apomorphic considerations are not convincing (see Hessler, this volume). The amphipods are more widely believed to be a highly advanced or apomorphic ordinal group within the Peracarida, particularly in the total absence of carapace, thoracic exopods, and antennal scale, the unique 3:3 subdivision of paired abdominal limbs, internal loss of the maxillary gland, and fusion of the urosomal ventral nerve ganglia.

The fossil record tends to support this view. Within the Eumalacostraca, Recent higher groups lacking a carapace (including Syncarida) occur as fossils in the Lower Carboniferous (Schram 1982). Corresponding taxa with a carapace (whole or in part) are found in Lower Carboniferous and Devonian deposits, and the fully bivalved phyllocaridan malacostracans are known from early Palaeozoic marine strata. Lastly, and significantly, the overwhelmingly dominant small- to medium-sized malacostracans in Recent aquatic and terrestrial environments, the Isopoda and the Amphipoda, totally lack a carapace.

In summary, the most modern and specialized amphipod subordinal and superfamily groups evolved during and since Cretaceous times; the more primitive modern groups very probably existed during the Jurassic, but existence of ancestral or prototype amphipods prior to the Triassic, and certainly late Paleozoic, is highly speculative.

In conclusion, whether amphipods are true peracaridans, or something else, the group is distinctive, monophyletic, and decidedly modern in aspect. The lack of a broadly significant fossil record is the single most important missing piece of the puzzle of the origin and evolution of the Amphipoda. Perhaps with the help of colleagues in related disciplines such as functional morphology, ultrastructural systematics, and the pooling of 'brainpower' of all amphipodologists, the picture will be completed and result in a universally acceptable phyletic classification.

# 7 ACKNOWLEDGEMENTS

Valuable commentary on many of the concepts presented here has already been made by colleagues, especially by Diana Laubitz, Anamaria Escofet, C.P.Staude, and Drs J.J.Dickinson, R.R.Hessler, J.K.Lowry and F.R.Schram. Assistance in preparation of the plates by Floy E.Zittin of Vancouver, Charles Douglas of Ottawa, and the Museum photo section is gratefully acknowledged. Field work in the southern hemisphere was made possible by: the C.S.S.Hudson in 1970 in southern South America; Drs John Morton and Graham Fenwick in New Zealand; and A.M.M.Richardson, A.J.Friend and T.M.Walker in Tasmania in 1978. Responsibility for the validity of the data and concepts and for errors and oversights that may occur, is my own. Some of this material is condensed and revised from Bousfield (1982b) and is used with the permission of McGraw-Hill.

# 8 APPENDIX

The text materials (Figure 2 and Table 4) encompass several names new to science or that require taxonomic redefinition or reconstitution. The following brief diagnoses will provide such until the publication of more detailed descriptions, since this information is currently unavailable elsewhere.

Superfamily Crangonyctoidea Bousfield 1977, emend Holsinger 1977, Bousfield 1979, 1982a
*Diagnosis:* Conforming with the original and subsequent diagnoses (above), with the following provisions: Gnathopods weakly or not sexually dimorphic, weakly or not amplexing (male grasps female laterally by coxal plates of one side, not dorsally as in Gammaridae); antennal calceoli usually present, of linear type (basal cup weakly or not developed), in males only, on peduncle and/or flagellum of antenna 2, rarely on antenna 1; mandibular left lacinia 4- or 5-dentate; coxal gills plate-like, usually present on peraeopod 7, or lacking; sternal gills usually present, of simple or lobate structure, usually paired variously on sternites 2-7, occasionally on pleon 1, or lacking; brood plates broad or medium broad, occasionally linear in some hypogean members.
*New Family Inclusion:* Family Niphargidae S.Karaman 1962, Bousfield 1979, 1982a
Family Crangonyctidae Bousfield 1973, emend 1977, Holsinger 1977
*Diagnosis:* Conforming with the emended diagnoses, with the following provisions: Mandibular left lacinia mobilis usually 5- (rarely 4-)dentate; gnathopod palmar margins with split-tipped (usually) or simple spines; coxal gill of peraeopod 7 occasionally lacking.
*New Inclusions:* Genus *Pseudocrangonyx* Akatsuka & Komai 1922. Type species: *P.yezoensis* Akatsuko & Komai 1922; Japan. Genus *Eocrangonyx* Schellenberg 1937. Type species: *E.japonicus* (Ueno 1931); Japan.
Family Paramelitidae Bousfield 1977
*New Inclusion:* Genus *Sternophysinx* Holsinger & Straskrabe 1973. Type species: *S.robertsi* (Methuen 1912); Transvaal.
Family Neoniphargidae Bousfield 1977
*New Inclusions:* Genus *Gininiphargus* Karaman & Barnard 1979. Type species: *G.pulchellus* Sayce 1899; S.E.Australia. Genus *Austroniphargus* Monod 1925. Type species: *A.bryophilus* Monod 1925; Madagascar. Genus *Sandro* Karaman & Barnard 1979. Type species: *S.starmühlneri* (Ruffo 1960); Madagascar. ·
Superfamily Eusiroidea Bousfield 1979
*New Family Inclusion: Paraleptamphopus* family group Bousfield 1980

*Diagnosis:* Smooth-bodied eusiroideans of epigean and hypogean freshwaters of New Zealand, characterized by: eyes small or lacking, inferior antennal sinus large to accommodate large peduncular segment 1 of antenna 2; cup-calceoli present on flagella of antennae, occasionally on peduncle of antenna 1, or not; gnathopods weakly sexually dimorphic and weakly amplexing, gnathopod 1 usually the larger; mandibular left lacinia 4-dentate; pleopods moderately developed, not powerful; uropods 1 and 2 lanceolate to sublinear; uropod 3, rami lanceolate, weakly foliaceous or spinose; telson lobes fused to shallow apical V-notch; coxal gills large, plate-like, usually present on peraeopod 7; sternal gills lacking; brood plates large, broad, marginal setae few, small. Type genus: *Paraleptamphopus* Stebbing 1899. Type species: *P.subterraneus* (Chilton 1882); New Zealand. Additional species: *P.ceruleus* (Thomson), and several new species (and genera) from surface and subterranean fresh waters of North and South Islands.

Superfamily Oedicerotoidea Bousfield 1979

Family Oedicerotidae Lilljeborg 1865, emend. Bousfield 1982a, Bousfield & Escofet in preparation
*Diagnosis:* Mainly marine members of superfamily Oedicerotoidea characterized by: body and appendages fossorial in form; peraeopods 3 and 4 fossorial, coxae usually broadened and deepened; peraeopods 5 and 6 subequal, fossorial, coxae usually deep, aequilobate; peraeopod 7 fossorial, elongate; gnathopods typically subequal, carpal lobes produced below; lower lip broadened, inner lobes well developed; urosome segments separate; brood plates sublinear or linear. Type genus: *Oediceros* Kroyer 1842, Stebbing 1906, Barnard 1969. Additional Genera: The type genus exemplifies a very large group of strongly fossorial genera, mainly subtidal, and mainly in the northern hemisphere, that usually possess strongly developed rostrum, dorsally contiguous eyes, 'pelagic' male sexual dimorphism, non- (or weakly) sexually dimorphic gnathopods with strongly developed carpal lobes, and narrow-lanceolate uropods 1-3.

Genus *Exoediceroides* new genus
*Diagnosis:* Eyes large, lateral; antennae elongate, calceolate; accessory flagellum minute; mandibular molar strong, palp segment 3 elongate; gnathopods subequal, weakly sexually dimorphic, carpal lobes weakly extended; peraeopods 3-6, dactyls minute, coxae deep; peraeopod 7 fossorial, dactyl not markedly elongate, 1-segmented; abdominal side plates 2 and 3 subquadrate; uropods 1 and 2 serially spinose; uropod 3, rami broad-lanceolate, strongly foliaceous; telson subquadrate; coxal gill on peraeopod 7; brood plates large, broadly sublinear.
*Type species: Exoediceroides maximus* new species
*Type material:* Ocean Beach, west coast Tasmania, surf zone at LW level, November 7, 1978, E.L.Bousfield & A.M.M.Richardson coll. Holotype female, ovigerous, 12.0 mm; allotype male, 9.0 mm. National Museum of Victoria, Melbourne. Paratypes. National Museum of Natural Sciences, Ottawa.
*Diagnosis:* With the characters of the genus. Antenna 1, peduncular segments 2 and 3 short, strongly setose posteriorly. Telson, apex broadly subtruncate, margin shallowly incised.

Genus *Parexoediceros* new genus
*Diagnosis:* Integument strongly pitted. Eyes large, fused dorsally on base of rostrum; antenna 1 much shorter than 2, flagella of both elongate, strongly calceolate; accessory flagellum lacking; mouthparts basic; mandibular molar strong, palp segment 3 short, tapering; gnathopods subequal, non-sexually dimorphic, carpal lobes moderately produced; peraeopods 3-6 strongly fossorial, dactyls well-developed, coxal plates medium deep, coxa of 6 posterolobate; peraeopod 7 very elongate; abdominal side plate 3 rounded behind; uropods 1-3 narrowly lanceolate, 3 non-foliaceous; telson subovate; coxal gill on peraeopod 7; brood plates large, sublinear.
*Type species: P.latimerus* new species.
*Type material:* Ocean Beach, west coast Tasmania, surf zone at LW level, November 7 1978, E.L.Bousfield & A.M.M.Richardson coll. Holotype female (brood plate II), 13.0 mm; Allotype male, 8.0 mm. National Museum of Victoria, Melbourne. Paratypes. National Museum of Natural Sciences, Ottawa.
*Diagnosis:* With the characters of the genus. Antenna 1, peduncular segments unarmed posteriorly, segment 3 short; peraeopods 5 and 6, segment 4 subtriangular, broadening distally; peraeopod 7, dactyl very elongate, multi-segmented; telson, apex subacute, with pair of fine setae.

Genus *Notoediceros* new genus
*Diagnosis:* A member of the *Exoediceros-Metoediceros* generic group, characterized by non-

pelagic reproductive morphology; sexually dimorphic, subequal, amplexing gnathopods (male); lateral eyes; weak rostrum; weakly (or not) calceolate antennae (male only); antenna 1, peduncular segment 3 not shortened, accessory flagellum vestigial; mandibular palp spatulate, or lacking; peraeopods 3-6, dactyls vestigial, coxal plates deep, coxa 6 weakly posterolobate; uropod 3 variously modified or reduced; telson short, rounded; coxal gill of peraeopod 7 vestigial or lacking; brood plates linear, or linear-spatulate. Distinguished from the other genera of this sub-group mainly by the total lack of antennal calceoli; the sublinear, weakly spinose rami of uropods 1 and 2; and the short, linear, unequally biramous and non-foliaceous uropod 3.

*Type species: N.tasmaniensis* new species

*Type material:* Ocean Beach, west coast of Tasmania, in freshwater stream outflow, near high-water level, November 7 1978, E.L.Bousfield & A.M.M.Richardson coll. Holotype male, 9.0 mm; allotype female, ovigerous, 7.0 mm. National Museum of Victoria, Melbourne. Paratypes. National Museum of Natural Sciences, Ottawa.

*Diagnosis:* With the characters of the genus. Antenna 1, peduncular segments strongly setose posteriorly; gnathopods (female), carpus elongate; uropod 3, inner ramus about half length of outer, margins with a few spine groups; telson, apex rounded, smooth; coxal gill of peraeopod 7 lacking.

*New Family Inclusion: Paracalliope* family group, Bousfield & Escofet in preparation

*Diagnosis:* Mainly marine-intertidal, estuarine, and freshwater members of the superfamily, of 'Gondwana' continental distribution (except South America), characterized by: body and appendages weakly (or not) fossorial in form; peraeopods 3 and 4 non-fossorial, coxae medium, not modified; peraeopods 5 and 6 weakly (or not fossorial, slightly unequal, coxae posterolobate; peraeopod 7 weakly fossorial, slightly elongate; gnathopods sexually dimorphic, subequal, carpal lobes little produced; lower lip 'tall', inner lobes weakly separated; urosome segments 2 and 3 usually fused; uropod 3, rami non-foliaceous; coxal gill of peraeopod 7 lacking; brood plates broad.

*Type species: P.fluviatilis* (Thomson 1879). New Zealand; fresh water.

*Remarks:* This generic group, and related members of the Australian regional Oedicerotidae are currently being studied by Barnard & Drummond (in preparation). Formal recognition of higher taxonomic categories are pending.

Superfamily Liljeborgioidea Bousfield 1979, Bousfield 1982a

*Diagnosis:* Conforming with the original diagnosis, with the following provisions: Mandibular left lacinia mobilis basically 5-(rarely 4-)dentate; sternal gills (when present) single, mid-ventral; gnathopods usually weakly or not sexually dimorphic, dactyls with elongate unguis.

*New Family Inclusion: Paracrangonyx* family group Bousfield 1980, Paracrangonyctidae, Bousfield 1982a

*Diagnosis:* Eyeless, hypogean fresh and brackish-water members, of southern continents, characterized by: bodies slender, elongate; abdomen well developed; antennae short, subequal; accessory flagellum minute, 1-2 segmented; gnathopods subequal, subsimilar, not noticeably sexually dimorphic; mouthparts superficially 'bogidielloid'; mandibular molar weak, palp weak, left lacinia 5-dentate; lower lip, inner lobes weakly developed; maxilla 1, outer plate with 9 apical spineteeth; coxae 1-4 often reduced and 5-7 conspicuously anterolobate; peraeopods 5-7 slender-homopodous; pleopods very reduced, usually uniramous; uropods 1 and 2 short, rami curvi-lanceolate; uropod 3 inaequiramous or uniramous, outer ramus usually 2-segmented; telson lobes fused to entire plate, with paired apical spines; coxal gills sac-like, weakly pedunculate; single rod-like sternal gills often present on peraeon segment 2-7, or fewer; brood plates elongate-linear. Type genus: *Paracrangonyx* Stebbing 1899. Type species: *P.compactus* (Chilton 1882). New Zealand, South Island. Additional genera: *Dussartiella* Ruffo 1979; *Pseudingolfiella* Noodt 1965 (both placed incorrectly in Bogidiellidae by Stock, 1981); and several new genera and species, mainly from hypogean fresh waters of New Zealand, currently being described by Sandro Ruffo (in preparation).

Superfamily Melphidippoidea Bousfield 1979

*New Family Inclusion: Phreatogammarus* family group Bousfield 1980, Phreatogammaridae Bousfield, 1982a

*Diagnosis:* Endemic to estuarine and fresh waters (epigean and hypogean) of New Zealand, characterized by: eyes reniform, pigmented, or lacking; antennae well-developed slender, lacking calceoli or brush setae; accessory flagellum well-developed; mouthparts basic; mandibular left lacinia 4-dentate; lower lip, inner lobes lacking or weakly developed; 'pelagic' male morphology lacking;

gnathopods subsimilar, weakly sexually dimorphic, dactyls toothed behind, with elongate unguis; uropods 1 and 2, rami linear; uropod 3, rami linear, subequal, spinose, non-foliaceous, outer ramus 1-segmented; telson lobes short, spinose, separated to base; urosome segments usually with dorsal groups of spinules; pleosome segments occasionally with postero-dorsal marginal denticles; coxal gill of peraeopod 7 small or lacking; paired sternal gills often present; brood plates usually large and broad. Type genus: *Phreatogammarus* Stebbing 1899. Type species: *P. fragilis* (Chilton 1882). New Zealand.

*Remarks:* Numerous new species, in several new genera, have recently been discovered in estuaries, and in both surface and subterranean fresh waters of North and South Islands of New Zealand (Bousfield 1980, Bousfield in preparation).

Superfamily Bogidielloidea Bousfield 1979, Bousfield 1982a, Stock, 1981 (part).

*Diagnosis:* Conforming with the original diagnosis, but with the following provisions: Exclusively hypogean in fresh and brackish (occasionally marine) waters of tropical-to-temperate regions, characterized by: antennae medium to short, peduncle of antenna 2 long; mouthparts reduced or modified; mandibular molar variously reduced, left lacinia mobilis 4-dentate; lower lip broad, inner lobes well-developed; gnathopods powerfully subchelate, subsimilar, weakly or not sexually dimorphic, carpus short, dactyls with short unguis; peraeopods 5-7 variously heteropodous, peraeopod 7 tending to elongation; coxae 5-7 anterolobate; pleopods variously reduced, usually biramous; uropods 1 and 2, rami linear, spinose; uropod 3 subequally biramous, foliaceous or spinose, outer ramus 1-segmented, telson lobes variously fused, apices and margins usually spinose; sternal gills lacking; brood plates short, linear.

*New Family Inclusion:* Artesiidae Holsinger 1980

*Type genus and species: Artesia subterranea* Holsinger 1980, *in* Holsinger & Longley 1980; Texas. Additional species: *Spelaeogammarus bahiensis* da Silva Brum 1975; Brazil.

*Remarks:* Holsinger's original diagnosis is expanded slightly to include the genus *Spelaeogammarus,* as follows: mandibular molar developed; coxal plates medium deep, increasing posteriorly; peraeopods 5 and 6 strongly antero-lobate; uropods 1 and 2, rami extending beyond peduncle of uropod 3; telson lobes variously fused, apex notched or deeply cleft. The plesiomorphic features of this family link the superfamily Bogidielloidea to a presumed marine epigean ancestral type, near the Melphidippoidea.


# DISCUSSION

ABELE: Is the evidence really strong that all those fresh water groups are really monophyletic?

BOUSFIELD: If you compare the subordinal diversity within the amphipods with that of the isopods the amphipods are almost all alike. There are nine suborders of isopods, including three totally parasitic groups, versus the compact amphipods with little parasitism and that only partially so as ectoparasites. There has just not been a great deal of diversity within the amphipods. The characters I am using may seem fairly strong to amphipod people, but to malacostracan phylogenists there is practically no change taking place within the amphipods. Degree of diversity is relative.

NEWMAN: I was intrigued by the variety of furcal conditions in amphipods which you outlined. I was wondering which is considered the most primitive.

BOUSFIELD: In my view, the nearly separated telson lobes are the plesiomorphic condition, fused at the base but separated through most of their length.

NEWMAN: Good.

BOUSFIELD: Evolution proceeds in two directions, total fusion or total separation.

MESSING: How does the current classification of hypogean amphipods concur with our understanding of plate tectonics?

BOUSFIELD: In my view it works very well. The hypogean amphipods are the dominant hypogean macrocrustacean group in the world. There are over 400 described species, and the world fauna will probably be well over 1 000. The amphipod fossil record is virtually useless. But here we have all these living fossils still alive in the groundwater. These taxa Holsinger and Longley are describing from the Texan artesian wells are fantastic animals. They have been preserved in that habitat probably since the Cretaceous.

# REFERENCES

Barnard, J.L. 1958. Index to the Families, Genera, and Species of the Gammaridean Amphipoda (Crustacea). *Occ. Pap. Allan Hancock Found.* 19:1-145.

Barnard, J.L. 1969. The families and genera of marine gammaridean Amphipoda. *Bull. US Nat. Mus.* 271:1-535.

Barnard, J.L. 1974. Evolutionary Patterns in Gammaridean Amphipoda. *Crustaceana* 27:137-146.

Barnard, J.L. & G.S.Karaman 1980. Classification of Gammarid Amphipoda. *Crustaceana, Suppl.* 6:5-16.

Bousfield, E.L. 1973. *Shallow-water Gammaridean Amphipoda of New England.* Ithaca: Cornell Univ. Press.

Bousfield, E.L. 1977. A new look at the systematics of gammaroidean amphipods of the world. *Crustaceana, Suppl.* 4:282-316.

Bousfield, E.L. 1979. A revised classification and phylogeny of amphipod crustaceans. *Trans. Roy. Soc. Canada* (4)16:343-390.

Bousfield, E.L. 1980. Studies on Freshwater amphipods of New Zealand and Tasmania. *Int. Conf. Biol. & Evol. Crustacea, Sydney, May 1980.* Abstract.

Bousfield, E.L. 1981a. Biogeographic indicators within the higher Crustacea of North America. *Ann. Meeting Can. Soc. Zool., Waterloo, Ont., May 1981.* Abstract.

Bousfield, E.L. 1981b. Evolution in North Pacific Coastal Marine Amphipod Crustaceans. In: G.Scudder & J.Reveal (eds), *Evolution Today. Proc. 2nd Int. Congr. Syst. & Evol. Biol.:*69-89. Pittsburgh: Hunt. Inst. Bot. Documentation.

Bousfield, E.L. 1982a. Amphipoda, Gammaridea, and Ingolfiellidea, In: *Synopsis and Classification of Animal Life. Vol.II:*254-285, 293-294. New York: McGraw-Hill.

Bousfield, E.L. 1982b. Amphipoda. In: *McGraw-Hill Yearbook of Science and Technology for 1982-1983:*96-100.

Bousfield, E.L. 1982c. The amphipod superfamily Talitroidea in the northeastern Pacific region. I. Family Talitridae: Systematics and Distributional Ecology. *Nat. Mus. Nat. Sci., Publ. Biol. Oceanogr.* No.11.

Bousfield, E.L. & A.N.Escofet in prep. Studies on the amphipod family Oedicerotidae in the southern hemisphere. *Nat. Mus. Nat. Sci., Publ. Biol. Oceanogr.*

Bowman, T.E. & H.-E.Grüner 1973. The Families and Genera of Hyperiidea (Crustacea: Amphipoda). *Smiths. Contr. Zool.* 146:1-64.

Dickinson, J.J. 1982. Studies on Amphipod Crustaceans of the Northeastern Pacific Region. I. 1. Family Ampeliscidae, genus *Ampelisca. Nat. Mus. Can., Nat. Mus. Nat. Sci. Publ. Biol. Oceanogr.* 10:1-40.

Dickinson, J.J. & R.L.Wigley 1981. Distribution of Gammaridean Amphipoda (Crustacea) on Georges Bank. *NOAA Tech. Rep. NMFS SSRF-746,* 25pp.

Friend, J.A. 1980. The taxonomy, zoogeography and aspects of the ecology of the terrestrial amphipods (Amphipoda:Talitridae) of Tasmania. PhD Thesis, Univ. Tasmania, Hobart. Vols.I & II:1-346.

Holman, H. 1981. Cuticular microstructure in the amphipod superfamily Talitroidea. *Am. Zool.* 21:956.

Holsinger, J.R. 1977. A review of the systematics of the Holarctic amphipod family Crangonyctidae. *Crustaceana, Suppl.* 4:244-281.

Holsinger, J.R. & Glenn Longley 1980. The Subterranean Amphipod Crustacean Fauna of an Artesian Well in Texas. *Smiths. Contr. Zool.* 308:1-62.

Howarth, M.K. 1981. Palaeogeography of the Mesozoic. In: P.H.Greenwood (ed.), *The Evolving Earth:* 1-264. Cambridge: Cambridge Univ. Press.

Karaman, G. & J.L.Barnard 1979. Classificatory Revisions in Gammaridean Amphipoda (Crustacea), Part 1. *Proc. Biol. Soc. Wash.* 92:106-165.

Kukalova-Peck, J. 1973. A Phylogenetic Tree of the Animal Kingdom (including Orders and Higher Categories). *National Mus. Nat. Sci., Ottawa, Publ. Zool.* 8:1-78.

Lincoln, R.J. 1979. *British Marine Amphipoda: Gammaridea.* London: Brit. Mus. Nat. Hist.

Lincoln, R.J. & D.E.Hurley 1981. The calceolus, a sensory structure of gammaridean amphipods. *Bull. B.M. (N.H.)* 40:103-116.

Moore, P.G. 1981. A Functional Interpretation of Coxal Morphology in *Epimeria cornigera* (Crustacea: Amphipoda:Paramphithoidae). *J. Mar. Biol. Assoc. UK* 61:749-757.

Myers, A.A. 1981. Amphipod Crustacea. I. Family Aoridae. *Mem. Hour Glass Cruises* 5:1-80.

Ruffo, S. 1979. Studi sui Crostacei Anfipodi XC. Descrizione di due nuovi anfipodi anoftalmi dell'Iran e del Madagascar. (*Phreatomelita paceae* n.gen. n.sp., *Dussartiella madegassa* n.gen. n.sp.). *Boll. Mus. Civ. Storia Nat. Verona* 6:419-440.

Sars, G.O. 1895. *Amphipoda: An Account of the Crustacea of Norway.* Christiana: Cammermeyers.

Schram, F.R. 1982. Fossil record and evolution of Crustacea. In: L.G.Abele (ed.), *The Biology of Crustacea. Vol.1:* 93-147. New York: Academic Press.

Stebbing, T.R.R. 1899. Revision of Amphipoda. *Ann. Mag. Nat. Hist.* (7)4:205-211.

Stebbing, T.R.R. 1906. Amphipoda I: Gammaridea. *Das Tierreich* 21:1-806.

Stebbins, G.L. & F.J.Ayala 1981. Is a new evolutionary synthesis necessary? *Science* 213:967-971.

Stock, J.H. 1980. Regression Model Evolution as Exemplified by the Genus *Pseudoniphargus* (Amphipoda). *Bidr. Tot Dierk.* 50:105-144.

Stock, J.H. 1981. The taxonomy and zoogeography of the Bogidiellidae, with emphasis on the West Indian taxa. *Bidr. tot Dierk.* 51:345-374.

Watling, L. 1981. An Alternative Phylogeny of Peracarid Crustaceans. *J. Crust. Biol.* 1:201-210.

MARTIN D. BURKENROAD
*Natural History Museum, San Diego, California, USA*

# NATURAL CLASSIFICATION OF DENDROBRANCHIATA, WITH A KEY TO RECENT GENERA

## ABSTRACT

A key to the living genera of Dendrobranchiata is presented. Comments on phylogenetic relationships and persistent problem areas in the taxonomy of the group are offered.

## 1 INTRODUCTION

Attempts at constructing natural taxonomic systems labor under the immediate disadvantage that symmetrical and simple diagnostic features are harder to achieve than in artificial systems. However, in the long run, attempts at arrangements that reflect evolutionary history usually increase the probability that new discoveries will be compatible with the system previously in use; whereas artificial arrangements are a source of unending confusion because they must often be fundamentally reorganized when new discoveries have to be accommodated.

The discrete non-interbreeding population units (species) might be compared with the separate filaments of a rope which has been unravelling as it extends through time. It is desired, for purposes of developing a natural system, to attempt to bundle the filaments of a given temporal cross-section of the rope together into definable groups, and combine these in turn into greater groups. Such hierarchies of filaments should be chosen to be conveniently distinguishable from each other. They should not be so numerous as to complicate identification at any given hierarchic grade. They should also be such that each bundle corresponds to an earlier filament distinct from those that give rise to other bundles of the same hierarchic grade at a given time. In practice these different desiderata are of course difficult to reconcile, as recognized by Schram (1981) in erecting a phylogenetic uncertainty principle. Even apart from the incompleteness of knowledge, inescapable difficulties result from actual differences in the rates and directions of unravelling of different parts of the rope. It is hoped that the arrangement proposed here for living dendrobranchiates is the best compromise at present attainable.

Burkenroad (1963, 1981) and Glaessner (1969) have recognized that the peneids form a natural rank, Suborder Dendrobranchiata, coequal with several other great groups of Decapoda. The dendrobranchiates are united by possession of the following advanced characters: (1) pleurobranchs appear later in the course of individual development than podobranchs and arthrobranchs; (2) pleura of the first pleomere are expanded to overlap the second pleura, and the hinges between the third and fourth pleomeres are covered;

279

(3) appendixes internae are absent, except for vestiges occasionally found on the male first and second pleopods; (4) gills are dendrobranchiae; and (5) first three pairs of pereiopods are chelate.

Following is a classification of the Dendrobranchiata, and an attempt at a natural key down to the level of genus. This outline is to be regarded as a summary of available information on relationships within the group. It is based on notes which have been accumulated over the years toward a definitive account, which this does not pretend to be. Treatment of the intricate legalistic problems of nomenclature, of which there are many still unresolved, has not been included; although it is hoped that most of the names employed are those which will in the end be retained. Many names of the present section are derived from Peneios, a Thessalian rivergod among whose descendants was Aristaios, the basis of several additional peneid names.

Suborder Dendrobranchiata Burkenroad 1963
  Superfamily Peneoidea Rafinesque 1815
    Family Aristeidae Wood-Mason 1891
      Subfamily Solenocerinae Wood-Mason & Alcock 1891
        *Haliporus* Bate 1881
        *Hymenopeneus* Smith 1882
        *Solenocera* Lucas 1850
      Subfamily Aristeinae Alcock 1901
        Tribe Aristeini Bouvier 1908
          *Aristeomorpha* Wood-Mason 1891
          *Hepomadus* Bate 1881
          *Plesiopeneus* Bate 1881
          *Hemipeneus* Bate 1881
          *Aristeus* Duvernoy 1840
        Tribe Benthysicymini Bouvier 1908
          *Benthysicymus* Bate 1881
          *Bentheogennema* Burkenroad 1936
          *Gennadas* Bate 1881
    Family Peneidae Rafinesque 1815
      Subfamily Peneinae, Burkenroad 1934a
        Tribe Peneini Burkenroad 1936
          *Funchalia* Johnson 1868
          *Heteropeneus* De Man 1896
          *Peneus* Fabricius 1798
        Tribe Parapeneini nov.
          *Parapeneus* Smith 1885
          *Artemesia* Bate 1888
          *Peneopsis* Bate 1881
          *Metapeneopsis* Bouvier 1905
        Tribe Trachypeneini nov.
          *Mangalura* Miers 1878
          *Macropetasma* Stebbing 1914
          *Trachypeneopsis* Burkenroad 1934
          *Atypopeneus* Alcock 1905
          *Protrachypene* Burkenroad 1934
          *Xiphopeneus* Smith 1869
          *Trachypeneus* Alcock 1901
          *Parapeneopsis* Alcock 1901
      Subfamily Sicyoninae Ortman 1898
        *Sicyonia* Milne-Edwards 1830
  Superfamily Sergestoidea Dana 1852

Family Sergestidae Dana 1852
    Subfamily Sicyonellinae nov.
        *Sicyonella* Borradaile 1910
    Subfamily Sergestinae Bate 1881
        *Petalidium* Bate 1881
        *Sergia* Stimpson 1860
        *Sergestes* Milne-Edwards 1830
        *Peisos* Burkenroad 1945
        *Acetes* Milne-Edwards 1830
Family Luciferidae nov.
    *Lucifer* Vaughan-Thompson 1830

## 2 KEY TO ADULT DENDROBRANCHIATA
(Superfamilies, families, subfamilies, tribes and genera)

1 A gill of the anterodorsal series (pleurobranch) present at least on the somites of the second and third maxillipedes. Some somites at least with three gills on each side. Total number of well-developed and conspicuous gills on each side at least 11 .................. *Superfamily* Peneoidea 2

– Pleurobranchs entirely absent. No somite with more than two gills on each side. Total number of well-developed gills on each side no more than seven ............. *Superfamily* Sergestoidea 28

2 (*Superfamily* Peneoidea) Anteromedian side of base of endopod of second pair of abdominal legs of male with two appendices (possibly representing appendix interna in addition to appendix masculina). Two well-developed gills of middle series (anteroventral and posterodorsal arthrobranchs) on penultimate thoracic somite ...................................... *Family* Aristeidae 3

– Male second pleopod with single appendix. Only one large arthrobranch on penultimate thoracic somite, although small rudimentary anteroventral one is sometimes present .. *Family* Peneidae 13

3 (*Family* Aristeidae) A spine (postorbital) on side of carapace behind orbital margin. Telson with tip appearing trifid, the well-developed point flanked by conspicuous pair of fixed lateral spines. Ocular somite with scale-like projection at base of eye-stalk. Posterolateral edge of base of endopod of male second pleopod with projection additional to the two appendices on the anteromedian edge ......
................................................................... *Subfamily* Solenocerinae[1] 4

– No postorbital spine. Telson tip not appearing trifid, its distal pair of lateral spines being moveable. Ocular somite without conspicuous scales. Base of endopod of male second pleopod without a posterolateral spur ............................................. *Subfamily* Aristeinae 6

4 Telson with several pairs of mobile lateral spines anterior to the fixed pair. Podobranchs on third maxillipede and anterior three pairs of walking legs. Projection of dorsomedian edge of basal segment of antennular peduncle (prosartema) short and rigid .......................... *Haliporus*

– Telson with only a pair of fixed spines. No podobranchs posterior to second maxillipedes. Prosartema elongate and flexible ..................................................... 5

5 Antennular flagella much flattened ..................................... *Solenocera*

– Antennular flagella cylindrical or somewhat flattened ...................... *Hymenopeneus*

6 Upper branch of antennule not much produced beyond its short, thickened basal part. Three or more rostral teeth ..................................................... *Tribe* Aristeini 7

– Both branches of antennule long. Rostral teeth usually two or less ..... *Tribe* Benthesicymini 11

7 (*Tribe* Aristeini) Carapace with hepatic spine ........................................ 8

– Hepatic spine absent ............................................... 9

8 Rostrum with more than three teeth. Well-developed epipodite on fourth leg. Well-developed podobranch on third leg ............................................... *Aristeomorpha*

– Rostrum with three teeth. Epipodite of fourth leg and podobranch of third minute or wanting ...
...................................................................... *Hepomadus*

9 Well-developed epipodite on fourth leg. Well-developed podobranch on third leg ... *Plesiopeneus*

– Epipodite of fourth leg and podobranch of third minute or wanting .................. 10

10 Cervical sulcus indistinct or absent. Pleurobranch of penultimate thoracic somite considerably less than half as long as that of ultimate somite ................................... *Aristeus*

– Cervical sulcus more or less distinct. Pleurobranch of penultimate thoracic somite around half the length of that of ultimate somite . . . . . . . . . . . . . . . . . . . . . . . . . . . . . . . . . . . . . . . . . . . . *Hemipeneus*

11 (*Tribe* Benthesicymini) Penultimate somite of abdomen carinated in dorsal midline. Exopodite of first maxillipede distally narrowed and segmented. Statolith normally with sand-grains . . . . . . . . .
. . . . . . . . . . . . . . . . . . . . . . . . . . . . . . . . . . . . . . . . . . . . . . . . . . . . . . . . . . . . . . . . . . . *Benthesicymus*
– Penultimate pleonic somite not carinated. Tip of exopodite of first maxillipede not set off from the rest. Statolith purely self-secreted . . . . . . . . . . . . . . . . . . . . . . . . . . . . . . . . . . . . . . . . . . . . 12

12 Podobranchs on third maxillipedes and first three pairs of walking legs . . . . . . . *Bentheogennema*
– Podobranchs wanting behind second maxillipede. . . . . . . . . . . . . . . . . . . . . . . . . . . . *Gennadas*

13 (*Family* Peneidae) Abdominal legs of third and fourth pairs with two branches. Exopod present on at least one thoracic appendage behind first maxillipede. Ocular somite with scale-like projection at base of eye-stalk. Scale-like projection on dorsomedian margin of basal segment of antennular stalk. Pleurobranch on at least three somites behind that of third maxillipede. Total number of conspicuous gills on each side at least 15 . . . . . . . . . . . . . . . . . . . . . . . . . . . . . *Subfamily* Peneinae 14
– Pleopods of third and fourth pairs single-branched, without endopodite. No trace of thoracic exopodites behind first maxillipede. Ocular somite and dorsomedian margin of base of antennular peduncle without scale-like projections. Pleurobranchs entirely absent behind somite of third maxillipede. Number of conspicuous gills on each side only 11 . . . . . . . . . . . *Subfamily* Sicyoninae, *Sicyonia*

14 Epipodite on the third maxillipede. Gill (pleurobranch) on the last thoracic somite *Tribe* Peneini 15
– No epipodite on third maxillipede. No gill on last thoracic segment . . . . . . . . . . . . . . . . . . . . . 17

15 (*Tribe* Peneini) Shell smooth and polished. Longitudinal ridge of sixth segment of abdomen discontinuous . . . . . . . . . . . . . . . . . . . . . . . . . . . . . . . . . . . . . . . . . . . . . . . . . . . . . . . . . *Peneus*
– Shell hairy. Longitudinal ridge of side of sixth pleonic segment continuous . . . . . . . . . . . . . . 16

16 Anteroventral corner of carapace rounded. Telson with four well-developed pairs of mobile lateral spines. Antennular flagella short. Incisor process of mandible not elongated, molar broad and heavy
. . . . . . . . . . . . . . . . . . . . . . . . . . . . . . . . . . . . . . . . . . . . . . . . . . . . . . . . . . . . . . *Heteropeneus*
– Anteroventral angle of carapace sharp. Telson with numerous inconspicuous lateral spinules. Antennular flagella long. Mandible scythe-like . . . . . . . . . . . . . . . . . . . . . . . . . . . . . . . . *Funchalia*

17 Telson with tip appearing more or less trifid, distal pair of lateral spines being fused to body of telson (although the line of junction is marked by a suture in *Artemesia*). Ventromedian margin of proximal segment of antennular peduncle usually with spine (absent in some individuals of *Parapeneus*, in which, however, tip of telson is obviously trifid) . . . . . . . . . . . . . . . . . . . . *Tribe* Parapeneini 18
– Distal pair of lateral spines of telson usually small and mobile, if present (fairly conspicuous, and more or less fused to telson in a few forms, as *Parapeneopsis stylifera*). Ventromedian margin of proximal segment of antennular peduncle without spine . . . . . . . . . . . . . . . . *Tribe* Trachypeneini 21

18 (*Tribe* Parapeneini) Carapace with longitudinal and transverse sutures. No more than one pair of minute lateral spines anterior to trifid tip of telson . . . . . . . . . . . . . . . . . . . . . . . *Parapeneus* 21
– Carapace without longitudinal or transverse sutures. Two or more well-developed pairs of lateral spines on telson, anterior to the distal fixed pair . . . . . . . . . . . . . . . . . . . . . . . . . . . . . . . . 19

19 Anteroventral angle of carapace sharp, tooth-like. Exopodite present on all maxillipedes and legs . .
. . . . . . . . . . . . . . . . . . . . . . . . . . . . . . . . . . . . . . . . . . . . . . . . . . . . . . . . . . . . . . . . . . . . 20
– Anteroventral angle of carapace rounded. Exopodite absent from second maxillipede and all legs . .
. . . . . . . . . . . . . . . . . . . . . . . . . . . . . . . . . . . . . . . . . . . . . . . . . . . . . . . . . . . . . . . *Artemesia*

20 Symmetrical petasma . . . . . . . . . . . . . . . . . . . . . . . . . . . . . . . . . . . . . . . . . . . . . . . . . *Peneopsis*
– Asymmetrical petasma . . . . . . . . . . . . . . . . . . . . . . . . . . . . . . . . . . . . . . . . . . . . . *Metapeneopsis*

21 (*Tribe* Trachypeneini) Pleurobranch on penultimate thoracic somite. Exopodite on maxillipedes and anterior four pairs of legs, none on fifth leg . . . . . . . . . . . . . . . . . . . . . . . . . . . . . . . *Mangalura*
– No pleurobranch on penultimate somite. Exopodite either present on all legs or absent from all four posterior pairs of legs . . . . . . . . . . . . . . . . . . . . . . . . . . . . . . . . . . . . . . . . . . . . . . . . . . . . 22

22 Exopodite on first maxillipede and first leg only. Petasma with enormous distolateral projection, longer than the rest of the structure . . . . . . . . . . . . . . . . . . . . . . . . . . . . . . . . . . . . *Macropetasma*
– Exopodite on third maxillipede at least, and on all legs; distolateral projection of petasma shorter than remainder of the organ . . . . . . . . . . . . . . . . . . . . . . . . . . . . . . . . . . . . . . . . . . . . . . . 23

23 Carapace without longitudinal or transverse sutures. Distal pair of lateral spines of telson mobile but mounted on elongate shoulders. Petasma with large many-lobed flap springing from distal end of ventral (posterior) face of each endopod, which partially hides short lateral spout . . . *Trachypeneopsis*

– Carapace with either longitudinal or transverse sutures or both. Spines of telson, if present and mobile, not mounted on projecting shoulders. Petasma without conspicuous distoventral flap . 24

24 Carapace with transverse but no longitudinal suture. Ischium of second pair of legs with spine . . . . . . . . . . . . . . . . . . . . . . . . . . . . . . . . . . . . . . . . . . . . . . . . . . . . . . . . . . . . *Atypopeneus*

– Carapace with longitudinal suture. Ischium of second legs unarmed . . . . . . . . . . . . . . . . . . . . . 25

25 Second maxillipede (alone among thoracic appendages) without exopodite. Chelae with slender palm around four times length of fingers . . . . . . . . . . . . . . . . . . . . . . . . . . . *Protrachypene*

– All thoracic appendages with exopodite. Palm of chelae not elongated . . . . . . . . . . . . . . . . . . 26

26 Carapace of adult without transverse suture. Fourth and fifth legs flagelliform, dactyl greatly lengthened and multiarticulate . . . . . . . . . . . . . . . . . . . . . . . . . . . . . . . . . . . . . . . *Xiphopeneus*

– Carapace with transverse suture. Fourth and fifth legs not flagelliform, dactyl not segmented . 27

27 No epipodite on third leg . . . . . . . . . . . . . . . . . . . . . . . . . . . . . . . . . . . . . . . *Parapeneopsis*

– Epipodite on third leg . . . . . . . . . . . . . . . . . . . . . . . . . . . . . . . . . . *Trachypeneus*[2]

28 (*Superfamily* Sergestoidea) Body moderately compressed. Chela (though sometimes with minute fingers) on second pair of thoracic legs at least. Antennule with two flagella. Mandible and maxilla with palp. First maxillipede with at least two well-developed distal branches (endopod and endite). Gills present . . . . . . . . . . . . . . . . . . . . . . . . . . . . . . . . . . . . . . . . . . *Family* Sergestidae 29

– Body extremely compressed, its depth several times its thickness. No chelae (third legs with microscopic dactyl but no fixed finger). Inferior antennular flagellum absent. Mandible and maxilla lack distal branches. First maxillipede lacks defined branches. No gills . . . *Family* Luciferidae, *Lucifer*

29 (*Family* Sergestidae) Telson with four well-developed pairs of lateral spines. Posterior two pairs of thoracic legs well-developed, seven-jointed. Chelae quite strong, with fingers about one-third as long as palm. The lamella representing posterior (dorsal) arthrobranch of somites of third maxillipede and first two pairs of legs well-developed, fringed with filaments. Statolith with sand-grains . . . . . . . . . . . . . . . . . . . . . . . . . . . . . . . . . . . . . . . . . . . *Subfamily* Sicyonellinae, *Sicyonella*

– Telson with three pairs of lateral spines at most. Posterior two pairs of thoracic legs six-jointed at most. Chelae weak or absent, with fingers at most much less than one-third of palm. Dorsal arthrobranch of somites of third maxillipede and first two legs a simple lamella without filaments, if present. Statolith entirely self-secreted . . . . . . . . . . . . . . . . . . . . . . *Subfamily* Sergestinae 30

30 (*Subfamily* Sergestinae) Posterior two pairs of thoracic legs fairly well-developed, but only six-jointed (dactyl wanting). First leg without chela. Maxilla with endite three-lobed at least. First maxillipede with exopod and endopod in addition to exite and endite . . . . . . . . . . . . . . . . . . . . . 31

– Posterior two pairs of legs small or absent, at most five-jointed. First leg with chela (palm elongate but fingers quite conspicuous). Maxilla with endite one-lobed. First maxillipede lacks exopod . 33

31 Petasma of male with processus ventralis deeply cleft. Best-developed gill with no more than about 13 branches, none of which bears more than 12 plates . . . . . . . . . . . . . . . . . . . . . *Petalidium*

– Petasma with processus ventralis very shallowly divided if at all. Some gills with numerous (20 or more) branches bearing numerous (25 or more) small plates . . . . . . . . . . . . . . . . . . . . . . . . . 32

32 Several large luminescent organs in the cephalothorax, developed from parts of gastro-hepatic gland. Ischium of first and second chelipeds with well-developed spine proximal to end of joint. Distal segment of antennular peduncle of females (and most males) slender, four times or more as long as its diameter (only about three times in males of *S.arcticus* and *S.similis*) . . . . . . . . . . . . . *Sergestes*

– Luminescent organs, if present, cuticular. Ischium of first and second legs without sub-terminal spine, although terminal angle may be weak or toothed; distal segment of antennular peduncle stout, not more than three times as long as its diameter . . . . . . . . . . . . . . . . . . . . . . . . . . . . . . . . . . . . *Sergia*

33 Fourth pair of walking legs present (though small), and some vestige at least of fifth pair. Tip of telson appearing trifid, both median and lateral teeth being well-developed, slender and sharp. Petasma of male with hook at tip of pars externa . . . . . . . . . . . . . . . . . . . . . . . . . . . . . . . . *Peisos*

– Fourth and fifth legs not represented. Tip of telson with lateral spines minute or absent, point at most a blunt triangle. Petasma without processus uncinatus . . . . . . . . . . . . . . . . . . . . . *Acetes*

# 3 DISCUSSION

Although it seems likely that adult ancestors of the Decapoda were in some respects more euphausiid-like than the modern forms, it is conceivable that modern euphausiids have lost

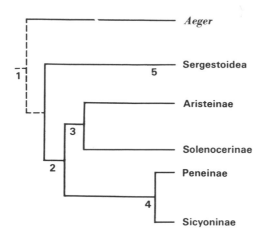

Figure 1. Possible phylogenetic relationships of dendrobranchiate superfamilies, families and subfamilies discussed in the text. 1. an *Aeger*-like stem from which the sergestoids and peneoids may have arisen; 2. a pre-solenocerine-like stock from which the peneids diverged very early; 3. a solenocerine-like stock from which the modern solenocerines and aristeines arose; 4. sicyonines appear to have been a relatively recent paedomorphic derivative of the peneines; 5. luciferids also exhibit marked paedomorphic features vis-à-vis the sergestids, but their exact affinities to the latter are difficult to assess for the present within the sergestoids.

decapod-like features, e.g. paedomorphic suppression of ancestral post-larval features in the course of specialization for a free-swimming life. The dendrobranchiates show more resemblances to the euphausiaceans than do other decapods (cf. Mauchline & Fisher 1969). Both groups generally shed their eggs freely into the water column or they adhere only a short time to the posterior thoracomeres; both groups are the only eumalacostracans with free nauplii; the larval development of both is marked by a succession of distinct phases; and there are similarities in the structure of the petasma, spermatophores, thelycum, and some of the photophores of euphausiaceans and dendrobranchiates (especially sergestids). However, many of these resemblances may be of kinds not necessarily indicative of close phylogenetic relationship.

A system of decapod classification which contrasts the long-tailed forms (Macrura) with the Anomura and the Brachyura is artificial, and has been discarded by Burkenroad (1963, 1981), Glaessner (1969), and de St Laurent (1979). The system (Boas 1880) which contrasts the Natantia (i.e. the decapod shrimp) with the Reptantia including Anomura and Brachyura also is artificial. The resemblances of the Dendrobranchiata to the Eukyphida and Euzygida are not indicative of a relationship between them closer than to the Reptantia. For example, the differences between the dendrobranchiates, the eukyphids, and the euzygids in the order of appearance of the arthrobranchs and pleurobranchs during development (Burkenroad 1939, 1945a, 1981) seem to connote a very ancient separation. A thorough study of the comparative morphology of eucarid spermatozoa may add further reasons for regarding the group Natantia as an artificial one.[3]

Peneoidea and Sergestoidea are no more than hierarchic promotions (Burkenroad 1981) of the groups Peneidae Rafinesque and Sergestidae Dana previously generally accepted. The purpose of this elevation in rank is to give room for formal recognition of what seems a closer relationship of the subfamilies Solenocerinae and Aristeinae to each other than to the Peneinae and Sicyoninae; and also to emphasize the distinctness of Sergestidae from the Luciferidae (Fig.1).

There are some faint hints that the common ancestor of Peneoidea and Sergestoidea may on the whole have been more like the Solenocerinae than like any other known groups, since the solenocerines are somewhat intermediate in form between the living aristeines and the Jurassic fossil *Aeger* (Burkenroad 1936, 1945a). A detailed comparison of a sufficient quantity of the various and well-preserved *Aeger* from the Jurassic Solenhofen fauna will probably help to clarify this question (and others).

It is likely that there are diagnostic differences between Recent adult Peneoidea and Sergestoidea additional to those mentioned in the key (such as the plate-like form in sergestids of the ultimate branches of gills, and the flattening of the last two pairs of their thoracic legs, when these are present); but such characters have not as yet been carefully checked across both groups.

## 3.1 *Peneoidea*

The second (posteriormost) appendix of the median side of the base of the endopod of the male second pleopod in the family Aristeidae (as here proposed; see Burkenroad 1934b) is thought to represent an appendix interna (Barnard 1950). As such, it would seem an important feature, presumably lost independently in the divergent groups Sergestidae and Peneidae. Aristeidae are also rather different from Peneidae in development (Burkenroad 1945a:569, 574, footnotes 5 and 7) in that the proximal endite of the maxilla is not reduced at the postmysis metamorphosis, and the thoracic exopods are merely reduced to their adult dimensions (while in peneids they disappear at that molt to reappear later). Thus it seems desirable to separate these two lines, even though some adult Solenocerinae resemble Peneinae rather than Aristeinae in general form (Fig.1).[4]

The subfamily Sicyoninae of the Peneidae differs from the Peneinae by becoming mature at a stage of development through which the larvae of the latter pass. This stage, marked in Peneinae by such features as temporary suppression of thoracic expodites, may well correspond to the adult mode of an ancestral stage in evolution of the Peneidae. However, the modern Sicyoninae have genitalia resembling those of certain Peneinae, (members of the Tribe Peneini) which are highly specialized in this feature. It thus seems possible that the Sicyoninae are paedomorphic derivatives of a relatively advanced peneine stock. Sicyonines may have rather recently acquired their distinctive features (Fig.1), even though these might also have been characteristic of some extinct derivative of an aristeoid-like form from which the Peneidae might have been derived by an earlier paedomorphosis.

### 3.1.1 *Solenocerinae.* *Haliporus* approaches Aristeinae, whereas *Hymenopeneus* and *Solenocera* approach Peneinae; but distinction of the two groups as tribes seems unnecessary at present [especially since the generic position of the form described as *Haliporus villosus* by Alcock & Anderson (1894) is uncertain, the figure showing some features reminiscent of true *Haliporus*].

*Hymenopeneus* is distinguishable from *Solenocera* largely by the single functionally significant but morphologically minor character of the antennal flagella cross-sections given in the key. It should be observed that separation of the pinnae of the respiratory exite of the first maxillipede somite into two (or more) groups, thought by Kubo (1949) to distinguish *Solenocera* from *Hymenopeneus,* is not diagnostic (since a similar separation proves to occur at least in *Hymenopeneus robustus* Smith. *Parahaliporus* Kubo (1949) seems synonymous with *Haliporoides* Stebbing, the latter being one of the several names available for subgenera of *Hymenopeneus* when these can be satisfactorily defined.

*Transolenocera,* proposed as a subgenus of *Solenocera* by Burkenroad (1934b) for *Parasolenocera maldivensis* Borradaile, evidently refers to a postlarva of the subfamily in the stage of temporary suppression of the postorbital spine. The generic position of Borradaile's postlarva seems uncertain.[1]

3.1.2 *Aristeinae.* Distinctions between certain genera of the Tribe Aristeini are unsatisfactory, since many of the available diagnostic characters are of a kind likely to vary from species to nearly-related species. Cross-resemblances between various species-groups placed in different genera suggest that actual phylogenetic relationships might be different from those usually recognized. The remarkable larvae of the tribe (*Cerataspis* H.Milne-Edwards, *Peteinura* Bate, etc.) are more varied in form than the adults; and, when assuredly identifiable with their parents, will doubtless supply useful clues to adult affinities (as well as prior names).

In the Tribe Benthesicymini, the genus *Benthesicymus* includes a number of well-defined species-groups which, when their interrelations are better understood, will probably invite subgeneric differentiation. *Benthonectes filipes* Smith seems to differ from some *Benthesicymus* only by a feature (elongation of posterior legs) less important than those distinguishing sub-groups of *Benthesicymus;* and *Benthonectes* is therefore regarded here as a synonym (possibly usable for one of the subgenera into which *Benthesicymus* may later be divided).

Some species of *Bentheogennema* now prove to resemble *Gennadas* in certain features previously thought diagnostic. Thus *B.pasithea* (De Man) has only a single pair of telson spines and only a minute gill on the somite of the first maxillipede. Since the genitalia of this species are remarkably like those of *Gennadas capensis* Calman, the question whether *Bentheogennema* is monophyletic has to be added to that of whether it is worth separating at the generic level. The answer to the latter question depends primarily on whether it would be convenient to divide *Gennadas* (as here used) into subgenera; which needs further consideration.

3.1.3 *Peneinae.* The tribe Peneini is a compact one, well-distinguished from other more advanced Peneinae by its retention of third maxillipede epipodites and last pereomere pleurobranchs. *Heteropeneus* seems to some extent intermediate between *Funchalia* and *Peneus,* since (in addition to the characters noted in the key) *Heteropeneus* has a thelycum much more like that of some *Funchalia* than that of the species of *Peneus* with an open receptacle. In turn, *Funchalia* is divisible into the subgenera *Funchalia* s.s and *Pelagopeneus* Burkenroad, the latter of which resembles *Heteropeneus* and *Peneus* in having ventral teeth on the rostrum.

Although no single feature that separates the Tribe Parapeneini from the Tribe Trachypeneini has yet to be found, they can be distinguished by a combination of various characters. Those individuals or species of Parapeneini which lack an antennular spine have an unmistakably trifid telson tip, while individual species of Trachypeneini with more or less fixed terminal telson spines have no antennular spine. Parapeneini seems a quite compact group of closely related forms. Trachypeneini as here defined is less compact; particularly by inclusion of the peculiar genus *Macropetasma.* However, all of the genera here included in the Trachypeneini have one or more features (additional to those of telson and antennular peduncle) which are unlike those of Parapeneini and suggestive of affinity with some of the other Trachypeneini (e.g. absence of pleurobranch from penultimate thoracic somite in *Macropetasma*). The arrangement suggested here probably follows natural lines, and may serve at least until the larval development becomes better known.

It seems likely that Parapeneini and Trachypeneini had a common ancestor with longitudinal and transverse sutures on the carapace, since these sutures are evidently related to peculiar features of the carapace of *Funchalia* of the Tribe Peneini (cf. Burkenroad 1936).

Among Parapeneini, the genus *Peneopsis* is very closely related to the genus *Metapeneop-*

*sis.* The members of the latter group are (so far as known) distinguished by possession of a tooth-like orbital angle and of a peculiar asymmetrical petasma. Some of the species are otherwise very like *Peneopsis*. However, *Metapeneopsis* contains a large number of species forming several distinct complexes. Although Kubo (1949) raised the group to generic rank for reasons which are quite mistaken, and although it is still not entirely certain that all species with a sharp orbital angle have an asymmetrical petasma, the generic separation will probably prove convenient, even though many relationships within *Metapeneopsis* are still unclear.

In the Tribe Trachypeneini, the genera *Atypopeneus, Protrachypene, Xiphopeneus, Parapeneopsis,* and *Trachypeneus* form a tight group from which *Trachypeneopsis, Mangalura* [with which *Metapeneus* Wood-Mason is synonymous (Burkenroad 1959)] and *Macropetasma* stand increasingly apart. Among the five most nearly related genera, *Xiphopeneus, Trachypeneus,* and *Parapeneopsis* seem especially close. A subgenus, *Trachysalambria,* has been separated from *Trachypeneus* s.s. by Burkenroad (1934a), but subsequent study of a wide range of Indo-Pacific material has revealed that some specimens with thelycum and petasma of the form found in *T.curvirostris* lack epipodites on the first or first two pairs of walking legs. No constant and important non-sexual distinctions between *T.curvirostris* and the *T.anchoralis* species complex have yet been found; hence the status of *Trachysalambria* must be regarded as uncertain. The American species-group centering on *T. constrictus* ('Section 1' of 'Division 2' of Burkenroad 1934b) seems clearly set apart from the rest of the genus, but the natural relationships of *Trachypeneus* to *Parapeneopsis* (and also relationships within the latter genus) require clarification.[2]

*Miyadiella* Kubo (1949), described as a unique adult peneine most nearly related to *Peneus,* seems in fact to have been based upon juveniles which might be developmental stages of *Atypopeneus.* I have seen a specimen from Amoy (unfortunately in poor condition) in the collection of the US National Museum, of a more advanced stage than Kubo's, which has eyes that apparently have not reached the middle of the second segment of the antennular peduncle. In this specimen (carapace 7.3 mm, total length 23 mm) the antennal flagellum is also more like that of adult Penaeinae (broken, but much longer than the antennule). The petasma of this specimen is obviously developing into the type found in *Atypopeneus.* Also, there is a transverse suture on the carapace (omitted by Kubo in his description of the *Atypopeneus* adult, hence perhaps also overlooked in his juveniles). Kubo's account of the legs of *Miyadiella* appears to be somewhat confused by errors in estimating magnification. His statement that *Miyadiella* has a gill on the last thoracic somite (and only two gills on the antepenultimate somite) seems suspect, since there are errors elsewhere in his branchial formulae (such as his report of three gills in addition to the podobranch on the somite of the second maxillipede in *Atypopeneus* and *Peneopsis*). It seems best to regard *Miyadiella* as a synonym of *Atypopeneus* until the existence of long-eyed adults has been demonstrated. If found, these might in any case not merit more than subgeneric separation.

3.1.4 *Sicyonia* is a large group as yet not clearly subdivisible, although the American species are on the whole rather distinct from those of the Old World (Burkenroad 1945b).

## 3.2 *Sergestoidea*

*Lucifer* is so highly specialized that its degrees of affinity to the various Sergestidae cannot be evaluated, and would seem to justify separation as a family.

Interrelationships among the genera of Sergestidae are also obscure. At most, it can be said that *Sicyonella* is primitive and clearly indicates a benthonic ancestry for the group; and that *Sergia* is very close to *Sergestes,* and *Peisos* to *Acetes. Petalidium* is hardly separable from *Sergia* and *Sergestes* in adult external characters, but seems to be set apart from both by peculiarities of its development. In a preceding discussion of this problem (Burkenroad 1945a), it was suggested that the remarkable larvae attributed to *Petalidium* by Gurney must belong to some other, as yet unknown sergestid adult. Confidence in this deduction can no longer be felt. Mastigopus larvae from South Africa, referred to in the preceding paper, have since been carefully compared by me with the two North Atlantic larvae identified as *Petalidium obesum* (Kröyer) by Hansen (1922). Although there are differences of detail (as previously noted in part), the South African larvae are not fundamentally distinguished, and may probably be attributed to *Petalidium foliaceum* Bate. In both the South African and North Atlantic larvae, the mouth parts of the younger individuals are congruent with those described by Gurney for the oldest stage of his larvae, while the mouth parts of the largest specimens are compatible with those of adult *Petalidium.* That is, the mandibular palp becomes slender and the palps of maxillule, maxilla, and first maxillipede, like the maxillary endites, appear to become well-developed. It thus seems probable that *Petalidium,* despite its adult resemblance to *Sergestes* (and especially to *Sergia*) may represent a distinct evolutionary line. A more penetrating search for adult differences is needed.

*Sergestes* seems naturally and conveniently divisible into the subgenera *Acheles* Cocco and *Sergestes* s.s. This division provides the chief reason for the present generic separation of *Sergestes* from *Sergia.* In the former subgenus the proximal spine of the inner margin of the dactyl of the third maxillipede is enlarged, the telson has a terminal point, the body of the pars media of the petasma is short; and the processus uncifer of the petasma has a terminal claw. Enlargement of the third maxillipedes has evidently occurred in some *Acheles* independently of the parallel development in *Sergestes* s.s., so that Hansen's (p.22) division of *Sergestes* (in the broadest sense, including *Sergia*) into forms with and without the enlargement seems an artificial one. *Acheles* as here defined includes: 'Tribu A' of Hansen's 1922 'Groupe II'; plus 'Tribu B' and a part ('II, 1') of 'Tribu A' of his 'Groupe I'. Its type, the spotted Mediterranean *A.arachnipodus* Cocco, appears to have been one of the species included by Hansen in '*S.corniculum* Kröyer'. The present subgenus *Sergestes* includes a part ('I') of 'Tribu A' of Hansen's 'Groupe I', plus 'Tribu B' of his 'Groupe II'.

*Sergia,* which comprises a part ('II, 2') of 'Tribu A' of Hansen's 1922 'Groupe I' of *Sergestes* s.l., has previously been distinguished as a subgenus (Burkenroad 1945a). It is divisible into distinct subgroups (as are also *Acetes* and *Lucifer*); but the application of subgeneric names to these groups requires further consideration.

# 4 ACKNOWLEDGEMENTS

As with my 1981 paper, in doing the work published here, I received help from the people acknowledged in my 1963 paper. Again, the efforts of Frederick R. and Joan M.Schram have allowed me to revise a manuscript begun more than 20 years ago. Without their efforts this product of my earlier years would never have reached print.

## 5 FOOTNOTES

1. This paper was begun many years ago, and was based on observations on all material available to me up to that time. I include only three genera for the Solenocerinae, omitting the additional taxa named by Pérez Farfante (1977). Her revision repeats in large part conclusions I have previously presented (Burkenroad 1936), with the addition of names for my subgroups. I am uncertain whether these names are needed.

2. Pérez Farfante (1972) erects a genus *Tanypeneus,* based on a newly collected species. Though it seems different from *Trachypeneus,* to which she closely compares it, I prefer not to add the generic name *Tanypeneus* to the key until its distinction from *Xiphopeneus* has also been treated.

3. The recent use by Bauer (1981) of a taxon Natantia is based on a single functional feature (mode of cleaning of the antennal flagella), of questionable phylogenetic importance. [Editor: see Felgenhauer & Abele, this volume.]

4. The manuscript of this present publication has been available for years (e.g. at the National Museum). The creation, with no formal diagnoses, of four families of peneoids by Pérez Farfante (1977) rather than two seems an overextension of my manuscript findings.

## REFERENCES

Alcock, A. & A.R.S.Anderson 1894. An account of a recent collection of deep sea Crustacea from the Bay of Bengal and Laccadive Sea. Natural History notes from H.M. Indian Marine Survey Steamer Investigations, Commander C.F.Oldham, R.N.Commanding. Series II, No.14. *J. Asiatic. Soc. Bengal* 63:141-185.

Barnard, K.H. 1950. Descriptive catalogue of South African decapod Crustacea. *Ann. S.Afr. Mus.* 38:1-837.

Bauer, R.T. 1981. Grooming behavior and morphology in the decapod Crustacea. *J. Crust. Biol.* 1:153-173.

Burkenroad, M.D. 1934a. Littoral Penaeidea chiefly from the Bingham Oceanographic Collection, with a revision of *Penaeopsis* and descriptions of two new genera and eleven new American species. *Bull. Bingham Oceanogr. Coll.* 4(7):1-109.

Burkenroad, M.D. 1934b. The Penaeidea of Louisiana, with a discussion of their world relationships. *Bull. Amer. Mus. Nat. Hist.* 68:61-143.

Burkenroad, M.D. 1936. The Aristaeinae, Solenocerinae, and pelagic Penaeinae of the Bingham Oceanographic Collection. Materials for a revision of the oceanic Penaeidae. *Bull. Bingham Oceanogr. Coll.* 5(2):1-151.

Burkenroad, M.D. 1939. Some remarks upon non-Peneid Crustacea Decapoda. *Ann. Mag. Nat. Hist.* 3 (11):310-318.

Burkenroad, M.D. 1945a. A new sergestid shrimp (*Peisos petrunkevitchi,* n.gen., n.sp.) with remarks on its relationships. *Trans. Conn. Acad. Arts Sci.* 36:553-591.

Burkenroad, M.D. 1945b. Status of the name *Sicyonia* H.Milne-Edwards, with a note on *S.typica* (Boeck) and descriptions of two species. *Ark. Zool.* 37(9):1-10.

Burkenroad, M.D. 1959. Decapoda Macrura I. Penaeidae. *Res. Sci. Miss. R.P.Dollfus en Egypte 1927-29. Paris* 25:67-92.

Burkenroad, M.D. 1963. The evolution of the Eucarida (Crustacea, Eumalacostraca) in relation to the fossil record. *Tulane Stud. Geol.* 2:3-16.

Burkenroad, M.D. 1981. The higher taxonomy and evolution of Decapoda (Crustacea). *Trans.San Diego Soc. Nat. Hist.* 19:251-268.

Glaessner, M.F. 1969. Decapoda. In R.C.Moore (ed.), *Treatise on Invertebrate Paleontology. Arthropoda 4, Part R:* R399-R533. Lawrence: Geol. Soc. Am. & Univ. Kansas Press.

Hansen, H.J. 1922. Crustacés décapodes (Sergestides) provenant des campagnes des yachts 'Hirondellé' et 'Princesse Alice' (1885-1915). *Res. Camp. Sci. Monaco* 64.

Kubo, I. 1949. Studies on penaeids of Japan and its adjacent waters. *J. Tokyo Coll. Fish.* 36:1-467.

Macuhline, J. & L.R.Fisher 1969. The biology of Euphausiids. *Adv. Mar. Biol.* 7:1-454.

Pérez-Farfante, I. 1972. *Tanypenaeus caribeas,* a new genus and species of the shrimp family Penaeidae (Crustacea, Decapoda) from the Caribbean Sea. *Bull. Mar. Sci.* 22:185-195.

Pérez-Farfante, I. 1977. American solenocerid shrimps of the Genera *Hymenopenaeus, Haliparoidea, Pleoticus, Hadropenaes* n.gen., and *Mesopenaeus* n.gen. *Fish. Bull.* 75:261-346.

Schram, F.R. 1981. On the classification of Eumalacostraca. *J. Crust. Biol.* 1:1-10.

St. Laurent, M.de 1979. Vers une nouvelle classification des Crustacés Décapodes Reptantia. *Bull. Off. Natu. Pêch. Tunisie* 3:15-31.

BRUCE E. FELGENHAUER & LAWRENCE G. ABELE
*Department of Biological Science, Florida State University, Tallahassee, USA*

# PHYLOGENETIC RELATIONSHIPS AMONG SHRIMP-LIKE DECAPODS

## ABSTRACT

Early classifications involving peneids, stenopodids, carideans, and procaridids are reviewed. These shrimp-like decapods have at various times been placed together in one suborder or in three separate suborders. The morphological characteristics of these groups are reviewed with particular reference to the gills, protocephalic skeleton, and foregut. We conclude that Dendrobranchiata (Peneoidea and Sergestoidea), Stenopodidea, and Procaridoidea represent independent evolutionary lines. We suggest that the Caridea is a heterogeneous group that should be re-examined.

## 1 INTRODUCTION

There is tremendous morphological diversity among decapod crustaceans. In size they range from pontoniid shrimp of a few millimeters to the giant Japanese spider crab, which can exceed 3 m with the legs extended. The shape of a decapod can vary from that of the coconut crab to that of planktonic sergestid shrimp. Despite this diversity there is a basic unity to the Decapoda that may be defined as follows: eucarid crustaceans with carapace fused dorsally to all thoracic segments and extending laterally to form a branchial chamber; exopod of the second maxilla with large lamellar expansion, the scaphognathite; eight pairs of thoracic appendages, the first three modified as maxillipeds; branchiae typically arranged in a series, (a) podobranchiae arise from epipodite with coxal insertion, (b) arthrobranchiae often in pairs and arise from body wall above coxae, and (c) pleurobranchiae arise above arthrobranchiae and never paired (Calman 1909; see Burkenroad 1981, for comments on homologies).

Although the group Decapoda has been relatively well-defined since Latreille the internal classification of the decapods remains a matter of some controversy (see Abele & Felgenhauer 1982, Bowman & Abele 1982). In this report, we will review earlier classifications and discuss the characters upon which they were based; and review the morphological diversity of decapods with special reference to four groups, the Peneoidea, Procaridoidea, Caridea, and Stenopodidea.

## 2 THE HISTORICAL DEVELOPMENT OF CLASSIFICATIONS

The very early classifications of the Decapoda are reviewed by Bate (1888), Calman (1909), Balss (1957a), and Glaessner (1969). Briefly, the early classifications (1700's, early 1800's)

291

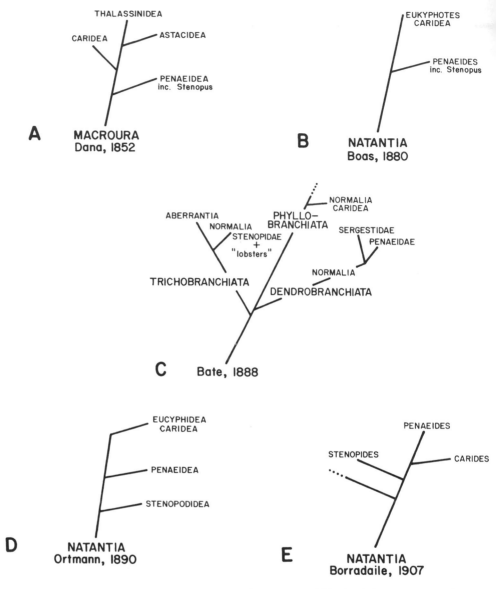

Figure 1. Phylogenetic relationships suggested by various authors within the shrimp-like Decapoda. Original spelling of taxa used by the various authors retained.

recognized two morphological types, a macrurous form with an elongated, subcylindrical body; and a brachyurous form with the cephalothorax greatly expanded and the abdomen reduced and folded beneath the cephalothorax. In Figures 1 and 2, we present the phylogenetic relationships suggested explicitly or implicitly by a number of authors representative of the various classifications that have been proposed. We have shown only those sections of the phylogenetic trees that deal with the groups under consideration here.

In 1852 Dana (Fig.1a) recognized Peneidea (including the genera *Stenopus* and *Spongicola*), Caridea, Astacidea, and Thalassinidea as the four major subdivisions of the Macroura. He considered the Peneidea as exhibiting the greatest amount of 'degradation', and placed them as the Macroura Inferiora. Dana based his classification on the form of the pereopods, the condition of the pleuron of the second abdominal somite and the form of the mandibular palp. Dana's search for the 'perfect' taxon may have been the result of his having undergone a deep religious conversion just prior to leaving on the Wilkes Expedition (Stanton 1975).

Boas (1880), in a major review, subdivided the decapods into the Natantia and Reptantia, and placed the Peneidea (still including *Stenopus*) and the Eukyphotes (Caridea) into the Natantia (Fig.1b). In the view of Boas, both the Natantia and Reptantia formed natural groups. In addition, Boas suggested that the peneid gill type (dendrobranchiae of Bate) gave rise to both the trichobranchiae and phyllobranchiae.

Huxley (1878) proposed a radical departure from earlier classifications based on an examination of gill structure. [Bate (1888) and others cite a classification of Huxley's (1883) but we have been unable to locate this reference.] He subdivided the decapods into the Trichobranchiata and the Phyllobranchiata. In the former group he created the Caridomorpha, containing the Peneidae, Stenopodidae, and Euphausiidae as independent taxa. The carideans were placed in the Phyllobranchiata. This classification represents the first rejection of the Natantia as a natural unit, but it was not widely followed (see Calman 1909).

In the *Challenger* report on the Macrura, Bate (1888) extended Huxley's use of the decapod gills as a basis for classification (Fig.1c). He recognized the gill of peneids as different from those of the other decapods and proposed the term Dendrobranchiata to include, under the subgroup Normalia, the Peneidae and Sergestidae. Bate also commented on the variability of the dendrobranchiate gill, and suggested that the trichobranchiate gill may have given rise to both the dendrobranchiate and phyllobranchiate gill types. Bate placed the Stenopidea as a tribe under the Trichobranchiata Normalia, which also included the Scyllaridae, Palinuridae, Eryonidae, Homaridae, and Astacidae. The Caridea (although the term was not used) were placed as a series of tribes under the Phyllobranchiata, subgroup Normalia. Bate, therefore, believed that the natantians were unrelated, and placed them in three separate divisions of Macrura.

Bates' classification was not followed by Ortmann (1890), who largely followed Boas (1880), placing the Stenopidea, Peneidea, and Eucyphidea (= Caridea) as three tribes within the Macrura Natantia (Fig.1d).

Borradaile (1907) published a widely cited and accepted scheme (Fig.1e). He recognized two suborders, the Natantia, containing three tribes (Penaeides, Carides, and Stenopides), and the Reptantia, containing all the other decapods. This classification doesn't follow Borradaile's own discussion of phylogeny where he suggests that Caridea and Reptantia arose independently from an early peneid stem. In fact Borradaile questions whether or not Natantia is a natural group. The position of the Stenopides is 'extremely doubtful' but is related to lower reptantians because 'it is trichobranchiate, has a curved mandibular palp and short endopodite to the first maxilliped, and lacks the copulatory apparatus of the male peneids and the spine (stylocerite) on the stalk of the antennule which is so characteristic of the Peneidea and Caridea'. However, some carideans have a curved mandibular palp and short endopod of the first maxilliped and a stylocerite is present in *Stenopus* and other stenopodidean genera (see de Saint Laurent & Cleva 1981, Holthuis 1946).

The classifications discussed thus far were proposed by neozoologists, and rarely con-

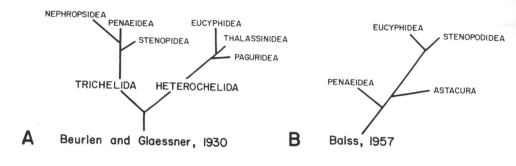

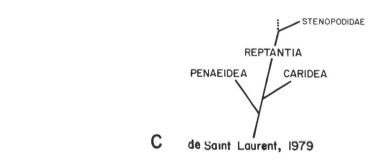

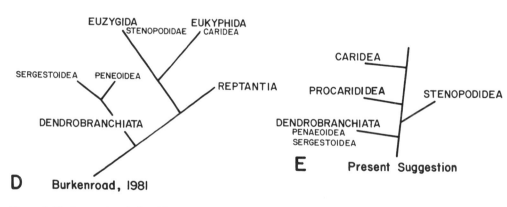

Figure 2. Phylogenetic relationships suggested by various authors within the shrimp-like Decapoda. Original spelling of taxa used by the various authors retained.

sidered fossils in any detail. Beurlen & Glaessner (1930, see also Glaessner 1960, 1969) proposed a classification (Fig.2a) that considered the known fossil record and the order of appearance of the various groups. They subdivided the Decapoda into two suborders, the Trichelida containing in one division (Nectochelida) the tribes Peneidea and Stenopidea, and in two additional divisions the fossil Paranephropsidea and the Recent Nephropsidea. The second suborder, Heterochelida, contained the remaining decapods, including the tribes Thalassinidea, Paguridea, and Eucyphidea (Caridea) in the division Anomocarida. The Caridea were derived from ancestral thalassinoids during the Jurassic. It was suggested that phyllobranchiate gills, variability of chelae, the main articulation of the pereopods being between merus and carpus, and the presence of appendix internae on the pleopods all supported this interpretation. Their classification of the Trichelida differs from their phylogeny which shows the Peneidea and Nephropsidea coming off together and later than the Stenopidea (Fig.2a). This classification suggests that Peneidea, Stenopodidea, and Astacidea had a common origin, and Beurlen & Glaessner suggested that the ancestor was probably trichelate with trichobranchiate gills. The above views were not widely accepted among neozoologists.

While Gurney (1924, see also 1936, 1942) did not review all aspects of decapod classification he did propose that the Stenopidea be placed as a separate section in the Macrura Reptantia. This placement was based on similarities between stenopodid larvae and certain Thalassinidea (e.g. Laomediidae). Gurney (1942) also pointed out that one stenopodid larva collected off Bermuda had a maxillule, which suggested a primitive condition.

In a major review of decapod systematics, Balss (1957a) utilized the concept of Natantia with three tribes, Peneidea, Stenopodidea, and Eucyphidea (Caridea) and the remaining decapods in the Reptantia. This is inconsistent with his suggested phylogeny (1957b, p.1801) which has Peneidea (Fig.2b) coming off a main stem followed by Astacura, and a third branch which bifurcates into the Eucyphidea (Caridea) and Stenopodidea. Balss apparently felt, as did Borradaile, that a practical classification and taxonomy were not necessarily congruent with phylogenetic relationships.

Burkenroad (1963) considered the evolution of the Eucarida in relation to the fossil record. He proposed that there were two major lines of decapod evolution, the suborder Dendrobranchiata, containing the Peneidea and Sergestidea, and the suborder Pleocyemata, containing the remaining decapods. Burkenroad recognized within the suborder Pleocyemata the supersection Natantia containing the Stenopodidea and Eukyphida (Caridea).

Burkenroad's recognition of the Dendrobranchiata as a distinct evolutionary line was accepted by de Saint Laurent (1979) in her revision of the classification of the Decapoda (Fig.2c). However, de Saint Laurent recognized three suborders of the decapods, the Peneidea (= Dendrobranchiata), the Caridea, and the Reptantia which include the Stenopodidea. She also listed 12 morphological features that, in various combinations, characterized the three suborders.

In 1981 Burkenroad revised his earlier (1963) classification proposing that four suborders of Decapoda be recognized: Dendrobranchiata, Euzygida (= Stenopodidea), Eukyphida (= Caridea), and Reptantia, apparently dropping the suborder Pleocyemata. This classification is somewhat inconsistent with Burkenroad's cladogram (Fig.2d), which indicates an independent origin for the Dendrobranchiata and Reptantia but a common origin for the Euzygida (Stenopodidea) and Eukyphida (Caridea), which would suggest that these latter two groups should be in the same suborder.

Neither Burkenroad (1981) nor de Saint Laurent (1979) dealt with the interesting shrimp genus *Procaris* described in 1972 by Chace & Manning. There are two known species, *P.ascensionis* from Ascension Island and *P.hawaiana* from Hawaii (Holthuis 1973). Chace & Manning (1972) erected a new superfamily Procaridoidea and family Procarididae for this shrimp, and included it in the Caridea. As discussed below we believe the procaridids represent an independent evolutionary line similar to both peneids and carideans (Fig.2e).

## 3 MATERIALS AND METHODS

The many decapod species examined in this study were selected from the authors' personal collections or obtained from private sources. Specimens were dissected either fresh or after preservation in various fixatives. Drawings and photographs of the endoskeletal system were prepared with the aid of a Wild-Heerbrugg camera lucida in addition to scanning electron micrographs.

### 3.1 *Preparation of endoskeleton*

Selected specimens were freed of as much muscle and membrane as possible without causing damage to the delicate endoskeleton. For removal of remaining adhering tissues, dissected specimens were placed in 15 % KOH at room temperature overnight. Each specimen was then placed in 5 % KOH and heated for approximately 15 minutes to complete the clearing process. After clearing, the endoskeletons were rinsed in distilled water for 5 minutes and then dehydrated to final storage in acid alcohol. Exuviae sometimes proved extremely useful in the study of the branchiae and complex endoskeleton features. Drawings were made from unstained endoskeletons using uneven illumination.

### 3.2 *Preparation of specimens for scanning electron microscopy*

The endoskeletons and foreguts examined with the scanning electron microscope (SEM) were cleared in KOH as described above. Before dehydration the specimens were postfixed in 2 % osmium tetroxide (in distilled $H_2O$) for two hours. Following fixation, endoskeletons were rinsed thoroughly in distilled water (3 changes five minutes each), dehydrated in a graded series of ETOH, and critical-point dried. Specimens were then mounted on stubs and coated with 20 nm of gold palladium for observation in a Cambridge S4-10 SEM at accelerating voltages of 5-30 kV using secondary and backscattered mode.

## 4 COMPARATIVE MORPHOLOGY

In Table 1, we have listed some morphological conditions that previous authors have indicated were important in characterizing the various groups. Although it is frustrating to state that the condition is 'unknown' or 'variable', the facts of the matter are that certain data are unavailable and that certain features are variable. Below we discuss each one of these characteristics in so far as data are available.

### 4.1 *Incubation of eggs and the first larval stage*

A basic difference among decapods is incubation of the eggs. Dendrobranchiate decapods release eggs into the water (although *Lucifer* carries them briefly on the pereopods, Burkenroad 1981), while all other decapods for which data are available carry the eggs on pleopodal setae. Nothing is known of the Procarididea.

The Dendrobranchiata hatch from the egg in the naupliar or protozoeal stage. Nothing is known of development in the Procarididea.

Table 1. Comparative morphological features among some Decapoda.

| Character | Dendro-branchiata | Procaridida | Stenopodida | Caridea |
|---|---|---|---|---|
| Incubation of eggs | None | Unknown | Pleopodal | Pleopodal |
| Eclosion from egg | Naupliar | Unknown | Zoeal | Zoeal |
| Gill type | Dendro-branchiate | Phyllo-branchiate | Tricho-branchiate | Phyllobranchiate |
| Epistomal condition | Membranous, articulated | Membranous, articulated | Membranous, articulated | Membranous, articulated |
| Cervical groove | Present | Present, weak | Present | Variable (present in *Glypho-crangon*) |
| Pleuron of second abdominal somite overlapping first and third | Never | Always | Never | Variable (not in some Psalidopidae, Glyphocrangonidae) |
| Pleonic hinges | Pleomeres 1-2, 2-3, 4-5, 5-6 | Pleomeres 1-2, 2-3, 4-5, 5-6 | Pleomeres 4-5, 5-6 | Variable, usually Pleomeres 1-2, 2-3, 4-5, 5-6 (all hinged in *Glyphocrangon*) |
| Form of telson | Narrowly tri-angular, uropods with-out diaresis | Subrectangular, lateral branch with diaresis | Subrectangular, uropods with-out diaresis | Subrectangular, uropods variable |
| Appendix internae | Absent | Absent | Absent | Usually present on pl.2-5 |
| Appendix masculinae | Absent | Absent | Absent | Present, sometimes reduced or absent, e.g. *Euryrhynchus*, *Synalpheus* |
| Form of first maxilliped (presence of caridean lobe) | No lobe on exopod | Very slight lobe on exopod | No lobe on exopod | Variable but usually with lobe |
| Form of third maxilliped | 7 segments, pediform | 7 segments, pediform | 7 segments, pediform | 3-5 segments, not pediform |
| Form of pereopods | 1-3 chelate, 4, 5 achelate | 1-5 achelate | 1-3 chelate, 4, 5 achelate | Variable (e.g. *Pseudocheles*), usually 1-2 chelae |
| Form of gastric mill | Well developed | Well developed | Developed | Usually reduced |

The development of some stenopodids has been described by Gurney (1924, 1936, 1942) and Williamson (1976). Based on development of *Stenopus* Gurney (1924) suggested that stenopodids are closely related to the Laeomediidae among the Reptantia. However, there are many unique features of stenopodid development according to Gurney (1924). The larva of *Stenopus* is unique among the decapods in hatching with four pairs of natatory limbs, while carideans hatch with three and reptants usually two. The uropods do not appear until stage IV and the last pair of pereopods do not develop until stage VI. In general the larval facies is distinct.

The larval development of carideans was reviewed by Gurney (1942 and citations therein). Williamson (1982) provides a more recent literature review as well as a key for the identification of crustacean larvae. Although larvae from most caridean families have been described, 'no concise definition which will be applicable to the whole group can be framed' (Gurney 1942). A review similar to that provided for the Brachyura by Rice (1980) is needed for the carideans.

## 4.2 *Branchiae*

Various authors (e.g. Huxley 1878, Bate 1888, Burkenroad 1963, 1981) have utilized the form and number of decapod gills as a basis for classification. Early authors (e.g. Milne Edwards 1837) recognized two basic gill morphologies: a trichobranchiate gill (Fig.4a,b) with a series of rather filamentous lateral branches arising from the main stem or branchial axis, and a phyllobranchiate gill (Fig.5) with paired lamellar branches arising from the branchial axis (see Huxley 1878). Bate (1888) pointed out that the so-called trichobranchiate gill actually included a rather distinct morphological type that is characteristic of peneids (Fig.4c,d) and sergestids (Fig.4f). This dendrobranchiate gill has paired lateral branches arising from the branchial axis with a series of subdivided secondary rami coming off each lateral branch. Examination of a relatively few species has convinced us that only the dendrobranchiate form is distinct. Trichobranchiate gills occur in a number of apparently unrelated taxa (e.g. *Stenopus, Aeglea, Palinura,* some Paguroidea), and we have had difficulty in distinguishing 'trichobranchiate' gills with paired flattened branches from phyllobranchiate gills. For example, the gills of *Upogebia* and axiids seem to us to be phyllobranchiate (see also Burkenroad 1963). Bouvier (1940) illustrated a series of gills from the dromiid genus *Dicranodromia* that appear to be transitional between trichobranchiae and phyllobranchiae. A re-evaluation of the morphologies and terms is necessary before a complete evaluation is possible.

Huxley (1878) and Bate (1888) suggested that the trichobranchiate type gave rise to both the dendrobranchiate and phyllobranchiate gills. As Bate considered the Dendrobranchiata to be the most primitive decapods, it would then follow that some ancestral decapod must have been trichobranchiate. However, it is not clear to us how a trichobranchiate form could give rise to a dendrobranchiate gill. Boas (1880) and Burkenroad (1981) both suggested that the dendrobranchiate gill could have given rise to the trichobranchiate and phyllobranchiate gills (Fig.3). An expansion of the two lateral branches of the dendrobranchiate gill results in a phyllobranchiate gill while loss of the secondary rami of the lateral branches would form a basic type of trichobranchiate gill. However, trichobranchiate gills usually have a large number of lateral branches rather than pairs, as would be the situation resulting from the loss of the secondary rami of a dendrobranchiate gill. The large amount of variation in dendrobranchiate gills (Bate 1888) suggests that reduction of the lateral branches and expansion of the secondary rami would result in a typical trichobranchiate gill. Whichever is correct, there is little doubt that phyllobranchiate gills represent a derived condition.

Burkenroad (1981) based his classification, in part, on the ontogenetic development of gills and the formulae of the adults. Briefly, pleurobranchs appear later in ontogeny than arthrobranchs in Dendrobranchiata, while pleurobranchs appear earlier than arthrobranchs in carideans and stenopodideans. The gills apparently appear simultaneously in Reptantia. Reptants also are unique in lacking a pleurobranch on the first pereonal somite even when one is found on the posterior somites. Comparison of gill formulae is, as Burkenroad points out, sometimes difficult because homologies are not immediately apparent when some of the gills are absent.

## 4.3 *The protocephalon*

In her discussion of decapod classification de Saint Laurent (1979) suggested that the form of the protocephalon is a major distinguishing feature among the suborders of the Decapoda. The protocephalon is the anterior portion of the endophragmal skeleton, consisting of the eyes, antennules, antennae, and associated skeletal elements (Snodgrass 1951, Young 1959). De Saint Laurent points out that a free ophthalmic segment is present within the Reptantia and that this is represented by an unpaired cavity in the protocephalon of this group. In addition, the epistome of peneids and carideans is said to be divided by a membranous invagination into two portions that move against each other. In Reptantia (except stenopodids, which de Saint Laurent includes in this group), the epistome is solid, without any subdivisions formed by membranes. We consider the epistome to be the skeletal elements that begin posteriorly at the site of attachment to the labrum. The epistome continues laterally and anteriorly as the skeletal elements forming the basal regions of the antennae, antennules, and ophthalmic area.

The morphology of the protocephalon is complex and requires numerous illustrations and photographs for an adequate description. Here we can only present a brief description of the following groups that are examined: Dendrobranchiata (e.g. *Peneus, Sicyonia*), Procarididea (*Procaris*), Stenopodidea (*Stenopus*), Caridea (e.g. *Atya, Potimirim, Palaemonetes, Macrobrachium, Oplophorus, Alpheus*), Reptantia Astacidea (*Cambarus*) [for a complete list, see Appendix I]. De Saint Laurent is essentially correct in her descriptions, but the variation within groups is much greater than perhaps she anticipated. In Figure 6a, the epistome of *Peneus* is shown. The mandibular palps have been removed from Figure 6a for clarity. The episome is membranous with a deep median invagination and membranes along the lateral extensions (Fig.6a). In *Peneus* then, the epistome consists of a series of medial and lateral skeletal elements separated by membranes (see Young 1959, fig.28). This morphology seems characteristic of all members of the Dendrobranchiata we examined.

The epistomal region of carideans is variable. In *Palaemonetes* (Fig.6d), there is a medial and, anteriorly, a lateral membranous region while in *Oplophorus* (Fig.6e) there is only a weak medial invagination suggesting a membranous region. The situation in large species of *Macrobrachium* is quite different and suggests that fusion of skeletal elements may be, in part, a function of size. In *Macrobrachium americanum* the epistome is rigid, heavily chitinized with no indication of membranes. Although the presence and degree of membranes in the epistome of carideans is variable all that we have examined have the epistome anterior to the base of the antennae.

The epistome of *Stenopus* is membranous, articulates with itself and is rather complex. It consists of a heavily armed anterior portion that is semicircular and to which the labrum attaches. Anterior to this portion ('ep' in Fig.6b) is a deeply recessed membrane that attaches to the anterior portion of the epistome that projects clearly between the antennal bases and the two lateral spines of the posterior portion of the epistome. This arrangement differs from anything we have seen in dendrobranchiates, carideans, or reptantians.

The epistomal region of *Procaris* (Fig.6f) has an anterior medial invagination that permits movement of the two lateral portions similar to both *Peneus* and some carideans.

In the reptantians that we have examined the epistomal region differs from that of the dendrobranchiates, carideans, procaridids, and stenopodids. In astacids (Fig.6c) the epistome is a large plate that is located between the mandibles and the antennae; it extends anteriorly between the bases of the antennae. The large posterior portion between the mandibles and antennae is always heavily sclerotized with no membranes present. However, even within a single family (e.g. Cambaridae) some species have a membrane separating the anterior from the posterior portion (Fig.6c) while in other species the entire epistome is a continuous plate. In general, reptantians (excluding stenopodids, which we do not consider reptantians) have the largest portion of the epistome as an extensive plate between the mandibles and the antennae. Perhaps the most extreme situation is found in the only extant representative of the Mesozoic Glypheoidea, *Neoglyphea inopinata,* where the epistome is extremely elongated, being almost twice as long as wide (Forest & de Saint Laurent 1981, Fig.9).

### 4.4 *The cervical groove*

Several authors have suggested that a cervical groove is absent in carideans, but it is present in some forms (e.g. alpheids, see Coutiere 1899) and is particularly well-developed in the genus *Glyphocrangon* (see Holthius 1971). A cervical groove is present in the other groups as noted in Table 1.

### 4.5 *Pleura*

A feature considered characteristic of carideans is the expanded pleura of the second abdominal somite which overlap those of the first and third (e.g. Holthius 1955). However, in some species of Psalidopodidae (see Chace & Holthuis 1978, *Psalidopus barbouri*) and Glyphocrangonidae (see Holthius 1971, *Glyphocrangon neglecta*) the pleura of the second abdominal somite are not expanded and do not appear to overlap those of the first and third somites. However, even in these groups the second pleuron overlaps the first when the abdomen is flexed. In dendrobranchiates and stenopodids the pleura never overlap while in procaridids it does (see Chace & Manning 1972).

### 4.6 *Pleonic hinges*

The pleural somites of many decapods are locked to each other by mid-lateral hinges. Each hinge is formed by an expansion of the pleuron into a knob-like structure that fits into a cavity formed by the adjacent pleuron. It is usually the posterior somite that locks into the anterior one. Burkenroad (1981) stated that pleural hinges are found on all somites of the Dendrobranchiata, but those of the junction 3 to 4 are hidden under the posterior margin of the third somite. We were unable to locate the hinge on the junction of 3 to 4 in specimens of *Solenocera*. In specimens of *Peneus* (and to a lesser extent *Solenocera*) there are muscle bundles under the margin of the third somite at the point where the hinge would be located but there are no obvious skeletal modifications that are so apparent on the other somites. In the Stenopodidea Burkenroad (1981) states that only the last three pleonic somites are hinged together, that is, hinges are present between somites 4 and 5, and 5 and 6. We have examined these in *Stenopus hispidus* and find them to be more ventral than in other groups, and they are not hinged in the same ball and socket morphology as in other groups. The hinges in most carideans are present on all somites but the junction of 3 to 4, presumably to permit greater flexing at that point. However, in *Glyphocrangon* (Holthius 1971), all somites appear to be hinged. Our single specimen of each species of *Procaris* has undergone some deterioration but hinges appear to be present on all somites except the junction of 3 to 4. Finally all somites are said to be hinged in Reptantia (Burkenroad 1981) and this was the case in those representatives that we examined although the morphology was variable. It seems clear that a much more detailed examination of these hinges is required before any strong phylogenetic conclusions can be drawn.

## 4.7 *Form of telson*

In their discussion of the systematic position of *Procaris* Chace & Manning noted that the telson of *Procaris* is more similar to that of carideans than to that of dendrobranchiates. In dendrobranchiates the telson tends to be narrowly triangular and the uropods lack a diaeresis, while in *Procaris* the telson is subrectangular and the lateral branch of the uropods has a diaeresis. However, as with other characters, carideans are variable in this regard. Some pontoniids have an almost oval telson (Holthuis 1951b) while *Psalidopus* has a narrowly triangular telson (Chace & Holthuis 1978).

## 4.8 *Appendices internae*

An appendix interna is absent from all pleopods in dendrobranchiates, procaridids, and stenopodids. The situation, however, is variable in carideans. An appendix interna is usually present on pleopods 2 through 5. However, the genus *Desmocaris* (usually in the family Palaemonidae, though we agree with Powell 1977, that it is probably a separate family) lacks appendices internae, and they are absent from the second pleopods in females of the palaemonid *Euryrhynchus* (Holthuis 1951b, Powell 1976).

## 4.9 *Appendix masculina and petasma*

The form of the male copulatory organ varies within the Decapoda. Dendrobranchiate males have the first pleopods modified as a petasma for sperm transfer (e.g. Farfante 1975). There is no major modification of male pleopods in stenopodids, though the first pleopod has some minor sexual modifications (Holthuis 1946), and though no one has sexed procaridids the pleopods are not modified (Chace & Manning 1972). In contrast, carideans usually have the second pleopod of the male with an appendix masculina which is involved in sperm transfer (Bauer 1976, Felgenhauer & Abele 1982). *Synalpheus* males lack an appendix masculina (Coutiere 1899). Holthuis (1951b) suggested that *Euryrhynchus* males lack an appendix masculina. However, Powell (1976) suggested that what has been called an endopod is actually an appendix masculina attached to a greatly reduced endopod. Powell (1976) also described a complex copulatory apparatus in the genus *Euryrhynchoides* that involves both the endopod and appendix masculina of the second pleopod. This structure appears to be unique among the carideans.

## 4.10 *Stylocerite*

As noted earlier, Borradaile (1907) suggested that stenopodids and reptantians lack stylocerites. However, a stylocerite is present in both of these groups as well as in peneids and carideans.

## 4.11 *Caridean lobe*

Numerous authors, including Burkenroad (1981), have stated that a lobe on the basal portion of the exopod of the first maxilliped is characteristic of carideans, hence its name, caridean lobe or α lobe of Boas (1880). This lobe is absent in dendrobranchiates and stenopodids. There is a weak, but distinct, lobe in *Procaris* and, again, its presence and development in carideans is variable. The most unusual first maxilliped is found among the Pasiphaeidae where it consists of a large, lamellar, 2-segmented appendage in *Pasiphaea semi-spinosa* (see Holthuis 1951a, fig.1). In another pasiphaeid, *Leptochela bermudensis,* the first maxilliped apparently consists of an endite, a small endopod, a club-like exopod with a distal indentation, and an epipod (Chace 1976, fig.6). In some crangonids (e.g. *Ponto-philus bidens,* see Holthuis 1951a, fig.33) there is no obvious lobe present. Among hippo-lytids it is weakly developed in *Latreutes parvulus* but well-developed in *Bythocaris cosme-tops* (see Holthuis 1951a, figs.20, 30). Among alpheids it is well-developed in some genera and weak in others (e.g. *Athanas,* see Coutiere 1899). In many families (e.g. Processidae, Palaemonidae, Oplophoridae, Physetocaridae) it appears to be well-developed in all mem-bers. Finally, there appears to be a 'caridean lobe' in some pagurids (see Kunze & Anderson 1979).

## 4.12 *Third maxillipeds*

The third maxillipeds in the Dendrobranchiata, Procarididea, and Stenopodidea are seven-segmented and pediform. The third maxillipeds in carideans never have seven segments and are often operculiform in shape. When the third maxillipeds are pediform they have five segments at most.

## 4.13 *Pereopods*

The form of the pereopods has been used as a basis for classification. The Dendrobranchiata and Stenopodidea have the first three pairs of pereopods chelate while the Procarididea are achelate. Among the carideans there are species with all five pereopods chelate, though the last three are modified chela (*Pseudocheles* in the family Bresiliidae, Chace & Brown 1978); or species with subchelate pereopods (e.g. *Glyphocrangon,* Holthuis 1971); and many species with the first two pereopods chelate. The chelae of carideans are extremely diverse.

## 4.14 *Morphology of the foregut*

A general decapod foregut (see Huxley 1880, Patwardhan 1935 and citations therein; Kaestner 1970, Schaefer 1970, Coombs & Allen 1978, Kunze & Anderson 1979) consists of two distinct regions. A J-shaped esophagus opens into a large anterior cardiac chamber and a smaller posterior pyloric region (Fig.7a, 8a) separated from the cardiac stomach by a ventral cardiopyloric valve of varying degrees of complexity. A ventral gland filter (am-pulla) (Fig.8a) is present in the floor of the pyloric chamber. This structure accepts only the smallest of particles and leads directly to the hepatopancreas. In general, both the cardiac and pyloric stomachs are chitinous and form a complex series of ossicles (Moc-quard 1883) and this is also true in *Atya* (Fig.7a). It is this chitinous, interior lining of the cardiac stomach that forms the gastric mill. The gastric mill typically consists of a large median tooth located on the urocardiac ossicle (Fig.7b,c) extending from the roof of the cardiac stomach, and a pair of lateral teeth borne on the zygocardiac ossicles (Fig.7d). Patwardhan (1935, 1936) (also Coombs & Allen 1978, de Saint Laurent 1979) suggested that the Stenopodidea and Peneidea have simple gastric mills, and from this con-dition two trends are apparent: a progressive reduction of the gastric mill in Caridea (said sometimes to be absent) from the Hippolytidae-Atyidae to the rest of the carideans, and, an increasing development of the gastric mill from the lower Reptantia-Astacura-Anomura to the Brachyura. The situation in reality, as might be expected, is much more complex.

4.14.1 *Dendrobranchiata*. The gastric mills of *Peneus* (Fig.9a), *Solenocera* (Fig.9b,c) and *Sergestes* (Fig.9d) exhibit striking similarities in their morphology. All have a well-developed median tooth (mt) armed with a series of teeth along their lateral margins (see Fig.9c). All three representatives also have strong lateral teeth arising from the zygocardiac ossicles.

4.14.2 *Stenopodidea*. The gastric mill of *Stenopus* (Fig.9e) has a moderately developed median tooth attached to a subcircular hastate plate. The teeth are knob-like and do not resemble those seen in other decapods. Well-developed peg-like lateral teeth are also present.

4.14.3 *Procarididea*. The foregut of *Procaris* has a well-developed median tooth (Fig.10a, b) armed with accessory teeth (Fig.10c). Large lateral teeth (Fig.10b,d) are also present. The floor of the cardiac chamber has the denticles guarding the entrance to the gland filter (gf) (Fig.10f).

4.14.4 *Caridea*. There is considerable diversity in the gastric mill of species referred to the Caridea. Perhaps one of the more unusual is found in the pasiphaeid genus *Leptochela* (Fig.11). The foregut itself is less than 1 mm in length with a well-developed gastric mill. The median tooth is magnificent and bifurcates medially into a large number of scaled teeth (Fig.11b). A large number of serrate petals surround the base of the median tooth (Fig.11f). Strong lateral teeth (Fig.11c,d) are present. The cardiopyloric valve is rather elaborate for a caridean (Fig.11e). (N.B. Fig.11e: cardiopyloric valve; X1500)

The chitinous foregut of *Atya* (Fig.7a) has a bifid, heavily armed median tooth (Fig.7b, c) with sharp teeth lining its interior (Fig.7c). A series of stout sclerotized teeth (lateral teeth) are present, arising from the broad zygocardiac ossicle (Fig.7d, arrow). A convoluted membrane borne on the median projection from the roof of the pyloric stomach (Fig.7e) is apparently unique to certain atyids.

The gastric mill of *Palaemonetes* (Fig.8b) consists of a very small median tooth with minute teeth (Fig.8c), extending from the roof of the cardiac chamber. There are no lateral teeth present; instead a lateral row of plumose setae is present (Fig.8b, arrow).

*Saron* (Fig.8a) lacks a gastric mill, the foregut consists of two sacs with no obvious chitinized regions. We also examined the foregut of *Crangon, Oplophorus, Gnathophyllum* and *Alpheus* and found no gastric mill present.

4.14.5 *Reptantia*. The massive gastric mill of *Cambarus* (Fig.9f) consists of a large, smooth bifid median tooth with strong lateral teeth.

The gastric mill region of *Upogebia* (Fig.8d,e,f) has a strong median tooth with approximately 16 stout lateral teeth present. The lateral teeth are robust, consisting of 20 or more movable plates (Fig.8d). The pyloric fingerlets are shown in Figure 8e, with a closeup in Figure 8f. Details of the morphology and function of the foregut of this species can be found in Powell (1974).

# 5 THE FOSSIL RECORD

The fossil record of the Decapoda has recently been reviewed by Glaessner (1960, 1969), Burkenroad (1963), and Schram (1982). The earliest decapod is *Palaeopalaemon newberryi,* known from the Upper Devonian in central North America (Schram et al. 1978). The species is reptant in form and shares some characters with the Astacidea and the Glypheoidea. Although *Palaeopalaemon* tells us little about the groups under consideration here, it is significant because it is the earliest decapod recognized and it is 100 million years older than any other known form.

Burkenroad (1963), following Brooks (1962), suggested that the Permian genus *Palaeopemphix* is an early decapod, but Glaessner (1969) suggested that it needs re-examination before this can be accepted. The problems center around interpretation of the carapace furrows.

Other than *Palaeopalaemon* the earliest accepted decapods are Permo-Triassic, the rare Peneidae and the extinct family Erymidae in the Astacidea (Glaessner 1969, Förster 1966, 1967). Among the Permo-Triassic Peneidae is the genus *Antrimpos,* very similar if not identical to *Peneus* (see Burkenroad 1963). Since the Peneidae are rather advanced dendrobranchiates, the group as a whole must have been present earlier than the Permo-Triassic.

Although the familial status is uncertain, carideans are known from the Middle Jurassic (e.g. *Udora*). The genus *Oplophorus* (family Oplophoridae) is known from the Upper Jurassic (see Glaessner 1969). Another interesting caridean is *Udorella* (family Udorellidae) known only from the Upper Jurassic. As reconstructed *Udorella* has five subchelate pereopods, long exopods, and a pediform third maxilliped which ends in a long, thin terminal segment. If the third maxilliped of *Udorella* has seven segments, there would be some similarity to *Procaris.* For the most part, the available fossils tell us little about caridean evolution.

There are no known fossils of either the Stenopodidea or the Procarididea.


## 6 THE PROPOSED CLASSIFICATION

Examination of the material listed in Appendix I and a survey of the literature has convinced us that the Dendrobranchiata, Procarididea, and Stenopodidea are natural taxa and should stand alone. The Caridea consists of a large and diverse group of species that may have to be reorganized in some other way. We do not wish to indicate the taxonomic level (suborder, infraorder, etc.) for these groups until more is known concerning the features in table 1 for all the decapods. For the present we recognize the Caridea as a taxon but believe that a revision is needed. Below we provide diagnoses for these taxa and then discuss their relationships.


### Dendrobranchiata Bate 1888

Eggs released free, hatch as nauplii or protozoeas. Gills consist of branchial axis, with paired lateral branches each with subdivided secondary rami (dendrobranchiate condition). Gastric mill well-developed, strong armed median tooth, well-developed lateral teeth. Protocephalon of an ocular plate and an epistomial region, the latter subdivided such that it can articulate with itself. Epistomal bars anterior to the labrum. Pleura of first abdominal somite overlap those of the second. Appendices internae and masculinae absent. First pleopod in males modified into a complex copulatory appendage, the petasma. Third maxillipeds pediform, with seven segments. First three pairs of pereopods chelate (Permo-Triassic to Recent).

**Procarididea** Chace & Manning 1972
Nothing is known of the eggs or larvae. Gills phyllobranchiate, branchial axis with pairs of lateral lamellae (Fig.12f). Gastric mill well-developed, with strong, armed median tooth, large lateral teeth. Protocephalon, with an occular plate and a subdivided epistome that articulates against itself. Pleura of the second abdominal somite overlap those of the first and third. Appendices internae and masculinae absent. Third maxillipeds pediform, with seven segments (Fig.12b,c). All pereopods achelate (Fig.12a). First four pereopods with very large epipods at right angles, extending into branchial chambers (Fig.12f). All maxillipeds and pereopods with well-developed exopods (Recent).

**Stenopodidea** Bate 1888
Eggs attached to pleopodal setae, hatch as zoeas. Gills trichobranchiate, branchial axis bearing numerous filaments irregularly arranged. Gastric mill with median tooth attached to subcircular hastate plate, knob-like teeth on margins of median tooth, well-developed lateral teeth present. Protocephalon with ocular plate and epistome, latter with heavily armed subcircular narrow portion attached to labrum connecting by a membrane anteriorly to a narrow portion between antennae. Pleura of second abdominal somite do not overlap those of first and third. First pleopod in both sexes uniramous, appendices internae and masculinae absent. Third maxillipeds pediform, with seven segments. First 3 pairs of pereopods chelate, third enlarged (Recent).

**Caridea** Dana 1852
Eggs attached to pleopodal setae, hatch as zoeas. Gills phyllobranchiate, branchial axis with pairs of lateral lamellae. Gastric mill variable.[well-developed in *Leptochela* (Pasiphaeidae) and *Atya* (Atyidae); greatly reduced to absent in Palaemonidae and Hippolytidae; absent in some members (at least all we examined) of Crangonidae, Alpheidae, Gnathophyllidae, and Oplophoridae]. Protocephalon of an occular plate and epistomal region, the latter usually subdivided so that it may articulate with itself (e.g. *Palaemonetes, Oplophorus*) or be as a solid plate (e.g. *Macrobrachium*). Pleura of second abdominal somite usually overlap those of the first and third (except in some species of *Glyphocrangon* and *Psalidopus*). Appendices internae and masculinae usually present (except *Desmocaris* lacks internae and *Euryrhynchus* may lack masculinae). First maxilliped usually with expansion of lateral border of exopod (absent in pasiphaeids and greatly reduced in many others). Third maxillipeds variable, with three to five segments. First and second pereopods usually chelate or achelate, but variable (Middle Jurassic-Recent).

A comment on nomenclature is probably in order as various names have been applied to the Dendrobranchiata, Caridea, and Stenopodidea. We agree with Burkenroad (1963) on the use of Dendrobranchiata Bate 1888 for the Peneidoidea and Sergestoidea. We believe that Caridea Dana 1852 was applied by Dana to families of shrimps that constitute this taxon today. We see no need to use the later terms of Boas (1880, Eukyphotes), Ortmann (1890, Eucyphidae) or Burkenroad's (1981) modification (Eukyphida) of Boas's term. Similarly we see no need to erect a new name for stenopodids in the manner of Burkenroad's (1981) Euzygida. We have cited Bate (1888) as the author of the higher taxon and Huxley (1878) as author of the family for stenopodids. Huxley (1878) included the stenopodids in the Trichobranchiata with peneids and it was Bate (1888) who separated the stenopodids as an independent group.

Finally, we believe that the evidence in favour of recognizing the Dendrobranchiata is overwhelming. The evidence in favor of other groups is not so decisive, but to include the procaridids with either the dendrobranchiates or the carideans would weaken the former and expand even more the already variable Caridea. We feel that until information becomes available on egg incubation and larval development in procaridids, it is best to separate them as an independent group. The taxonomic status of the stenopodids has changed several times in the past from a separate taxon to a subdivision of the Reptantia (Gurney 1924; de Saint Laurent 1979). However, we feel that inclusion of the stenopodids in the Reptantia would require too serious a modification of the definition of that group, especially in regard to the protocephalic region, pleonic hinges, and branchial formula.

# 7 DISCUSSION

It is now more than 100 years since Boas (1880) proposed a major revision of decapod classification. It would appear that interest in decapod phylogeny and classification has increased again. Guinot (1977, 1979) has proposed a reclassification of brachyurans. Rice (1980) re-examined brachyuran classification in light of larval characteristics. De Saint Laurent (1979) proposed a new higher classification of the decapods as well as some changes (1980a,b) in brachyuran classification. Burkenroad (1981) proposed a modification of his earlier (1963) classification, but emphasized modern rather than fossil forms. These studies and the present one are the result of both a renewed interest in comparative morphology and the discovery of important new species. A Recent representative of the previously thought to be extinct Glypheoidea, *Neoglyphea inopinata* (see Forest, de Saint Laurent, & Chace 1976, Forest & de Saint Laurent 1975) has stimulated a re-examination of the Reptantia, and the discovery of the shrimp *Procaris ascensionis* Chace & Manning 1972, has stimulated a study of the Caridea.

However, it is probably worthwhile to re-examine decapod morphology without any biases generated by previous classifications. As we have shown here, the concept of Caridea is based on a large number of variable characters. Similarly, the Reptantia is a diverse group of species that may not have enough shared derived characters to warrant unification as a single taxon.

The value of various characters in taxonomy should also be reconsidered. For example, the gills of axiids are often considered to be trichobranchiae, but some species have gills that appear to us to be phyllobranchiate. The presence and development of the caridean lobe is also a questionable character. It is absent in some pasiphaeids and little, if at all, developed in some alpheids. Additional characters that might have systematic value should also be evaluated and these are most likely to be discovered through extensive studies of comparative anatomy. For example, de Saint Laurent (1979) has called attention to the value of the endophragmal skeleton of the protocephalon as an important character, but very few species have been examined in this regard. We believe that an analysis of comparative morphology, internal and external, without attempting to fit species into any current classification would yield important results in sorting out the complexities of decapod evolution.

Finally, there is the problem of the fossil record. The oldest known decapod, *Palaeopalaemon newberryi,* may be characterized as follows (modified from Schram et al. 1978):

Rostrum present, unarmed, about one-third of carapace length; carapace with cervical, post-cervical, branchiocardiac, antennal and gastro-orbital grooves; median dorsal, antennal, branchiostegal, and lateral ridges present; first antennae small, second antennae with large broad scaphocerite; first pereopod large, chelate status unknown; pereopods two through five smaller than first, and two through four subchelate; first pleonic somite short, partially covered by carapace; second pleonic pleura not expanded; apparent pleonic hinges on all pleonic somites; telson broadly subtriangular, unarmed; uropods without diaeresis. This earliest known decapod is a mixture of astacidean and glypheoidean characteristics, and in addition, the antennal scale suggests a 'natantian' affinity. This juxtaposition of characters presents a problem because although most authors consider the dendrobranchiates to be the most primitive decapods, they do not appear in the fossil record until 100 million years after *Palaeopalaemon*. Although we recognize the capricious nature of the fossil record we believe that the available data would indicate that the Dendrobranchiata separated early from the other decapods, a conclusion reached by others. In contrast to most other authors, however, we believe that the origins and relationships of the Caridea and Stenopodidea (and possibly the Procarididea) are to be found among those groups traditionally considered reptants. In this regard our speculations are similar to those of Beurlen & Glaessner (1930), who derive the Caridea from ancestral thalassinoids.

# 8 ACKNOWLEDGEMENTS

We thank the following individuals for assistance, both allowing us access to specimens and commenting on the manuscript: Drs Raymond B.Manning, Robert H.Gore, Frederick R.Schram, Anthony J.Provenzano, Jr, Daryl L.Felder, Dr Sandra Gilchrist, Ms Elizabeth Woodsmall and Mr Kenneth Womble assisted with the graphics. Specimens were kindly provided by the San Diego Museum of Natural History, San Diego, CA, and Scripps Institution of Oceanography, La Jolla, CA. We also wish to thank Dr Anne Thistle for typing and editing the manuscript.

# 9 APPENDIX I

Material examined

Dendrobranchiata
  Peneidae
    *Peneus setiferus* (Linnaeus)
  Solenoceridae
    *Solenocera vioscari* Burkenroad
  Sergestidae
    *Sergestes similis* Hansen
  Sicyonidae
    *Sicyona* sp.
Caridea
  Palaemonidae
    *Palaemonetes kadiakensis* Rathbun
    *Palaemon floridanus* Chace
    *Macrobrachium acanthurus* (Wiegmann)
    *Pontonia* sp.

Hippolytidae
    *Thor floridanus* Kingsley
    *Lysmata wurdemani* (Gibbes)
    *Hippolyte zostericola* (Smith)
    *Tozeuma carolinense* Stimpson
    *Saron marmoratus* (Oliver)
Atyidae
    *Atya innocous* (Herbst)
    *Aty margaritacea* A.Milne Edwards
    *Micratya poeyi* (Guerin-Meneville)
    *Potimirim glabra* (Kingsley)
Oplophoridae
    *Oplophorus* sp.
    *Acathephyra* sp.
Alpheidae

*Alpheus lotini* Guerin
Crangonidae
   *Crangon crangon* (Linnaeus)
Gnathyphyllidae
   *Gnathyphyllum* sp.
Pasiphaeidae
   *Leptochela bermudensis* Bate
Procaridida
  Procarididae
   *Procaris ascensionis* Chace & Manning
   *P.hawaiana* Holthuis

Stenopodidea
  Stenopodidae
   *Stenopus hispidus* Oliver
Reptantia
  Cambaridae
   *Cambarus* spp.
  Upogebiidae
   *Upogebia pugettensis* (Dana)
  Axiidae
   axiid sp.

## 10 APPENDIX II

List of abbreviations

| | | | | |
|---|---|---|---|---|
| ab | branchial axis | | m | mandible |
| cch | cardiac chamber | | mp | mandibular palp |
| cm | convoluted membrane | | mpr | molar process |
| cpv | cardiopyloric value | | mo | mesocardiac ossicle |
| d | denticles | | mt | median tooth |
| e | esophagus | | p | pterocardiac ossicle |
| ep | epistome | | pc | pyloric chamber |
| epi | epipod | | pf | pyloric fingerlets |
| gf | gland filter | | po | pyloric ossicle |
| ip | incisor process | | sr | secondary rami |
| l | lamellae | | upg | uropyloric groove |
| la | labrum | | uo | urocardiac ossicle |
| lb | lateral branch | | upo | uropyloric ossicle |
| lt | lateral teeth | | zo | zygocardiac ossicle |
| lr | lateral ramus | | | |

## DISCUSSION

BAUER: You were talking about how some characters in defining carideans, such as the overlap of the second abdominal pleuron of the first and the third, may not be very good. Actually, on the slide you showed of *Glyphocrangon* it looked like the second was overlapping the first. Another point is whether or not there are articulations between the various abdominal segments. For example, in most carideans the hump is caused by the fact there is no articulations between the third and fourth pleura, and that is one of the characters Burkenroad used as a diagnostic feature for the carideans. I have found that in things like *Crangon,* which don't show the caridean hump, they lack that condyle between the third and fourth, whereas the peneids have a condyle between each of the segments. So that might be a better definition for some of these groups since the pleura, because of other selection pressures, are modified.

ABELE: *Glyphocrangon* does have condyles on the exterior surface of the second and third pleura.

BAUER: No, between the third and the fourth.

ABELE: They have them there as well. They are not as well-developed, but they are present there.

BAUER: Oh.

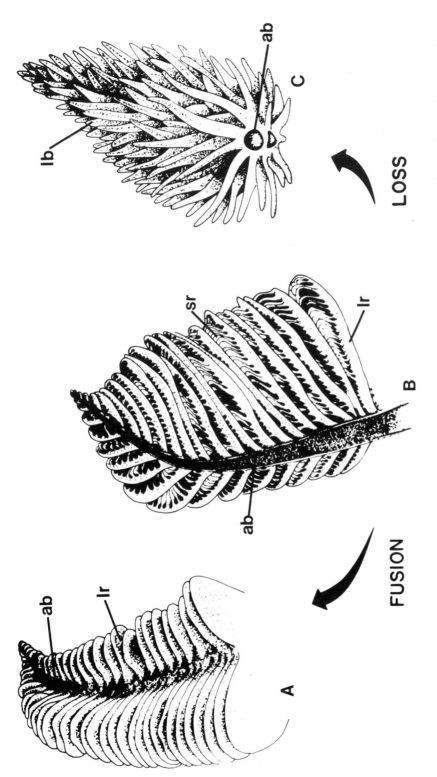

Figure 3. Hypothesis suggested by Boas 1880, and Burkenroad 1981, for the evolution of gill types among the Decapoda. (B) typical dendrobranchiate gill, consisting of lateral branches (lb) extending from the main branchial axis (ab) with a series of subdivided secondary rami (sr) from each lateral branch. Expansion of the lateral branches of the dendrobranchiate type would result in (A) phyllobranchiate gill; whereas loss of the secondary rami (sr) and/or reduction of the lateral branches would give rise to (C) trichobranchiate gill.

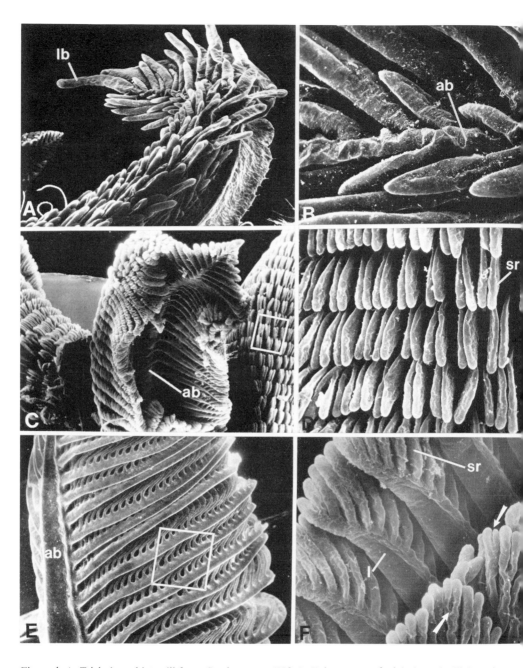

Figure 4. A. Trichobranchiate gill from *Cambarus* sp.; X50. B. Enlargement of trichobranch gill denoting the main branchial axis (ab) with lateral branches (1b) extending from the main gill axis; X100. C. Dendrobranchiate gill from *Peneus setiferus;* note the branchial axis with lateral branches extending from the main branch. The white box indicates the external view of the secondary rami (sr); X50. D. Enlargement of secondary rami (boxed area in C); X200. E. External view of dendrobranchiate gill of *Sergestes similis;* white box indicates branching secondary rami (sr); X60. F. Internal view of *S.similis* gill showing the secondary rami (sr) with white arrows denoting bifurcation of the secondary rami; X200.

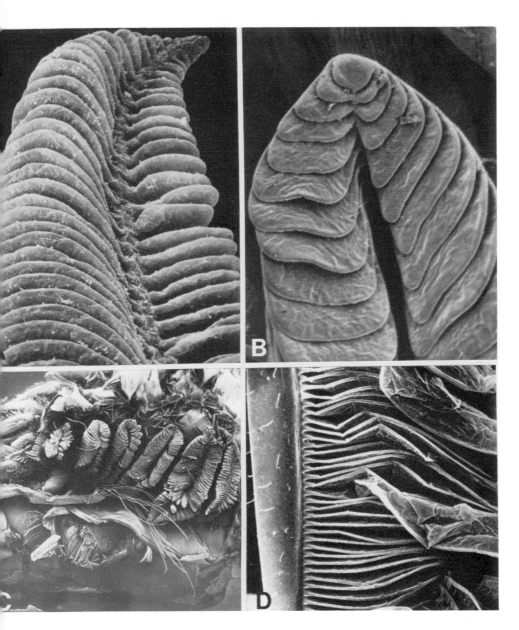

Figure 5. A. Phyllobranchiate gill type from *Palaemonetes kadiakensis;* X60. B. Phyllobranch gill plume of *Atya innocous,* showing the variation seen within this gill type; X80. C. Lateral view of entire gill region of *A.innocous,* indicating the arrangement of the phyllobranch gills; X20. D. Phyllobranch gill of *Oplophorus* sp.; note the thin, plate-like nature of the lateral rami; X50.

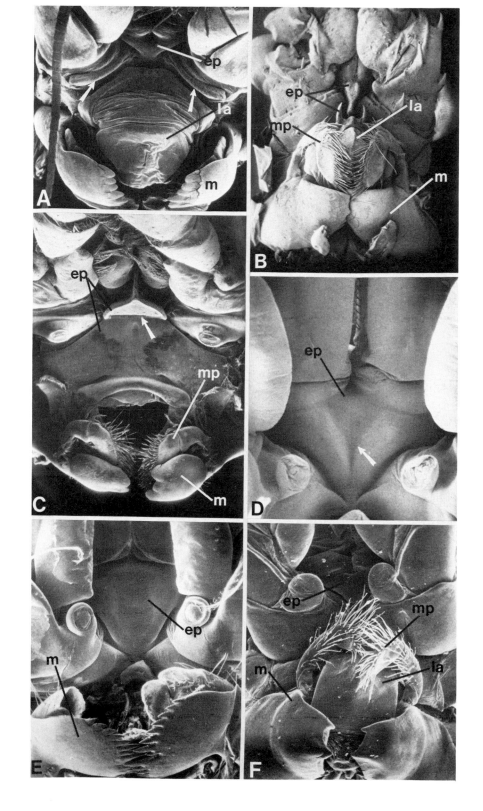

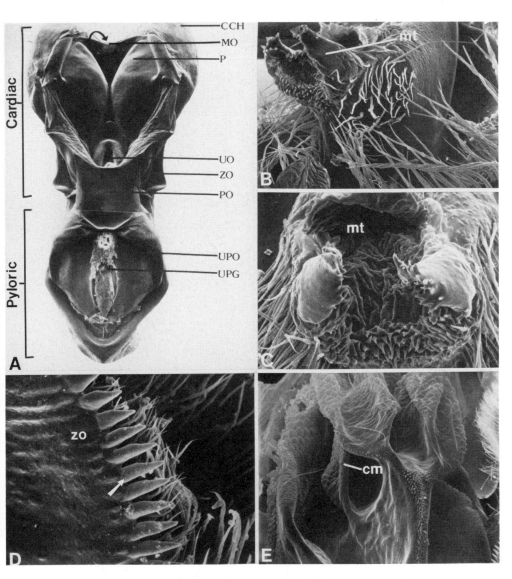

Figure 7. A. Dorsal view of the foregut of *Atya innocous* (membranes and musculature removed); note the distinct cardiac and pyloric regions along with the distinct chitinous regions (ossicles); X100. B. Lateral view of the bifid median tooth of *A.innocous* projecting from roof of cardiac chamber; X250. C. Close-up view of the median tooth of *A.innocous;* note presence of stout denticles on inner portion of tooth; X600. D. Zygocardiac ossicle of *A.innocous,* arrow indicates the lateral teeth; X600. E. Convoluted membrane (cm) of *Potimirim glabra* located within the pyloric chamber; X170.

Figure 6. A. Ventral view of the protocephalon of *Peneus setiferus;* note condition and location of epistome (ep) between the antennae; white arrow indicate epistomal bar; X25. B. Ventral aspect of protocephalon of *Stenopus hispidus;* note distinctive morphology and location of epistome (ep); X30. C. Protocephalon of *Cambarus* sp.; note location and morphology of the epistome; white arrow indicates membranous connection between anterior and posterior portions of rigid epistome; X20. D. Epistome (ep) of *Palaemonetes kadiakensis* in ventral view; white arrow denotes membranous points of articulation; X70. E. Protocephalon of *Oplophorus* sp.; large labrum has been removed to reveal nature of the epistome (ep); X40. F. Ventral aspect of the protocephalon of *Procaris ascensionis;* note condition and location of the epistome (ep); X40.

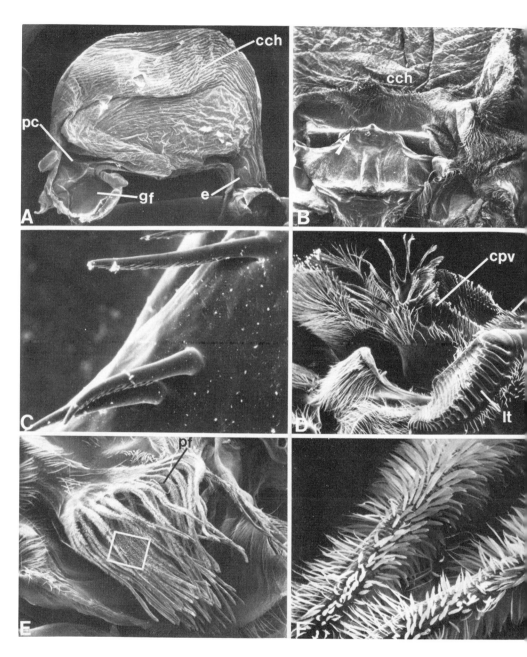

Figure 8. A. Foregut of *Saron marmoratus* (membranes and muscles removed); note lack of chitinized regions (ossicles and reduction of the pyloric chamber (pc); X20. B. Floor of the cardiac chamber (cch) of *Palaemonetes kadiakensis;* note the reduced median tooth (mt) (arrow), and lack of lateral teeth (lt); X1100. D. Lateral view of gastric armature of *Upogebia pugettensis;* note elaborate cardiopyloric valve (cpv); also shown are large lateral teeth (lt) and median tooth (mt) (the median tooth is obscured by the lateral teeth in this micrograph); X25. E. Pyloric fingerlets (pf) within the pyloric chamber of *Upogebia pugettensis,* X25. F. Close-up of pyloric fingerlets; X300.

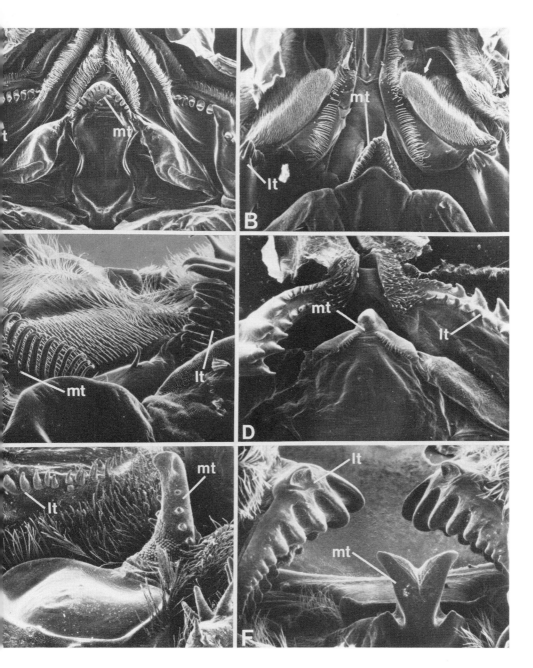

Figure 9. Cardiac chamber of *Peneus setiferus,* showing details of gastric mill; large median tooth is present (mt) flanked by large lateral teeth (lt); note rows of plumose setae anterior to median tooth which direct food to pyloric chamber (arrow); X20. B. Gastric mill of *Solenocera vioscari;* note large median tooth (mt) with robust lateral teeth (lt); dense pads of setae (arrow) direct food to pyloric chamber; X20. C. Lateral view of median tooth of *S.vioscari;* note long teeth borne on median tooth; X80. D. Gastric armature of *Sergestes similis;* note median tooth (mt) and lateral teeth (lt); X40. E.Gastric mill of *Stenopus hispidus,* elongate median tooth present with peg-like spines along its length (mt); lateral teeth also shown (lt); X50. F. Gastric mill of *Cambarus* sp.; note smooth bifid median tooth, massive lateral teeth (lt); X90.

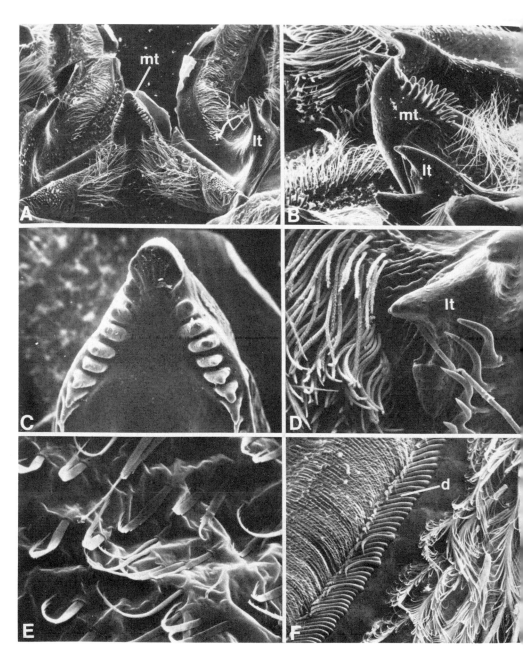

Figure 10. *Procaris ascensionis.* A. Gastric mill; note large median tooth (mt) and developed lateral teeth (lt); X50. B. Lateral view of gastric mill; X100. C. Close-up of median tooth (mt); X240. D. Details of lateral tooth; X200. E. Morphology of roof of cardiac chamber; X2000. F. Details of entrance to gland filter (gf); note denticles at (d); X765.

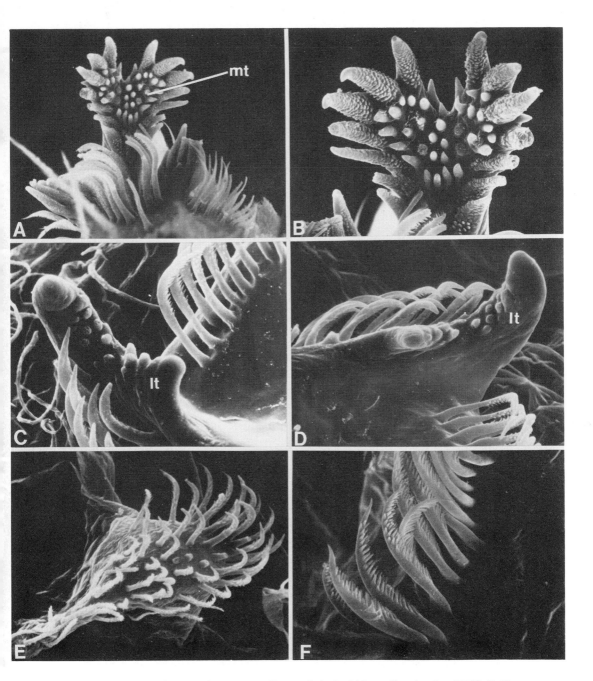

Figure 11. *Leptochela bermudensis*. A. Median tooth (mt) within cardiac chamber; X550. B. Close-up of median tooth; X2000. C and D. Robust lateral teeth; X1025. F. Serrate setae which surround median tooth (see Figure A); X1000.

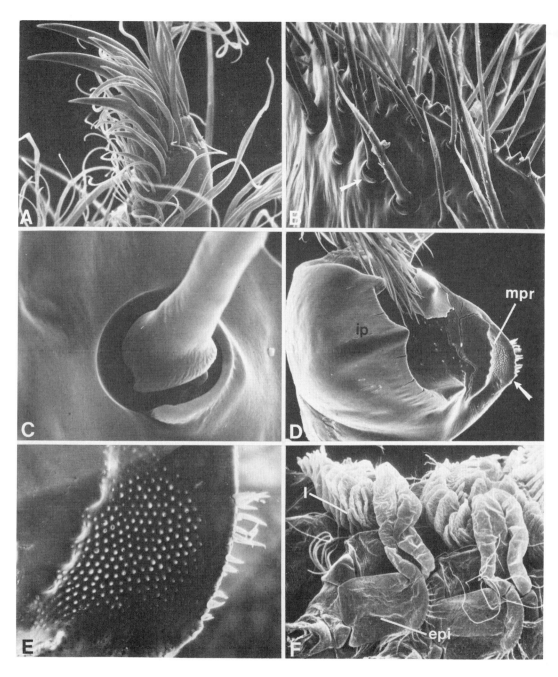

Figure 12. *Procaris ascensionis.* A. Second pereopod; X250. B. Third maxilliped (ischium); note the large movable setae (arrow) and gnathobasic-like projections medially (arrow); X1225. C. Enlargement of movable setae shown in B; X2000. D. Mandible; arrow indicates the reduced molar process; X75. E.Close-up of molar process; X280. F. Phyllobranch gills; note massive epipods (epi); X50.

# REFERENCES

Abele, L.G. & B.E.Felgenhauer 1982. Eucaridea. In: S.P.Parker (ed.), *Synopsis and classification of living organisms, Vol.2:* 294-326. New York: McGraw-Hill.

Balss, H. 1957a. Decapoda, Part VIII: Systematic. In: *H.G.Bronn, Klassen und Ordnungen des Tierreichs* 5(1):7(12):1505-1672. Leipzig: Akademische Verlags.

Balss, H. 1957b. Decapoda, Part XI: Stammesgeschichte. In: *H.G.Bronn, Klassen und Ordnungen des Tierreichs* 5(1):7(12):1797-1821. Leipzig: Akademische Verlags.

Bate, C.S. 1888. Report on the Crustacea Macrura collected by HMS Challenger during the years 1873-1876. *Challenger Rept. Zool.* 24:1-942.

Bauer, R.T. 1976. Mating behaviour and spermatophore transfer in the shrimp *Hepatacarpus pictus* (Stimpson) (Decapoda: Caridea: Hippolytidae). *J. Nat. Hist.* 10:415-440.

Beurlen, K. & M.F. Glaessner 1930. Systematik der Crustacea Decapoda auf Stammesgeschichtlicher Grundlage. *Zool. Jb. Abt. Syst.* 60:49-84.

Boas, J.E.V. 1880. Studier over Decapodernes Slaegtskabsforhold. *Vidensk. Selsk. Kristiania, Skrifter* (5)6:25-210.

Borradaile, L.A. 1907. On the classification of the decapod Crustacea. *Ann. Mag. Nat. Hist. London* (7):19:457-486.

Bouvier, E.L. 1940. Décapodes marcheurs. *Faune de France* 37:1-399, pls.I-XIV.

Bowman, T. & L.G.Abele 1982. Classification of the Crustacea. In: L.G.Abele (ed.), *Biology of the Crustacea, Vol.1:* 1-27. New York: Academic Press.

Brooks, H.K. 1962. The Paleozoic Eumalacostraca of North America. *Bull. Am. Paleontol.* 44:163-280.

Burkenroad, M.D. 1963. The evolution of the Eucarida (Crustacea, Eumalacostraca) in relation to the fossil record. *Tulane Stud. Geol.* 2:2-17.

Burkenroad, M.D. 1981. The higher taxonomy and evolution of Decapoda (Crustacea). *Trans. San Diego Soc. Nat. Hist.* 19:251-268.

Calman, R.T. 1909. Crustacea. In: R.Lankester (ed.), *A Treatise on Zoology, Vol.VII.* Adam & Charles Black, London.

Chace, F.A. 1976. Shrimps of the pasiphaeid genus *Leptochela* with descriptions of three new species (Crustacea: Decapoda: Caridea). *Smiths. Contr. Zool.* 222:1-51.

Chace, F.A. & D.E.Brown 1978. A new polychelate shrimp from the Great Barrier Reef of Australia and its bearing on the family Bresiliidae (Crustacea: Decapoda: Caridea). *Proc. Biol. Soc. Wash.* 91: 756-766.

Chace, F.A. & L.B.Holthuis 1978. *Psalidopus:* the scissor-foot shrimps (Crustacea: Decapoda: Caridea). *Smiths. Contr. Zool.* 277:1-22.

Chace, F.A. & R.B.Manning 1972. Two new caridean shrimps, one representing a new family, from marine pools on Ascension Island (Crustacea: Decapoda: Natantia). *Smiths. Contr. Zool.* 131:1-18.

Coombs, E.F. & J.A.Allen 1978. The functional morphology of the feeding appendages and gut of *Hippolyte varians* (Crustacea: Natantia). *Zool. J. Linn. Soc.* 64:261-282.

Coutiere, H. 1899. Les 'Alpheidae', morphologie externe et interne, formes larvaires, bionomie. *Ann. Sci. Nat. Zool.* (8)9:1-560.

Dana, J.D. 1852. Crustacea. *United States Exploring Expedition during the years 1838, 1839, 1840, 1841, 1842 under the command of Charles Wilkes, USN Vol.XIII:* 1-635.

Farfante, I.P. 1975. Spermatophores and thelyca of the American white shrimps, genus *Penaeus,* subgenus *Litopenaeus. Fish. Bull.* 78:463-486.

Felgenhauer, B.E. & L.G.Abele (in press). Aspects of mating behavior in the tropical freshwater shrimp *Atya innocous* (Herbst). *Biotropica.*

Forest, J. & M.de Saint Laurent 1975. Presence dans la fauna actuelle d'un représentant du groupe mésozoique des Glypheides: *Neoglyphea inopinata* gen. nov., sp.nov. (Crustacea Decapoda Glypheidae). *C.R. hebd. Seanc. Acad. Sci., Paris (D)*281:155-158.

Forest, J. & M.de Saint Laurent 1981. La morphologie externe de *Neoglyphia inopinata,* espècene actuelle de Crustacé Décapode Glyphéide. *Res. Camp. Musorstom. I – Philippines (18-28 Mars 1976) vol.1, Mém. ORSTOM 91:* 51-84.

Forest, J., M.de Saint Laurent & F.A.Chace 1976. *Neoglyphea inopinata:* A crustacean 'living fossil' from the Phillipines. *Science* 192:884.

Förster, R. 1966. Über die Erymiden, eine alte konservative Familie der mesozoischen Dekapoden. *Paleontographica* 125:61-175.

Förster, R. 1967. Die reptanten Dekapoden der Trias. *N.J. Geol. Paleontol. Abh.* 128:136-194.

Glaessner, M.F. 1960. The fossil decapod Crustacea of New Zealand and the evolution of the order Decapoda. *N.Z. Geol. Survey Paleontol. Bull.* 31:1-63.

Glaessner, M.F. 1969. Decapoda. In: R.C.Moore (ed.), *Treatise on Invertebrate Paleontology. Arthropoda 4. Part R, Vol.2:*R399-533. Lawrence: Geol. Soc. Am. & Univ. Kansas Press.

Guinot, D. 1977. Propositions pour une nouvelle classification des Crustacés Décapodes Brachyoures. *C.R. Adad. Sci. Paris* (D):285-1049-1052.

Guinot, D. 1979. Problèmes pratiques d'une classification cladistique des Crustacés Décapodes Brachyoures. *Bull. Off. Natn. Pêch. Tunisie* 3:33-46.

Gurney, R. 1924. Crustacea. Part IX – Decapod larvae. *Brit. Antarctic ('Terra Nova') Exped. 1910, Zool.* 8:37-202.

Gurney, R. 1936. Larvae of decapod Crustacea. Part I: Stenopidea. Part II: Amphionidae. Part III. Phyllosoma. *Discovery Repts.* 12:377-440.

Gurney, R. 1942. Larvae of decapod crustacea. *Ray Soc. London.*

Holthuis, L.B. 1946. Biological results of Snellius Expedition. XIV. The Decapoda Macrura of the Snellius Expedition I. The Stenopodidae, Nephropsidae, Scyllaridae and Palinuridae. *Temminekia* 7:1-178.

Holthuis, L.B. 1951a. The caridean Crustacea of tropical West Africa. *Alantide Rep.* 2:7-187.

Holthuis, L.B. 1951b. A general revision of the Palaemonidae (Crustacea Decapoda Natantia) of the Americas. I. The subfamilies Euryrhynchinae and Pontoniinae. *Occ. Paper Allan Hancock Found.* 11:1-332.

Holthuis, L.B. 1955. The recent genera of the caridean and stenopodidean shrimps (class Crustacea, order Decapoda, supersection Natantia) with keys for their determination. *Zool. Verhand.* 26:1-157.

Holthuis, L.B. 1971. The Atlantic shrimps of the deep-sea genus *Glyphocrangon* A. Milne Edwards 1881. *Bull. Mar. Sci.* 21:267-373.

Holthuis, L.B. 1973. Caridean shrimps found in land-locked saltwater pools at four Indo-West Pacific Localities (Sinai Peninsula, Funafuti Atoll, Maui and Hawaiian Islands), with the description of one new genus and four new species. *Zool. Verhandl.* 128:1-48, pls.1-7.

Huxley, T.H. 1878. On the classification and the distribution of the crayfishes. *Proc. Zool. Soc. London* 1878:752-788.

Huxley, T.H. 1880. *The crayfish, an introduction to the study of zoology.* London: C.Kegan Paul.

Kaestner, A. 1970. *Invertebrate zoology. Vol.III. Crustacea.* New York: Interscience.

Kunze, J. & D.T.Anderson 1979. Functional morphology of the mouthparts and gastric mill in the hermit crabs *Clibanarius taeniatus* (Milne Edwards), *Clibanarius virescens* (Krauss), *Paguristes squamosus* Mcculloch and *Dardanus setifer* (Milne Edwards) (Anomura: Paguridae). *Aust. J. Mar. Fresh. Res.* 30:683-722.

Milne Edwards, H. 1837. *Histore naturelle générale et particulière des Crustacés et insects. Tome Troisième.* Paris: Dufart.

Mocquard, F. 1883. Estomac des crustaceans podophthalmaires. *Ann. Sci. Nat.* 16:1-311.

Ortmann, A. 1890. Die Decapoden-Krebse de Strassburger Museums mit besonderer Berücksichtigung der von Herrn. Dr Döderlein bie Japan und bei den Lui-Kui-Inseln gesammelten und z. A. im Strassburger Museum aufbewahrten Formen. *Zool. Jb.* (Syst.) 5:437-540.

Patwardhan, S.S. 1935. On the structure and mechanism of the gastric mill in Decapoda. 5. The structure of the gastric mill in natantous Macrura-Caridea. *Proc. Indian Acad. Sci. (B)*1:693-704.

Patwardhan, S.S. 1936. On the structure and mechanism of the gastric mill in Decapoda. VI. The structure of the gastric mill in the Natantous Macrura – Penaeida and Stenopidea; conclusion. *Proc. Indian Acad. Sci. (B)*2:155-174.

Powell, R.R. 1974. The functional morphology of the fore-guts of the thalassinid crustaceans *Callianassa californiensis* and *Upogebia pugettensis*. *Univ. Calif. Pub. Zool.* 102:1-41.

Powell, R.R. 1976. Two new freshwater shrimps from West Africa: the first euryrhynchinids (Decapoda Palaemonidae) reported from the Old World. *Rev. Zool. Afr.* 90:883-902.

Powell, R.R. 1977. A revision of the African freshwater shrimp genus *Desmocaris* Solland, with ecological notes and description of a new species (Crustacea Decapoda Palaemonidae). *Rev. Zool. Afr.* 91:649-674.

Rice, A.L. 1980. Crab zoeal morphology and its bearing on the classification of the Brachyura. *Trans. Zool. Soc. London* 35:271-424.

Saint Laurent, M.de 1979. Vers une nouvelle classification des Crustacés Decapodes Reptantia. *Bull. Off. Natn. Pêch. Tunisie* 3:15-31.

Saint Laurent, M.de 1980a. Sur la classification et la phylogenie des Crustacés Décapodes Brachyoures. I. Podotremata Guinot, 1977, et Eubrachyura sect. no. *C.R. Acad. Sci. Paris (D)*290:1265-1268.

Saint Laurent, M.de 1980b. Sur la classification et la phylogenie des Crustacés Décapodes Brachyoures. II. Heterotremata et Thoracotremata Guinot 1977. *C.R. Acad. Sci. Paris (D)*290:1317-1320.

Saint Laurent, M.de & R.Cleva 1981. Crustacés décapodes: Stenopodidae. *Res. Camp. Musorstom I. Philippines (18-28 Mars 1976) Vol.1, Mem. ORSTOM* 91:151-188.

Schaefer, N. 1970. The functional morphology of the foregut of three species of decapod Crustacea: *Cyclograpsus punctatus* Milne-Edwards, *Diogenes brevirostris* Stimpson, and *Upogebia africana* (Ortmann). *Zool. Afr.* 5:309-326.

Schram, F.R. 1982. The fossil record and evolution of the crustacea. In: L.G.Abele (ed.), *Biology of the Crustacea. Vol.1:*93-147. New York: Academic Press.

Schram, F.R., R.M.Feldman & M.J.Copeland 1978. The late Devonian Palaeopalaemonidae and the earliest decapod crustaceans. *J. Paleontol.* 52:1375-1387.

Snodgrass, R.E. 1951. *Comparative studies on the head of mandibulate arthropods.* Ithaca: Comstock Publ.

Stanton, W. 1975. *The Great United States Exploring Expedition of 1838-1842.* Berkeley: Univ. Calif. Press.

Williamson, D.I. 1976. Larvae of Stenopodidea (Crustacea, Decapoda) from the Indian Ocean. *J. Nat. Hist.* 10:497-509.

Williamson, D.I. 1982. Larval morphology and diversity. In: L.G.Abele (ed.), *Biology of the Crustacea, Vol.2:* New York: Academic Press.

Young, J.H. 1959. Morphology of the white shrimp *Penaeus setiferus* (Linnaeus 1758). *Fish. Bull.* 145: 1-168.

A. L. RICE

*Institute of Oceanographic Sciences, Wormley, Godalming, Surrey, UK*

# ZOEAL EVIDENCE FOR BRACHYURAN PHYLOGENY

## ABSTRACT

The available zoeal evidence for brachyuran phylogeny is reviewed. Two main questions are addressed: first, the origin and early evolution of the primitive crab groups; and second, the subsequent phylogeny of the higher brachyurans. In both cases the evidence is equivocal, and, as in adult studies, its successful interpretation depends on the recognition of primitive as opposed to derived characters and the discrimination of resemblances resulting from common ancestry and convergence. Tentative phylogenetic schemes, based entirely on the zoeal stages, are presented for the heterotrematous eubrachyurans.

## 1 INTRODUCTION

Gurney's (1942) *Larvae of Decapod Crustacea* is still the most comprehensive review of the subject which has ever been undertaken. Despite the appearance of some hundreds of decapod larval papers since that time, Gurney's statement that 'It must be confessed that the evidence from development so far accumulated has not produced any very serious contribution to the systematics of the group' is still largely true. For although there are a number of cases in which larval characters have been useful in separating species, genera, or even families, and others in which the taxonomic position of a species has been clarified by the larvae (see Williamson 1982) there are very few examples in which the classification of the higher taxonomic categories has been significantly affected by a consideration of the larval stages. Burkenroad's (1981) most recent revision of the Decapoda is a possible exception since he acknowledges that it stemmed from 'an early collision with that stimulating work of Gurney . . .', and that it is '. . . directly dependent on the lead supplied by the peculiarities of branchial ontogenesis'. Nevertheless, although Burkenroad intimates that a review of decapod larvae would provide valuable phylogenetic information, his recognition of four decapodan suborders, each of which he believes is monophyletic, is based almost entirely upon adult characteristics.

A major problem in working on larval phylogenies is that fossil larvae are almost unknown. The available larval evidence for relationships within the decapods is therefore restricted to what can be gleaned from the extant forms. Nevertheless, as Williamson (1982) has pointed out, larvae are, in general, no more and no less difficult to separate than the adults; a fact which is interpreted as indicating that, as one would expect, selection has acted at all stages of decapod ontogeny from egg to adult. Consequently, it is 'just as legitimate to.discuss the phylogenetic relationships of nauplii, zoeas, and megalopas as of adult crustaceans, and the correct interpretation of the phylogeny of any species or group should be consistent with the evidence from all developmental phases' (Williamson 1982). Indeed, while the larval stages, being the principal dispersive phase in the decapodan life cycle, are almost without exception adapted for a planktonic life style, the habits of the adults include planktonic/nektonic, epibenthic and fossorial, and a wide variety of specializations within each of these. Consequently, the possibilities for confusing convergent evolution among adult decapods has been enormous, and presumably much greater than among the larvae.

Although there has been an increasing tendency in recent years for ontogenetic evidence to be considered in reviews of decapodan relationships, this has normally taken the form of brief references to the larval stages, more often than not in support of conclusions already reached on the basis of much more exhaustive considerations of adult features (e.g. Števčić 1971, 1973, Guinot 1978). In only one major decapodan group, the Brachyura, have the larval stages been used more generally to elucidate relationships (Aikawa 1929, 1937, Williamson 1965, 1976, Rice 1980a, 1981). This review will therefore be restricted to a consideration of the evidence for the inter-relationships of the various crab groups which has been adduced from the zoeal stages.

## 2 THE BRACHYURA

As an obvious pinnacle of crustacean evolution the typical adult crab is easily recognizable. Thus, the combination of a relatively short and broad carapace fused to the epistome, the last thoracic sternite fused with the more anterior ones, the short, flattened and symmetrical abdomen folded tightly beneath the thorax, the absence of biramous uropods, the first pereiopods always chelate and the third ones never so, clearly distinguishes the Brachyura from the other decapods, including those anomuran groups, such as the lithodids, porcellanids, and hippids, with a superficially crab-like form.

Nevertheless, the limits of the extant Brachyura are by no means clear, and a number of taxa, particularly the dromiaceans and the raninoids have been variously included in or excluded from the Brachyura by decapodan specialists during the past century or so. The reason for this is that these groups have a combination of primitive and specialized features which distinguish them from the higher Brachyura but do not clearly ally them with any of the other decapods.

One particularly significant feature which the dromioids and raninoids share with the homoloids and tymolids, and which distinguishes all of these groups from the higher brachyurans is the coxal position of the female sexual openings. This fundamental distinction led Gordon (1963) to suggest that all of the 'peditremen' crabs should perhaps be removed from the 'true' Brachyura, while Guinot (1978) united the dromioids, homoloids, raninoids, and tymolids in her section Podotremata, to distinguish them from her Heterotremata and Thoracotremata for which, together, de Saint Laurent (1980a) proposed the term Eubrachyura.

Much of the discussion of larval (and adult) evidence for brachyuran relationships in recent years has centred around the problems of whether any or all of the podotrematous groups should be included in the Brachyura, where these groups originated from, and whether any of them in turn gave rise to the higher crabs.

## 3 ORIGIN AND EARLY EVOLUTION OF THE BRACHYURA

On the basis of adult structure there have been three main suggestions for the origin of the crabs (see Števčić 1971): (1) that they arose from within the Astacidea, originally proposed by Huxley (1878), argued in great detail by Bouvier (1896, 1940), and supported by Balss (1957); (2) that they arose from within the Anomura, either from the Thalassinidea (Boas 1880) or from a form intermediate between the pagurids and the galatheids (Ortmann 1896); and (3) a suggestion mainly put forward by palaeontologists, that the Brachyura evolved directly from a *Pemphix*-like Triassic glypheoid ancestor (van Straelen 1928; Beurlen 1930, Glaessner 1930, 1960, 1969). But in all three theories the Brachyura are considered to be monophyletic; with the Dromiacea, that is containing both the Dromioidea and the Homoloidea, being the most primitive forms from which all the higher groups evolved.

The monophyletic view has not gone unchallenged by students of adult crabs. Thus, Bourne (1922) concluded from a study of the morphology, physiology and ecology of raninoids that they should be removed not only from the Oxystomata, in which they had usually been grouped together with the dorippids, leucosiids, and calappids, but from the whole of the Brachyura. Bourne established the separate tribe Gymnopleura for the raninoids, with a suggested origin from the Astacidea independent of that of the Brachyura. The larval evidence is totally opposed to Bourne's thesis since zoeal raninoids and higher brachyurans share so many common derived features that it is inconceivable that they could have resulted from separate phylogenies (Williamson 1965, Rice 1980a).

Nevertheless, the main criticisms of a monophyletic origin for the Brachyura as considered above have come from larval workers, but in relation to the Dromiacea rather than the raninoids. This alternative view began with Gurney (1924, 1942) and Lebour (1934) who were convinced that the dromiaceans should be excluded from the Brachyura and from the direct line of descent of the crabs, and has been extended by Williamson (1965, 1976, 1982) and by Rice (1980a, 1981), though both of these authors exclude only the Dromioidea from the Brachyura and consider the Homoloidea to be true crabs. This conclusion is based on a comparison of the larvae of the podotrematous crabs with those of the higher Brachyura and those of other decapodan groups, particularly the anomurans. Details of this rather complex argument have been discussed at length elsewhere (see above references). Sufficient details are given here to enable the main arguments to be followed. Further information, including details of larval morphology, are provided in the papers of Williamson & Rice.

The larvae of the podotrematous groups, either individually or collectively, have a number of features which distinguish them from those of the higher Brachyura and in which they resemble the typical anomurans. [This discussion is restricted to the dromioids, homoloids, and raninoids. The only known tymolid larva is the first zoea of *Cymonomus bathamae* Dell (Wear & Batham 1975) which has a strange combination of primitive phylogenetic and advanced ontogenetic features which Williamson (1976) has suggested may both result from the abbreviated development in this species.] These include the anteriorly directed rostrum (in the dromioids and homoloids), the flattened scale-like antennal exopod with numerous marginal setae, the setose antennal endopod in the early stages, generally more setose and segmented thoracic appendages, and the presence of setose uropods in the late zoeal stages. But although these characters are found in the larvae of many pagurids and galatheids, they do not necessarily indicate a close phylogenetic relationship between the anomurans and the podotrematous groups since they are plesiomorphic or primitive and could have been derived independently in both groups from a common ancestor.

The zoeal stages of the podotrematous groups are distinguished both from one another and from the higher brachyurans and anomurans by a number of specialized features, but in addition they exhibit quite distinct general evolutionary levels. Thus, the raninoids are the most advanced since the endopod of the second maxilliped is reduced to three segments, at least in the early stages, the third maxilliped never acquires a setose exopod, and the endopods of the uropods are very small or absent and are never setose. On these criteria the raninoids are rather more advanced than most anomurans and are at a level roughly equivalent to the hippids. In homoloid zoeas the endopod of the second maxilliped never consists of fewer than four segments in the early stages, setose exopods are developed on the third maxillipeds but not on the chelipeds, and the uropods have well-developed setose endopods. Homoloid zoeas are therefore at a similar evolutionary level to that of typical anomurans such as the pagurids and galatheids. Finally, the most primitive dromioid zoeas develop setose exopods on the third maxillipeds and the chelipeds and are thus more primitive than the accepted anomurans (and are at an evolutionary level equivalent to that of the thalassinids).

The primitive nature of the dromioid zoeas would seem to preclude the possibility that they originated from any forms as advanced as the modern pagurids or galatheids, but does not necessarily exclude them from the ancestry of the Brachyura. One feature of all dromioid zoeas, however, the hair-like second telson process, has been considered by Williamson (1965, 1976, 1982) and by Rice (1980a) to link them unequivocally with the anomurans and thalassinids, and at the same time separate them from the remainder of the podotremata and from the higher crabs. Apart from anomurans, thalassinids, and dromioids, the second telson process is reduced to a hair only in the zoeas of the Stenopodidea; in all other groups, including the euphausiids, peneids, sergestids, carideans, and nephropids, it is a short spine or absent, but never takes the form of a hair. Williamson & Rice have assumed that the hair-like condition is a derived or apomorphous character suggesting that no group possessing this feature, including the dromioids, could have been ancestral to any group in which the process is in the unreduced or primitive condition. [Though the second process is apparently not hair-like in some hippid zoeas (see Gurney 1942, Knight 1967).] Burkenroad (1981), however, has suggested that the hair-like condition is 'more likely to be a relic from the stem-form of the three incubatory suborders [his Euzygida (stenopodids), Eukyphida (carideans), and Reptantia], never present in dendrobranchiates [peneids and sergestids], and effaced in eukyphids, most Astacina, Palinura, and the higher Brachyura'.

This difference of interpretation is very difficult to resolve, although why such a primitive feature should have become altered independently, and in much the same way, in several quite distinct groups is not clear. However, if Burkenroad's view is correct, then I see no difficulty in deriving the higher brachyuran zoea from a dromioid ancestor and this, in turn, from a thalassinid — or perhaps from a glypheoid since the thalassinids and dromioids appear more or less simultaneously in the fossil record at the beginning of the Jurassic. Whether, under this view, the dromioids should then be included within the Brachyura is a separate question to which the larval stages do not provide a clear answer.

One possible source of evidence on this point is the ontogeny of the gills, for Burkenroad (1981) has pointed out that, whereas thalassinids and anomurans in which the bran-

chial formula is reduced generally lose the pleurobranchs from the posterior thoracic appendages, in the brachyurans it is the arthrobranchs which are lost and the pleurobranchs
which are retained. He therefore suggests that 'there was a brachyuran stem form with a
more or less complete branchial formula and coxal genital apertures which was distinguished
from other reptant lines by delayed appearance of posterior arthrobranchs (relative to posterior pleurobranchs) during individual development'. Unfortunately, insufficient attention has been given to the development of the gills in the larval literature, partly because
they are difficult to identify in zoeal material. Nevertheless, Burkenroad believes that the
accounts of the zoeas of *Dromia* by Gurney (1924) and Lebour (1934) suggest that the
developmental pattern in these larvae is, indeed, that to be expected of the brachyuran
stem form. A careful examination of branchial ontogeny in other dromioids would clearly
be valuable.

But while the available information on dromioid larvae does not clearly identify them
as brachyurans, that on homoloid and raninoid zoeas certainly does. Homoloid zoeas (e.g.
Rice 1964, 1970, Rice & Provenzano 1970, Rice & von Levetzow 1967, Rice & Williamson
1977, Williamson 1965) have a relatively short cephalothorax, often with a prominent dorsal carapace spine, a flexed abdomen, a second telson process which is never hair-like, and
a long antennal spine. They are consequently much more brachyuran in general appearance
than most dromioids. Nevertheless, when they are compared in detail with typical anomuran and typical branchyuran zoeas (Williamson 1965, Rice 1980) they have no characters
which obviously ally them with the higher crabs. Indeed, several homoloid features, particularly the distinctive carapace ornamentation, typically consisting of a series of denticulate
folds in the early stages which are replaced by a group of spines in the later zoeas, and the
form of the telson seems to separate them very clearly from the eubrachyurans. The raninoids, however, convincingly bridge this gap.

Typical raninoid zoeas (see Knight 1968, Sakai 1971) are very brachyuran, not only in
general appearance but also in several details of the appendage morphology, despite having
the primitive features of a flattened, scale-like antennal exopod with many marginal setae
and the acquisition of setose uropods in the later stages (Williamson 1965, 1976, Rice
1980a). The carapace in these zoeas carries a well-developed dorsal spine, a ventrally
directed rostrum and paired lateral spines, that is very similar to typical eubrachyurans but
quite different from the homoloids. Other described raninoid zoeas, however, such as
Aikawa's (1933) *Lithozoea serrulata* and Rice's (1970) raninoid larva C, have carapaces
which are much more reminiscent of the early stages of homoloids, with denticulate folds
or rows of small spines, indicating routes by which the homoloid features may have evolved
towards the more advanced brachyuran condition.

In attempting to bridge the considerable gap between homoloid and eubrachyuran
zoeas, Williamson (1976) saw another significant link in the strange zoeas which he had
originally (Williamson 1960) identified as majids but which he later tentatively assigned to
the Tymolidae. The major differences between these zoeas and the subsequent description
of the zoea of *Cymonomus* made such an identity very unlikely; they are now quite confi-
dently identified as *Dorhynchus thomsoni* Norman which has usually been included in the
sub-family Inachinae within the family Majidae. *Dorhynchus* has only two zoeal stages and
a number of other features, particularly in the setation of the maxillule and maxilla, which
quite clearly identify them as majids. They also have a curious combination of primitive
and advanced features which make their placement in the existing subdivisions difficult
and which suggest that the establishment of a new category is warranted (Rice 1980d).

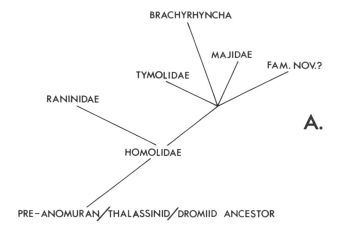

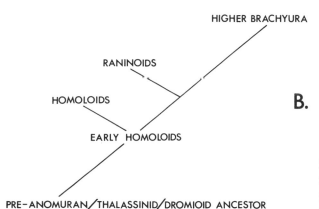

Figure 1. Representations of the phylogeny of the early Brachyura, A. according to Williamson (1976), B. according to Rice (1980a).

Their most remarkable features, however, are the ornamentation of the carapace, abdominal somites, and telson. The carapace has an anteriorly directed rostrum, three further spines or processes in the mid-line, a pair of supra-orbital spines, and a group of five lateral spines. Each abdominal somite carries one or two pairs of long spines and the telson has a single very long spine on each fork. In these characters, the *Dorhynchus* zoeas are remarkably similar to the later zoeal stages of homoloids (cf. Williamson 1976, fig.3d, i) and, with their obvious majid connections, seem to by-pass the raninoids in indicating an alternative possible route by which an ancestral homoloid zoea could have evolved into a spider crab zoea. Williamson therefore concluded that the raninoids and *Dorhynchus* (his fam.nov.?) had evolved from the homoloids by two distinct routes: (1) the raninoids showing no tendency to abbreviated development and having clear affinities with early stage homoloid zoeas, and (2) *Dorhynchus* showing clear affinities with late stage homoloids but with the zoeal phase abbreviated to only two stages. He then went on to discuss which of those two evolutionary routes had given rise to the remainder of the higher Brachyura.

The close affinity of the larvae of *Dorhynchus* with those of more conventional majids, including the presence of only two zoeal stages, indicated that this family, at least, had evolved from a stock close to the Homoloidea — fam.nov. evolutionary line. This left the problem of the origin of the rest of the eubrachyurans, which generally have five or more zoeal stages, either from the same line or independently from the raninoids. Williamson rejected this second possibility for two main reasons: (1) because majid zoeas are so similar to those of the other higher crab groups that a separate evolutionary history seems very unlikely; and (2) because he saw great difficulty in explaining the disappearance of the zoeal uropods if the non-majid eubrachyurans had been derived from the raninoids. Instead, he suggested that all the higher brachyurans have evolved from ancestors in which there were only two zoeal stages, a feature which has been retained in the majids and in the tymolids; whereas in most other brachyuran families the number of zoeal stages has been secondarily increased. The uropods were therefore lost at some evolutionary level between the homoloids and the majids, when the zoeal phase ended at a stage before the uropods would normally make their first appearance. Under this interpretation the raninoids represent a separate evolutionary 'cul-de-sac' originating from within the ancestral homoloids and acquiring the brachyuran zoeal characters by convergence (Fig.1a).

Having re-examined this question of the zoeal evidence for the phylogenetic relationships between the homoloids, the raninoids and the Eubrachyura (Rice 1980a), I came to a slightly different conclusion. For it seemed to me that raninoid zoeas share so many features with the advanced brachyurans in which both groups differ from the homoloids that completely separate evolutionary lineages from the homoloid ancestor were not feasible. Instead, I suggested that the modern larval condition in the homoloids, raninoids, and the higher brachyurans have all evolved from a more primitive ancestor which had larval characters common to all three, or from which the modern condition in all three taxa could be derived. The non-homoloid characters shared by the raninoids and higher brachyurans might therefore be explained by postulating that they had a common ancestral stem after separating from the homoloid evolutionary line, and only later diverged to produce their respective groups (Fig.1b).

This alternative scheme does not necessarily challenge Williamson's much more important suggestion that the higher Brachyura all evolved from forms in which there were only two zoeal stages, and may in fact support it. For despite the resemblance between the carapace in *Dorhynchus* zoeas and the late zoeas of homoloids, most higher brachyuran zoeas resemble early raninoids rather than the later stages, a situation which is more easily explained if the zoeal uropods were indeed lost by the elimination of these later stages. The main problem with this theory is to explain why *Dorhynchus* should resemble the homoloids if it had a much more raninoid-like ancestor, Williamson (1982) has perhaps supplied the answer when he points out that *Dorhynchus* cannot be regarded as being close to the evolutionary line from the Homoloidea to the majids. Far from having a generally primitive spider crab zoea, *Dorhynchus* has a rather advanced zoea with a few striking homoloid ancestral features. Williamson suggests that the modern Brachyura have perhaps retained the genes which could produce a 'homoloid' zoeal carapace and that the suppressive mechanism operative in most advanced crabs has been secondarily lost in *Dorhynchus.*

## 4 SUBSEQUENT EVOLUTION OF THE EUBRACHYURA

Whether this suggested phylogenetic scheme for the origin of the non-podotrematous crabs is correct or not, there can be little doubt that the 20 or so families (or superfamilies) of eubrachyuran crabs form a monophyletic group. For the 4 200 or more extant representatives share so many specialized features, both as adults and as larvae, that separate derivations through independent lineages are unthinkable. Nevertheless, within the Eubrachyura there has been an impressive degree of adaptive radiation, mainly during the Late Cretaceous and Early Tertiary, resulting in the exploitation of a remarkable range of habitats and life-styles and a corresponding wide range of adult morphological modifications.

During the last 150 years or so, attempts to categorize this eubrachyuran morphological diversity has resulted in a bewildering array of classifications revealing considerable disagreement among taxonomists and phylogeneticists. Nevertheless, until very recently some general groupings, varying in their specific constitutions, have been recognizable in many of the proposed systems. Thus, the dorippids, calappids, and leucosiids, often together with the raninoids, have usually been grouped in an oxystomatous section typified by an elongated mouth frame which narrows anteriorly, and by respiratory modifications usually associated with a burrowing habit. Similarly, the majids, parthenopids, and frequently the hymenosomatids form an oxyrhynchous group in which the carapace is generally more or less triangular with a well-developed rostrum, broad epistome, square buccal frame, and incomplete orbits. These forms are generally slow moving and appear to be mainly adapted for concealment. The remaining families have been variously subdivided, but often with some semblance of Milne-Edwards' (1834) recognition of a round-fronted group, the cyclometopes, and a square-fronted group, the catometopes. However, the general uncertainty about the validity of these divisions was reflected in Borradaile's (1907) widely accepted classification in which he created the Brachyrhyncha to rank alongside the Oxyrhyncha and to contain all the remaining non-oxystomatous crabs simply arranged in a single series of families.

The validity of these groupings, and particularly of the oxystomes and the oxyrhynchs, has been questioned many times in the past based on considerations of the morphology of both the adults and, more especially, the larvae (see Rice 1980a). The case for the removal of the raninoids from the oxystomes has already been dealt with, for their zoeal stages are so distinctive that they must clearly be separated not only from the other oxystomatous groups but also from the whole of the Eubrachyura. The zoeas of the remaining three constituent groups, that is the dorippids, calappids, and leucosiids, are also very distinctive, but they provide no clear grounds either for grouping them together or for separating them collectively or individually from the Brachyrhyncha. Similarly, the Oxyrhyncha seems to be an unnatural assemblage since while the zoeal evidence indicates that the majids form a homogeneous, monophyletic group, well-separated from the remainder of the Eubrachyura, the parthenopids have zoeas which are very similar to those of the xanthid-portunid section of the Eubrachyura. The placement of the hymenosomatids is much more difficult. Like the majids, they have an abbreviated development, but to three zoeal stages and no megalopa in this case compared with two zoeas and a megalopa in the spider crabs. Moreover, hymenosomatid zoeas do not develop pleopods and they show none of the precocious development typical of the majids. They do, however, have a number of apparently very advanced characters such as a trapezoidal-shaped telson with the loss of the outer three processes, reduced antennule and antenna, and a vestigial coxal endite on the maxilla armed with a single seta. Aikawa (1929), Gurney (1938, 1942), Wear (1965), and Rice (1980a) all felt that these characters allied the hymenosomatids with some of the most advanced eubrachyurans such as the pinnotherids and leucosiids, but de Saint Laurent (personal communication) believes that they may have had a separate evolutionary history from the remainder of the Eubrachyura since a very early stage.

In the most recent proposed general classification of the Brachyura (Guinot 1978) these groups have been fragmented and absorbed into new divisions based on the positions of the male and female genital openings. Beginning with the assumption that the primitive position for the decapodan sexual openings was coxal, on the third legs in females and on the fifth legs in males, Guinot suggested that there has been a general tendency within the crabs for the openings to move onto the corresponding sternites. This displacement has tended to occur initially in females and only later in males. The result is three distinct conditions, each represented in an extant group of crabs.

In the most primitive group both the female and male openings are coxal. This group, the Podotremata, corresponds to the Peditremen of Bouvier & Gordon and contains only the dromioids, homoloids, raninoids and tymolids; the zoeal evidence for the inter-relationships of these taxa have already been discussed. The remainder of the crabs (the Eubrachyura) all have sternal female genital openings and were divided by Guinot into two further groups based on the position of the male openings. The first of these, the Heterotremata, consists of a series of families (or superfamilies in Guinot's nomenclature) in which at least some representatives have coxal male openings. Finally, the most advanced group, the Thoracotremata, have both the male and female openings on the sternum in all cases. Guinot presented this classification as a practical solution to the problem of brachyuran systematics, and realized that it is somewhat unnatural since it groups the constituent families according to general evolutionary level rather than into phylogenetic lineages. Nevertheless, Guinot's treatment implied that if such lineages could be recognized they would be found to pass through the Heterotremata and into the Thoracotremata. The thoracotrematous condition was assumed to have evolved independently more than once, and Guinot suggested that some heterotrematous families, such as the Goneplacidae and Leucosiidae in which the male orifices are, in her terms, sometimes coxo-sternal, represent an intermediate stage towards this condition.

In reviewing the eubrachyuran zoeas (Rice 1980a) I found considerable agreement with the adult classifications at the family level and below, but could recognize only three major groups at a higher taxonomic level. Two of these, though not very clearly separated, corresponded reasonably closely to Guinot's Heterotremata and Thoracotremata (see below), but the most distinct group was the majid spider crabs which Guinot had included in the Heterotremata.

## 5 MAJIDAE

In this discussion I have retained the usual taxonomic ranking of the spider crabs as a single family divided into a series of sub-families, despite a wide variety of zoeal form which would support the establishment of a majoid superfamily as proposed by Guinot. The spider crabs are readily distinguishable from all other eubrachyurans in having only two zoeal stages, with at least nine marginal setae on the scaphognathite in the first stage, and with well-developed pleopods in the second stage. The many similarities between the larvae of the majids and the other eubrachyurans certainly indicate a common ancestry until after the typical brachyuran zoeal form had become established. The maxilla of first stage majids, however, distinguishes them so clearly from those of almost all other brachyurans, including those in which the development is similarly abbreviated, that a quite different developmental pathway seems to have been adopted by the spider crabs. The separation

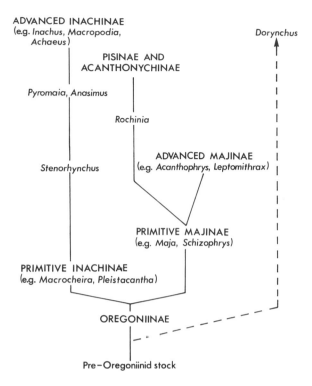

Figure 2. Suggested phylogenetic relationships among the extant Majidae, based entirely on zoeal features. See text for explanation.

of the majids from the other eubrachyurans seems to have taken place fairly early since the most primitive majid zoeas, that is those exhibiting the least reduction in spination, setation and segmentation, seem to be at an evolutionary level roughly equivalent to that of the most primitive xanthids. However, whereas the xanthids have a well-documented fossil history since the Upper Cretaceous, the existence of the majids prior to the Eocene is far from certain (Glaessner 1969).

There have been several attempts to devise phylogenetic schemes for the majids based on the larvae of the extant forms. Lebour (1928, 1931), for instance, suggested the sequence Majinae-Pisinae-Inachinae, and that *Hyas* should be transferred from the Pisinae to a new sub-family phylogenetically between the Majinae and the Pisinae. Similarly, Bourdillon-Casanova (1960) concluded that the inachinid genera *Inachus*, *Achaeus*, and *Macropodia* had been derived from a *Hyas*-like ancestor, in turn derived from the Majinae.

Aikawa (1937) read the phylogenetic sequence in the opposite direction, suggesting that the Inachinae are the most primitive majids and that the genera *Hyas* and *Chionoecetes* have the most advanced larvae, more so even than the Majinae.

Kurata (1969) produced a majid phylogenetic scheme based on a very careful examination of the degree of development of the carapace spines, the armature of the abdomen and telson, and the form and setation of the antenna and mouthparts in a large number of zoeas. Like Lebour and Bourdillon-Casanova, he believed that the reduced condition is advanced and therefore identified the genera *Maia*, *Macrocheira*, *Schizophrys*, and *Camposcia* as the most primitive majids from which he derived six distinct evolutionary lines. One of these lines led from *Maia* to the more advanced majinids such as *Acanthophrys* and *Lepto-*

*mithrax.* A second led from *Macrocheira* to *Stilbognathus* and then to *Herbstia.* Three separate lineages led from a form similar to *Camposcia,* one leading to *Micippa* and *Menaethius,* a second to *Acanthonyx,* and a third to the higher inachinids. A final lineage led from *Schizophrys* to the oregoniinids *Hyas, Chionoecetes,* and *Oregonia* and involved an enlargement of the carapace spines rather than a reduction as in the other lineages. Kurata nevertheless considered the oregoniinids to be more advanced because of the reduction in the telson fork spines and the loss or reduction of the carapace protuberances in the megalopa.

Kurata clearly felt that several of the majid subfamilies are heterogeneous groupings containing both primitive and advanced forms, and that a number of quite distinct evolutionary lines are represented in the family. From my own review of majid zoeas (Rice 1980a), I came to a similar general conclusion, and suggested a phylogenetic scheme for the family (Fig.2) which agrees quite closely with that proposed by Kurata, particularly in recognizing the genera *Macrocheira* and *Pleistacantha* as close to the ancestral stem form, in recognizing the remaining inachinids as among the most advanced majids, and in assuming a common and distinct ancestry for the Acanthonychinae and Pisinae. However, it disagrees with Kurata's scheme in a number of important points, particularly in considering the Oregoniinae, rather than the primitive Majinae or Inachinae, as being the closest extant spider crabs to the ancestral stock which gave rise to all the remaining groups except, perhaps, that represented by *Dorhynchus* (see below). From this oregoniinid-like ancestor I postulated two main lines, one leading to the primitive inachinid genera such as *Macrocheira, Pleistacantha, Camposcia,* and *Eurypodius* and the other leading to the primitive majinids such as *Maia* and *Schizophrys.* Two lines were suggested as originating from the primitive Majinae, one leading to the more advanced genera included in this sub-family, such as *Acanthophrys* and *Leptomithrax,* and the other leading to the Pisinae and Acanthonychinae via *Rochinia*-like zoeas which are very similar to those of other pisinids except in possessing well-developed lateral carapace spines. Finally, like Kurata, I suggested that the advanced group of inachinid zoeas, such as *Inachus, Macropodia,* and *Achaeus,* represent the end products of an evolutionary line originating in forms like the primitive inachinids. However, these primitive inachinid zoeas are so similar to those of the primitive majinids that it is possible that both groups are closely related to a single post-oregoniinid stock which gave rise to all three lines leading to the more advanced groups.

All of these suggested evolutionary lines are, of course, highly conjectural. But the least certain of all is that between the two groups of inachinid genera since the gap between them is so large. The zoeas of the primitive genera have all four carapace spines well-developed, dorso-lateral knobs on the third abdominal somite, at least two outer telson spines, the setae on the endopod of the maxilla in two clear groups, and the proximal segments of the endopods of both the maxillule and the second maxillipeds each carrying a single seta. In contrast, the advanced zoeas have neither rostral nor lateral carapace spines, no knobs on the third abdominal somite, only one outer spine on each telson fork, the setae on the endopod of the maxilla in a single group, and the proximal segments of the endopods of the maxillule and the second maxilliped both unarmed. At the time when this phylogenetic scheme was postulated the only described zoeas which seemed to be intermediate between these two extremes were those of *Stenorhynchus* (Yang 1976). Webber & Wear (1982) have subsequently described the zoeas of *Pyromaia* and *Anasimus* which seem to fall between *Stenorhynchus* and the more advanced Inachinae and lend a little more credibility to the existence of this lineage.

The zoeas of *Dorhynchus,* a genus which has usually been placed in the Inachinae, cannot be fitted easily into this phylogenetic scheme. The curious homoloid features of these larvae, including the forwardly directed rostrum, the presence of supra-orbital spines, the group of five lateral spines, the very prominent antero-dorsal carapace papilla and the long dorso-lateral spines on abdominal somites 2-5, all indicate a more primitive state than that found in the Oregoniinae. Most of the other features of the *Dorhynchus* zoeas, however, suggest a more advanced level than that of the Oregoniinae. Thus, the form of the antennal exopod is reminiscent of a typical majinid or of the primitive inachinids, whereas the retention of only one outer telson spine is a much more advanced character found otherwise only in the Pisinae, Acanthonychinae, higher Inachinae, and some Mithracinae. The terminal setation of the endopods of the maxillule and maxilla are both typical of the Acanthonychinae and Pisinae, while the unarmed proximal segment in the maxillulary endopod,, the extreme reduction of the setation of the basis of the second maxilliped and the failure of the sixth abdominal somite to separate from the telson in the second stage have been found in combination only in the most advanced inachinids. These zoeas are so distinct from all other described majids that they must surely belong to a hitherto unrecognized subfamily, and since they possess features which could not be readily derived from the extant members of the other sub-families I assumed that they have had a separate evolutionary history from a pre-oregoniinid spider-crab ancestor.

## 6 REMAINING HETEROTREMATA AND THE THORACOTREMATA

Compared with the majids, the other eubrachyuran zoeal groups which I thought I could recognize were much less easily distinguished. Each consisted of a series of families in which the zoeal stages are respectively relatively primitive or relatively advanced, judged by the degree of development of the armature of the carapace and abdomen, together with the setation and segmentation of the appendages. The only feature which seemed consistently to separate these two groups was the presence of six or more setae on the endopod of the maxilla in the primitive group, and five or fewer in the more advanced group. Distinguished in this way (and with the exception of the Majidae) the primitive group corresponded to Guinot's Heterotremata, except for the Leucosiidae, Dorippidae, and part of the Calappidae which have more advanced larvae seemingly allying them with the second group. With the addition of these families, this second group therefore corresponded to Guinot's Thoracotremata.

Encouraged by this correspondence, and despite the realization that evolution within the higher Brachyura has probably been quite complex, I attempted to identify possible phylogenetic lines running through the Heterotremata and into the Thoracotremata. However, although I was able to identify a few possible lineages within the primitive larval group I was unable to extend these into the more advanced families since many of these seem to have a confusing combination of advanced and primitive features which preclude their derivation from any of the extant primitive zoeal groups.

A possible solution, or at least an explanation, for these difficulties appeared with the publication of two very interesting notes on Guinot's classification by de Saint Laurent (1980a,b). While generally agreeing with Guinot's podotrematous, heterotrematous, and thoracotrematous groupings, de Saint Laurent disagrees fundamentally with her interpretation of the relationships between them. First, she sees great difficulty in deriving the eubrachyuran condition of the female genital apparatus from any of the groups of the present day Podotremata since, not only do no intermediate forms exist today, but she finds it difficult to see how such intermediate forms could ever have existed. Instead, she sees the distinction between the Podotremata and the Eubrachyura as cladistic and not simply a difference of evolutionary level. Second, de Saint Laurent believes that the separation of the heterotrematous and thoracotrematous forms is similarly cladistic, with no intermediate forms illustrating the transition between them. She redefined the Heterotremata (de Saint Laurent 1980b) as eubrachyurans in which the male genital ducts *always pass via the coxae* of the fifth legs before opening to the exterior. The supposedly intermediate forms (such as some Goneplacidae and Leucosiidae) in which the male orifices were referred to as coxo-sternal by Guinot are in fact, says de Saint Laurent, truly heterotrematous; the orifices only *appear* to be sternal because the tubular prolongations of the male ducts become encapsulated after they leave the coxae within an integumentary canal between sternites seven and eight to emerge finally in a sternal position. Although de Saint Laurent thus sees the division of the eubrachyuran stem into a heterotrematous group and a thoracotrematous group as a major event in crab evolution, she envisages a rather more complex situation in which several other phylogenetic lineages originated independently of this main division (de Saint Laurent 1980b, personal communication).

Adopting this view, the problem of tracing larval phylogenies from the Heterotremata to the Thoracotremata disappears, but is replaced by new difficulties. While the assumption of the thoracotrematous condition may, as de Saint Laurent suggests, have allowed this group to develop highly perfected locomotory mechanisms and to colonize a variety of terrestrial habitats, the heterotrematous crabs have retained the benthic habit or have become at least partly pelagic. The zoeal stages of both groups, on the other hand, have retained the primitive dispersive role so that convergent evolution is to be expected, and appears to have occurred. Thus, although the appendage setation of the most primitive thoracotrematous zoeas is much simpler (? advanced) than that of most heterotrematous zoeas, the same general trends are apparent in both groups. Consequently, both sections contain families with representatives having zoeas with some or all of the carapace spines absent, greatly reduced antennae, simplified setation of the endopods of the maxillule and maxilla, the segments of the endopod of the second maxilliped partly fused, and the sixth abdominal somite fused to the telson throughout the zoeal phase. Such features are found among the heterotrematous dorippids, leucosiids, and advanced majids, and among the thoracotrematous hymenosomatids and pinnotherids. This convergence results in spurious similarities between different groups of zoeas which have certainly confused me in the past (Rice 1980a,b). Nevertheless, it seems possible to identify at least some zoeal evolutionary lines amongst the more primitive groups and my present views on these are summarized below (Fig.3). Details of the criteria by which these lines have been identified are given in Rice (1981) and particularly in Rice (1980a), though some of the points in this summary differ slightly from these earlier statements.

1. Among majid zoeas, the combination of a wide range of general evolutionary levels from very primitive to very advanced, together with the possession of features which distinguish them from the zoeas of all other families, indicate that this family has had a quite distinct evolutionary history from that of the remaining Eubrachyura from a very early stage.

2. Of the remaining families, the most primitive zoeas belong to the Xanthidae. The ancestral stock, at least for most of the heterotrematous groups, is therefore assumed to have had zoeas similar to those of extant xanthid genera such as *Homolaspis, Ozius,* and *Eriphia.*

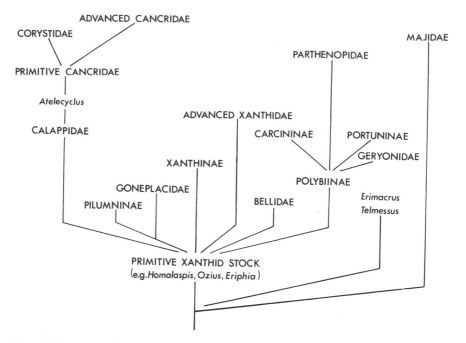

Figure 3. Suggested principal evolutionary lines among extant heterotrematous eubrachyurans, based entirely on zoeal features. See text for explanation.

3. From this stock, two major heterotrematous lines originated, one leading to the Portunidae, Geryonidae, and possibly the Parthenopidae, and the other to the Corystidae, Cancridae, and part of the Atelecyclidae.

4. Several groups of zoeas do not fit readily into this simple scheme and indicate the existence of several types of subsidiary evolutionary lines. (a) The sub-families Pilumninae and Xanthinae within the Xanthidae, and the Carcininae and Portuninae within the Portunidae seem to be offshoots from the main lines. (b) Some families seem to represent side branches which are evolutionary 'dead-ends'; the Geryonidae appear to have arisen in this way from a polybiinid ancestor, while the Goneplacidae possibly originated from the Pilumninae. (c) Several heterotrematous zoeas are so unusual that a quite separate lineage must be postulated: thus the Bellidae (*Corystoides, Heterozius,* and *Acanthocyclus*) could have originated from a primitive xanthid stock; the genera *Telmessus* and *Erimacrus* possibly originated from close to the same stock which gave rise to the majids; the origin of the genus *Orithyia* is completely unknown. The leucosiids and dorippids have zoeas which are so much more advanced than those of any of the other Heterotremata that they might have originated from almost anywhere within the group; at the same time, they have such unique telsons that they may have evolved quite independently of the other Heterotremata.

5. Within the thoracotrematous groups the Hymenosomatidae have very advanced zoeas at a level comparable with that of the Pinnotheridae (Rice 1980a). However, they are unique among the Thoracotremata in having three setae rather than two on the basal segment of the endopod of the first maxilliped. This suggests that they could not have evolved from any of the extant thoracotrematous groups and, in combination with their very spe-

cialized features, indicates that they have had a separate evolutionary history from a very early stage. Indeed, they may have acquired their thoracotrematous condition independently as de Saint Laurent (personal communication) has suggested.

6. The zoeas of the Gecarcinidae are the most primitive of the thoracotrematous groups. They may therefore be similar to the ancestral stock which gave rise to the rest of the Thoracotremata or they may have evolved from this stock before the more advanced zoeal characters of the other groups were acquired.

7. Within the Grapsidae the zoeas of the subfamily Grapsinae are the most advanced and could have been derived from those of any of the other subfamilies. The latter each have different combinations of primitive and advanced zoeal characters which are incompatible with a derivation one from another.

8. A similar situation exists in the Ocypodidae in which neither the Macrophthalminae nor the Scopimerinae could easily be derived from the other, but either subfamily could have given rise to the Ocypodinae.

9. The zoeas of the Ocypodinae are almost as advanced as those of the Pinnotheridae which could have originated from them. In turn, the pinnotherids have the most advanced thoracotrematous zoeas and, like the leucosiids within the Heterotremata, could not have been ancestral to any other group.


# 7 CONCLUSION

Brachyuran zoeas are all adapted for a relatively uniform pelagic existence, whereas the adults occupy a wide range of habitats and life-styles. Consequently, the zoeal stages may help to separate groups which have been classified together because of a superficial resemblance between the adults resulting from convergence. Examples of such clarification at relatively high taxonomic levels are provided by the separation of the constituent taxa of the Oxystomata and the Oxyrhyncha.

On the other hand, zoeal stages may suggest or confirm the essential unity of a taxon. Thus, the shared characteristics of all spider-crab zoeas which distinguish them from those of all other crabs strongly indicate that the majids form a monophyletic group.

Despite the absence of fossil material, zoeas may also indicate phylogenetic lineages, though this obviously depends upon the correct identification of primitive and derived characters. The importance of such identification is illustrated by the hair-like second telson seta of the stenopodids, thalassinids, anomurans, and dromioids. If this feature is derived or apomorphous then the dromioids seem to be closely related to the anomurans and excluded from a brachyuran ancestral role. If, on the other hand, the hair-like condition is primitive, then the dromioids could be close to the ancestral stem of the higher crabs.

Evolutionary trends within the eubrachyuran zoeas seem to be fairly easily recognizable; in general, advanced zoeas have reduced spination, segmentation and setation compared with the more primitive forms. However, these trends, which are presumably associated with the more efficient exploitation of the pelagic environment, may occur independently so that the zoeal stages are as susceptible to convergent evolution as the adults. The superficial resemblance of the zoeal stages of the Leucosiidae and Pinnotheridae, which as adults are fundamentally distinct, is an outstanding example of such convergence.

The general conclusion is that while students of adult decapodan morphology should

certainly take into account the larval evidence in attempting to establish phylogenetic relationships, it is equally important that larval workers should be aware of the potential errors
in assessing relationships on the basis of the developmental stages alone.

# REFERENCES

Aikawa, H. 1929. On larval forms of some Brachyura. *Rec. Oceanogr. Wks Japan* 2:17-55.

Aikawa, H. 1937. Further notes on brachyuran larva. *Rec. Oceanogr. Wks Japan* 9:87-162.

Balss, H. 1957. Decapoda. VIII. Systematik. *Bronn's Klassen und Ordnungen das Tierreichs* 5(1):7(12):
1505-1672.

Beurlen, K. 1930. Vergleichende Stammesgeschichte Grundlagen, Methoden, Probleme unter besonderer
Berücksichtigung der Höheren Krebse. *Fortschr. Geol. Paläont.* 8:317-586.

Boas, F.E.V. 1880. Studier over Decapodernes Slaegtskabs-forhold. *K. danske Vidensk. Selsk. Skr.* (6)1:
26-210.

Borradaile, L.A. 1907. On the classification of the Decapoda. *Ann. Mag. Nat. Hist.* (7)19:457-486.

Bourdillon-Casanova, L. 1960. Le méroplancton du Golfe de Marseille: les larves de crustacés décapodes.
*Recl. Trav. Stat. Mar. Endoume* 30:1-286.

Bourne, G.C. 1922. The Raninidae, a study in carcinology. *J. Linn. Soc. Lond. (Zool.)* 35:25-78.

Bouvier, E.-L. 1896. Sur l'origine homarienne des crabes. Étude comparative des Dromiacés vivants et
fossiles. *Bull. Soc. Philom. Paris* (8)8:34-110.

Bouvier, E.-L. 1940. Décapodes marcheurs. *Faune de France,* 37:1-404.

Burkenroad, M.D. 1981. The higher taxonomy and evolution of Decapoda (Crustacea). *Trans. San
Diego Soc. Nat. Hist.* 19:251-268.

Glaessner, M F 1930. Beiträge zur Stammesgeschichte der Decapoden. *Paläont. Z. Berlin,* 12:25-42.

Glaessner, M.F. 1960. The fossil decapod Crustacea of New Zealand and the evolution of the order
Decapoda. *N.Z. Geol. Surv. Paleont. Bull.* 31:1-63.

Glaessner, M.F. 1969. Decapoda. In: R.C.Moore (ed.), *Treatise on invertebrate paleontology, Part R,
Arthropoda 4, Vol.2:* R390-533. Lawrence: Geol. Soc. Am. & Univ. Kansas Press.

Gordon, I. 1963. On the relationship of the Dromiacea, Tymolinae and Raninidae to the Brachyura. In:
H.B.Whittington & W.D.I.Rolfe (eds), *Phylogeny and evolution of Crustacea:* 51-57. Cambrdige: Mus.
Comp. Zool.

Guinot, D. 1978. Principes d'une classification évolutive des Crustacés Décapodes Brachyoures. *Bull.
Biol. Fr. Belg.* 112:211-292.

Gurney, R. 1924. Decapod larvae. *Br. Antarct. Terra Nova Exped. (Zoology) Crustacea* 8:37-202.

Gurney, R. 1938. Notes on some decapod Crustacea from the Red Sea. VI-VIII. *Proc. Zool. Soc. Lond.*
108B:73-84.

Gurney, R. 1942. *Larvae of decapod Crustacea.* London: Ray Society.

Huxley, T.H. 1878. On the classification and the distribution of the Crayfish. *Proc. Zool. Soc. London.*
1878:725-788.

Knight, M.D. 1967. The larval development of the sand crab *Emerita rathbunae* Schmitt (Decapoda,
Hippidae). *Pac. Sci.* 21:58-76.

Knight, M.D. 1968. The larval development of *Raninoides benedicti* Rathbun (Brachyura, Raninidae),
with notes on the Pacific records of *Raninoides laevis* (Latreille). *Crustaceana, Suppl.* 2:145-169.

Kurata, H. 1969. Larvae of decapod Brachyura of Arasaki, Sagami Bay. IV. Majidae. *Bull. Tokai Reg.
Fish. Res. Lab.* 57:81-127.

Lebour, M.V. 1928. The larval stages of the Plymouth Brachyura. *Proc. Zool. Soc. Lond.* 1928:473-560.

Lebour, M.V. 1931. Further notes on larval Brachyura. *Proc. Zool. Soc. Lond.* 1931:93-96.

Lebour, M.V. 1934. The life history of *Dromia vulgaris. Proc. Zool. Soc. Lond.* 1934:241-249.

Milne-Edwards, H. 1834. Histoire naturelle des Crustacés, comprenant l'anatomie, la physiologie et la
classification de ces animaux. In: *Libraire Encyclopedique de Roret* 1:1-532, Paris.

Ortmann, A.E. 1896. Das System der Decapoden-Krebse. *Zool. Jb. (Syst.)* 9:409-453.

Rice, A.L. 1964. The metamorphosis of a species of *Homola* (Crustacea, Decapoda, Dromiacea). *Bull.
Mar. Sci. Gulf Caribb.* 14:221-238.

Rice, A.L. 1980a. Crab zoeal morphology and its bearing on the classification of the Brachyura. *Trans.
Zool. Soc. Lond.* 35:271-424.

Rice, A.L. 1980b. The first zoeal stage of *Ebalia nux* A.Milne Edwards 1883, with a discussion of the zoeal characters of the Leucosiidae (Crustacea, Decapoda, Brachyura). *J. Nat. Hist.* 14:331-337.

Rice, A.L. 1981. Crab zoeae and brachyuran classification: a re-appraisal. *Bull. Br. Mus. Nat. Hist. (Zool.)* 40:287-296.

Rice, A.L. & K.G.von Levetzow 1967. Larvae of *Homola* (Crustacea: Dromiacea) from South Africa. *J. Nat. Hist.* 1:435-453.

Rice, A.L. & A.J.Provenzano 1970. The larval stages of *Homola barbata* (Fabricius) (Crustacea, Decapoda, Homolidae) reared in the laboratory. *Bull. Mar. Sci.* 20:446-471.

Rice, A.L. & D.I.Williamson 1977. Planktonic stages of Crustacea Malacostraca from Atlantic Seamounts. *'Meteor' Forsch.-Ergebn.* 26:28-64.

Saint Laurent, M.de 1980a. Sur la classification et la phylogenie des Crustacés Décapodes Brachyoures. I. Podotremata Guinot 1977, et Eubrachyura sect. nov. *C.r. hebd. Seanc. Acad. Sci. Paris,* (D)290: 1265-1268.

Saint Laurent, M.de 1980b. Sur la classification et la phylogenie des Crustacés Décapodes Brachyoures II. Heterotremata et Thoracotremata Guinot 1977. *C.r. hebd. Seanc. Acad. Sci. Paris* (D)290:1317-1320.

Sakai, K. 1971. The larval stages of *Ranina ranina* (Linnaeus) (Crustacea, Decapoda, Raninidae) reared in the laboratory, with a review of uncertain zoeal larvae attributed to *Ranina. Publs Seto Mar. Biol. Lab.* 19:123-156.

Števčić, Z. 1971. The main features of brachyuran evolution. *Syst. Zool.* 20:331-340.

Števčić, Z. 1973. The systematic position of the family Raninidae. *Syst. Zool.* 22:625-632.

Straelen, V.van 1928. Sur les Crustacés Décapodes triassiques et sur l'origine d'un phylum des Brachyoures. *Mem. Acad. R. Cl. Sci.* Ser.5, 14:496-516.

Wear, R.G. 1965. Zooplankton of Wellington Harbour, New Zealand. *Zool. Publs. Victoria Univ. Wellington* 28:1-31.

Wear, R.G. & E.J.Batham 1975. Larvae of the deep-sea crab *Cymonomus bathamae* Dell 1971 (Decapoda, Dorippidae) with observations on larval affinities of the Tymolinae. *Crustaceana* 28:113-120.

Webber, W.R. & R.G.Wear 1982. Life history studies on New Zealand Brachyura 5: Larvae of the family Majidae. *N.Z. J. Mar. Freshw. Res.* 15:331-383.

Williamson, D.I. 1960. A remarkable zoea, attributed to the Majidae (Decapoda, Brachyura). *Ann. Mag. Nat. Hist.* (13)3:141-144.

Williamson, D.I. 1965. Some larval stages of three Australian crabs belonging to the families Homolidae and Raninidae, and observations on the affinities of these families (Crustacea: Decapoda). *Aust. J. Mar. Freshw. Res.* 16:369-398.

Williamson, D.I. 1976. Larval characters and the origin of crabs (Crustacea, Decapoda, Brachyura). *Thalassia Jugosl.* 10:401-414.

Williamson, D.I. 1982. Larval morphology and diversity. In: L.G.Abele (ed.), *Biology of Crustacea,* vol.1: New York: Academic Press.

Yang, W.T. 1976. Studies on western Atlantic Arrow crab genus *Stenorhynchus* (Decapoda, Brachyura, Majidae). I. Larval characters of two species and comparison with other larvae of Inachinae. *Crustaceana* 31:157-177.

FREDERICK R. SCHRAM
*San Diego Natural History Museum, San Diego, California, USA*

# METHOD AND MADNESS IN PHYLOGENY

'And . . . the Lord God . . . brought them unto Adam to see what he would call them: and whatsoever Adam called every living creature, that was the name thereof. And Adam gave names to all the cattle, and fowl of the air, and to every beast of the field.' (Genesis 2:19-20)

## ABSTRACT

Five operational principles utilized in phylogenetic analysis are outlined: Hennig's Axiom, the Principle of Relativity, the Conundrum of Convergence, Popper's Law, and the Phylogenetic Uncertainty Principle. The constraints imposed by these principles require that we come to understand and recognize the limits to certainty in phylogenetic analysis. This uncertainty complements current ideas on the operation of punctuational and stochastic factors in macroevolutionary processes. A method of phylogenetic analysis, stochastic mosaicism, is applied to several invertebrate groups. This technique suggests a conceptual model of evolution which, when contrasted to Sewell Wright's adaptive landscape model, more clearly reflects the evolution of *Baupläne* in space-time.

## 1 INTRODUCTION

Though taxonomy is the world's oldest profession, there has been little agreement on just how 'to do it'. Any volume of *Systematic Zoology* in the last decade bears witness to the taxonomic controversies which rage in the night, often not too gently, over matters, which to the outsider, may seem indeed much ado about nothing.

There has never been any shortage of conflicting ideas about just how to advance the frontiers of phylogenetic knowledge. Most recently, like Tories and Whigs, evolutionary and phylogenetic systematists have campaigned among the rank-and-file scientific community for converts to 'the cause'. Although the 'Hennigian Whigs' seem to hold the balance of power at present, this does not insure that the 'Simpsonian Tories' might not make a stunning comeback with the next swing of the pendulum in 'public opinion'. However, though we have been subjected to intense argument in recent years on just how to judge characters and assess biotic relationships, there has been little concerted effort in the course of this debate to outline the limits of our ability to know.

In point of fact, systematics and phylogeny operate under a set of 'laws' no less rigid than those which govern sciences like physics or chemistry. These determine the manner

331

in which we analyze data and constrain the limits of any set of conclusions. The rules of logic are not repealed when one leaves the 'empirical purity' of the physics laboratory for the more complex realms of biologic habitats. There are a series of principles utilized in the performance of phylogenetic analyses which can be clearly delineated. Our failure in recent years to arrive at a consensus on 'how to' do phylogeny and taxonomy stems, however, from the fact that not all of these principles have been equally recognized. Before we can understand method and recognize madness in crustacean (or any other phylogeny) it behooves us to outline these 'rules of the game'.

## 2 PHYLOGENETIC PRINCIPLES

2.1 *Hennig's axiom – Only synapomorphies determine degree of relationship between taxa and define cladistic proximity* (primitive characters are only useful to help define taxa).

This axiom has become, within the last decade, *the* dominant principle of phylogenetic analysis and has served to put systematics on a sounder conceptual foundation than it has probably ever had previously. Phylogenetic systematics has forced even the skeptics to carefully consider matters of what are advanced and what are primitive characters before they propose phylogenetic scenarios. Though cladistic analysis is not advocated nor practiced *in toto* by all taxonomists, the most vocal Hennigian advocates have forced the discipline as a whole to be more analytical and more careful in the definition of characters in order to attain any degree of sympathetic reception for newly proposed phylogenies.

In the original proposition of phylogenetic systematics Hennig (1966) clearly outlined a method by which everyone can effectively 'play' the phylogenetic 'game'. His ultimate and lasting accomplishment may be that he more carefully clarified the roles advanced (apomorphic) and primitive (plesiomorphic) characters can play in phylogenetic analysis. At one time, it was possible to place an inordinate amount of emphasis on the possession of shared primitive characters among taxa as a means of establishing common 'ancestors' (hypothetical creatures which were a sort of sum total array of primitive characters). Though the invention of such 'ancestors' may have a role in the design of scenarios to explain presumed phyletic events, they have little to do with the actual determination of degree of relationship among taxa. Gould (1980b) specified the pernicious effects of rigid preconceptions determining phylogeny when he observed that such have: '. . . led to a reliance on speculative storytelling in preference to the analysis of form and its constraints; and, if wrong, in any case, it is virtually impossible to dislodge because the failure of one story leads to invention of another rather than abandonment of the enterprise'.

Let us consider an example of this from the realm of crustacean studies. Calman (1909) established a facies theory to classify all eumalacostracans and to relate a phylogenetic scenario for the groups. He established a supposed ancestor, 'the caridoid animal', which was a compilation of primitive features he called the caridoid facies. These facies (as somewhat modified by Hessler 1982) are:

1. Carapace covers and encloses thorax.
2. Moveably stalked compound eyes.
3. Biramous antennule.
4. Scale-like exopod on antenna as a single joint.
5. Thoracopods with well-developed exopods.

6. Abdomen with complex massive musculature to achieve strong ventral flexion of tail fan.

7. Internal organs mainly excluded from abdomen (a consequence of 6).

8. Paddle-like uropods and telson forming a tail fan.

9. Pleopods I-V alike, natatory.

However, all of the above characters, except for 6 and 7, can be considered plesiomorphic. Well-developed carapaces are found in many of the crustacean orders, stalked eyes occur in phyllocarids, biramous antennules are also seen in remipedes, thoracic or trunk exopods are a common occurrence throughout crustaceans, and natatory pleopods (or posterior trunk appendages) also occur in remipedes. What therefore appears to be a long and impressive list to unite the eumalacostracans actually tells us very little about eumalacostracan kinships. The only synapomorphies that can provide any argument at all in favor of recognizing the legitimacy of Eumalacostraca as a valid taxon are 4, 6-7, and 8 (see Hessler, this volume).

The apomorphies of one-joint antennal scale, caridoid musculature, and thoracic restriction of viscera also serve to exclude one group from the Eumalacostraca which Calman (1904, 1909) placed there, viz. the Hoplocarida. Hoplocaridans appear to possess (at least in the stomatopods) an antennal scale of two joints, their abdominal musculature is specialized in a manner different from that of caridoids, and they retain a more primitive arrangement with viscera throughout the entire length of the body. Schram (1969a,b) separated the hoplocaridans from eumalacostracans on the basis of a series of hoploid facies, but Burnett & Hessler (1973) argued to maintain the plesiomorphic caridoid facies in the sense of the taxonomy of Calman (1904) to retain mantis shrimp as a superorder of Eumalacostraca. In fact, the Hoplocarida (both fossil and recent orders) possess a series of synapomorphic features which argue for their separate status (Schram 1982, see also Knuze, this volume):

1. Triramous antennules.

2. Antennal scale of two joints.

3. Thoracopods with three-segment protopod, and four-segment endopod.

4. Abdomen with distinct ('quasi-caridoid') musculature.

5. Pleopod gills dendrobranchiate.

6. Possible fusion of anterior somite to achieve a six-segment from a seven-segment abdomen.

A strict consideration, therefore, of just what might be plesiomorphic and apomorphic among malacostracans leads to a rejection in the mode of Gould of the 'speculative storytelling' of Calman, which indeed has been heretofore 'virtually impossible to dislodge' and which still commands a small but determined pocket of loyalists. The Calman caridoid animal may be permitted to survive only as a hypothetical creature useful in weaving a 'just so story' of what may have happened in the course of eumalacostracan history. However, it can no longer be used in classification and phylogenetic analysis in order to tell us to what degree any taxa, e.g. syncarids and mysidaceans, might be related.

2.2 *Principle of relativity — The assessment of plesiomorphies and apomorphies is a subjective decision* (Robert Hessler most aptly encapsulated the idea for me in a private conversation — 'one man's plesiomorphy is another man's apomorphy').

If Hennig's Axiom were in fact a law, or even a postulate, i.e. an eternal but unprovable verity, then there would be no need to argue about cladistic schemes. Hennig's Axiom, however, only outlines an empirical technique. Conclusions derived from application of the technique to individual cases hold only if one can agree to the initial limiting conditions established by the cladistician when he or she defines what is to be considered plesiomorphic and apomorphic.

In practice, the determination of whether or not a feature is derived is frequently constrained by a more or less 'universal consensus' of workers on a group rather than the independent judgement of individual taxonomists. As a result of such consensus, a certain amount of stability is achieved, which in turn insures, for better or worse, that all phylogenetic theorizing operates within certain 'acceptable limits'. Though such consensuses can persist for long periods of time, they can and do undergo changes in status, often quite dramatically.

For example, until the middle of the 1900's, it was widely assumed that the primitive crustacean limb was biramous. The peculiar mixopodial, polyramous anterior limbs of the Devonian *Lepidocaris rhyniensis* were explained as modifications of a condition represented by the more primitive posterior biramous paddles and biramous larval appendages (Scourfield 1926, 1940). With the discovery of *Hutchinsoniella macracantha* and the subsequent phylogenetic analysis that began with Sanders (1957), and proceeding through Hessler & Newman (1975) and even Schram (1982, actually written in 1978), a new consensus was 'firmly' established, viz. one that abandoned the biramous theory and established the polyramous mixopodial limb as the most primitive type. The anterior mixopodia of *Lepidocaris* were easily fitted into the resultant phylogenetic schemes based on cephalocarids as an ancestral type, and the posterior adult and larval biramous limbs of *Lepidocaris* were conveniently ignored. However, the recent discovery (Yager 1981) of an even more primitive crustacean, *Speleonectes lucayensis,* with homonomous biramous limbs on all body segments would seem to lend some support to the old biramous theory (see Schram, this volume). The biramous appendage may again return to its original position as a model of the most primitive crustacean appendage type, and the polyramous mixopodium may thus be reinterpreted as an apomorphic form. *Lepidocaris* now appears to be a crucial transitional type which displays how anterior limbs initially become specialized as mixopodia while posterior and larval appendages retained the more generalized structure. Purely polyramous types like cephalocarids are thus seen to be derived forms, which provides some morphological confirmation of Schram's (1982) interpretation of the biology of cephalocarids as rather specialized.

Another example of a changing consensus comes from the study of isopods. A long prevalent view (e.g. Calman 1909) was that either the flabelliferans or asellotes were the most primitive of isopod groups. However, Schram (1974) suggested that an analysis of characters indicated phreatoicid isopods to be more primitive. A new consensus on isopod evolution may be emerging, since Watling & Kensley (personal communication) derive a stem form for isopods similar to that implied by Schram's conclusions.

A final example of consensus in flux may be supplied by peracarids. Calman (1904) grouped into a superorder Peracarida a series of orders possessing oöstegite brood pouches in females and a lacinia mobilis on the mandibles. This arrangement has become well entrenched in eumalacostracan phylogeny (e.g. Siewing 1956, Fryer 1964, Hessler 1969). However, a series of papers has now begun to appear (Schram 1981; Watling 1981, this volume) which question for different reasons Peracarida as a valid taxon. What the form of the new consensus will be is not yet clear, but it would appear that our views of what is primitive or advanced, or what is indeed useful at all, in regards to 'peracarid' types is changing.

In a period of 25 years the pendulum of consensus swung away from biramous limbs as plesiomorphs, to biramous limbs as apomorphs, and now appears to be swinging back again; from flabelliferans or asellotes as primitive isopods towards some other groups; from Peracarida as a useful taxon to questioning of such. Should all this be disturbing to crustacean phylogenists? Not at all, these are all merely examples of validity of the corrolary of relativity.

2.3 *The conundrum of convergence – Similarities of form suggest affinity, but the development of form is mediated by the needs of function* (for arthropodologists of any persuasion this might just as well be termed Manton's Conundrum, since so many of the controversial and startling phylogenetic analyses of that lady actually arose from her attempts to recognize and deal with convergence between various arthropod groups).

This is perhaps the most disturbing of the five principles for phylogenists and taxonomists to deal with. The riddle lies in this: how can things appear to be alike and yet be different? Hessler, in House (1979:488) argued 'that if one no longer give reliance on similarity as an indicator of affinity then taxonomy should be given up: similarity as a criterion should only be abandoned after proof'. However sympathetic we may feel towards this statement, we must remember first, that similarities come in two 'flavors', synapomorphies and symplesiomorphies, and only the former are useful in phylogenetic analysis; and second, form is rather meaningless without an understanding of function and developmental history. Complete and adequate phylogenetic judgements can be made only when all three are adequately understood.

For example, Manton (1964) provided a most valuable service to carcinology by cutting the Gordian Knot of the 'Mandibulata'. The supposed synapomorphy of mandibles between uniramians and crustaceans was shown, after exhaustive analysis of function and embryonic derivation, to be convergent. Entomologists and myriapodologists (students of the Uniramia) seem determined to maintain the connection (see e.g. Gupta 1979). However, so many autapomorphic conditions of adult locomotor functional morphology (Manton 1977) and early development (Anderson 1973) now so effectively intervene, that the appearance of mandibular structures in several disparate arthropodous groups has to be considered a convergent phenomenon. If mandibles are a unifying feature, they can only be so as a symplesiomorphy, or at best a synapomorphy back to such a distant ancestral type as to be absolutely meaningless.

The Principle of Relativity tells us the determination of apomorphy is subjective, thus any ultimate recognition of convergence is dependent on initial analysis and definition of characters. The problem is rampant throughout the Crustacea, and Arthropoda as a whole, since there seems to be no way to classify arthropodous taxa so as to eliminate all cases of convergence (see e.g. Schram 1981). A working rule to apply here is that one should be reluctant to invoke convergence without a sound reason; i.e. one should assume apparent synapomorphies are in fact such, but always be alert for conflicting evidence from other apomorphies or from ontogeny which can question initial conclusions.

2.4 *Popper's Law – A statement can never be definitively proved, only disproved* (the level of 'proof' in an argument is only a function of probability, whereas 'disproof', once achieved, is certainty).

Popper's Law is a principle of logic, but has emerged as an important element in contemporary debate over whether or not phylogenetic analysis is a real science. That Popper's

Law applies to all in phylogenetic analysis is a result of the operation of the Principle of Relativity and the Conundrum of Convergence. 'Relative proof' or disproof under the strictures of Hennig's Axiom arises from the basic subjectivity of character state determination, as well as the ever present possibility of convergence being recognized in evolving lines as functional need is understood.

For example, as discussed above, we might reject Calman's views on eumalacostracan history and taxonomy. Concepts of what is plesiomorphic and what is apomorphic among 'peracarids' are being clarified, and a new period of stability appears to be in the offing. Calman relied on primitive and convergent characters to determine relationships among 'peracarids', and as a result his methods and conclusions are now being questioned. Complete consensus on an alternative explanation for eumalacostracan evolution has yet to be achieved, however, since the Calman scheme can and is being questioned in various different ways and on different grounds. The convergence of structure in subthoracic brood pouches (Schram 1981), the varying apomorphic derivations of the carapace and its nonplesiomorphism (Dahl 1977, Watling 1981), and the possibility of the caridoid facies being nothing but convergences (Tiegs & Manton 1958) all illustrate the operation of Popper's Law vis-à-vis Calman's facies theory. Though Calman's speculations were considered 'laws carved in stone', it was unreasonable to expect them to be anything but temporary expedients to classify the knowledge of that time (early 20th century). Likewise, current systems being advanced will undoubtedly be modified or rejected by future workers after having served their purpose. Calman never *proved* anything; he merely made a statement equivalent to saying 'all swans are white'. Nor have the scores of carcinologists who have used Calman's system since its inception done anything to *prove* it, other than make an indirect statement as to the probability of its usefulness. These are all 'white swans' in the sense of Popper. However, one man's rejection of facies theories is enough to serve as the 'black swan' which allows the rejection of the general proposition that the caridoid facies are useful.

*2.5 Phylogenetic uncertainty principle – In delineating lines of evolution one can either group them in proximity to each other, and be uncertain as to the degree of proximity; or, one can postulate archetypes to unite the lines, and be uncertain such hypothetical animals ever existed and connected the lines one seeks to link.*

This principle (first outlined in Schram 1979, 1981) formalized the constraints placed on all phylogenetic analysis by the four principles discussed above. The consequence for phylogenetic analysis is that the most one can do with certainty is to establish morphologic *Baupläne,* using characters that are both plesiomorphic as well as apomorphic. These *Baupläne* are the basic structural plans, each forming a nexus, about which the evolution of a group appears to take place.

The actual grouping and/or linking of phyletic lines is dependent on assessment of apomorphic conditions. One can attempt to increase the probabilities of likelihood of apomorph determination by attempting outgroup comparisons to assess polarity (e.g. Watrous & Wheeler 1981) or by using a somewhat more mathematical approach (e.g. Marx et al. 1977). However, no matter what steps are taken to increase the probability of 'success', one cannot escape the basic subjectivity engendered by the Principle of Relativity and the Conundrum of Convergence. Furthermore, the Hennigians tell us recognition of ancestors is impossible (e.g. Boudreaux 1979), or, at best, futile (Engelmann & Wiley 1977).

Does the Uncertainty Principle completely negate the rationale for phylogenetic analysis? Not at all. We can continue to try to connect lines of evolution into cladograms, but the principle of uncertainty warns us that the farther the dichotomous branchings are from the basic morphotypes we seek to interrelate, the greater the level of uncertainty. We can continue to manufacture hypothetical ancestors on demand so long as we realize these imaginary creatures are *always* subject to rejection (disproof) for any number of reasons arising from their basic subjectivity.

The Uncertainty Principle clarifies the conceptual role phylogeny plays in evolutionary theory. Like the 'laws' of physics, a phylogeny and/or classification can stand only so long as it adequately serves to direct data collection and interpretation. For example, Calman's caridoid facies served to provide a framework in which anatomical systems could be analyzed for their phyletic usefulness (e.g. Siewing 1956). However, when a phylogenetic scheme proves to be inadequate in this role [e.g. attempting to reconcile synapomorphies and convergences among eumalacostracan types (Schram 1981) or applying a strict cladistic analysis to the Peracarida (Watling 1981, this volume)] it must be rejected in favor of some 'new synthesis'. The new scheme possesses a lesser degree of uncertainty than the old one it replaces for a time, but it does not approach a level of absolute certainty, which is effectively unachievable.

## 3 IMPLICATIONS OF PRINCIPLES

The lack of certitude that is engendered by the principles discussed above may constrain the kinds of activities a phylogenist can engage in if he wishes to remain within the limits of some high degree of certainty. This appears to require the recognition of *Baupläne,* or morphotypes, (at whatever level one is working, be it phyla and classes or down to genera and species), and the elucidation of the manner and extent of radiations within those structural plans. This is to say that the lowest levels of uncertainty seem to lie in recognizing form and seeking to understand function. Cladogram erection can, of course, continue to be an activity for a phylogenist/taxonomist provided the degrees of uncertainty are explicitly or implicitly recognized and accepted. Such a position would imply that while some real advantage may be gained by arguing definitions of basic structural types and elucidation of function, there can be little meaningful debate on the advantages of one cladogram over another. In the end, only purely pragmatic considerations will serve to award accolades to that one of several contending schemes that best serve to facilitate the accumulation of more data and the interpretation of basic form and function. There is nothing particularly new in such a position. D'Arcy Thompson (1917) and Severtsov (1912) were early advocates of considering the importance of form and function in phylogenesis independent of any theories of organic evolution (for summary, see Adams 1980).

The limitations engendered by the five principles above lead Schram (1981, 1982) to utilize an approach I termed *stochastic mosaicism.* During this attempt to identify basic forms for eumalacostracans so a cladistic analysis of relationships could be performed, it was found that there was no way to arrange characters in phenograms without encountering problems with convergences. It seemed that all could be done without too much uncertainty was to outline the *Baupläne,* that is, to consider a series of alternative states for certain basic features of the eumalacostracan body plan (uniramous or biramous thoracopods, presence or absence of subthoracic brood structures, absence or degree of development of a carapace). These produced a series of 16 possible *Baupläne,* morphotypes, or 'paper animals' about which eumalacostracan evolution could have taken place (Fig.1). When the

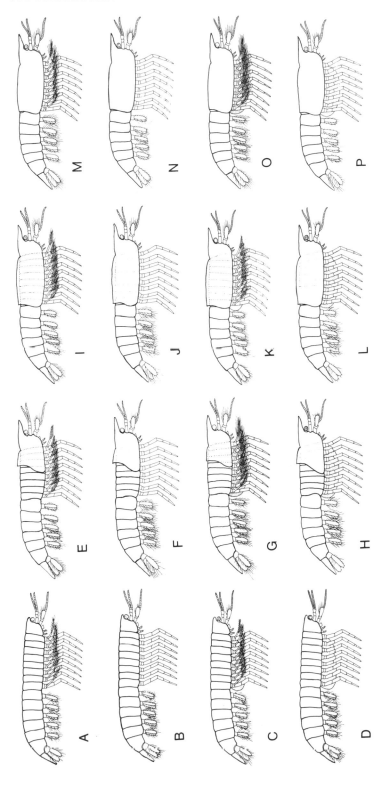

Figure 1. Matrix of paper animals illustrating the possible *Baupläne* within the Eumalacostraca. Column A-D, without carapace; E-H, short carapace; I-L, full carapace not completely fused to thorax; M-P, carapace completely fused to thorax. Rows A-M and C-O, with biramous thoracopods; B-N and D-P, with uniramous thoracopods. Rows C-O and D-P, with subthoracic broodpouches. The status of each *Bauplan* is as follows: A. syncarid; B and C. no known groups developed; D. acarideans (isopods and amphipods); E. thermosbaenaceans; F. no known group developed; G. hemicarideans (cumaceans, tanaids, and spelaeogriphaceans); H. no known group developed; I. waterstonellideans; J. belotelsonideans; K. mysidaceans; L. and M. no known groups developed; N. decapods; O. euphausiaceans; P. amphionidaceans.

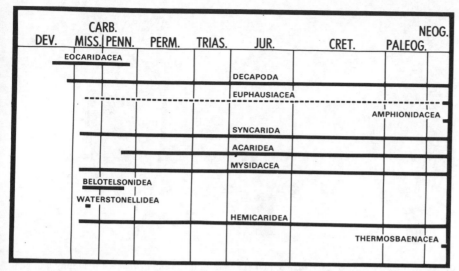

Figure 2. Presently recognized stratigraphic ranges for eumalacostracan orders as delineated by Schram (1981).

fossil and recent record was then examined, it was noted that 10 of the 16 morphotypes actually existed either now or in the past. It was as if nature had played a cosmic cottie game, experimenting with random combinations of alternative morphologic states in a matrix of all possible animals.

Furthermore, the fossil record of the realized morphotypes indicated that they all arose about the same time, within the late Devonian to early Carboniferous (Fig.2). Not all of the possible morphotypes were actually developed (or at least have yet to be recognized), nor have all *Baupläne* been equally successful in terms of the extent of the adaptive radiation achieved. However, in a truly stochastic system, there is no reason to suppose all or any of the possible morphotypes need to have actually appeared at all. The implication is that whichever did appear and whichever did achieve a greater radiation was due to a large element of chance.

As a result of this analysis, Schram (1981) rejected the traditional Calman taxonomy of Eumalacostraca and proposed an alternative system recognizing ten orders, and a cladogram utilizing cohorts that reflected what appeared to be a likely grouping of those orders.

The most promising thing about the stochastic mosaic approach is that it has applications in many other (and by implication all) animal groups. It is rooted in a tradition based on the importance of morphological analysis in discerning animal relationships, with historical roots contemporaneous with Darwin (Zangerl 1948). Although I do not pretend expertise in a variety of higher taxa outside of crustaceans, I do feel that some of the recent literature on the higher taxonomy of major animal groups indicates that greater attention should be paid to the determination of *Baupläne* in a stochastic mosaic framework. This would prove quite productive towards understanding group origins and interrelationships.

For example, Haugh & Bell (1980a,b) have asserted '. . . that the high-rank phyletic taxonomy of echinoderms must reflect the morphology of alternative coelomic body plans'. In an attempt to incorporate new information about the internal soft anatomy of echino-

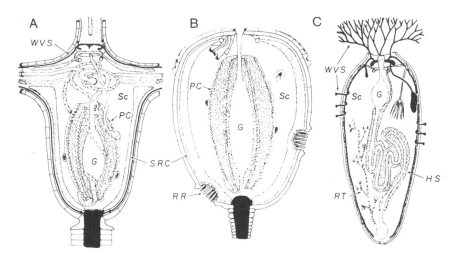

Figure 3. Paper animals proposed by Haugh & Bell (1980a,b) illustrating *Baupläne* for subphyla of echinoderms. A. with both water vascular and subdermal respiratory coeloms (e.g. camerate crinoid); B. with subdermal respiratory coelom only (e.g. rhombiferan cystoid); C. with water vascular coelom only (e.g. holothurian). G: gut, HS hemal system, PC: perigastric coelom; RR: respiratory rhomb, RT: respiratory tree, Sc: somatocoel, SRC: subdermal respiratory coelom, WVS: water vascular system.

|  | WVS present | WVS absent |
|---|---|---|
| **SRC present** | subphylum I | subphylum II |
| **SRC absent** | subphylum III |  |

Figure 4. Matrix of coelomic character states for echinoderms. WVS: water vascular system; SRC: subdermal respiratory coelom.

derms, Haugh & Bell (1980a) recognized a series of three 'subphyla' in which most of the fossil and recent echinoderm classes can be placed (Fig.3).

In fact, Haugh & Bell unwittingly employed a stochastic mosaic analysis to achieve their end by utilizing coelomic morphology (presence or absence of a water vascular system, presence or absence of a subdermal respiratory coelom) which delineate four *Baupläne*, only three of which Haugh & Bell realized (Fig.4). Their Subphylum I has both coelomic systems present (and of which only the Crinoidea survive); Subphylum II has only a subdermal respiratory coelom (and, though there was a tremendous Paleozoic radiation of some ten classes, is now extinct); and Subphylum III has only a water vascular system (the most successful subphylum with all three of its classes, Echinoidea, Stellaroidea, and Holothuroidea, still extant). One morphotype derived from the matrix (neither water vascular nor subdermal coeloms, i.e. a 'paper animal' with a crystalline calcite endoskeleton and a visceral somatocoel only) has not been recognized by Haugh & Bell.

There is a fairly close correspondence in terms of time or origin of the three actualized echinoderm subphyla, similar to that seen in eumalacostracans. Subphyla I and II have their earliest occurrence in the Cambrian; and while Subphylum III currently has its earliest

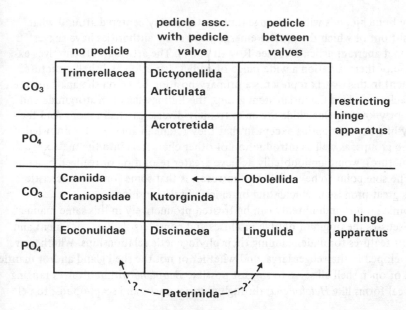

| | no pedicle | pedicle assc. with pedicle valve | pedicle between valves | |
|---|---|---|---|---|
| CO₃ | Trimerellacea | Dictyonellida Articulata | | restricting hinge apparatus |
| PO₄ | | Acrotretida | | |
| CO₃ | Craniida Craniopsidae | Kutorginida | ◄----Obolellida | no hinge apparatus |
| PO₄ | Eoconulidae | Discinacea | Lingulida | |

Note: The table above uses $CO_3$ and $PO_4$ as the left-column labels. The "restricting hinge apparatus" and "no hinge apparatus" labels appear at the right. Below the matrix: `?----Paterinida----?`

Figure 5. Matrix of character states for brachiopods utilizing nature of pedicle, development of skeletal hinge apparatus, and type of mineralization in shell. Obolellida typically have the pedicle between valves, though some associate it with the pedicle valve; and Paterinida have forms with and without pedicles.

record in the lower Ordovician, a recent trend has been to extend the range of many invertebrate groups back into the Cambrian as knowledge of that earlier period improves. It is probable that the three recognized echinoderm *Baupläne* extend back and arose together in earliest Paleozoic times.

Yet another example of how stochastic mosaicism provides a clearer view of a major taxon's origins may be seen among brachiopods. An overview by A.D.Wright (1970), augmented by data from Williams & Hurst (1977), indicates that an array of characters basic to the brachiopod plan appears to yield a matrix of 'paper animals' which correspond in many instances to the actual groups Wright and other authorities (see e.g. Rowell 1981) seek to separate taxonomically.

Brachiopoda are currently arrayed into two classes, Inarticulata and Articulata. The articulates at present, and through most of Phanerozoic time, have been the dominant group. However, the inarticulates are a much more diverse taxon than the traditional view, derived from *Lingula* as the 'typical inarticulate' would allow. A random arrangement of certain basic brachiopod features yields a series of *Baupläne*: carbonate or phosphatic shells; lack of a pedicle, pedicle emerging between shells, or pedicle emerging in some groove or foramen in the pedicle valve; and presence or absence of some locking or restricting hinge mechanism. These can be arranged into a matrix in which Wright's 'orders' can be nicely fitted (Fig.5).

Wright sought to demonstrate the independent origin of these several brachiopod lines in earliest Paleozoic times from a suite of somewhat phoronid-like 'brachiophorates'. The matrix achieved reveals some overlap of his tentative groups, but in all cases these represent problematic taxa poorly understood by brachiopod authorities. For example, the

Dictyonellida are branchiopods with a strange umbonal anatomy centered around what appears to be a slit out of which the pedicle emerges, and some authorities have suggested these may be in fact aberrant articulates (see Rowell 1965). The articulates themselves exhibit in the course of their radiation a wide range of pedicle variation and shell structure, and their placement in this matrix represents a primary position based on the anatomy seen in the generally acknowledged Cambrian stem group, the Billingsellacea. Kutorginida and Paterinida are poorly known. Obelellida are an extremely diverse catchall taxon. And the Craniida are very like the craniopsids except in that they retain an anus. Better knowledge of several of these groups, as well as introduction of other characters into the matrix (such as shell microstructure), would undoubtedly achieve greater resolution of realized brachiopod *Baupläne*. The sole point to be made here, however, is that some form of stochastic mosaicism offers great promise in elucidating brachiopod relationships.

Other taxonomic groups undoubtedly can be treated productively in this same manner. Morris (1979) discusses the origin of bivalve molluscs and considers shell development and form as important features for understanding their phylogenetic relationships. Whether or not shells are developed in the veliger larva, and whether or not the shell gland and/or mantle secretes a closed or open shell all serve to suggest possible *Baupläne* within bivalves ranging from problematical forms like *Helcionella,* through rostrochonchs and scaphopods, to pelecypods.

The strange array of Cambrian arthropod forms might be better understood if a stochastic mosaic analysis were attempted. Such is hinted at by Briggs (in House 1979:487). Although promising beginnings have been made in delineating the components of the Burgess Shale fauna (Briggs, this volume), completion of all redescriptions of these Cambrian arthropods must be awaited before any effective analysis can begin. At that time, analysis of basic structural features such as body and appendage form, numbers and types of cephalic segments, numbers and locations in relation to the mouth of sensory (antennal-like) appendages, post-cephalic tagmatization, and gross internal anatomy will undoubtedly prove important in the delineation of arthropodous *Baupläne,* and perhaps suggest interrelationships.

Finally, even vertebrate history lends itself to such an analytic approach as discussed here. Houde & Olson (1981) recognized a series of early Tertiary birds which in an unexpected way combined certain features of avian anatomy that have been of central concern in discussions of bird evolution. Advanced birds could be viewed as being divided on the basis of four structural types: (1) with a paleognathous palate and flying ability, the new group of birds of Houde & Olson; (2) with a paleognathous palate and no ability to fly, Ratitae; (3) with a neognathous palate and an ability to fly, Carinatae (excluding the Impenes or penguins); and (4) a neognathous palate and no ability to fly, Impenes. Such might suggest an approach to bird taxonomy much like that of Wetmore (1960).

## 4 APPLICATION TO EVOLUTION THEORY

What has been discussed above deals with a manner and method for engaging in phylogenetic analysis. What, if anything, can this tell us about evolutionary processes involved in the origin of major groups?

In recent years, a critical examination of various aspects of the general fossil record has led to some new and stimulating insights into the evolutionary process (Raup 1978, Gould

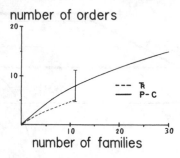

number of orders

number of families

Figure 6. Rarefaction analysis of the number of families in orders of Permo-Carboniferous, with 95 per cent confidence limits, and Triassic malacostracans. The limits indicate that the faunal extinction for malacostracans across the Permo-Triassic, though dramatic in terms of taxa changeover, is not greater than that which might be expected due to chance.

1980b). Ideas have advanced along several fronts that suggest, at least in the realm of macroevolution, that strict reductionist, gradualist, and Darwinian explanations cannot supply all the answers.

A series of evaluations has led to the conclusion by some of these workers that species origins occur relatively quickly rather than gradually through time and that these periods of 'punctuation' are followed by long periods of species stasis (e.g. Eldredge & Gould 1972, Gould & Eldredge 1977). The patterns discussed above for stochastic mosaicism in major groups, for example in Eumalacostraca (Figs.1 and 2), suggests a punctuational or quantum evolutionary mode also operates in conjunction with a simultaneous origin of higher taxa as well.

As for the extinction process, causal factors can and do operate, but when viewed as a whole, purely stochastic factors seem to be equally important (Raup 1978). Again, such patterns are seen in an analysis of the Permo-Triassic extinction of Malacostraca (Schram in press) where, in a rarefaction analysis of families in orders, a purely random pattern of disappearance of taxa across the boundary could not be rejected (Fig.6).

Finally, the random generation of lineages (Raup et al. 1973) and of artificial morphologies (Raup & Gould 1974) by computer program produced patterns that were found to correspond with those observed in the fossil record. This might suggest that deterministic explanations of a group's actual history, as seen in the fossil record, may not be justified if stochastic processes cannot first be rejected. And indeed, the matrices of stochastic mosaic analyses used here seem to imply that the quick evolution of several alternative *Baupläne* in the early evolution of a group is determined by a large element of chance. Just what morphologies do appear and which expand into major radiations seem due as much to chance as anything definable. Gould (1980b) stated it succinctly: 'Just as mutations are random with respect to the direction of change within a population, so too might speciation be random with respect to the direction of macroevolutionary trend'.

Gould (1980a) posed an important question relevant here: 'Why is morphological space so sparsely populated, but so clumped where it is occupied?' One answer, derived from the analysis presented above, is that morphology is 'clumped' because it starts out that way. *Baupläne* arise randomly from a set of potential possibilities that are essentially either/or propositions. For example, in eumalacostracans, thoracopods can be either biramous or uniramous, given the basic constraints of the malacostracan stenopod. *Baupläne* within eumalacostracans will then cluster about these, and only these, alternatives. So while basic anatomical features combine to form a morphotype or structural plan, constraints are placed on the animals by the limits of overall morphology within any particular phylum. The *Baupläne* of echinoderms (Fig.3) revolve around coelomic cavities, because these are

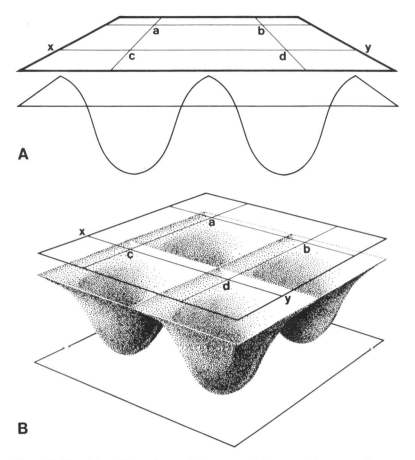

Figure 7. Pictorial rendition of an evolutionary model proposed here as an alternative to the 'adaptive landscape' model of Wright (1932). A structural morphotype can be viewed as occupying a place in space-time similar to a pit, i.e. its presence is like a curvature of a plane. If there were no manifested *Baupläne,* the plane would be a flat smooth surface. Transitions to adjacent *Baupläne* are achieved when a character varies over a singularity at the rim of the pit, essentially expressed as an 'either/or' condition. Variations on a *Bauplan* would tend to clump deep in the pit at least curvature, while extreme variants would be found near the rim. A is a cross-section through the curved surface in B, a-d define the centers of *Baupläne* manifestation, comparable to the morphotypes expressed in the matrices of Figures 1 or 3.

basic features of importance to Echinodermata. *Baupläne* of eumalacostracans can never involve coelomic variants, since coeloms within the Crustacea are not important structural elements; nor do they involve dermal calcite endoskeletons, since possession of such are not part of being an arthropod.

Part of the difficulty with attempting to understand the unevenness of morphology has been due to a peculiar distortion in our outlook engendered by a conceptual model originally advanced by Sewell Wright (1932). He proposed an 'adaptive landscape', where manifested morphologies occupy peaks of 'high adaptation'; and these peaks were separated by valleys, which represent poorly adaptive gene combinations. Furthermore, it was proposed that the topography of the landscape in this model could shift through time as differing gene combinations developed greater or lesser adaptive potential. Under the strictures of such a model, constructed phylogenies pose problems to an analyst because of the temptation to ask why *that* form (taxon), i.e. *that* adaptive peak, and not some other one slightly different from it, developed at a particular time and place.

The analytical approach of stochastic mosaicism, set in the context of contemporary evolutionary theorizing (i.e. toward what Gould calls a 'new and general theory of evolution' reframed in stochastic ideographic terms) suggests an alternative model to that of Wright. Schram (1980) posited that evolution is a phenomenon better understood in a relativistic-quantum mechanical context. Such a view suggests that a more productive approach to phylogenetic studies would be to focus on adaptive types, i.e. assess how the course of a radiation is constrained by the initial *Bauplan* upon which it is based, rather than to try to judge strict cause and effect, i.e. engage in the subjective evaluation of characters to manufacture ancestors or stem groups which might have given rise to descendent taxa. Relating structural types to each other in such a context would serve to provide some historical perspective, but would not be a necessary adjunct to assessing adaptive types.

In such a situation, manifested morphologies would be better visualized as occupying 'pits' (i.e. multidimensional curvatures in space-time co-ordinated by character states). Each pit is demarcated from another by a rim which defines the limits of a particular *Bauplan* (Fig.7). Transitions from one morphotype to another (one pit to another) are more or less abrupt since one gets to a point, or singularity, on the rim where a particular character state is 'either/or'. It is easiest for variations on a structural plan to cluster in the depths of the 'pit' at least curvature, and more difficult to maintain themselves at extremes near the rim at maximum curvature. *Baupläne* are not 'adaptive peaks' separated by 'non-adaptive valleys' which somehow must be bridged in the course of evolution. Rather, one morphotype can give rise to another when at the extremes of character expression a shift takes place to a different character state, propelled by either chance, epigenetic factors, or selective advantage to yield another *Bauplan* upon which a separate adaptive radiation can be built. This is why it is most important to understand the functional limits of an adaptive type; for if we do so, the phylogenetic relationships will sort themselves out.

The paper animals of Figures 1 and 3 are hypothetical constructs to help us focus on the nexus of possible *Baupläne*. Adaptive factors may play a role to maintain a particular species at a point on the curvature. However, the development at all of any real organisms within a particular, but only possible, structural plan is largely due to two things: first, whether a possible *Bauplan* may or may not become manifest; and second, occupation of particular *Baupläne* affecting the probability of a potential adjacent morphotype's development. But the central evolutionary precept still applies: that it is out of pre-existing forms that new forms develop. Phylogenesis is akin to the generation in a star of the elements on the periodic table; what is actually formed within a stellar interior at a particular point in time depends on the physical parameters of the star and what elements have appeared previous to that time. The evolution of *Baupläne* is dependent on what has evolved up to a point, i.e. it is essentially random but Markovian.

# 5 CONCLUSIONS

The view of phylogenetic analysis advocated here is one of several perfectly legitimate and effective approaches for classifying organisms with a goal of reflecting our knowledge concerning their evolution. Cladistics can be used in phylogenetic analysis, though I suspect it is more profitably employed with higher taxonomic categories. It allows us to be a bit more objective in regards to taxonomic and phyletic judgements. However, in applying cladistics too rigidly in the erection of higher taxa one can cause considerable trouble. As an example, the analysis of potential eumalacostracan phenograms (Schram 1981) indicates that slight differences of emphasis as to the relative importance of characters in determining hierarchy causes widely divergent and conflicting groupings of structural types in cladograms.

The difference of approach is essentially one of conceptual outlook. Evolution need not necessarily be viewed as a directional phenomenon. A static view might be preferable under some circumstances for, at least as regards higher taxonomic catagories, a *Bauplan* 'X' is not necessarily 'better' than a *Bauplan* 'Y', merely different. The mode of change from one type to another is modulated in large part by selection, but in large part by chance as well. The appearance of basic forms appear to be clustered in time in punctuational events.

## 6 ACKNOWLEDGEMENTS

It is very difficult to adequately acknowledge the source of ideas which have generated in the course of discussions with colleagues over a number of years. Certain people deserve special mention, however. Professors R.R.Hessler, W.A.Newman, of Scripps Institution, and myself have an ongoing dialogue on matters of crustacean phylogeny, made all the more valuable since we frequently disagree with each other. Dr A.P.Rasnitsyn, of the Paleontological Institute of the Soviet Academy of Science, has been most stimulating in matters dealing with evolution theory. Dr D.M.Raup, Field Museum, performed the rarefaction on data supplied by me. The following individuals read earlier drafts of the manuscript and offered valuable comment and criticism: B.M.Bell, A.J.Boucot, M.J.Grygier, and G.K.Pregill. Research was supported by NSF grant DEB 79-03602.

## DISCUSSION

BRIGGS: I can't help observing, that the reason your *Baupläne* appear suddenly in the case of the echinoderms and branchiopods is at that stage in time they were evolving more heavily mineralized skeletons and therefore appearing at that time. My other comment is that what would be really interesting in terms of stochastic mosaicism would be if you could workout some sort of deterministic explanation why certain combinations don't actually appear, or, indeed if they do, why they don't survive.

SCHRAM: [As for the first point, whatever the reason, the significance lies in it happening all at once. And for the second:] I wonder if causal explanations are at all possible. One might be able to develop a selectional explanation, but one might equally have to conclude these are purely random events. Unless there is a good reason to reject the null hypothesis that these events are random, one would be in a difficult position to defend an explanation that claims there is a selective advantage of one *Bauplan* over another.

WATERMAN: Compound eyes may be useful in the study of phylogeny in Crustacea.

SCHRAM: I would wonder in constructing cladograms on single characters of this sort whether you are focussing on something of very restrictive usefulness when dealing with phylogeny. By focussing on single features [subject as they are to adaptive pressures] one overlooks the total body plan in which these single characters are a part.

MESSING: It seems what we really need *is* more fine detailed studies in order to determine whether characters we are looking at are really the same. We are only going to find out about true homologies if we look at structure very closely.

SCHRAM: You have to be careful when you are doing this. The temptation is to suddenly start building phylogenies on single characters. However, Manton's argument is appropriate here, that it is absolutely important to understand function of structures, the phylogenies will sort themselves out later.

WATLING: You commented about not seeing the *Baupläne* for the fine structure. What I have been searching for is *Baupläne* that include the fine structures. It seems the first step in the analysis is to look at all these structures for patterns, and then see where patterns overlap. Then perhaps one can recognize *Baupläne* from sequences of overlapping patterns. Animals evolved as units, changes in one part of the body often necessitate changes on other parts of the body. When we deal with origin of taxa at the level of order or superorder, we have to be concerned with total functional plans for an animal's body. All of the features of the animal then have to be put in the context of life habit. Then, one's job when looking at modern animals is to subtly survey the range of anatomical features so that one can distill that information which may have some phylogenetic meaning. I've tried to take schemes of Calman and others and dismantle them as a starting point. We should not get hung up on that particular paradigm of phylogenetic relationships which has a caridoid ancestor. That may prove to be the correct answer when all is said and done, I personally don't think so. However, until we take a detailed look at our previous suppositions the issue cannot be settled.

GORE: Evolution occurs at the species level, and what we group as higher taxonomic levels, like families and orders, is our own construct. What constitutes a *Bauplan* by its very definition, precludes too fine structure. *Baupläne* only afford an overall view. To understand the fine structure one has to effectively destroy the *Baupläne*, that is, to pull it apart and understand its makeup. It is like a glass box which has to be broken in order to examine its contents.

SCHRAM: I have deliberately used few characters to outline morphotypes or *Baupläne*, for a pedagogical device as much as anything else. However, there is a fine line that exists between what I've done and what some may be tempted to do. How many characters can one throw into the matrix before it becomes impossible to analyze? I should hate to have to analyze *Baupläne* for Ed Bousfield's amphipods; it staggers the mind, and what would it yield us. A consideration of morphotypes can only serve to help us focus our attention and subsequently sort out which of the finer characters might be of further use in phylogeny [and which merely introduce extraneous 'noise' if one tries to deal with them in outlining *Baupläne*].

BOUSFIELD: Computers can help to evaluate characters and opposing phylogenies.

GRYGIER: Computers can be programmed to come up with the most parsimonious explanation given certain data.

SCHRAM: There is an important point missing in that. What is a phylogeny for? The unending debate about testing phylogenies [or producing 'perfect' phylogenies] largely escapes me. Phylogenies are pragmatic instruments. They function as long as they allow us to collect data and analyze it. As soon as they cease to be useful in that regard they have been in effect 'tested' and found wanting. We then develop another to take its place. Parsimony may not always be the best way — is there a parsimonious way to look at amphipods?

HESSLER: Parsimony is always the best answer! You try and find an explanation that includes all the available data, and that is the most parsimonious explanation. If you have an explanation that sets a piece of data aside for no apparent reason, that is not a parsimonious explanation.

SCHRAM: But if you use all the data, sometimes what you end up with is chaos.

MESSING: A large number of characters in an analysis can yield fresh insight. One of the protozoology talks at these meetings was assessing relationships in flagellates using 114 unweighted characters. Some traditional and familiar groups emerged in the resultant cladogram, others disappeared revealing they had been founded only one or two characters, and several entirely new postulates for relationships emerged. The use of a wide range of characters can provide one with a useful first order approximation which can then suggest further lines of investigation.

BRIGGS: One has to use as much as one can get one's hands on in the initial analysis. I've been thinking today that if I only had sperm in the Cambrian all my problems would be solved.

NEFF: If you abandon parsimony what else is there? If the word parsimony bothers people, I suggest we substitute something else called 'common sense', that is, what explains most about the world. I suggest further that if you look at all the data and you get chaos, that suggests there is something wrong with the character analysis.

HESSLER: This business of getting chaos and having to re-examine characters — maybe that would suggest merely reweighting characters.

NEFF: Experience would seem to indicate that in problems of that sort, it is not a matter of weighting, but a question of character identification. These are generally problems of homology determination. In this I would agree with the speaker that insight into function, and let us not forget development, allow some insight into the nature of the characters.

WATERMAN: Isn't it true that you are trying to reach an explanation that describes the natural world? If the natural world is not parsimonious you are distorting it if you try to force a parsimonious explanation.

NEFF: Parsimony only applies to our explanations, the real world is what it is.

SCHRAM: My argument is that in phylogenetic analysis one has to also pay attention to the forest and not worry too much about the trees, or even the leaves on the trees. I have very little patience with the issue forced by Cracraft this morning which insisted Dr Boxshall's analysis of maxillopodans wasn't any good because one is not allowed to pull a group apart unless one sticks the pieces elsewhere. The methodology is so sacred to pure cladists that the important issues raised by looking at functions are viewed as inconsequential. That is ridiculous.

BOXSHALL: I, of course, agree. But on another related issue, the trouble I always find in constructing phylogenies is do you accept a solution with eight incongruencies or nine. If you strictly apply the methodology you take the one with eight incongruencies. But it doesn't necessarily follow. Function can help. That group of copepods [misophrioids] with the carapace-like structure has some very strange features. They retain the antennary gland in the adult, they are the only free living group of copepods with one nauplius instead of six, they have an expandable gut, and they lack the nauplius eye. These are a great complex of characters, but if you consider them as a whole they can be desdribed as one character. They are adapted to a deep-sea existence. They are opportunistic macrophages. They have moved everything out of the middle of the body so the gut can expand, they have a cuticle hinging mechanism to accommodate the expansion, they have a carapace to protect the body, in the deep-sea they don't need eyes, and they have lecithotrophic development rather than planktotrophic development like all other free-living copepods. Now what do you call a character, is this one character or is it several? This is the problem of strict application of parsimony.

NEFF: A strict application of cladistics, and I consider myself a cladist, has been more destructive than helpful in the sense that it argues that it has to be either one hypothesis or another. There is still a problem if there are hypotheses that are that close. We really cannot judge characters that well. A strict application of parsimony can be counter productive in the sense that it can consider problems as closed which may not be closed.

# REFERENCES

Adams, M.B. 1980. Severtsov and Schmalhausen: Russian morphology and the evolutionary synthesis. In: E.Mayr & W.B.Provine (eds), *The evolutionary synthesis:*193-225. Cambridge: Harvard Univ. Press.

Anderson, D.T. 1973. *Embryology and phylogeny in Annelids and Arthropods.* New York: Pergamon Press.

Boudreaux, H.B. 1979. *Arthropod phylogeny with special reference to insects.* New York: Wiley.

Burnett, B.R. & R.R.Hessler 1973. Thoracic epipodite gills in the stomatopods: a phylogenetic consideration. *J. Zool., Lond.* 169:381-392.

Calman, W.T. 1904. On the classification of the Crustacea Malacostraca. *Ann. Mag. Nat. Hist.* (7)13: 144-158.

Calman, W.T. 1909. Crustacea. In: E.R.Lankester (ed.), *A Treatise on Zoology, Vol.VII.* London: Adam & Charles Black.

Dahl, E. 1977. The amphipod functional model and its bearing upon systematics and phylogeny. *Zool. Scripta* 6:221-228.

Engleman, G.F. & E.O.Wiley 1977. The place of ancestor-descendant relationships in phylogeny reconstruction. *Syst. Zool.* 26:1-11.

Eldredge, N. & S.J.Gould 1972. Punctuated equilibria: an alternative to phyletic gradualism. In: T. Schopf (ed.), *Models in Paleobiology:*82-115. San Francisco: Freeman, Cooper.

Fryer, G. 1964. Studies on the functional morphology and feeding mechanism of *Monodella argentarii. Trans. Roy. Soc. Edinb.* 66:49-90.

Gould, S.J. 1980a. The promise of paleobiology as a nomothetic, evolutionary discipline. *Paleobiol.* 6: 96-118.

Gould, S.J. 1980b. Is a new and general theory of evolution emerging? *Paleobiol.* 6:119-130.

Gould, S.J. & N.Eldredge 1977. Punctuated equilibria: the tempo and mode of evolution reconsidered. *Paleobiol.* 3:115-151.

Gupta, A.P. 1979. *Arthropod Phylogeny.* New York: Van Nostrand Reinhold.

Haugh, B.N. & B.M.Bell 1980a. Fossilized viscera in primitive echinoderms. *Science* 209:653-657.

Haugh, B.N. 1980b. Classification schemes. *Univ. Tenn., Studies in Geol.* 3:94-105.

Hennig, W. 1966. *Phylogenetic Systematics.* Urbana: Univ. Illinois Press.

Hessler, R.R. 1969. Peracarida. In: R.C.Moore (ed.), *Treatise on Invertebrate Paleontology, Part R, Arthropoda 4, Vol.I:*R360-393. Lawrence: Geol. Soc. Am. & Univ. Kansas Press.

Hessler, R.R. 1982. Evolution within the Crustacea. In: L.G.Abele (ed.), *Biology of Crustacea, Vol.I:* 149-185. New York: Academic Press.

Hessler, R.R. & W.A.Newman 1975. A trilobitomorph origin for the Crustacea. *Fossils and Strata* 4:437-459.

Houde, P. & S.L.Olson 1981. Paleognathous carinate birds from the early Tertiary of North America. *Science* 214:1236-1237.

House, M.R. 1979. *The origin of major invertebrate groups.* London: Academic Press.

Manton, S.M. 1964. Mandibular mechanisms and the evolution of arthropoda. *Phil. Trans. Roy. Soc. (B):* 247:1-183.

Manton, S.M. 1977. *The Arthropoda.* Oxford: Clarendon Press.

Marx, H., G.B.Rabb & H.K.Voris 1977. The differentiation of character state relationships by binary coding and the monothetic subset method. *Fieldiana: Zool.* 72:1-20.

Morris, N.J. 1979. On the origin of Bivalvia. In: M.R.House (ed.), *The Origin of Major Invertebrate Groups:* 381-413. London. Academic Press.

Raup, D.M. 1978. Approaches to the extinction problem. *J. Paleontol.* 52:517-523.

Raup, D.M. & S.J.Gould 1974. Stochastic simulation and evolution of morphology – towards a nomothetic paleontology. *Syst. Zool.* 23:305-322.

Raup, D.M., S.J.Gould, T.J.M.Schopf & D.S.Simberloff 1973. Stochastic models of phylogeny and the evolution of diversity. *J. Geol.* 81:525-542.

Rowell, A.J. 1965. Order uncertain – Dictyonellida. In: R.C.Moore (ed.), *Treatise on Invertebrate Paleontology, Part H, Brachiopoda* 1:H359-361. Lawrence: Geol. Soc. Am. & Univ. Kansas Press.

Rowell, A.J. 1981. The origin of brachiopods. *Univ. Tenn., Studies in Geol.* 5:97-109.

Sanders, H.L. 1957. The Cephalocarids and crustacean evolution. *Syst. Zool.* 6:112-129.

Schram, F.R. 1969a. Polyphyly in the Eumalacostraca? *Crustaceana* 16:243-250.

Schram, F.R. 1969b. Some middle Pennsylvanian Hoplocarida and the phylogenetic significance. *Fieldiana: Geol.* 12:235-289.

Schram, F.R. 1974. Late Paleozoic Peracarida of North America. *Fieldiana: Geol.* 33:95-124.

Schram, F.R. 1979. Book review – Arthropod phylogeny with special reference to insects. *Syst. Zool.* 28:635-638.

Schram, F.R. 1980. O relyativistsko-kvantovo-mechkanicheskom podhkode k evalyutsii. *Zh. obshchey biol.* 41:557-573.

Schram, F.R. 1981. On the classification of Eumalacostraca. *J. Crust. Biol.* 1:1-10.

Schram, F.R. 1982. The fossil record and evolution of Crustacea. In: L.G.Abele (ed.), *The Biology of Crustacea, Vol.I:* 93-147. New York: Academic Press.

Scourfield, D.J. 1926. On a new type of crustacean from the Old Red Sandstone – *Lepidocaris rhyniensis. Phil. Trans. Roy. Soc., Lond. (B)* 214:153-187.

Scourfield, D.J. 1940. Two new and nearly complete specimens of young stages of the Devonian fossil crustacean *Lepidocaris rhyniensis. Proc. Linn. Soc., Lond.* 152:290-298.

Severtsov, A.N. 1912. *Etindy po teorii evolyutsii: individyualnoe razvitie i evolyutsiya.* Kiev.

Siewing, R. 1956. Untersuchungen zur morphologie der Malacostraca. *Zool. Jb., Abt. Anat. Ontog.* 75: 29-176.

Tiegs, O.W. & S.M.Manton 1958. The evolution of Arthropoda. *Biol. Rev.* 33:255-337.

Thompson, D.W. 1917. *On growth and form.* Cambridge: Cambridge Univ. Press.

Watling, L. 1981. An alternative phylogeny of peracarid crustaceans. *J. Crust. Biol.* 1:201-210.

Watrous, L.E. & Q.D.Wheeler 1981. The out-group comparison of character analysis. *Syst. Zool.* 30:1-11.

Wetmore, A. 1960. A classification for the birds of the world. *Smith. Misc. Coll.* 139(11):1-37.

Williams, A. & J.M.Hurst 1977. Brachiopod evolution. In: A.Hallam (ed.), *Patterns of Evolution as Illustrated by the Fossil Record:* 79-121. Amsterdam: Elsevier.

Wright, A.D. 1979. Brachiopod radiation. In: M.R.House (ed.), *The Origin of Major Invertebrate Groups:* 235-252. London: Academic Press.

Wright, S. 1932. The roles of mutation, inbreeding, crossbreeding selection in evolution. *Proc. VI Int. Cong. Gen.* 1:356-366.

Yager, J. 1981. Remipedia, a new class of Crustacea from a marine cave in the Bahamas. *J. Crust. Biol.* 1:328-333.

Zangerl, R. 1948. The methods of comparative anatomy and its contribution to the study of evolution. *Evolution* 2:351-374.

# LIST OF DISCUSSANTS

Lawrence G.Abele, Florida State University, Tallahassee, Florida 32306 USA
Raymond T.Bauer, University of Puerto Rico, Rio Pedras, Puerto Rico, USA
Edward L.Bousfield, National Museum of Natural Science, Ottawa, Ontario, K1A 0M8, Canada
Geoffrey A.Boxshall, British Museum (Natural History), London, SW7 5BD, UK
Derek E.G.Briggs, Goldsmith's College, University of London, London, SE3 3BU, UK
Joel Cracraft, University of Illinois Medical Center, Chicago, Illinois 60680 USA
Bruce E.Felgenhauer, Florida State University, Tallahassee, Florida 32306 USA
Robert H.Gore, Smithsonian Institution, Fort Pierce, Florida 33450 USA
Mark J.Grygier, Scripps Institution of Oceanography, La Jolla, California 92093 USA
Robert R.Hessler, Scripps Institution of Oceanography, La Jolla, California 92093 USA
Janet C.Kunze, Scripps Institution of Oceanography, La Jolla, California 92093 USA
Charles G.Messing, University of Miami, Coral Gobles, Florida 33124 USA
Nancy A.Neff, American Museum of Natural History, New York, New York 10024 USA
William A.Newman, Scripps Institution of Oceanography, La Jolla, California 92093 USA
Frederick R.Schram, San Diego Natural History Museum, San Diego, California 92112 USA
Jürgen Sieg, Universität Osnabrück, D-2848 Vechta, German Federal Republic
Talbot Waterman, Yale University, New Haven, Connecticutt 06520 USA
Les Watling, University of Maine, Walpole, Maine 04573 USA
Mary K.Wicksten, Texas A & M University, College Station, Texas 77843 USA

# TAXONOMIC INDEX

The orthography of *Peneus* and derivative names was an editorial decision to establish conformity for volume with spelling used in Burkenroad's chapter.

# SUBJECT INDEX

abbreviated development 320
abdomen ´2, 4, 5, 12, 18, 36,
  37, 61, 77, 83, 85, 86, 89,
  90, 100, 107, 111, 115,
  125, 147, 151, 152, 156,
  177, 182, 184, 190-192,
  196, 206, 207, 261, 271,
  282, 292, 314, 317, 322-
  324, 333
abdominal flexion 168
abdominal gills 167, 182
abdominal musculature 145,
  147, 150-152, 156, 162,
  168, 193
abdominal segments 31, 43,
  61, 152, 156, 317
abdominal segment fusion 167
abdominal viscera 152
abdominalization 124-128, 130
abyssal fauna 247
accessory flagellum 261
accessory teeth 303
acron 61
adaptation 49, 56
adaptive landscape model 344,
  345
adaptive potential 345
adaptive radiation 339
adaptive types 345
adductor muscle 32, 35, 37,
  128, 184
adductor muscle scars 37, 86,
  87
advanced thoracotrematous
  zoeas 327
advanced zoeas 323, 326
aeschronectids 166, 167, 177
aglaspids 9
Alan of Lille 67
α lobe 301
alpheids 300, 302, 306
ameirids 60
ammonites 64
amphipod ancestral prototype
  257, 270, 271
amphipod characters 258

amphipod classification 267,
  268
amphipod distribution 257
amphipod fossil record 257,
  269, 271, 275
amphipod geologic ranges 270
amphipod taxonomic diagnoses
  272-275
amphipods 56, 57, 59, 64, 114,
  153-156, 158, 161, 191,
  193, 195, 196, 198, 203-
  205, 215, 222-225, 347
ampullae 179, 181, 183, 302
Amoy 287
anamorphic development 105,
  108
anamorphic larval sequence 105
anamorphy 107
anaspidaceans 107, 148, 149,
  153, 156, 159, 172, 193,
  202, 215
anaspidids 196
anatomical reductions 41
Anaxagoras 66
Anaximander 66
Anaximines 66
ancestral cirriped 122, 123
ancestral copepod 122, 123
ancestral crustacean 31, 43,
  79, 140
ancestral eumalacostracan 153,
  207
ancestral homoloid zoea 318
ancestral homoloids 319
ancestral malacostracans 189,
  191, 201
ancestral ostracods 8
ancestral peracarid 223
ancestral post-larval features
  284
ancestral tanaidacean 229
ancestral types 332, 334, 337
androgenic glands 253
annulate exopods 35
annulate pleopod 196
angiosperm leaf litter 269

anomuran zoeas 317
anomurans 314-316, 327
anostracans 15, 31
Antarctic 246
antennae 2, 5, 6, 10, 16, 18,
  19, 35, 36, 60-62, 81, 85-88,
  90, 94, 135, 137, 146, 147,
  161, 166, 193, 261-263,
  265, 299, 300, 305, 307,
  320, 322, 325
antennal bases 299
antennal endopod 315
antennal exopod 23, 81, 315,
  317, 324
antennal flagella 285
antennal glands 85-87, 89, 91,
  95, 223, 226
antennal scale 114, 115, 147,
  150-152, 156, 157, 161,
  162, 185, 192, 193, 200,
  271, 307, 332, 333
antennal spine 317
antennary muscles 132
antennary gland 348
antennular attachment 79
antennular flagella 153, 281-
  283, 287, 289
antennular peduncle 233, 281-
  283, 286, 287
antennular spine 286
antennular segments 166
antennular statocysts 203
antennulary muscles 132
antennules 10, 23, 35, 36, 59-
  63, 79, 81, 83, 85-88, 94,
  135, 140, 146, 147, 166,
  185, 192, 238, 281, 283,
  287, 293, 299, 307, 320
antizoeae 177
anthozoans 85
anus 61, 342
aortae 221, 222, 226
appendages (general) 3, 4, 23,
  232, 342
apomorphies 146, 147
appendix interna 280, 281,

361